T0141850

Studies in Big Data

Volume 122

Series Editor

Janusz Kacprzyk, Polish Academy of Sciences, Warsaw, Poland

The series "Studies in Big Data" (SBD) publishes new developments and advances in the various areas of Big Data- quickly and with a high quality. The intent is to cover the theory, research, development, and applications of Big Data, as embedded in the fields of engineering, computer science, physics, economics and life sciences. The books of the series refer to the analysis and understanding of large, complex, and/or distributed data sets generated from recent digital sources coming from sensors or other physical instruments as well as simulations, crowd sourcing, social networks or other internet transactions, such as emails or video click streams and other. The series contains monographs, lecture notes and edited volumes in Big Data spanning the areas of computational intelligence including neural networks, evolutionary computation, soft computing, fuzzy systems, as well as artificial intelligence, data mining, modern statistics and Operations research, as well as self-organizing systems. Of particular value to both the contributors and the readership are the short publication timeframe and the world-wide distribution, which enable both wide and rapid dissemination of research output.

The books of this series are reviewed in a single blind peer review process.

Indexed by SCOPUS, EI Compendex, SCIMAGO and zbMATH.

All books published in the series are submitted for consideration in Web of Science.

Farhad Hosseinzadeh Lotfi · Tofigh Allahviranloo ·
Morteza Shafiee · Hilda Saleh

Supply Chain Performance Evaluation

Application of Data Envelopment Analysis

 Springer

Farhad Hosseinzadeh Lotfi
Department of Mathematics
Science and Research Branch
Islamic Azad University
Tehran, Iran

Morteza Shafiee
Department of Industrial Management
Faculty of Economics and Management
Shiraz Branch, Islamic Azad University
Shiraz, Iran

Tofigh Allahviranloo
Department of Software Engineering
Istinye University
Istanbul, Turkey

Hilda Saleh
Department of Mathematics, Faculty
of Science
Central Tehran Branch, Islamic Azad
University
Tehran, Iran

ISSN 2197-6503   ISSN 2197-6511 (electronic)
Studies in Big Data
ISBN 978-3-031-28249-2   ISBN 978-3-031-28247-8 (eBook)
https://doi.org/10.1007/978-3-031-28247-8

© The Editor(s) (if applicable) and The Author(s), under exclusive license to Springer Nature
Switzerland AG 2023
This work is subject to copyright. All rights are solely and exclusively licensed by the Publisher, whether
the whole or part of the material is concerned, specifically the rights of translation, reprinting, reuse
of illustrations, recitation, broadcasting, reproduction on microfilms or in any other physical way, and
transmission or information storage and retrieval, electronic adaptation, computer software, or by similar
or dissimilar methodology now known or hereafter developed.
The use of general descriptive names, registered names, trademarks, service marks, etc. in this publication
does not imply, even in the absence of a specific statement, that such names are exempt from the relevant
protective laws and regulations and therefore free for general use.
The publisher, the authors, and the editors are safe to assume that the advice and information in this book
are believed to be true and accurate at the date of publication. Neither the publisher nor the authors or
the editors give a warranty, expressed or implied, with respect to the material contained herein or for any
errors or omissions that may have been made. The publisher remains neutral with regard to jurisdictional
claims in published maps and institutional affiliations.

This Springer imprint is published by the registered company Springer Nature Switzerland AG
The registered company address is: Gewerbestrasse 11, 6330 Cham, Switzerland

Preface

The progress of management, especially in the fields of commerce and the fields of engineering and industrial management, due to the industrialization movement in today's world, brings forward new methods of management. Being aware of these new achievements of management science and synchronizing with them as much as possible to use them in practice is a requirement for progress and dynamism in similar fields. In addition, the use of new management methods is also important from another point of view. In today's dynamic environment, the use of classical management concepts not only does not help the management of organizations but also causes a lack of progress and dynamism in the organization.

What helps managers in this regard is strategic thinking. Therefore, managers should try to use strategic thinking as the dominant thinking; and use this attitude in all their management decisions. This book has also tried to use this thinking as the main guide in presenting its content and providing supply chain topics with the perspective of management and strategic thinking.

One of the principles of management that play an important role in management science and industries is the issue of supply chain management. Unfortunately, there are few books in the field of supply chain performance evaluation. This book has tried to practically discuss the fundamental principles of supply chain evaluation with a strategic view.

In recent years, researchers have paid a lot of attention to issues related to the supply chain. By reviewing information sources, we can easily see the growing trend of supply chain studies in various fields of organizational literature. But the noteworthy point is that along with the emergence of any new system and procedure in the organization, we need to measure the performance of this new system, in other words, its efficiency and effectiveness, so that we can improve the conditions of the organization by identifying opportunities for improvement.

In today's global markets, companies are not units with a unique brand that can operate independently. The complexity of goods and services in today's world is such that it rarely happens that an organization or an institution can produce a product or provide a service on its own without the help and cooperation of other organizations. In this uncertain and competitive environment, companies have faced extensive and serious challenges, and to face these challenges, we see the emergence of new methods and approaches in the business world every day.

The supply chain management approach is one of the new methods in the business world that has emerged to face and deal with these challenges. The conditions that led to the definition and design of such an attitude are the ever-increasing competition and the efforts of companies to survive. Companies consider the secret of this survival in reducing costs and satisfying customer needs. Customers' needs and interests can include price reduction, timely delivery, and good quality.

In supply chain management, to achieve customer satisfaction and maximize profitability, there is a need to measure performance and continuous improvement. That is before an improvement process can be started in the supply chain, a clear perspective of the existing supply chain structure and how it works must be obtained. It should be noted that the goal of all those who work in the fields of supply is to increase competitiveness or increase customer service. The reason is that today, from the point of view of the final customer, an organizational unit is not responsible for its competitiveness or services, and the supply chain considers all organizations at once. Therefore, competition has shifted from companies to supply chains. The services provided and their incompatibility with the costs and revenues, the lack of sufficient information about the services provided to respond to the complaints and demands of the customers, the lack of strategic focus in the supply chain, and the need to measure performance and continuous improvement shows in the supply chain. Also, before an improvement process can be started, a clear picture of the structure of the existing supply chain and how it works must be obtained through supply chain analysis.

This book intends to identify the current performance and competitive position of that supply chain in comparison with other supply chains by presenting and reviewing the techniques and models for measuring the efficiency and performance of the supply chain. Determining the performance of a supply chain is a good description of the existing situation (what is). Determining the performance of a supply chain is useful for describing the past and present of the supply chain processes, and on the other hand, it can be used to determine performance goals and start the improvement process. To achieve this, a strategic framework or model is needed, so that it is possible to extract the indicators related to the efficiency of the supply chain and design the appropriate model.

Having a strategic attitude and system is necessary to measure performance and efficiency. Until now, the most common method to measure performance and efficiency in the field of supply chain management has been to use pre-designed models by institutions or people engaged in the field of supply chain management, and fewer of these measurements. A strategic and quantitative approach has been used to determine performance goals and start the improvement process. Although research has been done in this field, it is still not possible to find comprehensive research that centrally measures the performance of the supply chain by combining quantitative and qualitative approaches with a strategic view. On the other hand, with the emergence of unexpected crises such as the Corona epidemic, the need to evaluate, control and manage the supply chain is much more important than before.

Therefore, the authors of this book tried to make these experiences available to those interested, considering the experience of several years of training, research, and implementation of projects in the supply chain performance evaluation field. It should be noted that the writing of Chaps. 1–3, which are mostly about the management of the supply chain, started under the supervision of Morteza Shafiee. The responsibility of Chaps. 4 and 5 were entrusted to Tofigh Allahviranloo. After that, Chaps. 6 and 7 were written under the supervision of Farhad Hosseinzadeh Lotfi, and finally, the book ended with the writing of Chaps. 8–10 under the supervision of Hilda Saleh.

This book is one of the few books that can be used academically and industrially. Students of management, applied mathematics, and industrial engineering at all levels, especially doctoral level, as well as managers and industrialists of various industries, and supply chain activists, are the main target audience of this book.

Tehran, Iran   Farhad Hosseinzadeh Lotfi
Istanbul, Turkey   Tofigh Allahviranloo
Shiraz, Iran   Morteza Shafiee
Tehran, Iran   Hilda Saleh

Contents

1   **Supply Chain Management**   1
   1.1   Introduction   1
   1.2   Concepts of Supply Chain Management   3
         1.2.1   Evolution of Supply Chain Management   3
         1.2.2   Definitions   5
         1.2.3   Reasons for the Importance and Need of Supply
                 Chain Management   9
         1.2.4   Components of Supply Chain Management 12
         1.2.5   Objectives of Supply Chain Management 13
         1.2.6   Scope of Supply Chain Management Performance 15
         1.2.7   Principles of Supply Chain Management 20
         1.2.8   Supply Chain Management Processes 21
         1.2.9   Decision-Making Steps in Supply Chain
                 Management 28
   1.3   Strategic Supply Chain Management 33
         1.3.1   The Role of Strategic Thinking in Supply Chain
                 Management 34
         1.3.2   Supply Chain as a Strategic Asset 36
         1.3.3   How to Start Turning Your Supply Chain
                 into a Strategic Asset 40
         1.3.4   Benefits of Turning the Supply Chain
                 into a Strategic Asset 41
   1.4   The Structure of the Supply Chain Network 42
         1.4.1   Identification of Supply Chain Members 42
         1.4.2   The Structural Dimensions of the Supply Chain
                 Network 43
   1.5   Conclusion 44
   References 44

2   Performance Evaluation of Supply Chain Management   47
    2.1   Introduction   47
    2.2   Evaluation and Measurement of Supply Chain Performance   48
        2.2.1   Reasons and Importance of Supply Chain
                Performance Measurement   48
        2.2.2   Definitions of Supply Chain Performance
                Measurement System (SCPMS)   50
        2.2.3   Principles and Characteristics of a Suitable
                Supply Chain Performance Measurement System
                (SCPMS)   51
        2.2.4   Evolution of SCPMS Models   52
        2.2.5   Reasons for Changes in SCPMS Models   53
        2.2.6   Review and Classification of Common SCPMS
                Frameworks and Models   57
    2.3   The Role of the Supply Chain Performance Evaluation
        System (SCPMS) in Achieving the organization's Goals   63
    2.4   Supply Chain Performance Evaluation Criteria   64
        2.4.1   Limitations of Supply Chain Performance
                Evaluation   68
        2.4.2   Appropriate Feature of Supply Chain Performance
                Evaluation Criteria   69
    2.5   Conclusion   70
    References   70

3   Main Models and Approaches in Supply Chain Evaluation   75
    3.1   Introduction   75
    3.2   Basic Models in Supply Chain Evaluation and Its Different
        Approaches   76
        3.2.1   Supply Chain Operation Reference Model
                (SCORE)   79
        3.2.2   New Supply Chain Paradigms or Approaches   92
    3.3   Network Models in Supply Chain Evaluation   97
        3.3.1   Definitions of Supply Chain Network (SCN)   98
        3.3.2   Difference Between Supply Chain Network
                and Supply Chain   100
        3.3.3   Supply Chain Network Structure   102
        3.3.4   How to Analyze the Supply Chain Network?   104
        3.3.5   Common Types of the Supply Chain Network   106
    3.4   Conclusion   110
    References   111

4   Supplier Performance Evaluation Models   117
    4.1   Introduction   117
    4.2   Existing Models of Supplier Selection and Evaluation   118
        4.2.1   Total Cost of Ownership (TCO) Models   118
        4.2.2   Multiple Attribute Decision Making (MADM)   121

4.2.3 Mathematical Programming Models 133
4.3 Conclusion 141
References 142

5 Examining Supply Chain Crises and Disruptions 149
5.1 Introduction 149
5.2 Definitions of Supply Chain Disruptions 150
5.3 Examples of Crises (Disruptions) in the Supply Chain 151
5.4 Vulnerability of the Supply Chain and Its Aggravating
Factors Against Disruption 153
5.5 Identification of Supply Chain Risks 156
5.5.1 Risk Definition and Supply Chain Risk
Management 156
5.5.2 Different Forms of Supply Chain Risks 157
5.5.3 Supply Chain Risk Management Processes 162
5.5.4 Disruption Management Solutions in the Supply
Chain 163
5.6 The Concept of Reliable Supply Chain 173
5.6.1 Reliability of the Supplier 174
5.6.2 Reliability of Producer 174
5.6.3 Reliability of the Distributor 175
5.6.4 Reliability of the Entire Supply Chain 175
5.7 Conclusion 175
References 176

6 Data Envelopment Analysis 179
6.1 Introduction 179
6.2 Basic Concepts and Definitions 180
6.3 Multiplier Models 182
6.4 Envelopment Models 186
6.4.1 Radial Models 188
6.5 Non-radial Models 201
6.5.1 Additive Model 202
6.5.2 RAM Model 204
6.5.3 SBM Model 205
6.5.4 Enhanced Russell Model 208
6.6 Hybrid Models 210
6.7 Cost and Revenue Efficiency 212
6.8 Ranking 213
6.8.1 Anderson and Peterson Method 214
6.8.2 Cross Efficiency 215
6.9 Return to Scale 216
6.10 Network DEA 218
6.10.1 Series Network 219
6.10.2 Parallel Networks 231
6.10.3 Mix Networks 233

    6.11 Conclusion 236
    References 238

7 Supplier Selection 243
    7.1   Introduction 243
    7.2   Evaluation of Suppliers' Performance to Select the Best
           Supplier Using DEA Models 244
    7.3   Supplier Selection in the Presence of Weight Restriction 249
    7.4   Supplier Selection with the Special Type of Data 253
          7.4.1   Supplier Selection with Imprecise Data 253
          7.4.2   Supplier Selection with Non-discretionary Data 255
          7.4.3   Supplier Selection with Undesirable Inputs
                and Outputs 257
          7.4.4   Supplier Selection with Dual-Role Factors 265
          7.4.5   Supplier Selection with Fuzzy Data 269
    7.5   Supplier Selection by Cross-Efficiency Models 280
    7.6   Supplier Selection Using Cost and Profit Efficiency 290
    7.7   Conclusion 295
    References 296

8 Performance Evaluation of the Supply Chains Using DEA 301
    8.1   Introduction 301
    8.2   Performance Evaluation of the Supply Chains Using DEA
           Models 303
          8.2.1   Performance Evaluation of the Healthcare Supply
                Chain 304
          8.2.2   The International Biomass Supply Chain: The
                Reconfiguration of the Chain Based on Different
                Processing Scenarios and Transportation Methods
                in the Presence of Undesirable Output 306
          8.2.3   Evaluating the Performance and Determining
                the Regression and Progress of Supply Chains
                Using the Malmquist Index Based on Success
                Indicators in the Financial Market in the Presence
                of Ratio Measures 312
          8.2.4   Oil Import Supply Chain: Performance Evaluation
                of Supply Chain Based on Supply Chain Risk 316
          8.2.5   Evaluating the Impact of Different
                Information-Sharing Scenarios for Reconfiguring
                the Supply Chain Using Cross-Efficiency 319
    8.3   Aggregation of Indicators in the Evaluation of Supply
           Chain Performance 322
    8.4   Benchmarking for Supply Chain 324
    8.5   Conclusion 325
    References 326

9   Supply Chain Evaluation by Network DEA 329
    9.1   Introduction 329
    9.2   Performance Evaluation of Supply Chains Using Radial
           Models in NDEA 330
           9.2.1   Definition of New PPS for Performance
                    Evaluation of Supply Chain 330
           9.2.2   Performance Evaluation of Centralized
                    and Decentralized Supply Chains 338
    9.3   Performance Evaluation of Supply Chains Using
           Non-radial Models in NDEA 346
    9.4   Performance Evaluation of the Supply Chains Using
           Hybrid Models in NDEA 348
           9.4.1   Performance Evaluation E of Two-member Series
                    Supply Chain Based on Short-term and Long-term
                    Decision-making 349
           9.4.2   Performance Evaluation of Mixed Multi-member
                    Supply Chains 357
           9.4.3   Performance Evaluation of Supply Chain Using
                    Network Epsilon-based DEA Model 364
    9.5   Performance Evaluation of Supply Chains Using Multiplier
           Models 367
           9.5.1   Performance Evaluation of Supply Chains Using
                    Frontier-shift 367
           9.5.2   Performance Evaluation of Three-member Supply
                    Chains with a Focus on the Education Supply Chain ... 374
           9.5.3   Performance Evaluation of Mixed Supply Chains 376
           9.5.4   Performance Evaluation of Supply Chains
                    in the Presence of Returnable Relations
                    and Identifying the Unit of Relative Efficient 377
    9.6   Performance Evaluation of Supply Chains Using Game
           Theory 381
           9.6.1   Performance Evaluation Using NDEA Models
                    with the Approach of Non-cooperative Games 382
           9.6.2   Performance Evaluation Using NDEA Models
                    with the Approach of Cooperative Games 388
    9.7   Performance Evaluation of Supply Chains in the Presence
           of Fuzzy Data Using FNDEA 390
           9.7.1   Performance Evaluation of Supply Chains
                    in the Presence of Fuzzy Data Using Envelopment
                    Form of NDEA Models 391
           9.7.2   Performance Evaluation of Supply Chains
                    in the Presence of Fuzzy Data Using Multiplier
                    Form of NDEA Models 397

9.8   Determining the Type of Return to Scale in the Supply Chain ...   408
9.9   Conclusion   413
References   414

10   Performance Evaluation of Supply Chains by Bi-Level DEA   419
10.1   Introduction   419
10.2   Bi-Level Programming   420
10.3   Cost Efficiency in the Two-Member Supply Chain
        (Leader–Follower) Using Bi-Level DEA   421
        10.3.1   BPL Problem-Solving Method   424
10.4   Cost Efficiency Assessment in a Supply Chain
        with Multiple Followers Using Bi-Level DEA   426
        10.4.1   BPL Problem-Solving Method   429
10.5   Profit Efficiency Using Bi-Level DEA   430
        10.5.1   BPL Problem-Solving Method   432
10.6   Resources Reallocation in the Supply Chain Using
        Bi-Level DEA   434
        10.6.1   BPL Problem-Solving Method   438
10.7   Conclusion   441
References   441

Chapter 1
Supply Chain Management

1.1 Introduction

Over the last few decades, managers have witnessed dramatic global change due to technological advances, the globalization of markets, and the stabilization of the political economy. As the number of world-class Competitors increases, organizations are forced to improve internal processes to stay ahead of the competition quickly. In the 1980s, companies sought techniques and strategies to reduce production costs and compete in different markets. Some of these techniques were: the "just-in-time" system, "Kanban system," "lean production," "total quality management," and so on. Companies could reduce their production costs as much as possible by using these techniques. Still, competing companies also reduced their production costs as much as possible by using the same methods. Other opportunities to reduce costs should be found and stay in the competitive market. One of these options, which offers many potential opportunities to reduce costs to organizations and companies, is the supply chain and its management [44, 52]. At the same time and in the 1990s, besides improving production capabilities, industry managers realized that the materials and services received from different suppliers significantly increased the organization's qualifications to deal with customers' needs, competition, and activities in the international arena. This trend, in turn, doubled the organization's focus, supply bases, and sourcing strategies.

On the other hand, the managers realized that it is not enough to produce a good quality product. Providing products with the criteria desired by the customer and with the quality and cost expected by them created new management challenges for today's organizations. Therefore, with such an attitude, the "supply chain" and "supply chain management" approaches came into existence [45]. However, the emergence of the supply chain concepts also raised the issue of its management. Supply chain management is an interdisciplinary subject comprising disciplines such as marketing, operations management, purchasing, and support [29]. Supply chain management at the operational level includes inventory management, forecasting, replacement, and planning. Still, at the strategic level, supply chain strategies are

© The Author(s), under exclusive license to Springer Nature Switzerland AG 2023
F. Hosseinzadeh Lotfi et al., *Supply Chain Performance Evaluation*, Studies in Big Data 122, https://doi.org/10.1007/978-3-031-28247-8_1

proposed, which include defining and prioritizing supply chain goals and policies. Therefore, the purpose of supply chain management is to improve the efficiency of the supply chain process so that the right product reaches the customer at the lowest cost [50]. The thought that supply chain management can lead to better responses to customers and ultimately more profitability has made many managers pay attention to the issue of supply chain management [29]. In general, the factors driving organizations toward supply chain management can be expressed as follows [48]:

- Need for improvement activities
- Increasing the level of outsourcing
- Increase in the shipping cost
- Competitive pressures
- Increasing globalization and the importance of world trade
- The complexity of the supply chain and the need to manage inventories.

With the attention and focus of today's organizations and companies on supply chain management issues, we can examine supply chain management at three levels:

- Macro level: This includes the fundamental visions and goals of the supply chain and is the framework of the strategic management of the supply chain.
- Strategic level: which seeks to realize the vision of the supply chain in different organizational fields with the help of supply chain strategies. Strategic management of the supply chain supports the organization's competitive strategy by empowering the supply chain.
- Operational level: which uses supply chain strategies designed at a strategic level and includes planning, control, and execution of processes [46].

Efficient supply chain management allows a company to coordinate the production and transfer of products throughout the production and distribution channel, from the supply of raw materials and parts to placing the finished product in the hands of the customer. Supply chain management includes all activities related to the flow and transformation of goods from the raw material (extraction) stage to delivery to the final consumer and information flows. Therefore, the scope of supply chain management is beyond an organization. It considers everything involved in producing and delivering a product or service and connects them all in such a way that it works as an efficient and integrated team. It means that manufacturers, customers, suppliers, transport companies, and even recently, commercial competitors are united and act in the form of a united and integrated network to make the best use of the time and resources used. Therefore, having an efficient and responsive supply chain is a critical factor for competition in today's market [45]. And in this regard, in this chapter of the book, the concepts of the supply chain are discussed first. Then, how the supply chain is described as a strategic asset and the concepts of the supply chain management network are also discussed.

1.2 Concepts of Supply Chain Management

In this section, the supply chain management concept is reviewed generally. The authors try to create a comprehensive perspective of the issue.

1.2.1 Evolution of Supply Chain Management

In the 1950s and 1960s, most manufacturers focused on mass production or production using economies of scale to reduce production and organization costs. They were primarily looking for the balance of output and preventing production bottlenecks using their internal strength and technology. During this period, product development and diversification were slow, keeping extensive inventories of raw materials and finished goods to meet demand. This process and strategy for decades led to increased operational, investment, and production costs. During this period, sharing the company's internal information, technology, and expertise with customers and suppliers was usually unacceptable and high risk. Little by little, managers of industries and organizations realized the importance and role of inventory types in controlling company costs. Therefore, in the 1970s, MRPII was introduced, and managers realized the impact of inventory during manufacturing on production cost, quality, new product development, and delivery time. This decade introduced new concepts in material management and organizational performance, and managers realized its importance and role in controlling administrative costs. With the emergence of intense competition in the 1980s, organizations had to market a product with a low price, quality, reliability, and higher flexibility in design and production. Therefore, this decade formed concepts such as just-in-time production (JIT), flexible production systems (FMS), lean production, and other management innovations.

One of the side benefits of using "just-in-time" production systems, which significantly contributed to the emergence and development of supply chain management, is understanding the critical role of the buyer and the seller in different positions such as producer-supplier, producer-customer, etc. The strategic benefits of direct communication with suppliers established the foundation for forming supply chain management and its concepts. This relationship caused the change and development of ideas such as logistics management, transportation, and in a word, the development of material management by integrating new concepts into it. Finally, with material management's action, integrated logistics became the first indication of supply chain management. Later, this concept of integrated logistics was developed as supply chain management.

In fact, in the 1990s, the evolution of the concept of supply chain management began, and other essential concepts of supply chain management were expanded, which will be further analyzed in this book. Diagrams 1.1, 1.2, and 1.3 show the evolution stages of supply chain management in different decades. A crucial point that needs to be mentioned here is that the supply chain components have undergone fundamental changes with the development of information technologies or the emergence of new phenomena and crises, such as the Corona epidemic.

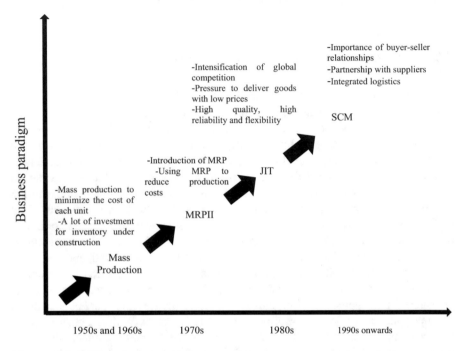

Diagram 1.1 Different business paradigms

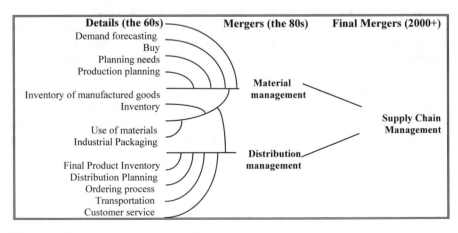

Diagram 1.2 Evolution of the supply chain

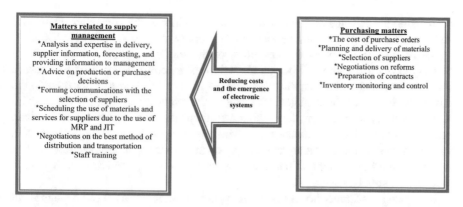

Diagram 1.3   The history and evolution of supply management from purchasing to supply

The evolution of management shows that companies have paid attention to supply chain management first to deal with severe environmental changes and improve organizational performance, then to gain more market share and competitive advantage [46].

1.2.2   Definitions

Various definitions of supply chain management have been provided, but before that, it is necessary to explain the concept of the supply chain. According to an explanation, the supply chain includes all activities related to the flow and goods transformation from the stage of raw material (extraction) to delivery to the final consumer, as well as information, flows related to them, and consists of the following components:

1. Upstream supply chain: It includes primary suppliers (who can be assemblers or manufacturers) and their suppliers; all these routes originate from materials. The main activities of this sector include purchasing and transportation.
2. Internal supply chain: It includes all the procedures used by an organization in converting inputs (materials, human resources, etc.) provided by suppliers into outputs, from the time the materials enter the organization until the time the final product moves outside the organization for distribution. The activities of this department include material handling, inventory management, manufacturing, and quality control.
3. Downstream supply chain includes all the procedures for distributing and delivering products to final customers. It is often observed that the supply chain ends when the product is handed over or consumed; however, the activities of this part include packaging, storage, and transportation [37].

In general, the supply chain is a chain that includes all activities related to the flow of goods and the transformation of materials, from the stage of raw material preparation to the stage of delivery of the final product to the consumer. Figure 1.1 shows the flow of goods and materials during supply chain management [46].

Therefore, by accepting this definition of the supply chain, it can be said that "supply chain management" is a combination of art and science used to improve access to raw materials, manufacture products or services, and transfer them to the customer. Supply chain management includes the integration of supply chain activities as well as information flows related to them through improvement in chain relationships to achieve a reliable and sustainable competitive advantage. Therefore, supply chain management integrates supply chain activities, and related information flows through improving and harmonizing actions in the production supply chain and product supply.

According to Christopher, the powerful solution to achieve a cost advantage is not necessarily the volume of products and economic scale but the supply chain management. According to him, the supply chain is a network from upstream to downstream organizations involved in diverse processes and activities that create value in the form of products and services in the hands of the final customer. Many people have explained and analyzed the concept of supply chain management. Some have understood it with concepts such as logistics, operations management, procurement, or a combination of these three [12].

However, we can refer to the comprehensive definition provided by the World Supply Chain Association: "Supply chain management is the integration of key business processes from the end user to the main supplier that provides products, services, and information that create value." [31, 56].

Also, the Association of Council of Supply Chain Management Professionals (CSCMP) has provided the following definition of supply chain management: Supply chain management contains the planning and management of all activities related to the supply of resources, their transformation into products, and support management. It also includes supply chain management, communication, and cooperation between chain members. These members can be suppliers, intermediaries, customers, etc. Therefore, supply chain management is responsible for managing supply and demand between one or more organizations [16].

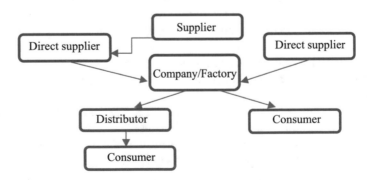

Fig. 1.1 Flow of goods and materials during supply chain management

And finally, to achieve the highest value, the supply chain requires efficient and effective management of all flows and processes between all members at all stages, including product design, procurement, purchase of raw materials, forecasting, planning, production, distribution, and after-sales service. The supply chain management will include all the assets, information, and processes that lead to the supply of these items [11, 57].

Other definitions of supply chain management are mentioned in Table 1.1.

We can conclude a comprehensive definition by reviewing the above descriptions of supply chain management. Supply chain management is the management of all activities related to the transfer of goods from raw materials to the end user, which includes source selection and supply, production scheduling, order processing, inventory management, transportation, Storage, and customer service. Also, it consists of the information systems needed to monitor and coordinate activities. According to the above definitions, the supply chain network can be presented in Fig. 1.2.

In another definition, it can be said that supply chain management includes Customer Relationship Management (CRM) systems and Enterprise Resource Planning (ERP). Customer Relationship Management (CRM) seeks to convey customers' demands and needs to the organization and create a strong relationship between the organization and customers. Undoubtedly, one of the goals of customer relationship management is to transfer market demands to the organization. Then, through the Enterprise Resource Planning (ERP) system, the organization distributes the market demands in all the internal processes of the organization and communication processes with suppliers. The success of supply chain management depends on the integration and relationship between Customer Relationship Management (CRM) and the Enterprise Resource Planning (ERP) system. In Fig. 1.3, this position of dependence is given.

From the set of definitions above, it can be seen that the concept of supply chain management is different from the traditional description of logistics (support). Logistics usually refers to activities that occur within the boundaries of a single organization. In contrast, a supply chain is a network of companies that work with each other and coordinate their actions and activities to deliver a product (service) to the market. Also, traditional logistics focuses on procurement, distribution, maintenance and repairs, and inventory management [18].

While emphasizing the importance of traditional logistics, supply chain management also covers other activities such as marketing, new product development, finance, and customer service. From a broader view of supply chain thinking, these additional activities are now required as part of the work to complete the customer's request.

Supply chain management looks at the supply chain and its organizations as a single entity. It provides a systematic approach for better understanding and managing various activities and coordination between them. This systemic approach offers a framework to show the best response (answer) to the needs of business in the case that without it, it seems that there will be contradictions and paradoxes between these needs. For example, the need to maintain customer service (at a high level) leads to

Table 1.1 Definition of supply chain management

Definition	Year	Researcher
Supply chain management represents planning, organizing, and controlling activities, which include activities related to the transfer and flow of goods and services. Their information flows from the source of raw materials to the final consumers	(2000) [5]	Ballou et al.
Conceptually, supply chain management means dimensions such as strategic cooperation of suppliers, communication with customers, and the level of information sharing	(2006) [34]	Li et al.
A supply chain means shaping the processes of physical, informational, financial, and knowledge flows to satisfy the final consumer's needs through products and services related to suppliers. In this case, supply chain management also includes designing, maintaining, and operating supply chain processes to meet end-user needs	(2000) [3]	Ayers
Supply chain management is a set of techniques effectively integrating suppliers, manufacturers, warehouses, and stores. With this approach, the goods are produced and distributed in the right volume, place, and time to minimize the system's total cost and the service level requirements. Supply chain management as a business attitude has revolutionized business by increasing the business skills and performance of all members in the supply chain	(2006) [49]	Shepherd and Gunter
Supply chain management is a set of perspectives that seeks integration and efficient cooperation between materials, information, and financial flows along the supply chain. This integration ensures that the product is in the right place on time, with the right amount, at the lowest possible cost, and satisfies the customer's needs. In other words, supply chain management coordinates the supply chain processes and the flow of materials from the supplier to the customer	(2009) [22]	Hilletofth
Supply Chain management means integrating and coordinating the transfer of materials, information and, financial flow among several organizations (and within each organization) that are legally independent, but all are members of the chain. In other words, the establishment of companies is such that it provides continuity within the chain from the origin to the destination for the supply of products or services	(2011) [18]	Gold and Seuring
Supply chain means the systematic and strategic coordination of traditional business functions and tactics between these functions in a company with other businesses in the supply chain to improve the company's long-term performance (separately) and the entire supply chain. In other words, supply chain management means coordination in production, inventory (warehouse), location, and transportation between players in a supply chain, to achieve the best combination of responsiveness and efficiency for success in the market	(2018) [25]	Hugos

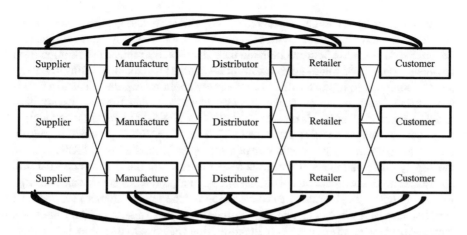

Fig. 1.2 Supply chain network [11]

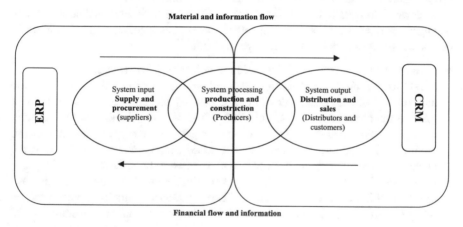

Fig. 1.3 Position of CRM and ERP with supply chain management (SCM)

maintaining a high volume of inventory if activity and production at a high-efficiency level require a reduction in inventory levels [11].

The process can be managed in an integrated and coordinated manner only when these contradictions are resolved, and a balance is created between them.

1.2.3 Reasons for the Importance and Need of Supply Chain Management

With the expansion of the globalization of markets, the only way to continue the survival of companies and organizations depends on increasing competitiveness and gaining and maintaining a sustainable competitive advantage. Customer service should be at the forefront of activities to achieve the supply chain's competitiveness. Today, the competition between individual companies has replaced the competition between supply chains, and supply chains link suppliers to a manufacturing company and the company to its customers. Proper supply chain management ensures excellent customer service, low costs, and short cycle times. Today, the practical solution to achieve a cost advantage is not necessarily the volume of outcomes and economic scale but the supply chain management [4]. Therefore, the importance of supply chain management lies in the definition of the supply chain. Stadler and Kilger define the supply chain as the supply chain is a set of organizations that are divided into upstream and downstream organizations and with a product or service and through various activities and processes to create value for the final customer [29, 53]. The supply chain of a product can be tiny. Still, the number of supply chain links has increased exponentially due increase in production capacity, mass production, extensive storage, and different product supply and distribution channels. In this situation, supply chain management and its relations are essential [49].

According to Handfield and Nichols [21], the three main factors that have made managers realize the importance of supply chain management and thoughtfully pursue the issue of supply chain management are:

- Traditional and electronic information revolution: The structure in traditional SCM has two buying and selling sides, where information flows from one side and goods from the other side, and after passing through certain hierarchies, it reaches the consumer. In electronic SCM, with the introduction of the Internet, the above traditional structure has been changed, and new parts are added to it, including the addition of databases. The SCM process has changed due to the change in its structure, and parts of it are expected to be removed.
- Customers demand to buy products and services with good quality, lower cost, more suitable delivery conditions, and more modern technology with a longer lifespan, ultimately leading to increased competition among producers and manufacturers.
- The necessity of creating a new structure in inter-organizational relations: a significant factor that is perhaps the most critical and complex part of effective supply chain management is the management of links in the supply chain. This factor in the analysis of supply chain management has a tremendous effect on the performance level of different parts of the chain, in such a way that in the supply chain, many failures and successes are caused by how relationships are formed and managed in the chain, and without a doubt, addressing this issue in the field of the supply chain, seems very important and vital [47]

Also, in the past, most organizations under-managed their supply chains; instead, they focused on their operations and immediate suppliers. But several important factors are driving the need for supply chain management for business organizations that actively manage their supply chain. These essential factors are:

- Need to improve operations

Over the last decade, organizations have implemented lean manufacturing and total quality management activities. They will be able to achieve improved quality while eliminating a large number of overhead costs of their system. There is still room for advancement, and the opportunity exists mainly in procurement, distribution, and supply chain support.

- Increasing the level of sourcing from abroad

Organizations are increasing their levels of outsourcing; it means buying goods and services instead of producing or providing them by themselves. As the level of outsourcing increases, organizations spend significant time and money on supply-related activities (wrapping, packing, handling, loading, unloading, and set-up). Much of this cost and time is spent on these and other unrelated activities that may not be necessary.

- Increase in the shipping cost

Worldwide shipping costs are increasing and need to be carefully managed.

- Competitive pressures

Competitive pressures have led to increasing new products, shorter product creation and development periods, and growing demand for custom manufacturing. In addition, adopting quick response strategies and trying to reduce delivery time are also among these pressures.

- Globalization

Expanding globalization expands the physical length of supply chains.

- The increasing importance of e-commerce

The increasing importance of e-commerce has added new dimensions too commercial buying and selling and provided new challenges.

- Complexity of supply chains

Supply chains are complex, dynamic, and have many inherent uncertainties that can adversely affect the supply chain. Such as incorrect forecasts, late deliveries, substandard quality, equipment failure, and changed or canceled orders.

- The need to manage inventories

Inventory plays a meaningful role in the success and failure of the supply chain, so coordinating inventory levels throughout the supply chain is critical. Shortages can disrupt the timely flow of work and damage supply chain performance, while excess inventories increase unnecessary costs. It is not unusual to have a lack of inventory in one part of the supply chain and excess inventory in another.

1.2.4 Components of Supply Chain Management

In a systemic view, the supply chain consists of components that are related to each other, and the interaction between them will lead to the success of the entire system. A supply chain is made up of three parts as a system.

1. System input: types of resources needed by the organization, information, and knowledge related to the organization's resource suppliers.
2. System processing: production, assembly, and conversion of input resources into products.
3. System output: distribution of products (goods and services) and their transfer to customers.

Also, four leading players play a role in a supply chain system in a supply chain which includes [23]:

1. Suppliers that are located at the entrance of the system.
2. Producers that are located in the processing section of the system.
3. Distributors that are located at the end of the system.
4. Customers that are located at the end of the system.

By considering these factors, the different states of the supply chain can be drawn as in Figs. 1.4, 1.5 and 1.6.

Therefore, the components of supply chain management are the variables by which business processes are integrated and managed by the supply chain. Lambert and cooper [31] state the following components for a successful supply chain:

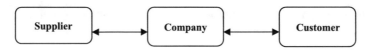

Fig. 1.4 Simple supply chain [25: 45]

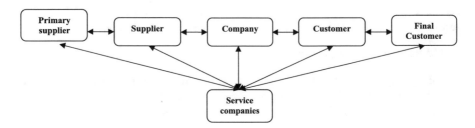

Fig. 1.5 Expanded supply chain [25: 45]

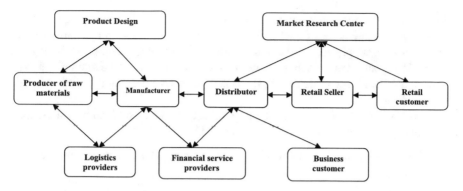

Fig. 1.6 Advanced supply chain [25, 45]

- Planning and Control
- Work structure, which determines how activities and tasks are performed.
- The organizational structure of an independent company or supply chain.
- Product flow structure. Supply, production, and distribution networks throughout the supply chain.
- Structure of the flow of information, which defines the type of information and how often it is updated.
- Management methods based on philosophy or their absence in the supply chain can affect the level of commitment of other supply chain members.
- Shared rewards and risks.
- Attitude and culture.

1.2.5 Objectives of Supply Chain Management

The goal of any supply chain is to maximize the total value of the product, which is the difference between the worth of the final product for the customer and the cost that the supply chain spends to satisfy the customer's demand. In most commercial supply chains, value is closely related to the profitability of the supply chain. Supply chain profitability is the difference between what is received from the customer and the total cost spent throughout the supply chain. The final profit is shared between all stages of the supply chain. The greater this capability, the greater the success of the supply chain. Of course, the success of the supply chain is measured based on the profitability of the supply chain, not based on the individual profit of each stage, because the maximum gain of each step does not necessarily guide to the maximization of the profitability of the supply chain.

Also, the mission of supply chain management is to obtain customer satisfaction, which is a factor that causes the survival and continuity of the company. The goals of supply chain management are to minimize the costs related to the flow of materials and information so that the suitable goods and services reach the right customer (intended) in the right place and time, in the right amount, with the right quality, and in the right conditions. This goal will not be achieved by integrating these activities and improving supply chain relations. In addition, with this integration, the flow of materials and information will move more smoothly, regularly, and dynamically and provide the possibility of achieving the strategic and operational demands of chain management.

Therefore, the goals of supply chain management can be divided into short-term goals and long-term goals, which are described below:

(A)   Short-term goals: The short-term goals of supply chain management are:

- Efficiency
- Inventory reduction
- Reducing the time delay of product production compared to the schedule
- waste reduction
- More detailed planning for product production through information feedback

(B)   long term purposes: The long-term goals of supply chain management are:

- Obtain customer satisfaction
- Maintain market share
- Gaining profit for all organizations involved in the supply chain
- And in general, it can be said that SCM has a significant impact on the company and the consumer.

Supply chain management activities can improve customer service. They can ensure customer satisfaction by providing the necessary products at the right place and time. By increasing customer satisfaction, companies can build and improve customer loyalty.

SCM also provides a significant benefit to companies by reducing operating costs. SCM activities can reduce the cost of purchasing, production, and the entire supply chain. Reducing costs improves a company's financial position by increasing profits and cash flow.

Perhaps the vital role of SCM in society is less known and understood. SCM can contribute to human survival by improving health care, protecting people from climate extremes, and preserving life. People rely on supply chains to provide necessities such as food, water, medicines, and healthcare. The supply chain is also critical to delivering electricity to homes and businesses, providing lighting, heating, air conditioning, and refrigeration energy.

SCM can also enhance the quality of life by boosting job creation, providing a basis for economic growth, and improving living standards. It offers many career opportunities as supply chain professionals design and control all supply chains in a community and inventory, warehousing, packaging, and logistics. In addition, one of the common characteristics of most developing nations is the absence of a well-developed resource supply chain. Societies with strong and well-developed supply chain infrastructure, such as large rail networks, intercity highway systems, and airports and seaports, can exchange goods at lower costs, allowing consumers to purchase more products and economic growth.

1.2.6 Scope of Supply Chain Management Performance

In today's competitive and diverse market, for the production and supply of a product, if the supply chain is used, experience has proven that it should be applied in five main and decisive areas, which are called the five primary drivers of the chain. In using the supply chain, the main attention is paid to the five drivers simultaneously and with each other, and there is no precedence or priority among them. And companies, organizations, and different industries making decisions. In particular, five factors define their supply chain capabilities as the capabilities required by the supply chain will be achieved through their management. Figure 1.7 shows these five factors.

In the effective management of the supply chain, it is necessary to know each of the drivers and how they work because each driver can influence the supply chain and create specific capabilities, and in the next stage of the supply chain management, it should be able to develop presumably be obtained from the combination of these stimuli. First, it is necessary to examine each of these stimuli independently:

- Production

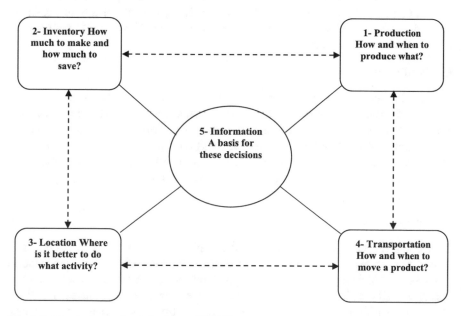

Fig. 1.7 Five main supply chain drivers [25, 35]

This variable expresses the capability and ability of the supply chain in manufacturing and storing the product in factories and warehouses. Managers must determine the balance between responsiveness and efficiency because creating capacity is costly, and excess capacities are unused resources that are not used and naturally do not generate income. Therefore, the higher the new capacity, the lower the efficiency. By choosing one of the two production approaches (product-oriented or task-oriented) or combining the two mentioned methods, the industry can better respond to customer demands. In the first approach, the direction is to expand expertise and knowledge of operations related to a specific category of products. At the same time, the second approach is to acquire expertise about a particular process instead of a specific desired product. In production, the market needs for the type, quality, price of the product and required periods, etc., must be carefully coordinated with the production and storage facilities and capacities in transportation and considered.

• Inventory

In the supply chain, inventory includes everything from raw materials and equipment in the production process to finished products. Inventories are kept by manufacturers, distributors, and retailers. In this case, managers must decide between responsiveness and efficiency. Maintaining a large stock or the supply chain makes it possible to respond quickly to market fluctuations. However, creating and maintaining inventory is costly, and to reach the highest efficiency level, the inventory should be kept as low as possible. In this manner, the organization must balance the cost of maintenance and the cost of creating more flexible production capabilities by using three methods: periodic inventory, safety inventory, and seasonal inventory depending on the conditions and facilities.

• Location and location

The location refers to the geographical location of the drivers of the supply chain. In addition, the type of activities of each driver is determined in this section. Creating a balance between responsiveness and efficiency by deciding on limited locations or locations close to scattered customers at this stage, considering equipment costs, labor costs, taxes, duties, and proximity to suppliers, is done by customers. Location and location decisions are very strategic since they profoundly impact a supply chain's performance and cost characteristics.

• Transportation

Transportation refers to the movement of anything from raw materials to finished products between components of a supply chain. Creating a balance between responsiveness and efficiency depends on the type of transportation method directly. Transportation methods, including shipping, trains, pipelines, trucks, airplanes, and electronic transportation, each have advantages and limitations. Since transportation costs are the third cost factor. They are in a supply chain; Therefore, the decisions taken are crucial. As a general rule, it can be stated that with the higher value of the products and the smaller volume (such as electronic devices and components or medical equipment and materials), the transportation network should focus on the speed of response. On the other hand, the lower the value of the products and the larger their volume (such as bulky goods such as beans, wood, timber, etc.), the network's focus should be more on efficiency.

- Information

The basic information for decision-making is about the four drivers of the supply chain and acts as a bridge of communication between them. The depth and accuracy of information help managers make the right decisions and ensure the supply chain's success. The mission of information systems is created based on activities in the following three areas:

1. Receiving and exchanging data
2. Data storage and retrieval and data analysis
3. Reporting.

Also, in the supply chain, there are various information systems with different combinations of the above three areas, the most important of which are:

(1) Enterprise resource planning (ERP)
(2) Logistic systems (Ls)
(3) Advanced planning schedules (APS)
(4) Transport planning schedules (TPS)
(5) Demand planning (DP)
(6) Revenue management (RM)
(7) Customers Relationships Management (CRM)
(8) Sales Force Automation (SFA)
(9) Manufacturing Executing Systems (MES)
(10) Warehouse Management Systems (WMS).

The above systems are the systems that have a combination of the three mentioned areas. These systems are designed to achieve different supply chain goals. It can be said that the supply chain management system is an integrated set that includes the applications of scheduling and advanced planning, transportation planning, demand planning, and inventory planning. This tool needs all existing information systems in the form of inventory management systems, transportation scheduling systems, and warehouse management systems to strengthen its analysis and plan. It should also be acknowledged that today the demands and expectations of the global economy have forced companies and industries to make general improvements in the supply chain in the field of competition and struggle to enter the market and maintain market share. Four new and promising information technologies can be used to complete supply chain management systems. And in this regard, four new information technologies can be used to complete supply chain management systems.

1. Radio frequency identification (RFID)

This technology tracks pallets, boxes, and individual items moving through the supply chain from the manufacturer to the end customers. Today, with this technology becoming cheaper, its general use in the form of active and passive tags and radio scanners to collect information about products and their movement throughout a supply chain has become common. Currently, there is an organization called GSI for developing product standards, which has introduced a new term for numbering patterns, which only includes a single 14-digit numbering pattern called the Global Trade Item Number (GTIN) electronic product code. The product's electronic code, or 14-digit electronic code, contains the following four critical pieces of information:

1. Editing code: It shows the type of editing of the electronic code of used products.
2. Manager code: The organization that creates the electronic code introduces the product.
3. Product code: specifies the type of product or service.
4. Serial number: identifies a specific product or service sample.

Organizations and companies can register a global number for their goods by registering in the GSI organization and obtaining a manager code by adding a serial number. Currently, most companies use systems based on electronic product code information standards combined with active radio frequency identification tags. In these systems, a radio scanner on an inactive sticker is active. It sends the product's electronic code number to an application system that uses the Internet to communicate with a technique called the global network of the product's electronic code. The electronic product code network is a system designed by the GSI organization. It enables the company to identify the type of product to which the electronic product code number is assigned and more information about that particular item, such as the date of manufacture and the history of movement during Get the supply chain.

2. Business process management (BPM)

A process is a set of consecutive actions that lead to creating a product or providing a specific service. Also, the method based on which an organization uses people's opinions, products, and information are called the business process performance method of that organization. Business process management is a method companies apply to create continuous improvement and a growing trend to improve the performance of company activities. To manage the business process, the companies must first identify the critical processes of the respective organization and determine the steps that each process must go through; by using the BPM software, the method of categorizing and continuously displaying the data that moves and trades at each stage, it shows the knowledge and process of organizing their functions. In addition to the above, BPM software can automate many routine activities, such as transferring information from one activity to another, or it can be used to detect specific errors and automatically send warnings to people who should. It was used to respond to these conditions because the people involved in the activities of a process will observe it as soon as an event occurs and will be able to react effectively to respond to problems and improve the productivity of the hood. Also, information about the performance of each process by BPM software. It is collected and creates an important database that companies and industries can use to design new strategies when existing processes can no longer meet business needs.

3. Business intelligence (BI)

The speed of changes in the supply chains is always the reason for the eagerness of people and organizations to understand the direction of the changes to keep their interests and position together. Using the business intelligence system in this connection helps them discover what is happening inside the organization and the markets they serve. When information is collected from different sources by BI software that can analyze data and display its results, the direction of changes will be revealed. This software can analyze various data from spreadsheets and simple charts.

4. Modeling using simulation

With the speed of business changes, companies are forced to make decisions that significantly affect their operations and profitability. For this purpose, they use simulation software that allows them to create a model of a factory, a supply chain, or a product delivery route or service provision and expose them to inputs. Put different positions and observe what happens. A plan is implemented on paper until it is modeled, and its performance in various conditions is not simulated; its problems cannot be estimated in practice. Through simulation, these problems are revealed much faster and cheaper than doing them in the real world. In the application of business process management, by using information collected by business intelligence management as simulation inputs, business processes are modeled under different conditions, and companies can model and Test new business models before they are implemented and choose the best model for implementation.

It is necessary to remember that the four described technologies will be effective and efficient when they are used in interaction with each other so that successful supply chains can be designed and implemented. After identifying, bottlenecks and deficiencies in a supply chain, business intelligence databases and analytical software can first identify the root of the problems. Then with the help of simulation systems, they can model the potential reforms in the supply chain and the possible effects of each change. Observe and finally select and implement the most effective reforms that will produce the expected results with a high level of confidence [58, 11].

1.2.7   Principles of Supply Chain Management

Obtaining the appropriate and expected results in the areas of supply chain performance requires the application of a set of principles that are known as the principles
of supply chain management, including:

1. Planning

The strategic part of supply chain management is planning. To carry out planning,
one must first determine the status of the company or organization concerning the
three subcategories that affect planning, which are as follows:

• Demand Forecasting

Demand forecasting is based on knowledge of the four main market variables (supply,
demand, product specifications, and competitive environment) and is obtained using
qualitative, causal, time series, or simulation methods. Companies react to the
demand forecast based on the production capacity, utilization, and the amount of
their inventory in planning.

• Product pricing

Companies and supply chains influence the demand by using the price, and most
tend to price the product during peak consumption seasons to increase the demand
to earn maximum income, a. Thempany's cost structure can be organized in such a
way. It is necessary to establish a balance between demand and production capacity.

• Inventory management

The goal of inventory management is to reduce inventory as much as possible while
maintaining customer service. Based on the forecast of demand and product price, it
creates savings due to the increase in warehouse volume to reduce cost and achieve
the best product price.

2. Provision of resources (supplies)

Choosing the suppliers who deliver the goods and services needed to make the product
or provide the service has been one of the main activities for buying goods and
services with the lowest price and the most suitable quality since ancient times. Even
today, although this work is still essential, other activities in this sector are becoming
equally important and have added to the importance of this sector. The activities
carried out in the supply department include purchase management, consumption
management, vendor selection, contract negotiation, and contract management.

3. Manufacturing

The activities to realize the production, testing, packaging, and preparation for
the delivery of goods are in this stage. They include product design, production
scheduling, and production unit management.

4. Delivery management (orders)

The primary mission of order management is to collect information on the estimation of customers' needs and the consumption market from retailers and distributors and transfer it to service companies and manufacturers, in which order delivery dates, product replacements, and back orders decline serious attention seriously due to the complexity of the supply chain and changes in market demand, order and delivery management is being combined with customer relationship management, which is presented as marketing or sales operations.

5. Feedback (feedback processing)

In the goods field, most end customers, retailers, distributors, and manufacturers all return incorrect manufactured products or products that were damaged during transportation or in the factory. Companies and supply chains should track the types of returns and react appropriately to them.

1.2.8 Supply Chain Management Processes

Before dealing with the issue of supply chain processes, the concept of the process will be briefly explained. A process is a series of related activities that are carried out to achieve a specific goal. The process shows how the work is done during the internal task areas of the organization. The process can be considered a value chain where each stage (each chain seed) adds value to the previous step. In other words, the inputs of a process are generally the outputs of different processes. Therefore, processes in an organization are typically planned and implemented under controlled conditions to achieve added value. A process is a set of interrelated activities/processes that turn a set of inputs into a set of outputs (products/services) in line with the organization's business activity and create added value. Finally, the process approach considers flows from the entire organization in a way that includes all activities and makes them integrated and oriented, so it can be said that supply chain processes are a set of different activities that include the entire value chain from customers to suppliers.

Many think organizations should focus on processes to evaluate their business performance. In general, the organization's effectiveness depends on its processes' effectiveness. When the organization can be considered successful, the performance process indicators are in appropriate conditions [49].

According to Chen and et al.'s opinion [10], the effectiveness of supply chain processes is a suitable criterion for measuring supply chain performance and measuring supply chain performance plays a vital role in gaining competitive power. For this purpose, to achieve a model for measuring organizations' competitiveness level based on supply chain processes, it is crucial to identify supply chain processes and supply chain performance measurement indicators. In the rest of the book, along with introducing different attitudes about supply chain processes, a list of the methods mentioned in the subject literature is also presented.

1.2.8.1   Process Approach to the Concept of Supply Chain Management

In the supply chain discussion, there are two different process approaches:

1. Staged attitude

The phased approach aims at the processes formed and divided within a supply chain course at the intersection of two consecutive points. And the purchase is divided. In this view, all the process in a supply chain are divided into one or two categories of related components, which ultimately depends on the customer's demand.

2. Tensile/Compressive attitude

The pull/push approach is concerned with processes divided into two categories within the supply chain to fulfill a customer's order or anticipate customer orders. Pulling methods are triggered by a customer's order, while push processes are triggered, and a forecast of customer orders is formed. Stretching techniques are implemented to fulfill a customer order's response. While during the implementation of a push process, the demand is unknown and must be predicted [11].

As stated, there are two types of attitudes about supply chain processes. According to these two types of perspectives, some of the models mentioned in the supply chain management literature are mentioned:

(1)   **Handfield and Nichols [21] process model**

This model of supply chain management processes based on the nature of actions or the type of operations that are carried out during the chain; it is divided into three categories, which are:

(a) **Information management**: In discussing the supply chain, coordination in activities is essential. This point is also valid in discussing information management in the chain, management of information systems, and information transfer. Coordinated and appropriate information management between partners will cause increasing effects in speed, accuracy, quality, and other aspects. Correct management of information will lead to more coordination in the chain, and in general, in the supply chain, information management will be effective in different parts [53].

(b) **Logistics management**: In the analysis of production systems, the issue of logistics includes the physical part of the supply chain. This department, which consists of all physical activities from the raw material preparation stage to the final product, including transportation, warehousing, production scheduling, etc., occupies a relatively large part of the supply chain activities. The scope of logistics is not only the flow of materials and goods but the focus of supply chain activities, where relationships and information are the supporting tools for improvement activities. In one of the definitions, logistics is called 7R,[1] which

---

[1] Right product, Right quantity, Right Place, Right condition, Right quality, Right cost, Right customer.

includes correcting product supply in the right amount, time, place, conditions, and cost for the customer [51].

(c) **Relationship management**: Relationship management tremendously impacts all aspects of the supply chain and its performance level. In many cases, the information and technology systems required for supply chain management activities are readily available and can be completed and deployed in a relatively short period. However, many initial failures in the supply chain result from the poor transfer of expectations and the resultant behaviors between the parties involved in the chain. In addition, the most critical factor for successful supply chain management is reliable communication between partners in the chain in such a way that partners have mutual trust in each other's capabilities and operations. In other words, in developing any integrated supply chain, trust and confidence among partners and their reliability plan are critical and vital elements for achieving success [27, 31].

(2) **The Process Model of the Supply Chain Association (2005)**

According to this model, a supply chain consists of five processes. The five main processes of the supply chain include planning, sourcing, manufacturing, delivery, and return. They broadly describe these efforts, including supply and demand management and finding sources related to raw materials and parts. Manufacturing and assembly, warehousing and inventory tracking receiving orders and order management, distribution in all relevant channels, and delivery to the customer. This model was examined in detail on the principles of supply chain management. This model is considered a process reference model by the Supply Chain Association as a formal diagnosis and problem-solving tool for the supply chain management. This model regarded only a supply chain framework that provides detailed performance criteria, industry best practices, and software requirements in a business process model. The focus of this model is on describing the management processes within the supply chain and the relationships between them, compiling standard indicators to measure the performance of the supply chain processes and evaluating and modifying the configuration of the supply chain to support the continuous improvement and promotion of the competitive management of the supply chain in the business environment [55, 42–44].

(3) **Chan's process model [9]**

Based on this model, the supply chain can be divided into six categories based on the type of operations that take place along the chain:

(a) **Relationship management and customer service**: A business process that looks at the customer and develops a structure to create and maintain customer relationships.

(b) **Supply chain planning and demand management**: This process tries to prevent excess supply or demand in different parts of the chain by balancing supply and demand in the entire chain.

(c) **Manufacturing and logistics flow management**: A process that includes all the necessary activities for the production and transfer of goods from factories to the next stage of the chain, and the task of obtaining, implementing, and managing flexibility in manufacturing is the responsibility of this process.

(d) **Procurement and management of relationships with suppliers**: This process is responsible for purchasing and procuring materials and parts in different categories of the supply and well as developing and maintaining effective relationships with suppliers of materials and components.

(e) **Product development and commercialization:** This process provides a structure for developing and placing products in the market and cooperating with customers and suppliers.

(f) **Order fulfillment and distribution process:** This process includes all the necessary activities to define the customer's needs, design the distribution network and fulfill the customer's requests with the minimum delivery cost.

Among the above six processes, the two processes of relationship management with suppliers and relationship management with customers are the processes that communicate with the chain's external members, i.e., customers and suppliers. The other four supply chain processes are also related to suppliers and customers but establishing and maintaining the relationship of different processes is the responsibility of the two mentioned. It should be kept in mind that supply chain management processes play a critical role in the organization's competitiveness. Environmental stimuli and intra-organizational pressures influence processes. On the other hand, the type of relationships between them in terms of integration and cohesion can play a critical role in increasing the competitiveness of organizations [9, 534].

(4) **Process model of** [11]

In this model, all supply chain management processes in the organization are divided into three main process categories. It manages the primary operations of the flow of information, product, and capital needed to receive, produce and fulfill the customer's request, which will be explained further.

(a) **Customer Relationship Management (CRM)**: All processes focus on interactions and communications between the organization and its customers. These processes are also called downstream supply chains, including after-sales service, delivery of goods, sales, and marketing.

(b) **Internal Supply Chain Management (ISCM)**: All the processes inside the organization. These processes are actions that take place inside the organization to make goods. Like product design, it includes product production and manufacturing, warehousing and inventory control, quality management, etc.

(c) **Supplier Relationship Management (SRM)**: All processes focus on interactions and communications between the organization and its suppliers. These processes are also referred to as upstream supply chains. Supplier relationship management is defined as commercial activities necessary to record and understand interactions with different suppliers in the profitability of an economic enterprise. Supplier relationship management has the task of reducing the costs of providing goods and services, managing relationships with suppliers, and, most importantly, increasing profits.

It should be noted that the purpose of all three processes is to serve the same customer and achieve the same goal; therefore, for the supply chain to be successful, all three approaches must be well integrated. Companies' organizational structure dramatically impacts the success or failure of efforts to integrate these three processes [11].

(5) Global Supply Chain Forum (GSCF)

From a process point of view, supply chain management integrates critical business processes in a chain consisting of the final customer to the primary supplier that provides products, services, and information and creates added value for customers and stakeholders. According to this point of view, the eight processes of the supply chain are:

(a) **Customer relationship management (CRM)**: The customer relationship management process determines how to create and continue the relationship with the customer. Management identifies customers and a key group of them that should be targeted. The purpose of this process is the strategic level is identifying customer segmentation, preparing criteria for customer classification, providing guidance for customers to provide products and services according to each customer's taste, preparing a standard framework, and preparing guidelines for sharing the benefits of improving processes with customers. This part of the agreement about the product and service is written and implemented.

(b) **Customer Service Management (CSM)**: The customer service management process is the organization's perspective vis-a-vis the customer. This department provides a unified source of information from the customer, including access to goods, shipping date, and order status. This department is responsible for implementing agreements. In short, the purpose of this process at the strategic level is to create the infrastructure and cooperation tools needed to enforce contracts on services and products and provide the critical point of contact with the customer. At the functional level, this process is responsible for responding to events. It is internal and external.

(c) **Demand Management (DM)**: The process of demand management must balance the demands of customers with the supply capabilities of the organization. This process includes forecasting and coordinating demand with production, purchase, and distribution. Also, the task of this process is to create and implement a plan for possible events when a problem occurs in the performance. At the strategic level in this process, the forecasting approach, information flow plan, synchronizing procedure, the management system of possible events, and preparing a standard framework for this work are determined.

(d) **Order Fulfillment (OF)**: The process of fulfilling strategic orders pays attention to the requirements of manufacturing, logistics, and marketing for the design of the distribution network. At the strategic level, marketing goals,

supply chain structure, and customer service goals are reviewed, order fulfillment requirements are defined, the distribution network is evaluated, a plan for
order fulfillment is designed, and a standard framework is prepared.

(e) **Manufacturing Flow Management (MFM)**: This process deals with creating
the product and creating the flexibility needed in manufacturing to provide
the product to the target market. The method includes all activities required for
product flow management, availability, Implementation, and flexibility management. At the strategic level, the production, sourcing, marketing, and logistics
strategies are examined, and the degree of flexibility required in production, the
stretching/push limits of the organization, production constraints, requirements,
and production capabilities are determined.

(f) **Product Development and Commercialization (PDC)**: Product development
is vital for the organization's continued success. Developing new products in a
short time and presenting them to the market is an essential part of the organization's success. When the product reaches the market is one of the crucial things
focused on in this process. Supply chain management includes the integration
of customers and suppliers in the product development process to reduce the
time the product reaches the market. By decreasing the creation's life cycle
time, the product's development and presentation to the market should be done
quickly to compete with other similar products. At the strategic level of this
process, sourcing, production, and marketing strategies are examined, the idea
generation process is created, guidance is prepared for the members of the inter-
departmental Product development teams, product limitations and new product
guidelines are provided, and finally, a standard framework is provided for this
work.

(g) **Supplier Relationship Management (SRM)**: The supplier relationship process
determines how the organization interacts with suppliers. As the organization
needs to develop its relationships with its customers, it should have a close relationship with some suppliers and a proper relationship with others. This process
is related to defining and managing agreements about services and products. At
the strategic level, in this process, reviewing the organization's production and
sourcing strategies, determining criteria for categorizing suppliers, preparing
guidelines for the degree of customization in agreements about services and
products, creating a standard framework, and creating guidelines for sharing
the benefits of improving processes with suppliers.

(h) **Return Management (RM):** Effective referral management is vital to supply
chain management. While many organizations ignore this process because they
do not believe its essential, this process can help the organization achieve a
sustainable competitive advantage. If this process is implemented correctly, it is
easy to identify the opportunities to improve the project's production and failure
points. At the strategic level, this process includes reviewing legal and environmental guidelines, creating a referral network and credit laws, identifying the
secondary market, and preparing a standard framework for this work [32].

(6) Porter's value chain process model [41]

In this model, supply chain processes are discussed based on the value chain concept of Michael [44]. He believes that the best way to describe the operations of any company is to tell them according to the value chain. All companies in a particular industry have a similar value chain, including raw material procurement, product design and manufacturing, and customer service.

The main discussion in value chain management is that the costs of each production stage with the value it creates are examined and compared with competitors, and evaluation and cost management are done. This chain relates to the most influential environmental factors on both sides, i.e., suppliers of raw materials and parts and factors after production to customers. The relationship between the value chain of the organization and the value chain of suppliers and customers forms the chain that Porter calls the "value system." Of course, this concept is called other titles such as value network, extended value chain, and supply chain. According to Porter's point of view, supply chain processes include two following categories:

(a) Main processes

The main activities are those activities that are called value-adding activities. Doing them increases the product's value and moves it towards the customer. For example, raw materials are imported, received, stored, Etc. (inbound logistics), then production operations are carried out on these materials that upgrade them to the manufactured product. Next, the manufactured product is packaged, transported, and stored (outbound logistics), then marketing activities increase the product's value, and its sale turns the product into money. After-sales service is another value-adding activity applied to the product at the end. All these activities are a direct source of profitability for the organization.

(b) Support processes

Management support processes that provide the necessary support from the supply chain processes and determine their direction, such as organizational infrastructure (management financial accounting), human resource management, research and development, procurement, and procurement [41, 220].

(7) The American Quality and Productivity Center process model

This center presented a process framework in 1991. In designing this model, more than 80 organizations interested in being pioneers in using modeling (from all over the world) participated. Companies such as Boeing, BT, Ford, IBM, the US Navy, Etc., officials believe that a common vocabulary not dependent on a specific industry is necessary to classify information based on processes and helps organizations transcend the limitations of the terminology within the organization. The process classification framework can be valuable for understanding and mapping business processes.

This center has also divided the business processes of the organization into two categories, operational procedures and support and management processes, which are mentioned in Table 1.2.

According to the mentioned processes from the different researchers' views, the presented techniques can be summarized in Table 1.3.

1.2.9   Decision-Making Steps in Supply Chain Management

Decision-making in supply chain management has three main stages which are [11]:

- Step 1—Designing the supply chain strategy: deciding how to create a coherent structure in line with the organization's strategies and specifying the combinations and processes needed at each stage.
- Step 2—Supply chain planning: adopting operational decisions and policies in the way of activities of the organizations in the chain without making changes in the strategic decisions adopted in the previous step.
- Step 3—Supply chain operations: making decisions and taking measures to respond as best as possible to customers' orders and specific needs, which are usually reviewed and decided daily or weekly.

1.2.9.1   Obstacles and Challenges in Making Decisions in Supply Chain Management

Among the essential obstacles in creating coordination in supply chain management, the following 5 cases can be mentioned [11]:

1. Incentive obstacles: Obstacles that cause non-cooperation of members with each other and cause noticeable changes and decrease in profits of the chain, for example, inappropriate treatment of the organization with suppliers or distributors.
2. Information processing obstacles: Obstacles that cause the information to be distorted in the whole chain, and as a result, the organization is unable to respond to the real needs of the customers, for example, sending incorrect information from the distributors to the organization.
3. Operational obstacles: Obstacles to fulfilling orders include the supplier's failure to send goods on time to the organization.
4. Price barriers: Factors that cause a lack of price stability in the final product; these factors can be applied to the chain from within or outside the organization; for example, the lack of stability in the price of imported raw materials can be mentioned.
5. Behavioral barriers: Refer to internal issues involving the organization.

Table 1.2 Components of supply chain processes according to the process model of the American Center for Quality and Productivity

Support and management processes		Operational processes	
Process components	Process name	Process components	Process name
– Creation and management of human resource strategies, – Policies and programs Sourcing, selection, and recruitment of employees - Developing and improving employees – Appreciating and maintaining employees – Relocating and retiring employees – Management of employee information	1. Management and development of human capital	– Definition of business concept and long-term vision – Preparation and preparation of business strategy – Management of strategic initiatives	1. Preparation and compilation of vision and strategy
– IT business management – Creation and management of customer relationship information technology – Management of business volatility and risk – Creating and maintaining information technology solutions – Implementation of information technology solutions – Providing and supporting information technology services – Organization information management – IT knowledge management	2. Information technology (IT) management	– Designing products and services – Development of products and services	2. Designing and developing products and services

(continued)

Table 1.2 (continued)

Support and management processes		Operational processes	
Process components	Process name	Process components	Process name
– Planning (budgeting) and management accounting – The process of accounts payable and expense savings – General accounting and financial reporting – Accounting of incomes – Management of fixed assets - Salary process – Management of treasury operations – Management of internal controls – Management of taxes (fiscal affairs)	3. Management of financial resources	– Preparation and preparation of marketing, distribution, and channel strategy – Preparation and compilation, and management of customer strategy – Management of advertising, pricing, and activities – Management of sales partners and allies – Management of opportunities and sales lines – Management of sales orders - Sales promotion	3. Marketing and selling products and services
– Destruction of assets and workplaces – Design and construction of assets – Maintenance of assets and workplaces – Physical risk management – Management of capital assets	4. Acquisition, construction, and management of assets (property)	– Planning the supply of required resources (planning the supply chain) – Production/manufacturing/delivery of products – Logistics and warehousing management – Providing services to customers – Buying materials and services	4. Delivery of products and services

(continued)

Table 1.2 (continued)

Support and management processes		Operational processes	
Process components	Process name	Process components	Process name
– Determination of health, safety, and environmental effects – Compilation and implementation of health, safety, and environmental programs – Education and training of employees (about health, safety, and environment – Monitoring and management of health, safety, and environmental programs – Ensuring compliance with laws and regulations – Management of therapeutic activities	5. Environmental health and safety management	– Preparation and compilation of service/customer care strategy – Customer service management – Providing installation and repair services after sales – Measuring and evaluating customer satisfaction – Customer service workforce management	5. Customer service management
– Establishing relations with investors – Managing relations with the government and industries – Management of relations with the board of directors and the assembly – Management of ethical and legal issues - Management of public relations programs	6. Management of extra-organizational relations		
– Creation and management of organizational performance strategy – Development of trans-organizational knowledge management capacity - Modeling performance – Change management	7. Knowledge management, improvement, and transformation		

Table 1.3 Components of supply chain processes

Reference	Components of supply chain processes	Row
Handfield and Nichols	Information management, logistics management, relationship management	1
Quality and Productivity Association of America	Operational processes (preparation and development of vision and strategy, design and development of products and services, marketing and sales of products and services, delivery of products and services, customer service management), support processes (governance and development of human capital, Information technology (IT) management), management of financial resources, acquisition, construction and management of assets (property), environmental health and safety management, extra-organizational relations management, knowledge management, improvement, and transformation)	2
Porter's value chain	Inbound logistics, manufacturing and production, outbound logistics, marketing and sales, after-sales services, support processes (organizational infrastructures (management financial accounting), human resources management, research and development, procurement and procurement)	3
Chopra and Meindl	Customer relationship management, internal supply chain management, supplier relationship management	4
SCORE	Planning, sourcing, manufacturing, delivery, return	5
GSCF	Customer relationship management, customer service management, product development and commercialization, demand management, manufacturing flow management, supplier relationship management, order fulfillment process, return process	6
Chan et al.	Customer service and communication management, supply chain planning and demand management, manufacturing and logistics flow management, procurement and supplier relationship management, product development and commercialization, order fulfillment and distribution process	7

1.2.9.2   Management Functions Against Obstacles and Challenges in the Way of Supply Chain Management

To achieve effective supply chain management, the following five functions help to organize and coordinate more effectively. These functions are briefly described below [21]:

1. Structure of supply chain partners: This function is determined to gain customer satisfaction to create competitiveness in the organization and effectively increase the desirability of the chain and create a strong chain. As a result, it makes coordination in producing new products, improves quality, and reduces costs.
2. Implementation of collaborative communication: This function focuses on sharing information between partners in the chain. In general, this function expands the communication of the supply chain to participation with factors outside the company.
3. Supply chain design for strategic profitability: This management function has been guided and aligned regarding the effective participation of factors outside the organization. It considers the demands of the members in the chain to increase strategic profitability.
4. Supply chain management information: This function uses new technologies to transfer knowledge and improve the supply chain.
5. Reducing supply chain costs: The leading indicator in measuring the effectiveness of a supply chain is the reduction of prices, which is one of the organization's highest goals in creating this chain.

1.3   Strategic Supply Chain Management

Before addressing this issue, it is necessary to define the strategy concept. According to Ahlstrand and his colleagues [1], event strategy is an emergency response to an unpredictable situation, not something planned. They define *strategy* as "a pattern of flow of decisions and actions." In their definition, "pattern" refers to both intention (intended) and emergency (unplanned). According to them, organizations in the real world employ a combination of contingency-based strategies and pre-intention-based strategies. The realized strategy of each company is a combination of measured and planned strategy and emergent or nascent strategy (reactive responses without prior planning to changing conditions).

After defining the strategy, it is necessary to determine the strategic management, which is mentioned below:

- From the perspective of Fred and his colleagues, strategic management is the art and science of formulating, implementing, and evaluating multiple task decisions that enable the organization to achieve its long-term goals [17].

- From Pearce and Robinson's point of view, strategic management is a set of existing decisions and activities to formulate and implement a strategy to achieve the organization's goals [40].
- According to Hunger and Wheelen, strategic management is a set of management decisions and activities that determine a company's future performance and emphasizes monitoring and evaluating external opportunities and threats and the strengths and weaknesses of a company to develop and implement a new strategic direction for it [26].

Also, the distinctive characteristic of strategic management is its emphasis on strategic decision-making. According to Grant [20], strategic decisions have three essential characteristics, including:

- Are important.
- They require the allocation of significant resources.
- They are not easily reversible.

On the other hand, strategic thinking is another concept used to describe strategic management's innovative and creative aspects. Strategic thinking is one of the new reforms in the world of management. Graetz [19] considers the purpose of strategic thinking to maintain a competitive advantage in today's complex and ambiguous environment and believes that the role of strategic thinking is to pursue innovation and imagine a new and different future.

1.3.1 The Role of Strategic Thinking in Supply Chain Management

Liedtka [33] is one of the experts who gave a comprehensive definition of strategic thinking, and by presenting a model, he showed the components of strategic thinking and the relationships between them, which is shown in Fig. 1.8.

These primary elements can be introduced in the supply chain management below. The role of strategic thinking in this discussion will be discussed by examining these items in supply chain management.

- The first element in Liedtka's model is the systemic approach. In supply chain management, a complete pattern of value creation factors should always be considered from the beginning to the end of the chain. In this model, the critical chain competencies, opportunities and threats, and the relationship between these factors are correctly drawn, and the concept of integration in the chain is evoked.
- The second element in Liedtka's model is focusing on the goal. This pillar recommends concentrating the members' resources in the chain to achieve the goal. These resources include the physical facilities of the members and technological capabilities, which are considered for the achievement of all the members.

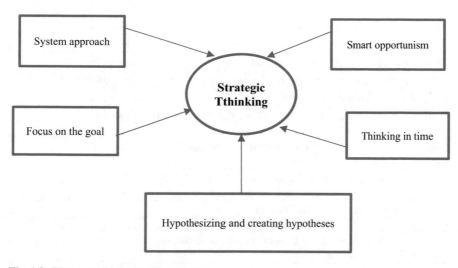

Fig. 1.8 Elements of strategic thinking [33]

- The third element in Liedtka's model is intelligent opportunism. This pillar considers the attention to the body of the chain and the creative capacities hidden in it as a workable matter. It is essential to pay attention to the surrounding environment, welcome new ideas, and discover emerging opportunities.
- The fourth element in Liedtka's model is hypothesis creation and hypothesis creation. This element relies on the scientific method of "hypothesis and hypothesis testing" and evaluates the product of new and prudent hypotheses to deal with the surrounding environment and interact with environmental developments.
- The last element of Liedtka's model is the thought of continuity in time. In this view, the chain is considered as a single organization that does not live only in the present, and the thinking is that today's success of the chain is the result of yesterday's capabilities chain. To succeed in the future, the foundation must be laid today.

Bonn [6] considered "strategic thinking" as "competency-oriented" in an organization. Also, Bonn [7] considered the lack of "strategic thinking" as the biggest weakness of senior managers and considered spreading the vision and turning it into the organization's daily work as the strategic thinkers' primary task. He defines *strategic thinking* as a solution to strategic problems by combining a rational and convergent strategic approach with a creative and concurrent thinking process. He proposes a conceptual framework of strategic thinking that consists of systemic thinking, creativity, and foresight. Table 1.4 summarizes the standard features mentioned for strategic thinking in the research literature.

Based on this, and according to the literature and characteristics identified about strategic thinking, it can be used in supply chain management. The supply chain can be turned into a strategic asset, described below.

Table 1.4 Characteristics of strategic thinking identified in the research literature

Characteristics of strategic thinking	Researchers
Vision, creativity, flexibility, and entrepreneurship	Rowe [42]
Being creative, combining, and insightful in thinking	Mintzberg [35]
Integrative, creative, and divergent thinking	Heracleous [24]
Systems vision, goal-oriented, timely thinking, hypothesis-driven, and intelligent opportunism	Liedtka [33]
Integrative thinking, creativity, divergent thinking, insight in thinking, and innovative thinking	Graetz [19]
Analytical thinking, conceptual, far-sighted, based on knowledge and integration	Thompson Jr. [54]
Systemic thinking, creativity, and foresight	Bonn [7]
The ability to combine knowledge according to the resources and activities of the organization, the ability to find opportunities and conceptualize new markets	Kim [30]
Organizational systematic thinking, insight reflection, and redrafting	Karğın and Aktaş [28]
Conceptual thinking, visionary thinking, creativity, analytical thinking, learning ability, combining ability, and objectivity	Nuntamanop [38]
Systemic thinking, creative thinking, vision-oriented thinking, and market-based thinking	Moon [36]
Systemic thinking, innovative thinking, and a thinking-oriented view	Al-Zu'bi [2]

1.3.2 Supply Chain as a Strategic Asset

If, like many companies and only think about changing the supply chain when something has gone wrong, such as:

- Very high inventory level and meager sales
- Customers' dissatisfaction with the company's goods and services
- Poor shipping delays from suppliers
- Unpredictable and difficult conditions
- Lowering performance compared to other competitors.

Furthermore, when thinking about changing the supply chain, you probably do not see it as a valuable, strategic asset that can give the company a competitive advantage. Moreover, if so, companies that use their supply chain as a strategic weapon are at risk.

Successful managers have concluded that the supply chain can be a strategic differentiator. They are constantly looking for new ways to add value and push the boundaries of performance. Moreover, they continue refining their supply chain to stay one step ahead of the competition. Therefore, to strategically manage the supply chain and creates a unique and effective supply chain network that will advance the strategic goals, five key elements are required, which are described below [13]:

- Operational strategy

It shapes all the decisions about how to produce goods and services, the operations, and all the critical decisions that affect the entire supply chain and the investments. For example, operational strategy determines how employees, factories, and warehouses manage their order desks and how they design their processes and information systems.

- Distribution channel strategy

The distribution channel strategy relates to presenting the products and services to buyers or end users. These decisions address issues such as whether sell to customers indirectly through distributors or retailers or directly through the Internet or a direct sales force. The market and geographic segments considered will guide the decisions in this regard. Since profit margins vary depending on the channels used, we must decide on the optimal distribution channel mix. These decisions drive asset and cost performance within a company. They should be part of the company's overall distribution channel strategy, along with decisions about pricing, vendor financing policies, advertising, and other terms and conditions.

- Outsourcing strategy

Consider outsourcing activities with low strategic importance or that a third party can do better, faster, or cheaper. Outsourcing allows companies to scale up or down quickly, create new products, or change themselves in the market by taking advantage of the expertise and capacity of other companies. This flexibility and agility can make a big difference in today's competitive global markets. Most importantly, outsourcing allows companies to focus on their core competencies and improve their competitive position. However, considering outsourcing decisions' risks and strategic implications before introducing new products, managing inventory levels, Etc., are strategic activities that cannot be outsourced to a third party.

- Customer service strategy

Customer service strategy should be based on two things:

1. The overall volume and profitability of customer accounts
2. Understanding what customers want.

   Both areas are integral to supply chain strategy because they help prioritize and focus a company's or organization's capabilities.

- Property network

The final component of supply chain configuration involves decisions about the network of company assets, factories, warehouses, production equipment, order desks, and service centers that make up the business. These assets' location, size, and mission significantly impact supply chain performance. Most companies choose one of the following three network models based on factors such as business size, customer service requirements, tax benefits, suppliers, local content laws, and labor costs:

1. Global model: Production of a given product line is done in one location for a global market. This model is influenced by factors such as the need to align manufacturing with research and development (R&D), control unit production costs for highly capital-intensive products, or highly specialized manufacturing skills.
2. Regional model: Production takes place primarily in the region where the products are sold, although there may be some interregional flow based on the specialization of the production center. The regional model is often chosen based on various factors, including customer service levels, import duty levels, and the need to adapt products to specific regional requirements
3. Country model: Production takes place mainly in the country where the market is located. This model is the choice for goods that are very expensive to transport. Other factors include duties and tariffs and access to the market, which is conditional on domestic production.

Due to price competition, many companies produce in low-cost countries to reduce unit production costs. When choosing such a location, key considerations include production costs, corporate tax rates, export incentives, presence of critical suppliers or duty-free imports, infrastructure, and skilled labor. Supply chain flexibility and total supply chain cost are important considerations when designing an asset network, especially for products with highly variable demand and short life cycles.

1.3.2.1   Four Criteria for the Right Strategy in the Supply Chain

Configuration components of operations strategy, distribution channel strategy, outsourcing strategy, Customer service strategy, and asset network; are the basic building blocks of the strategic structure of the supply chain. However, to advance strategic business goals and gain a competitive advantage, these components and the choices made about each one is also needed, followed by four criteria to have a suitable strategy in the supply chain and gain a competitive advantage [13]:

* Alignment with business strategy

A Supply chain strategy should directly support and drive business strategy. An effective business strategy begins with a master strategic vision that defines the boundary conditions for the business. The core strategic vision clarifies the answers to crucial business strategy questions, for example:

(1)   What are overall strategic goals?
(2)   What value does it offer to customers?
(3)   How does the company differentiate itself in the market?

- Alignment with customer needs

Do you know what customers want? Are there opportunities not taken advantage of simply because you cannot imagine them?

Answering these questions can be challenging. Most assumptions about what customers want are wrong, and most customers are unaware of their needs or cannot fully express them. Voice of the customer (VoC) is a method that can be used. It also helps customers discover their needs and causes them to transform customer needs into the needs of new products and services and provide new business opportunities. Aligning with customer needs means identifying specifics. The company or organization will lose market share if customers do not get what they need. Therefore if the company/organization does not know what the customers want, ask them or use VoC and review this issue periodically to change according to the customers' needs.

- Alignment with the position of power

A proper supply chain strategy is based on understanding the power and influence of the company/organization towards customers and suppliers. Moreover, this is important because the relative strength of the company/organization determines what can realistically be achieved in reconfiguring the supply chain to achieve the overall strategic goals of the company/organization. Companies often underestimate their power because they think globally rather than limiting their scope to a country or market segment. Even relatively small companies can strategically partner with selected suppliers or customers and gain a competitive advantage. The following strategies can increase the influence and power in the job market:

(a) If you are not a priority for your suppliers, your cooperation and service are insufficient, reconsider your supplier relationships.
(b) Consider shifting the power equation by focusing on a few smaller suppliers and allowing them to grow their business in exchange for working more closely with you over time to reduce costs, increase efficiency, and improve overall performance.
(c) The brand name can also be the primary source of supply chain power, especially in consumer markets. If your products are popular with consumers, you will have more leverage with retailers and other channel partners.

- Adaptability, because the competitive advantage is temporary and market conditions change.

Change is necessary. Market conditions change, business strategies evolve, and new technologies emerge. If you do not pay attention to these things, your supply chain can get out of sync. Supply chain strategy, like business strategy, must be consistent. Although change is constant, the frequency of significant change will vary across industries. For example, in the PC industry, companies make significant changes to their supply chain roughly every three to five years, driven by their constant pursuit of rapid, cost-effective new product introductions.

These four criteria may seem basic, but in the reality, few companies follow them. The practice of strategic supply chain development and management is not shared, and many users have only the most essential supply chain strategy process, indicating that these concepts are either poorly understood or challenging to implement.

1.3.3 How to Start Turning Your Supply Chain into a Strategic Asset

Payne [39] describes three strategies to create and transform the supply chain into a strategic asset, which are mentioned below:

(1) An honest assessment of the current performance of the company/organization

Before making changes, conduct a thorough and fair evaluation of the supply chain, including suppliers, the overall supply chain architecture (including structural, human, and technological components), and customers.

(2) Having an in-depth analysis of the various sources of supply chain capital of the organization/company

In general, there are several types of "capital" that an organization needs to understand and develop to turn its supply chain into a strategic asset, including:

- Social capital: refers to the capital created by relationships with suppliers, customers, banks, and governments. Poor relationships are a vital source of supply chain failure and can lead an organization to be reactive in its approach to supply chain management.
- Intellectual capital: That includes patents, copyrights, innovations, and trade secrets that drive both the current and future supply chain, as well as the supplier and customer behaviors.
- Human capital: It includes the people, skills, and competencies of the supply chain talent, as well as the team's ability to meet and exceed expectations.
- Financial capital: includes cash, payments, and margin implications for all supply chain interactions. Creating a robust single data warehouse capable of aggregating data generated throughout the supply chain is one approach required to accurately model the impact of the supply chain on the organization's finances. Unfortunately, many organizations fail to put in place the governance structure (putting the right teams in the right place) needed to implement this type of initiative.
- Physical capital: It includes plants, equipment, products, and any other physical element that may affect the supply chain.
- Information capital: includes measurements and key performance indicators taken throughout the supply chain, data collected from supply chain processes, and

measurement devices that exist to collect information from the supply chain (especially in the busy world of the Internet of Things).
- Technological capital includes technical engineering and analytical resources needed to support the operations and mass measurement of a strategic global supply chain.

(3) Drawing a correct path for the organization/company

To draw this path and move towards strategic supply chain management, several points are needed, which are mentioned below:

- Correct ownership structure for the supply chain: Run a small but strong mix of key stakeholders, including logistics leadership, technology and analytics leadership, and customer and supplier-focused leadership and knowledge. This structure can take different forms in different organizations. However, putting the right people in the right place at both the analysis and execution stages is critical.
- Build credibility incrementally using data: As you begin working to transform the supply chain into a more strategic asset, focus on reliability and take a close look at the data you collect.
- Dealing with critical costs, such as supplier costs: In many cases, small but innovative changes may be necessary to improve supplier costs. To figure out where to start, try benchmarking the supply chain costs and performance against competitors in the industry and beyond. Even adopting a collaborative approach with the suppliers to jointly eliminate waste using supply chain strategies such as "lean" or "theory of constraints" can be essential. Finding ways to increase volume in the supply chain through optimization (e.g., route selection optimization) can also be critical.
- Innovate with critical suppliers: Collaborating with crucial suppliers as innovation partners can be essential in realizing massive change. This is especially true in the technology and analytics sector and represents a great way to build credibility.
- Apply analytics, technology, and automation in a coherent and integrated manner: The debates and developments taking place in the Internet space continue to present enormous opportunities for the supply chain world. However, many companies are not ready to take advantage of these changes. One reason is that finding people with cross-functional expertise in supply chain management, technology, and analytics is challenging.

1.3.4 Benefits of Turning the Supply Chain into a Strategic Asset

If an organization can turn its supply chain into a strategic asset, it will have the following benefits:

- Reduce costs,
- Increase in income,

- Improving the use of assets,
- Transportation efficiency,
- Improving efficiency,
- Eliminate waste and avoid the cost
- Increasing safety and reducing injuries,
- Flexibility (which can be competitively used by your organization in other areas),
- Increasing speed and agility on a global scale,
- Improving the efficiency of employees,
- The ability to react faster to market changes [39].

1.4 The Structure of the Supply Chain Network

The supply chain network structure consists of all companies participating in the service production chain from raw materials to final consumption and the relationships between them (through which business activities or business processes are carried out). According to Lambert and Cooper [31], this structure consists of a central company (controlling company) and several related companies (suppliers and customers). Dimensions to be considered include the length of the supply chain and the number of suppliers and customers at each level. The noteworthy point is that the supply chain does not appear in the form of a chain but is more like the branches of a tree whose roots and branches are a symbol of a network [14]. For this reason, finding only one participating company in the chain is unusual. With all these details in mind, a dilemma arises: how many of these branches and roots need to be managed? The most common factors determining the number of companies that must be addressed in the supply chain concept are the product's complexity, the number of suppliers, and the availability of raw materials [31]. Managers and officials say that it is not necessary to coordinate and integrate all relationships throughout the supply chain in management because the level of communication between relationships is very different. In supply chain management, choosing the most appropriate community level for each connection is necessary. The most appropriate relationship is the most important thing to the company. To learn and know how to plan the supply chain network, Lambert and Cooper [31] state the analysis of three structural aspects of the network:

- Members of the supply chain
- Structural dimensions of the network
- Different types of process-building communication.

1.4.1 Identification of Supply Chain Members

To determine the network's structure, it is necessary to identify the supply chain members. The chain members should be classified according to the level, and their criticality for the company's success should be evaluated. In most cases, the integration and coordination of all process relationships can have opposite, complex, and even impossible results [8].

Supply chain members include all companies and organizations with which the central company interacts bilaterally, directly or indirectly, through its suppliers or customers, from the point of origin to the point of sale. Nevertheless, to make a complex network more manageable, it is essential to distinguish the leading members from the supporting members [15].

The leading members of the supply chain are independent companies or strategic business units that perform value-added activities and operational or management activities in business processes that produce a specific output for a particular customer and market. In contrast, supporting members are companies that merely provide resources, knowledge, and tools to the core members of the supply chain. For example, helping companies include shipping companies, lending banks, warehouse building owners, and companies supplying equipment to produce and print business brochures. A company can perform both core and support activities. Also, a company can perform main activities in one process and support activities in another process.

It should be noted that the difference between the main and supporting members of the supply chain is not evident in all cases. Nevertheless, the proposed definition, at least, suggests an acceptable operational simplification that covers the essential aspects for those who may be considered important members of the supply chain. The description of the primary and supporting members makes it possible to define the supply chain's starting point and the point of sale. In general, there is no immediate member at the starting point of the supply chain because all the supporting members are considered [8].

1.4.2 The Structural Dimensions of the Supply Chain Network

The three critical structural dimensions for describing, analyzing, and implementing a supply chain are:

- Horizontal structure
- Vertical structure
- Horizontal position within the chain.

Horizontal structure refers to the number of levels in the supply chain. Which can be big or small according to the number of available levels. For example, the network structure is too large for the automotive industry. Automotive parts are manufactured in multiple locations by a large number of suppliers around the world who send their products to the main hubs of Khoro's assembly subsystems and make the long journey to final assembly. Vertical structure refers to the number of suppliers or customers available at each level. A company can have a narrow vertical structure with very few companies at each level or a broad vertical structure with many suppliers or customers at each level.

The third structural dimension is the horizontal position of the company within the supply chain. A company can be located far from or close to the primary source of supply or far from or close to the final customer in a place between these supply chain boundaries [31].

1.5   Conclusion

In this chapter, three essential parts of supply chain management were described, which were explained in the first part to inform the readers about the primary and critical concepts of supply chain management. Then, in the second part, how to turn the supply chain into a strategic asset and achieve a more competitive advantage was described for the readers' knowledge. Moreover, in the third part, the structure of the supply chain network was described so that the readers can learn about the planning of the supply chain network and its members and dimensions, and in this way, they can know the most appropriate level of society in each of the different layers of the supply chain network.

References

1. Ahlstrand, B., Lampel, J., Mintzberg, H.: Strategy Safari: A Guided Tour Through the Wilds of Strategic Management. Simon and Schuster (2001)
2. Al-Zu'bi, H.A.: Strategic thinking competencies and their impact on strategic flexibility. J. Curr. Res. Sci. 4(1), 35 (2016)
3. Ayers, J.B.: Handbook of Supply Chain Management. CRC Press (2000)
4. Azevedo, S.G., Prata, P., Fazendeiro, P.: The role of radio frequency identification (RFID) technologies in improving process management and product tracking in the textiles and fashion supply chain. In: Fashion Supply Chain Management Using Radio Frequency Identification (RFID) Technologies, pp. 42–69. Woodhead Publishing (2014)
5. Ballou, R.H., Gilbert, S.M., Mukherjee, A.: New managerial challenges from supply chain opportunities. Ind. Mark. Manage. 29(1), 7–18 (2000)
6. Bonn, I.: Developing strategic thinking as a core competency. Manag. Decis. 39(1), 63–71 (2001)
7. Bonn, I.: Improving strategic thinking: a multilevel approach. Leadersh. Org. Dev. J. 26(5), 336–354 (2005)
8. Campuzano, F., Mula, J., Peidro, D.: Fuzzy estimations and system dynamics for improving supply chains. Fuzzy Sets Syst. 161(11), 1530–1542 (2010)
9. Chan, F.T.: Performance measurement in a supply chain. Int. J. Adv. Manuf. Technol. 21(7), 534–548 (2003)
10. Chen, I.J., Paulraj, A., Lado, A.A.: Strategic purchasing, supply management, and firm performance. J. Oper. Manag. 22(5), 505–523 (2004)
11. Chopra, S., Meindl, P.: Supply chain management. Strategy, planning & operation. In: Das Summa Summarum des Management, pp. 265–275. Gabler (2007)

12. Christopher, M.: Logistics and Supply Chain Management: Strategies for Reducing Cost and Improving Service. Financial Times: Pitman Publishing, London (1998). ISBN 0 273 63049 0 (hardback) 294+ 1× pp. (1999)
13. Cohen, S., Roussel, J.: Strategic Supply Chain Management: The Five Disciplines for Top Performance. McGraw-Hill Education (2013)
14. Cooper, M.C., Ellram, L.M., Gardner, J.T., Hanks, A.M.: Meshing multiple alliances. J. Bus. Logist. 18(1), 67 (1997)
15. Davenport, T.H.: Process Innovation: Reengineering Work Through Information Technology. Harvard Business Press (1993)
16. Defee, C.C., Stank, T.P.: Applying the strategy-structure-performance paradigm to the supply chain environment. Int. J. Logist. Manage. 16(1), 28–50 (2005)
17. Fred, R.D., Forest, R.D.: Strategic Management: A Competitive Advantage Approach (2016)
18. Gold, S., Seuring, S.: Supply chain and logistics issues of bio-energy production. J. Clean. Prod. 19(1), 32–42 (2011)
19. Graetz, F.: Strategic change leadership. Manage. Decis. 38(8):550–564 (2000). Quoted in Chrusciel, D., Field, D.W.: Success factors in dealing with significant change in an organization. Bus. Process Manage. J., 503–516
20. Grant, R.M.:Contemporary strategy analysis. Wiley (2021)
21. Handfield, R.B., Nichols, Jr., E.L.: Introduction to Supply Chain Management. Prentice Hall, Englewood Cliffs, NJ (1999)
22. Hilletofth, P.: How to develop a differentiated supply chain strategy. Ind. Manag. Data Syst. 109(1), 16–33 (2009)
23. Holmberg, S.: A systems perspective on supply chain measurements. Int. J. Phys. Distrib. Logist. Manag. 30(10), 847–868 (2000)
24. Heracleous, L.: Strategic thinking or strategic planning? Long Range Plan. 31(3), 481–487 (1998)
25. Hugos, M.H.: Essentials of Supply Chain Management. Wiley (2018)
26. Hunger, J.D., Wheelen, T.L.: Essentials of Strategic Management, vol. 4. Prentice Hall, New Jersey (2003)
27. Johnson, M., Mena, C.: Supply chain management for servitised products: a multi-industry case study. Int. J. Prod. Econ. 114(1), 27–39 (2008)
28. Karğın, S., Aktaş, R.: Strategic thinking skills of accountants during adoption of IFRS and the New Turkish Commercial Code: A survey from Turkey. Procedia Soc. Behav. Sci. 58, 128–137 (2012)
29. Ketchen, D.J., Jr., Giunipero, L.C.: The intersection of strategic management and supply chain management. Ind. Mark. Manage. 33(1), 51–56 (2004)
30. Kim, W.C.: Blue ocean strategy: from theory to practice. Calif. Manage. Rev. 47(3), 105–121 (2005)
31. Lambert, D.M., Cooper, M.C.: Issues in supply chain management. Ind. Mark. Manage. 29(1), 65–83 (2000)
32. Lambert, D. M. (2008). Supply chain management: processes, partnerships, performance. Supply Chain Management Inst.
33. Liedtka, J.M.: Strategic thinking: can it be taught? Long Range Plan. 31(1), 120–129 (1998)
34. Li, S., Ragu-Nathan, B., Ragu-Nathan, T.S., Rao, S.S.: The impact of supply chain management practices on competitive advantage and organizational performance. Omega 34(2), 107–124 (2006)
35. Mintzberg, H.: The fall and rise of strategic planning. Harv. Bus. Rev. 72(1), 107–114 (1994)
36. Moon, B.J.: Antecedents and outcomes of strategic thinking. J. Bus. Res. 66(10), 1698–1708 (2013)
37. Nikam, M., Satpute, S.: RFID: changing the face of supply chain management. Presentational slides, Welingkar Institute of Management and Development Research, Mumbai, India (2004)
38. Nuntamanop, P., Kauranen, I., Igel, B.: A new model of strategic thinking competency. J. Strateg. Manag. 6(3), 242–264 (2013)
39. Payne, N.: How to Turn Your Supply Chain into a Strategic Asset. Linkedin (2016)

40. Pearce, J., Robinson, R.: Strategic Management. McGraw Hill (2014)
41. Porter, M.E.: Competitive Advantage, pp. 33–61. Free Press, New York (1985)
42. Rowe, A.J.: Strategic Management: A Methodological Approach. Addison Wesley Publishing Company (1989)
43. Schnetzler, M.J., Sennheiser, A., Schönsleben, P.: A decomposition-based approach for the development of a supply chain strategy. Int. J. Prod. Econ. **105**(1), 21–42 (2007)
44. Shafiee, M., Saleh, H., Ghaderi, M.: Benchmarking in the supply chain using data envelopment analysis and system dynamics simulations. Iranian J. Supply Chain Manage. **23**(70), 55–70 (2021)
45. Shafiee, M., Rafatmah, M.: Supply chain analysis via the queuing theory approach. J. Prod. Manage. **13**(50), 205–234 (2019)
46. Shafiee, M., Tarmast, P.: The impact of supply chain management processes on competitive advantage and organizational performance (Sapco company case study). Q. J. Quant. Stud. Manage. **5**(2), 108–128 (2014)
47. Shafiee, M., Fallahi Ghiyasabadi, L., Rezaee, Z.: A study and survey of social capital influence on supply chain integration (case study: food industry in Fars Province). ORMR **3**(2), 43–65 (2013)
48. Shafiee, M., Rezaee, Z., Ebrahimi, A.: Strategic Supply Chain Management. Terme Publications (2009)
49. Shepherd, C., Günter, H.: Measuring supply chain performance: current research and future directions. In: Behavioral Operations in Planning and Scheduling, pp. 105–121 (2010)
50. Si, Y.W., Edmond, D., Dumas, M., Chong, C.U.: Strategies in supply chain management for the trading agent competition. Electron. Commer. Res. Appl. **6**(4), 369–382 (2007)
51. Somuyiwa, A.O.: Modeling outbound logistics cost measurement system of manufacturing companies in Southwestern, Nigeria.Eur. J. Soc. Sci. **15**(3), 382–395 (2010)
52. Stavrulaki, E., Davis, M.: Aligning products with supply chain processes and strategy. Int. J. Logist. Manage. **21**, 127–151 (2010)
53. Stadtler, H.: Supply chain management and advanced planning—basics, overview and challenges. Eur. J. Oper. Res. **163**(3), 575–588 (2005)
54. Thompson Jr, A.A., Strickland III, A.J.: Strategic Management: Concepts and Cases (1978)
55. Tan, K.C.: A framework of supply chain management literature. Eur. J. Purchasing Supply Manage. **7**(1), 39–48 (2001)
56. Van Hoek, R.I.: "Measuring the unmeasurable"—measuring and improving performance in the supply chain. Supply Chain Manage. Int. J.
57. Wang, R., Lin, Y.H., Tseng, M.L.: Evaluation of customer perceptions on airline service quality in uncertainty. Procedia Soc. Behav. Sci. **25**, 419–437 (2011)
58. Zuckerman, A.: Supply Chain Management: Operations 06.04 (Express Exec. Publisher: Capstone, 1 (2002).

Chapter 2
Performance Evaluation of Supply Chain Management

2.1 Introduction

In today's highly volatile market environment, companies face significant challenges in meeting customer needs. In addition, competition has shifted from a single firm to the entire supply chain (SC). In this context, supply chain management (SCM), by effectively organizing activities from the supplier to the final customer, plays a vital role in maintaining the company in the global market. It leads to business management from procurement of raw materials to production to distribution, customer service, and finally, the products are reprocessed and disposed of. Hence, according to the role of supply chain management, indicators and performance measures are needed to measure the effectiveness and efficiency of SC to improve SC's performance to meet customer expectations.

Furthermore, the success of any SC or business depends on an effective performance measurement system (PMS). Therefore, an effective PMS in the context of SC is needed for the proper measurement at the right time [34]. Also, supply chain performance measurement (SCPM), as an essential management tool and a means to achieve success, enables the supply chain to strategically manage and continuously control the achievement of goals [2]. On the other hand, one of the most complex decision-making problems for managers is evaluating the adequate performance of the supply chain, which includes various criteria [71]. Because this issue depends on the performance of the supply chain, the performance of the supply chain depends on the following activities:

- Supply chain-wide activities in meeting end-customer needs, including product availability, on-time delivery, and the necessary inventory and capacity in the supply chain to deliver that performance in a responsive manner
- Extensive supply chain activities outside the company's borders, including procuring raw materials, components, sub-parts, and finished products and distribution through different channels to the end customer.

© The Author(s), under exclusive license to Springer Nature Switzerland AG 2023
F. Hosseinzadeh Lotfi et al., *Supply Chain Performance Evaluation*,
Studies in Big Data 122, https://doi.org/10.1007/978-3-031-28247-8_2

- Extensive supply chain activities outside traditional organizational lines include procurement, production, distribution, marketing and sales, and research and development.

Therefore, to win in this new environment, supply chains need performance measures or "Metrics" that support the improvement of global supply chain performance and do not support only limited companies or specific performance measures that hinder chain improvements [26]. Therefore, the right design and evaluation of the supply chain are essential issues for managers and researchers, doubling the process's complexity due to the internal relationships of the units involved. Therefore, in the design of the system, all communications, interactions, priorities, influences, and limitations should be taken into account as much as possible so that the evaluation result of a supply chain provides more accurate feedback on performance for improvement [68]. Therefore, the primary purpose of this chapter is to identify the concepts of supply chain performance evaluation as well as the role of the performance evaluation system in advancing the supply chain goals and analyzing the key performance indicators in the field of supply chain management.

2.2 Evaluation and Measurement of Supply Chain Performance

2.2.1 Reasons and Importance of Supply Chain Performance Measurement

Performance measurement is generally defined as quantifying the efficiency and effectiveness of action [61]. Effectiveness; is the extent to which customer needs are met. While efficiency measures the economical use of company resources to achieve a predetermined level of customer satisfaction [2]. Therefore, performance evaluation is an important activity for the survival and growth of any company [71]. The old saying goes, "You cannot improve what you cannot measure." Therefore, organizations may need to measure for different purposes for the following reasons [63, 71]:

- Identifying the successes of the organization
- Diagnosing and solving customer needs
- Help to understand the processes of the organization and identify the location of problems, bottlenecks, wastes, etc.
- Where improvement is required, it ensures that decisions are objective rather than subjective and shows whether the planned improvement occurred.

In general, the importance of performance measurement systems can be summarized for the following reasons [1, 2]:

- **Directing organizational actions**

Performance measurement guides actions in two ways:

(1) Measurement criteria are seen in the organization, and people are trying to achieve high performance according to these criteria.
(2) Measurement criteria take action by identifying areas for improvement. In fact, after identifying weaknesses, management should take corrective measures to address such issues.

- **A framework for decision making**

Performance measurement provides a basis for evaluating options and determining decision criteria. The structure of the measurement system guides decisions and actions at the strategic, tactical, and operational levels. Hence, a relevant performance management system targets performance optimization in multiple objectives.

- **Closed loop control**

Feedback is an integral part of any process. An effective performance management system provides the necessary feedback to reveal progress, identify problems, identify potential opportunities for improvement, facilitate understanding and communication between supply chain members, and test the effectiveness of different strategies.

Meanwhile, the main reason for poor supply chain performance is the lack of a measurement system [58]. The performance measurement system provides a set of actions for management that can be used to improve performance, plan efforts, and increase competition [82]. Organizations should measure not only the final output but also the processes involved in reaching the final output to find the problem that causes the variance between the target and the actual specification of the final product [83].

Also, performance measurement systems (PMS) are essential for companies, supply chains, and society because they are the primary mechanisms for implementing supply chain management [24, 42, 82].

With a proper PMS embedded in the supply chain, supply chain strategy implementation, control, decision-making, communication, and improvement can be realized [4, 10, 19, 24, 53, 56].

Therefore, the performance evaluation of the supply chain as a necessary management tool helps to improve the performance of the supply chain in the direction of its excellence [12]. Also, Chen and Paulraj [14] believe that the measurement of supply chain performance causes the following:

- Facilitating a better understanding of the supply chain
- Positive impact on the behavior of supply chain agents
- Improving the overall performance of the supply chain

Also, performance measurement is one of the critical aspects of successful SCM [2]. Gunasekaran et al. [20] described effective performance measurement as essential for SCM. Lai et al. [41] also stated that inadequate performance measurement is one of the main obstacles to effective SCM.

In general, when we need to manage, we should be able to measure it with the help of which we can control the performance of the entire functions involved in the SC of a company. Therefore, a dynamic, accurate, and integrated Supply Chain Performance Measurement System (SCPMS) is needed [13].

However, not all measures and performance measurement system designs are equally effective in achieving these beneficial results [25]. Therefore, it is crucial to understand how to design and use Supply Chain Performance Measurement Systems (SCPMS) and how they evolve when they are embedded as a practice in the supply chain [24].

2.2.2 Definitions of Supply Chain Performance Measurement System (SCPMS)

Looking at the topical literature on Supply Chain Performance Measurement Systems, we notice that different authors defined SCPMS depending on the purpose they had during their studies. Here are some functional definitions in this regard:

- SCPMS is a system that evaluates SC performance precisely after considering all SC performance criteria [38].
- SCPMS is a measurement system that can evaluate supply chain performance [65].
- SCPMS is an essential organizational system that paves the way to gaining and maintaining competitiveness [67].
- SCPMS are management and central mechanisms that are efficient and effective in achieving supply chain management [24].
- Maestrini et al. [51] also provided the following general definitions for the Supply Chain Performance Measurement System:

  (a) Supply chain performance measurement systems (SCPMS) breathe new life into business operations thanks to new technologies that enable the collection, integration, and sharing of information across multiple supply chain partners.
  (b) SCPMS is a set of metrics used to quantify the efficiency and effectiveness of supply chain processes and relationships, spanning multiple organizational functions and companies and enabling SC coordination.
  (c) SCPMS is a set of metrics used to quantify the efficiency and effectiveness of intercompany processes shared between multiple buyers and suppliers.

- SCPMS is a system to set goals, evaluate performance and determine future paths [81].
- SCPMS identifies and minimizes the gap between planning and execution [8].

- SCPMS is a system that determines overall goals and provides a set of performance measures to achieve these goals [8].
- SCPMS is an essential management tool that provides the necessary assistance to improve performance in line with SC excellence [35].
- SCPMS is essential to an effective plan, control, role, and decision-making [9].
- SCPMS is a system for effective communication and coordination of SC elements and functional areas [32].
- SCPMS is a tool capable of combining single organizational actions with inter-organizational and SC actions [55].
- SCPMS is a system to achieve the following goals [84]:

  (a) Support goal setting
  (b) Performance evaluation
  (c) Determining future directions at the strategic, tactical, and operational levels.

- Supply chain performance measurement systems (SCPMS) allow the adoption of performance measures that span different companies and processes. Hence, SCPMSs represent a way to improve SC governance by ensuring value-added, informed, and timely decisions [19].
- SCPMS is a system to control the improvement of processes in SC [20].
- SCPMS is a system that analyzes available resources and measures output with flexibility in operations [7].

2.2.3 Principles and Characteristics of a Suitable Supply Chain Performance Measurement System (SCPMS)

SCPM is a critical tool for designing and developing organizational change systems. Therefore, SCPM has been discussed in several types of research so that the authors and experts have proposed various desirable features for an SCPMS. However, they all agreed that an effective SCPM should have the following features [2, 5, 7, 19, 40, 66, 79]:

- Inclusiveness: Cover all aspects and processes of a supply chain.
- Universality: To provide the possibility of comparison under different operating conditions.
- Measurement capability: Quantitative and measurable output.
- Compatibility: The criteria are compatible with the supply chain objectives.

Charkha and Jaju [13] also believe that a good Supply Chain Performance Measurement framework should focus on the customer and measure the right things. Therefore, the following criteria were listed for a suitable SCPM system:

- Be meaningful, unambiguous, and widely understood.
- Be owned and managed by teams within the organization.
- Be based on a high level of data integrity.

- Collect data in such a way that it is embedded in routine procedures.
- Be able to prioritize actions, establish connections between actions, and create improvements related to the organization's critical goals and key drivers.

In general, Krmac [39] consider the following as the principles of supply chain performance evaluation:

- The main point of performance appraisal is that emphasis should be placed on key performance indicators. Therefore, mathematical or informational tools should be used to analyze these indicators.
- Use indicators that can reflect SC business processes.
- Indicators should reflect not only the performance of an individual company but also the subsystem and the entire SC. This is even more important for achieving sustainable SC management (SSCM).
- Immediate assessment methods should be used as much as possible, as it is more meaningful to do so than to analyze afterward.
- Ensure that the SC performance assessment objective is consistent with the overall SC strategy objective. Otherwise, the SC performance will not contribute significantly to the strategic goal.
- Indicators that support the strategic objective should be selected.
- One of the ultimate goals of every SC is to satisfy customers. Managers must understand what the demands are to meet actual customer demand. Meanwhile, Managers should also clearly understand SC performance from the customer's point of view.

2.2.4 Evolution of SCPMS Models

As previously stated, measurement is essential because it directly affects behavior that affects supply chain performance. Thus, SCPM provides a tool by which a company can assess whether its supply chain has improved or degraded. To trace the history and performance evaluation models, it should be said that in the past, companies relied only on financial accounting measures, many of which date back to the nineteenth century and even earlier. Then, at the beginning of the twentieth century, diversification led to the refinement of performance measurement. In 1903, the DuPont Company implemented the "Rate of Return on Investment (ROI)" to evaluate the performance of various units and developed the "DuPont System Scale." After that, it was widely accepted. Since then, financial indicators have become systematic [63]. After the Second World War, the company environment changed and became full of uncertainty and diversity, which required establishing a balance between marketing, research, development, and human and financial resources [40]. Therefore, the companies changed their priorities and started using financial and non-financial indicators, i.e., the combined approach.

Moreover, with the evolution and maturity of the business organization concept in the late 1990s, performance measurement systems wholly changed to a balanced,

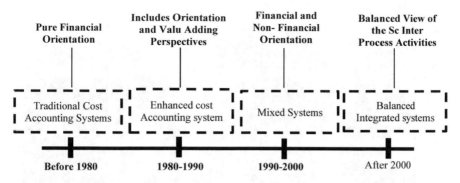

Fig. 2.1 SCPM evolution timeline [2: 4]

integrated approach [63]. Figure 2.1 shows the evolution of SCPM systems in a timeline divided into four periods. Before the 1980s, traditional cost accounting systems
used only financial indicators, relying solely on quantitative and general financial
measures, ignoring other strategic and non-financial measures such as customer
loyalty or service quality. Subsequently, cost accounting systems were improved,
whereby the range of economic indicators was expanded to cover different functions
and specific operations in the supply chain. In the early 1990s, Kaplan and Norton
[36] developed the balanced scorecard (BSC) approach, which first introduced the
concept of hybrid systems. Their method addresses the importance of monitoring and
evaluating non-financial indicators. During the last decade, the concept of integrated
online systems and e-commerce has been strongly evolving to enable information
sharing and facilitate the entire measurement process from different perspectives of
the supply chain.

Also, after reviewing the literature and SCPM models, the characteristics of SCPM
models are shown in Fig. 2.2.

2.2.5   Reasons for Changes in SCPMS Models

Traditional performance evaluation models only paid attention to financial indicators. Financial indicators are critical in evaluating whether operational changes
will improve the financial health of an organization or not. However, as highlighted
and emphasized in the literature, they are not sufficient to measure supply chain
performance for the following reasons [40, 43]:

1. Financial indicators tend to be short-term, internally focused, and historically
   oriented.
2. Financial indicators are not related to important strategic and non-financial
   performance indicators such as customer satisfaction and product quality.
3. Financial indicators are not directly related to operational effectiveness and
   efficiency.

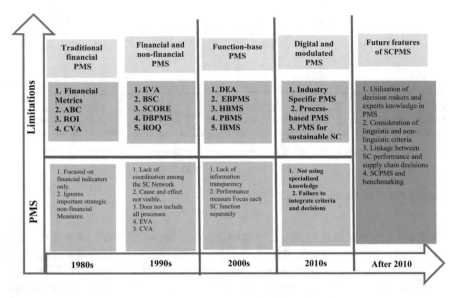

Fig. 2.2 Characteristics of supply chain performance measurement systems (SCPMS) over time
[38]

Various systems and approaches have been developed in response to the shortcomings of traditional accounting methods for measuring the per supply chain
performance. In addition, the reasons for changes in SCPMS models are shown
in Fig. 2.3.

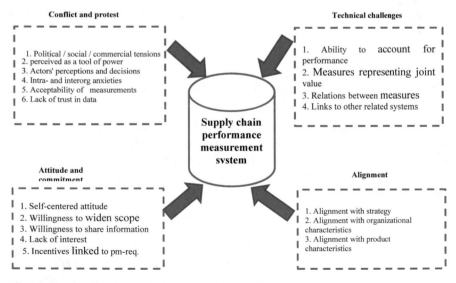

Fig. 2.3 Reasons for changes in SCPMS models [24: 52]

2.2.5.1 Technical Challenges

Reviewing the literature shows that technical challenges are one of the reasons for the changes in SCPMS models. These challenges include:

1. Due to complexity, many of the achievements of supply chain management, such as customer satisfaction and the quality of relationships with suppliers, cannot be easily tracked through numbers [29].
2. The actions included do not represent the creation of shared value [33], are short-sighted and non-strategic and do not have specific long-term goals [9].
3. The lack of industry standards may also create challenges [50].
4. Lack of recognition of interdependencies between supply chain criteria [30] and also between organization-specific performance criteria and SCM-based criteria [80]: Linking key metrics to performance drivers using cause and effect relationships is essential to sustain SCPMS [9]. In addition, the availability of accurate information [57] also improves the stability of SCPMS.
5. Lack of trust in data reliability [50].
6. Finally, if communication with other systems, such as cost accounting systems and information technology systems, does not work as intended, it is an indication that this affects the design and challenges the performance and lifetime of the embedded SCPMS kills [23, 85].

2.2.5.2 Alignment

Reviewing the literature shows that one of the reasons for changes in SCPMS models is related to the alignment between SCPMS and the environment in which these systems are embedded. These types of SCPMS changes are related to organizational characteristics such as:

1. Organizational structure [23] and organizational maturity [48]: These affect the design, implementation, and use of SCPMS.
2. The relationship between product criteria and features [62] with strategies [48, 69]: When the metrics are not representative of the strategies or products that are supposed to help management, this affects the sustainability of the embedded SCPMS.
3. Dynamic Environment: it affects the SCPMS design [74].

Therefore, the everyday use of performance measurement varies in different contexts and even in the operations of large companies [33].

2.2.5.3 Attitudes and Commitment

The literature and research background on various Supply Chain Performance Evaluation Systems indicate that the actors and users involved are also potentially influential and can influence the design, implementation, and use of SCPMS [42], which creates challenges such as:

1. Avoiding a self-centered attitude in the companies involved [30]:

This attitude may lead to the inability or unwillingness to expand the scope of measurement activities [80] and look beyond the company's boundaries [30]. However, the literature and research background often show that the desire for Information sharing [30, 42] and measurement information communication between buyer and supplier is limited in practice [33]. Therefore, the actors involved should try to understand the value and potential of this information and intra-organizational and inter-organizational communications [42].

2. Specific SCPMS content and logic should be in place [9]:

In general, when feelings of disinterest in the system and performance measures are low [50] and also when supplier ownership of performance measurement requirements is low [85], creating incentives related to performance measurement requirements has been suggested as an essential strategy [50, 85].

2.2.5.4 Conflict and Objection

Another reason for changes in SCPMS is to create concern, protest, or active resistance [80]. These cause the following challenges:

1. Political, social, and commercial tensions within and between companies affect the design, implementation, and use of SCPMS [72].
2. These tensions affect the understanding and importance of all components and decision-makers involved in and around the SCPMS [23].
3. The sense of intra-organizational and inter-organizational anxiety affects SCPMS activity [72].

Therefore, a key question is whether the information and performance indicators embedded in the SCPMS are acceptable to buyers and suppliers [57]. When SCPMS is challenged, the literature and past research background show that the ability to resolve conflicts between organizational strategies and the supply system is essential [57]. For this purpose, the following solutions work:

- The active participation of supply chain partners in the design stage and using an authentic strategy for conflict resolution is essential [48, 50].
- Training, ownership, responsibility, financial support, openness, and cooperation within and between supply Chain participants help overcome barriers [72, 80].

2.2.6 Review and Classification of Common SCPMS Frameworks and Models

Supply chain frameworks are categorized based on size and complexity, from a simple model representing independent decision-making to complex corporate interactions and behaviors model. Therefore, appropriate performance measurements that measure supply chain characteristics and exchanges should be designed. In general, the larger and more complex the supply chain is, the more challenging it will be to measure the efficiency of the supply chain [15]. Since customers, as the final judges, can determine how much value is created at the supply chain level, supply chain performance can be measured through customer satisfaction levels and costs [16]. Various models and frameworks have been presented for supply chain measurement, which is a classification [6] have divided the performance evaluation models based on three perspectives, process, and hierarchical approaches:

- Perspective-oriented approach

This approach integrates generic performance measures and causal hypotheses that determine the relationship between performance measures. The perspective approach is a unique supply chain performance measurement perspective based on researchers' perspectives.

- Process-oriented approach

Due to the critical dimensions of the supply chain, identifying a supply chain's key activities and processes is necessary to evaluate an effective performance measurement system. Among the key processes, we can mention internal procurement, manufacturing core, external procurement, marketing, and sales.

- Hierarchical approach

Hierarchy approach; A hierarchical performance measurement system evaluates supply chain performance through different hierarchical levels. Analyzing supply chain performance metrics and indicators at strategic, tactical, and operational levels helps managers make the right decisions.

Also, the literature and background related to SCPM are relatively huge. Several attempts have been made to measure supply chain performance using conventional approaches. The literature review showed that there are generally two categories of SCPM systems: financial and non-financial [2, 16, 40, 44]. Each category is explained below [2]:

2.2.6.1 Financial Performance Measurement Systems (FPMS)

Financial performance measurement systems are generally referred to as traditional accounting methods for measuring supply chain performance. As explained in the previous section, they mainly focused on economic indicators and were always criticized for their inadequacy as they ignored necessary non-financial strategic measures. In the literature and background of past research, there are two methods as the most popular methods for financial performance measurement systems (FPMS), which are [43]:

1. Activity Based Costing (ABC)

The ABC approach was developed in 1987 by [37] to link financial measures with operational performance. Unlike the traditional cost accounting methods, where the overhead costs are divided equally, this method divides the activities into individual tasks or cost drivers. It then estimates the resources, such as the time and costs required for each activity. Overhead costs are then allocated based on these cost drivers. This approach is designed to allow a better assessment of the productivity and expenses of a supply chain process. However, it still suffers from the significant limitation of relying on purely financial measures [37].

2. Economic Value Added (EVA)

EVA is an approach to estimating the return on capital or the economic added value of the company. It was developed in 1995 by Stern et al. To correct the shortcomings of traditional accounting methods that focused only on short-term financial results; it provides little insight into a company's success in creating long-term value for shareholders. The EVA approach assumes that shareholder value increases when a company earns more than its cost of capital. The EVA measurement attempts to determine the value created by a company based on the excess of operating profit over capital employed (through debt and equity). Although helpful in evaluating high-level executive partnerships and long-term shareholder value, EVA metrics cannot reflect operational supply chain performance because they only consider pure financial indicators [76].

2.2.6.2 Non-financial Performance Measurement Systems (NFPMS)

Researchers believe that existing non-financial SCPM approaches can be classified into nine types of models based on measurement criteria, and these nine groups are:

1. Supply Chain Balanced Scorecard (SCBS)

In 1992, Kaplan and Norton introduced the Balanced Scorecard (BSC) as an essential performance management tool. Since then, it has been recognized as the leading performance measurement tool in research and industry. Because it enables managers to have a balanced view of operational and financial measures at a glance, the authors proposed four essential perspectives that managers should monitor as follows:

- Financial perspectives
- Customer

- Internal business processes
- Perspectives on innovation and learning.

By considering these four perspectives, managers can transform strategies into specific metrics that can monitor the overall impact of a strategy on the company. However, as discussed in the literature, it suffers from two significant limitations:

(a) It is a top-down approach. Therefore, it is not collaborative and may fail to detect interactions between different process metrics. According to [47], BSC is a static approach that does not provide an opportunity to develop, communicate, and implement a strategy when applied in a corporate environment.
(b) Although powerful and widely used in industry, the BSC provides only a conceptual framework. It lacks an implementation methodology and therefore deviates from the concept's merit.

2. Supply Chain Operations Reference Model (SCOR)

The Supply Chain Council developed the SCOR model [31, 46, 75]. Its first version was created in 1996. It is a framework for the detailed examination of the supply chain by defining and classifying the processes that make up the chain, assigning criteria to such processes, and examining comparable criteria. The SCOR model establishes a supply chain as consisting of five main integrated processes:

- Planning
- Source
- Construction
- Delivery
- Performance return.

Also, most processes are measured from 5 perspectives:

- Reliability
- Accountability
- Flexibility
- Cost
- Property.

3. Dimension-based Measurement Systems (DBMS)

The concept of DBMS is based on the assumption that any supply chain can be measured based on dimensions [66]. Initially, in 1999, Beamon identified three types of metrics as essential components in supply chain performance measurement systems: resources (R), outputs (O), and flexibility (F). Examples of resource performance measures include production cost, inventory cost, and return on investment (ROI). Output metrics include total sales, on-time delivery, and fill rate, while flexibility metrics measure volume changes and new product introductions. Another example of a DBMS is the one identified by [27], who suggests that a supply chain should operate in the following three key dimensions:

- Services

Services are related to anticipating, attracting, and fulfilling customer demands.

- Assets

Assets include anything of financial value, such as inventory and cash.

- Speed

Speed includes metrics that are time-related to track responsiveness and speed of execution.

DBMS are generally simple, flexible, and easy to implement. However, they do not reflect the performance and internal operations in the chain because they only focus on high-level measures [2].

4.   Interface-based Measurement Systems (IBMS)

IBMS was initially proposed in 2001 by Pohlen and Lambert [64]. They offered a framework in which the performance of each stage in the supply chain is related. This link-to-link approach provides a means to align performance from the point of origin to the end of consumption to maximize shareholder value for the entire supply chain and each company. The IBMS approach seems good in theory, but in the business environment, it requires openness and complete sharing of information at each stage, which is ultimately tricky to implement [66].

5.   Perspective-based Measurement Systems (PBMS)

PBMS looks at the supply chain from all possible perspectives and provides measures to evaluate each [66]. This system was developed in 2003 by Otto and Kotzab [61] and included six main points of view as follows:

- System dynamics
- Research in operations
- Logistics
- Marketing
- Organization
- Strategy.

6.   Hierarchical-based Measurement Systems (HBMS)

In 2004, Gunasekaran et al. developed the HBMS, in which actions are classified as strategic, tactical, or operational. The main idea was to allocate activities at that management level to deal with them best, thus facilitating quick and appropriate decision-making [66]. However, it is impossible in such systems to create a clear guide to place actions at different levels that can reduce conflict levels between various supply chain partners [2].

7. Function-based Measurement Systems (FBMS)

FBMS is a tool that combines measures to cover different functions in a supply chain [66]. It was initially developed in 2005 by Christopher [17] to protect detailed performance measures applicable to various parts of the supply chain. Although it is easy to implement and objectives can be assigned to individual departments, it does not provide high-level measures to cover the entire supply chain. FBMS is criticized for viewing individual supply chain functions in isolation, leading to local benefits that may harm the whole supply chain.

8. Efficiency-based Measurement Systems (EBMS)

EBMS are systems that measure supply chain performance in terms of efficiency. Most EBMS are based on DEA. Despite being very useful, it can sometimes be misleading for managers and stakeholders due to relative returns. However, in recent years, the combination of this model with other supply chain measurement approaches has become popular, which has had very acceptable results in the field of supply chain evaluation, which will be discussed in the following chapters.

9. Generic Performance Measurement Systems (GPMS)

Since the early 1980s, several general performance measurement models and frameworks have been developed, i.e., not necessarily specific to supply chains. Each of these has its advantages and limitations. However, a literature review shows that few of them have been widely cited. Some of them are discussed below [2, 40, 79]:

- Performance Prism

The performance prism is a performance measurement framework that suggests that performance should be measured in five distinct, but related, perspectives, which are illustrated by Neely et al. [60] below:

- Stakeholder satisfaction
- Strategies
- Processes
- Capabilities
- Stakeholder participation

A Performance Prism has a much more comprehensive view of different stakeholders than other frameworks. The main strength of this conceptual framework is that it first questions the company's existing strategy before starting the performance selection process. Hence, it ensures that the performance measures have a strong foundation. The performance charter also considers new stakeholders (employees, suppliers, alliance partners, or intermediaries).

It is usually neglected when formulating performance measures. Although the performance charter goes beyond traditional performance measurement, the main drawback is that it provides little information on identifying and selecting performance measures [40, 79].

- Performance Pyramid

One of the needs of any performance evaluation system is a clear relationship between performance indicators at different hierarchical levels of the organization so that each unit strives to achieve the same goals. One model that includes how to create this relationship is the performance pyramid model. The purpose of the performance pyramid is to link the organization's strategy with its operations through the translation of goals from top to bottom (based on customer priorities) and measurements from bottom to top [40, 49]. This framework includes four levels of goals:

- The creation of an organizational performance pyramid begins with defining the organization's vision at the first level, which then turns into the business units' goals.
- At the second level, business units regulate short-term goals such as profitability and cash flow and long-term goals such as growth and improvement of the market situation (financial and market).
- External effectiveness of the organization (left side of the pyramid)
- Internal efficiency of the organization (right side of the pyramid)

This framework reveals the difference between external groups' indicators (customer satisfaction, quality, and on-time delivery) and internal business indicators (such as productivity, time cycle, and waste). Business operational systems bridge high-level and daily operational indicators (customer satisfaction, flexibility, and productivity). Finally, four key performance indicators (quality, delivery, work cycle, and waste) are used daily in units and work centers [79].

The most important strength of the performance pyramid is its effort to integrate the organization's goals with operational performance indicators. However, this approach does not provide any mechanism for identifying key performance indicators, and this model has no concept of continuous improvement [17].

- Medori and Steeple

In 2000, Medori and Steeple [52] presented a comprehensive and integrated framework for auditing and improving performance evaluation systems. The graphical framework of their approach is presented under the name Circuit and Steeple. This framework consists of six precise steps:

- In the first step, like most other frameworks, the starting point of this model is the definition of the organization's strategy and its success factors.
- In the second step, the organization's strategic requirements are matched with six competitive priorities: quality, cost, flexibility, time, timely delivery, and future growth.
- In the third step, selecting appropriate indicators begins with using a checklist containing 105 indicators with complete definitions.

- In the fourth step, the existing performance evaluation system is audited to identify the company's current indicators.
- In the fifth step, indicators are discussed, and each indicator is explained with eight components: title, goal, model, equation, Frequency, source of information, responsibility, and improvement.
- In the sixth and last stet deals with periodic revisions of the company's performance evaluation system.

2.3 The Role of the Supply Chain Performance Evaluation System (SCPMS) in Achieving the organization's Goals

According to the definitions of the supply chain performance evaluation system in Sect. 2.2, In general, the goals and importance of the performance measurement system can be listed as follows:

- Khan et al. [38] consider the primary goal of SCPMS to improve the efficiency and effectiveness of SC.
- In addition, Tangen [78] also considers the following objectives of SCPMS:

  (a) Providing guidelines for better management and decision making
  (b) Provide guidelines to support analysis
  (c) Providing guidelines to improve the overall performance of the organization

- Also, Behrouzi and Wong [8] consider SCPMS as a system to achieve the following four main goals:

  (a) waste reduction
  (b) Time compression
  (c) Flexible response
  (d) Efficient use of all types of resources and unit cost reduction.

- Moreover, Charkha and Jaju [13] also consider the following six steps as practical steps and goals of the SCPMS system:

  (a) Creating strategic goals
  (b) Transforming strategic goals into a set of desired standards/parameters/criteria.
  (c) Prioritizing parameters and criteria according to their impact on performance.
  (d) Making connections between criteria
  (e) Understand overall performance
  (f) Finding remedial actions

  These steps are shown in Fig. 2.4.
  Therefore, a simple framework for measuring supply chain performance is a framework that includes not only measurement but also defining and understanding

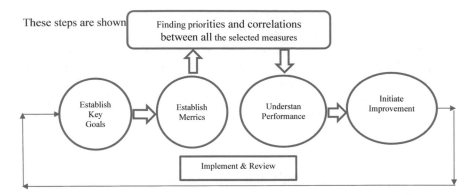

These steps are shown

Finding priorities and correlations between all the selected measures

Establish Key Goals

Establish Merrics

Understan Performance

Initiate Improvement

Implement & Review

Fig. 2.4 Effective SCPMS steps [13: 6]

criteria, collecting and analyzing data, then prioritizing and creating a relationship between actions, and taking improvement actions [13].

2.4 Supply Chain Performance Evaluation Criteria

Improving supply chain performance is a continuous process that requires performance measurement systems. Different variables measure the effects of supply chain performance evaluation on the revenue and costs of the entire system. After identifying key performance indicators, managers plan, control, and continuously implement them according to the selected vital indicators' results to improve performance. It is necessary to measure the performance of the supply chain in specific time intervals to improve the efficiency of the supply chain over time. To make an accurate assessment, in this case, it is necessary to design some indicators in this field. The performance evaluation criteria of supply chains are categorized into the following groups:

1. Quantitative and qualitative [7]
2. Cost and non-cost [20]
3. Quality, cost, delivery, and flexibility [70]
4. Cost, quality, resource utilization, flexibility, visibility, trust, and innovation [12]
5. Resources, staff, flexibility [7]
6. The efficiency of collaboration, coordination, and supply chain configuration [28]
7. Focusing on strategic, tactical, operational levels [20]
8. Score model (design, sourcing, manufacturing, delivery, return, or customer satisfaction) that can measure quantitative and qualitative criteria such as cost, time, quality, flexibility, and innovation [73]
9. Input and output and combined criteria [12]

10. Modeling scales of lean, agile, lean-agile chain [3]
11. Scales and critical performance criteria in the supply chain [22]
12. Balanced scorecard approach [11]
13. Tangible/intangible [20]
14. Sustainability/being green [62]
15. Financial/non-financial [21].

Most of the evaluation criteria of supply chain performance are economical and quantitative (cost, customer, responsiveness, and productivity). In general, supply chain performance causes growth rates and the effects of new markets and products [18].

Many measurement systems are weak in strategic settings and balanced approaches and systematic thinking, so these systems face problems in systematically identifying appropriate criteria. Many researchers have used the balanced scorecard and cost-based activities to solve this problem to evaluate the organization's performance. From the process view, the SCOR model deals with the construction of a facility to assess the performance of the systematic supply chain. However, this model is an improvement tool to identify, evaluate, and control the supply chain's performance.

Identifying key performance indicators in the organization differs depending on the activities they perform and are identified accordingly. In some evaluation systems, these criteria are classified into five categories:

- Sources
- output
- flexibility
- Innovation
- Information.

These indicators have a complex relationship with each other. These relationships are equal, hierarchical (string or chain), and paired. In Table 2.1, performance criteria based on supply chain processes are categorized.

In another research, the performance of the supply chain has been classified into five categories, which are:

- The flexibility of delivery systems to meet customer needs
- Empowering supplier participation
- Cost competitiveness
- Shorter order cycle
- Responding to customers flexibly [47].

In today's customer-centric markets, where the customer is the basis for achieving better financial results, the competence and spirit of the marketing effort are considered one of the most important sources of financial performance. Market share and sales growth in the shadow of increasing the added value of the price and sales revenue and reducing the cost of each unit can help achieve financial goals, leading to a significant increase in total profitability. Research literature considers four competitive

Table 2.1 Supply chain performance criteria based on processes [54]

Category	Performance evaluation criteria
Sources	Total supply chain costs, distribution cost, warehouse cost, production cost, total sales cost, management cost, information, warranty cost, return on investment (net profit rate to total assets), and added value
Output	Sales (profit), inventory shortage rate (lost sales), stocking rate (achieving target stocking rate, average item replenishment), the time between order and item arrival, on-time delivery percentage, order fulfillment, customer churn, complaint rate, Customer, planned process cycle time, cash to cash cycle time
Flexibility	Supply chain responsiveness, delivery flexibility, production flexibility, new product flexibility, purchasing flexibility, logistics flexibility, information flexibility
Technology	New product sales rate, number of new products launched, supply chain stability and process improvement
Information	Information accuracy, information suitability, information availability, information sharing

priorities, speed, quality, flexibility, and cost, important in supply chain measurement. Also, some researchers believe the critical parameters in evaluating supply chain performance include cost, delivery reliability, quality improvement, conformance to specifications, the time between order and receipt, product arrival time to the market, process improvement, and responsiveness.

Strategic objectives include vital elements that have resource, staff, and flexibility criteria. The criteria of resources (generally cost) and output (general responsiveness to the customer) are widely used in the supply chain model. However, the application of flexibility should emphasize three types of performance evaluation criteria. Resource criteria, output criteria, and flexibility. These criteria are shown in Table 2.2, as it is clear that they have different goals.

These criteria are critical to supplying chain success. Therefore, the supply chain performance evaluation system should evaluate these criteria [7] (Fig. 2.5).

The following supply chain performance evaluation criteria at three strategic, tactical, and operational levels are presented in Table 2.3.

These criteria are shown schematically in Fig. 2.6.

Table 2.2 Objectives of performance evaluation criteria [7]

Performance measure type	Goals	Purpose
Resources	High level of efficiency	Efficient resource management is critical to profitability
Output	High level of customer service	Without acceptable output, customers will turn to other supply chains
Flexibility	Ability to respond to a changing environment	In an uncertain environment, supply chains must be able to respond to change

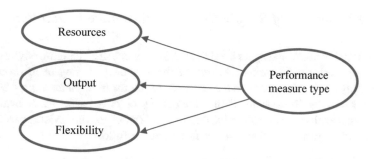

Fig. 2.5 Performance evaluation criteria from Beamon's perspective [7]

Table 2.3 A list of supply chain performance evaluation indicators [20]

Non-financial	Financial	Performance metric	Level
✓		Total cash flow time	Strategic
	✓	Rate of return on investment	
✓		Flexibility to meet particular customer needs	
✓		Delivery lead time	
✓		Total cycle time	
✓	✓	Level and degree of buyer-supplier partnership	
✓		Customer query time	
✓		The extent of co-operation to improve quality	Tactical
	✓	Total transportation cost	
✓		The truthfulness of demand predictability/forecasting methods	
✓		Product development cycle time	
	✓	Manufacturing cost	Operational
✓		Capacity utilization	
	✓	Information carrying cost	
	✓	Inventory carrying cost	

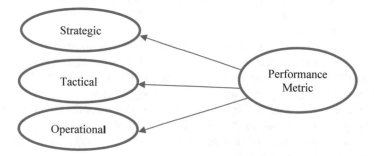

Fig. 2.6 Performance evaluation criteria from the perspective of Gunasekaran et al. [20]

2.4.1 Limitations of Supply Chain Performance Evaluation

Considering that many managers need to make decisions in a changing environment, determining performance criteria is one of the challenges facing decision-makers in the supply chain. Therefore, managers face problems in implementing a suitable performance evaluation system, so to solve these problems, performance evaluation systems have evolved. Among the limitations and criticisms that exist regarding the supply chain performance evaluation methods are as follows:

- Lack of connection with the organization's strategy
- Focusing on cost criteria and not considering non-cost criteria
- Lack of balanced methods
- Lack of focus on customers and competitors
- Lack of attention to supply chain integration and, as a result, local optimization
- Lack of systemic thinking.

In addition to these criticisms, performance evaluation systems in the field of production systems have problems and limitations, the most important of which are:

- Expecting results in a short period
- Without centralized strategic systems, the evaluation system does not correspond correctly to the strategic goals, organizational culture, or reward systems.
- In such systems, instead of progressing and improving the entire system, they seek to optimize the internal situation by reducing deviations from the existing standards.
- Failure to obtain competitor information through comparative evaluation.

In implementing a performance evaluation system, differentiating criteria according to the business process are necessary to identify and determine the items suitable for each of the strategic, tactical, and operational levels and the organization's goals.

Considering the difference between cost and non-cost measures (such as time, quality, innovation, and flexibility) is of great importance because if only cost indicators are used, it will give a misleading picture of supply chain performance. Gives. Time and quality criteria reflect the ability of the supply chain to provide a high level of service to customers.

Flexibility and innovation represent the ability of the supply chain to meet demand and supply with rapid changes. Two indicators of innovation and flexibility have been considered strategic drivers in future supply chain development. Therefore, it is necessary to check the performance of the supply chain using five evaluation indicators (cost, time, quality, flexibility, and innovation). For the effective competition, a comprehensive analysis should be done to check the results with the performance of the supply chain. In general, he listed the main problems and limitations of performance evaluation as follows:

- Lack of communication and understanding

The optimal feedback system ensures the life and survival of the system. Many evaluators keep the evaluation results to themselves for some reason. Naturally, the evaluator's lack of knowledge about the strengths and weaknesses of his performance leads to the lack of growth and development of the evaluator's capacity and abilities.

- Problems related to assessment outcomes

Among the other problems in the evaluation system is the lack of optimal use of the evaluation results in the organizational and individual dimensions. Suppose the systems of promotion, payment of salaries and wages, rewards, identification of training needs, and such things are based on performance evaluation. In that case, it can be undoubtedly said that creating a positive attitude towards the assessment and motivation in the evaluator increases effort in the direction of growth and improvement of their capacities.

- Increased cost to profit

It should always be considered whether the evaluation cost is reasonable and appropriate compared to the profit obtained from the result. Because, in some cases, the measurement price in some activities is much higher than their results.

- Lack of information

The lack of information in many organizations prevents them from setting appropriate goals.

2.4.2 Appropriate Feature of Supply Chain Performance Evaluation Criteria

Gunasekaran et al. [19] have recently emphasized that supply chain performance measures should be mainly balanced (financial vs. non-financial) and classified into strategic, tactical, and operational management levels. Tan et al. [77] stated that performance measures should be measurable, non-conflicting, and clearly defined throughout the chain, along with many other characteristics. In general, it can be said that the performance evaluation criteria of the supply chain should have the following characteristics [13]:

- It should be simple and easy to use
- It should have a clear purpose
- Must provide prompt feedback
- It should be about performance improvement, not just monitoring
- Can strengthen the company's strategy
- Able to relate it to the long-term and short-term goals of the organization
- It should match the company's organizational culture
- They should not contradict each other

- It should be integrated both horizontally and vertically into the company structure
- It should be compatible with the company's existing recognition and reward system
- Focus on what is important to customers
- Focus on what the competitor is doing
- It should lead to the identification and removal of waste
- It helps to accelerate organizational learning
- Evaluate groups, not individuals, in terms of performance for timing
- Establish specific numerical standards for most purposes
- It should reflect relevant non-financial information based on the critical success factors of each business
- Financial and non-financial measures should be aligned and proportionate in a strategic framework
- There should be a minimal deviation between organizational goals and measurement goals.

2.5 Conclusion

In this chapter, three essential parts about the importance of evaluating and measuring supply chain performance were described, and in the first part, the concepts of evaluating and measuring supply chain performance were explained. Then, in the second part, the role of the performance measurement system (SCPMS) in achieving the organization's goals and the importance of the performance measurement system were discussed. Moreover, in the last part, the supply chain performance evaluation criteria and the appropriate features of these criteria and limitations in the field of performance evaluation were described so that in addition to the readers getting to know the concepts of supply chain performance evaluation and its importance, they also get to know the performance evaluation criteria. So that they can improve the efficiency of their supply chain and their continuous control and implementation because; Improving supply chain performance is a continuous process that requires the use of performance measurement systems. In addition, there are different criteria for performance measurement, and it is necessary to identify them.

References

1. Abu-Suleiman, A., Boardman, B., Priest, J.W.: A framework for an integrated supply chain performance management system. In: IIE Annual Conference. Proceedings, p. 1. Institute of Industrial and Systems Engineers (IISE) (2004)
2. Agami, N., Saleh, M., Rasmy, M.: Supply chain performance measurement approaches: review and classification. J. Organ. Manage. Stud. **2012**, 1 (2012)
3. Agarwal, A., Shankar, R., Tiwari, M.K.: Modeling agility of supply chain. Ind. Mark. Manage. **36**(4), 443–457 (2007)

4. Anthony, R., Govindarajan, V.: Management Control Systems (2007)
5. Arzu Akyuz, G., Erman Erkan, T.: Supply chain performance measurement: a literature review. Int. J. Prod. Res. **48**(17), 5137–5155 (2010)
6. Balfaqih, H., Nopiah, Z.M., Saibani, N., Al-Nory, M.T.: Review of supply chain performance measurement systems: 1998–2015. Comput. Ind. **82**, 135–150 (2016)
7. Beamon, B.M.: Measuring supply chain performance. Int. J. Oper. Prod. Manage. **19**(3), 275–292 (1999)
8. Behrouzi, F., Wong, K.Y.: An investigation and identification of lean supply chain performance measures in the automotive SMEs. Sci. Res. Essays **6**(24), 5239–5252 (2011)
9. Bhagwat, R., Sharma, M.K.: Performance measurement of supply chain management. Int. J. Comput. Ind. Eng. **53**, 43–62 (2007)
10. Bhagwat, R., Sharma, M.K.: An application of the integrated AHP-PGP model for performance measurement of supply chain management. Prod. Plan. Control **20**(8), 678–690 (2009)
11. Bullinger, H.J., Kühner, M., Van Hoof, A.: Analysing supply chain performance using a balanced measurement method. Int. J. Prod. Res. **40**(15), 3533–3543 (2002)
12. Chan, F.T., Qi, H.J., Chan, H., Lau, H.C., Ip, R.W.: A conceptual model of performance measurement for supply chains. Manag. Decis. **41**(7), 635–642 (2003)
13. Charkha, P.G., Jaju, S.B.: Supply chain performance measurement system: an overview. Int. J. Bus. Perform. Supply Chain Model **6**(1), 40–60 (2014)
14. Chen, I.J., Paulraj, A.: Understanding supply chain management: critical research and a theoretical framework. Int. J. Prod. Res. **42**(1), 131–163 (2004)
15. Chen, C., Yan, H.: Network DEA model for supply chain performance evaluation. Eur. J. Oper. Res. **213**(1), 147–155 (2011)
16. Estampe, D., Lamouri, S., Paris, J.L., Brahim-Djelloul, S.: A framework for analysing supply chain performance evaluation models. Int. J. Prod. Econ. **142**(2), 247–258 (2013)
17. Ghalayini, A.M., Noble, J.S.: The changing basis of performance measurement. Int. J. Oper. Prod. Manag. **16**(8), 63–80 (1996)
18. Gopal, P.R.C., Thakkar, J.: A review on supply chain performance measures and metrics: 2000–2011. Int. J. Product. Perform. Manag. **61**(5), 518–547 (2012)
19. Gunasekaran, A., Patel, C., McGaughey, R.E.: A framework for supply chain performance measurement. Int. J. Prod. Econ. **87**(3), 333–347 (2004)
20. Gunasekaran, A., Patel, C., Tirtiroglu, E.: Performance measures and metrics in a supply chain environment. Int. J. Oper. Prod. Manag. **21**(1/2), 71–87 (2001)
21. Gunasekaran, A., & Ngai, E. W.: Build-to-order supply chain management: A literature review and framework for development. J. Oper. Manag. **23**(5), 423-451 (2005)
22. Gunasekaran, A., & Kobu, B.: Performance measures and metrics in logistics and supply chain management: A review of recent literature (1995–2004) for research and applications. Int. j. Prod. Res. **45**(12), 2819-2840 (2007)
23. Hald, K.S., Ellegaard, C.: Supplier evaluation processes: the shaping and reshaping of supplier performance. Int. J. Oper. Prod. Manag. **31**(8), 888–910 (2011)
24. Hald, K.S., Mouritsen, J.: The evolution of performance measurement systems in a supply chain: a longitudinal case study on the role of interorganisational factors. Int. J. Prod. Econ. **205**, 256–271 (2018)
25. Hanson, J.D., Melnyk, S.A., Calantone, R.A.: Defining and measuring alignment in performance management. Int. J. Oper. Prod. Manag. **31**(10), 1089–1114 (2011)
26. Harrison, T.P., Lee, H.L., Neale, J.J.: The Practice of Supply Chain Management: Where Theory and Application Converge. Springer Science & Business Media (2005)
27. Hausman, W.H.: Supply chain performance metrics. In: The Practice of Supply Chain Management: Where Theory and Application Converge, pp. 61–73. Springer, Boston, MA (2004)
28. Hieber, R.: Supply Chain Management: A Collaborative Performance Measurement Approach, vol. 12. vdf Hochschulverlag AG (2002)
29. Hofmann, E., Locker, A.: Value-based performance measurement in supply chains: a case study from the packaging industry. Prod. Plan. Control **20**(1), 68–81 (2009)

30. Holmberg, S.: A systems perspective on supply chain measurements. Int. J. Phys. Distrib. Logist. Manag. **30**(10), 847–868 (2000)
31. Huan, S.H., Sheoran, S.K., Wang, G.: A review and analysis of supply chain operations reference (SCOR) model. Supply Chain Manage. Int. J. **9**(1), 23–29 (2004)
32. Huang, E.: A systematic approach for supply chain improvement using design structure matrix. J. Intell. Manuf. **18**(2), 285–299 (2007)
33. Jääskeläinen, A., Thitz, O.: Prerequisites for performance measurement supporting purchaser-supplier collaboration. Benchmark. Int. J. **25**(1), 120–137 (2018)
34. Ka, J.M.R., Ab, N.R., Lb, K.: A review on supply chain performance measurement systems. Procedia Manuf. **30**, 40–47 (2019)
35. Kamalabadi, N., Bayat, A., Ahmadi, P., Ebrahimi, A., Kahreh, M.S.: Presentation a new algorithm for performance measurement of supply chain by using FMADM approach. World Appl. Sci. J. **5**(5), 582–589 (2008)
36. Kaplan, R.S., Norton, D.P.: The balanced scorecard: measures that drive performance. Harv. Bus. Rev. **70**(1), 71–79 (1992)
37. Kaplan, R.S., Bruns, W.: Accounting and Management: A Field Study Perspective. Harvard Business School Press, Massachusetts (1987). ISBN 0-87584-186-4
38. Khan, S.A., Chaabane, A., Dweiri, F.: Supply chain performance measurement systems: a qualitative review and proposed conceptual framework. Int. J. Ind. Syst. Eng. **34**(1), 43–64 (2020)
39. Krmac, E. (ed.): Sustainable Supply Chain Management. BoD–Books on Demand (2016)
40. Kurien, G.P., Qureshi, M.N.: Study of performance measurement practices in supply chain management. Int. J. Bus. Manage. Soc. Sci. **2**(4), 19–34 (2011)
41. Lai, K.H., Ngai, E.W., Cheng, T.C.E.: Measures for evaluating supply chain performance in transport logistics. Transp. Res. Part E: Logist. Transp. Rev. **38**(6), 439–456 (2002)
42. Laihonen, H., Pekkola, S.: Impacts of using a performance measurement system in supply chain management: a case study. Int. J. Prod. Res. **54**(18), 5607–5617 (2016)
43. Lapide, L.: What about measuring supply chain performance. Achieving Supply Chain Excellence Through Technology **2**(2), 287–297 (2000)
44. Lauras, M., Lamothe, J., Pingaud, H.: A business process oriented method to design supply chain performance measurement systems. Int. J. Bus. Perform. Manage. **12**(4), 354 (2011)
45. Li, S., Ragu-Nathan, B., Ragu-Nathan, T.S., Rao, S.S.: The impact of supply chain management practices on competitive advantage and organizational performance. Omega **34**(2), 107–124 (2006)
46. Lockamy, A., McCormack, K.: Linking SCOR planning practices to supply chain performance: an exploratory study. Int. J. Oper. Prod. Manag. **24**(11–12), 1192–1218 (2004)
47. Lohman, C., Fortuin, L., & Wouters, M.: Designing a performance measurement system: A case study. European journal of operational research, 156(2), 267-286, (2004)
48. Luzzini, D., Caniato, F., Spina, G.: Designing vendor evaluation systems: an empirical analysis. J. Purch. Supply Manag. **20**(2), 113–129 (2014)
49. Lynch, R.L., Cross, K.F.: Measure Up! The Essential Guide to Measuring Business Performance. Mandarin (1991)
50. Maestrini, V., Luzzini, D., Caniato, F., Maccarone, P., Ronchi, S.: Measuring supply chain performance: a lifecycle framework and a case study. Int. J. Oper. Prod. Manag. **38**(4), 934–956 (2018)
51. Maestrini, V., Luzzini, D., Maccarone, P., Caniato, F.: Supply chain performance measurement systems: a systematic review and research agenda. Int. J. Prod. Econ. **183**, 299–315 (2017)
52. Medori, D., Steeple, D.: A framework for auditing and enhancing performance measurement systems. Int. J. Oper. Prod. Manag. **20**(10), 1119–1145 (2000)
53. Melnyk, S.A., Stewart, D.M., Swink, M.: Metrics and performance measurement in operations management: dealing with the metrics maze. J. Oper. Manag. **22**(3), 209–218 (2004)
54. Mintzberg, H., Ahlstrand, B., Lampel, J.B.: Strategy Safari. Pearson, UK (2020)
55. Molnár, A., Gellynck, X., Kühne, B.: Conceptual framework for measuring supply chain performance: an innovative approach. In: 1st International European Forum on Innovation and System Dynamics in Food Networks, pp. 261–271. Universität Bonn-ILB Press (2007)

56. Mondragon, A.E.C., Lalwani, C., Mondragon, C.E.C.: Measures for auditing performance and integration in closed-loop supply chains. Supply Chain Manage. Int. J. **16**(1), 43–56 (2011)
57. Morgan, C., Dewhurst, A.: Using SPC to measure a national supermarket chain's suppliers' performance. Int. J. Oper. Prod. Manag. **27**(8), 874–900 (2007)
58. Morphy, E.: Measuring up. Export Today **15**(6), 52–57 (1999)
59. Najmi, A., Makui, A.: Providing hierarchical approach for measuring supply chain performance using AHP and DEMATEL methodologies. Int. J. Ind. Eng. Comput. **1**(2), 199–212 (2010)
60. Neely, A., Adams, C., Crowe, P.: The performance prism in practice. Measur. Bus. Excell. (2001)
61. Otto, A., Kotzab, H.: Does supply chain management really pay? Six perspectives to measure the performance of managing a supply chain. Eur. J. Oper. Res. **144**(2), 306–320 (2003)
62. Park, J.H., Lee, J.K., Yoo, J.S.: A framework for designing the balanced supply chain scorecard. Eur. J. Inf. Syst. **14**(4), 335–346 (2005)
63. Parker, C.: Performance measurement. Work Study (2000)
64. Pohlen, T.L., Lambert, D.M.: Supply chain metrics. Int. J. Logist. Manag. **12**(1), 1–19 (2001)
65. Putri, Y.D., Huda, L.N., Sinulingga, S.: The concept of supply chain management performance measurement with the supply chain operation reference model. IOP Conf. Ser. Mater. Sci. Eng. **505**(1), 012011
66. Ramaa, A., Rangaswamy, T.M., Subramanya, K.N.: A review of literature on performance measurement of supply chain network. In: 2009 Second International Conference on Emerging Trends in Engineering & Technology, pp. 802–807. IEEE (2009)
67. Ramezankhani, M.J., Torabi, S.A., Vahidi, F.: Supply chain performance measurement and evaluation: a mixed sustainability and resilience approach. Comput. Ind. Eng. **126**, 531–548 (2018)
68. Saleh, H., Hosseinzadeh Lotfi, F., Rostmay-Malkhalifeh, M., Shafiee, M.: Provide a mathematical model for selecting suppliers in the supply chain based on profit efficiency calculations. J. New Res. Math. **7**(32), 177–186 (2021)
69. Schmitz, J., Platts, K.W.: Supplier logistics performance measurement: indications from a study in the automotive industry. Int. J. Prod. Econ. **89**(2), 231–243 (2004)
70. Schönsleben, P.: Integral Logistics Management: Planning and Control of Comprehensive Supply Chains. CRC Press (2003)
71. Shafiee, M., Lotfi, F.H., Saleh, H.: Supply chain performance evaluation with data envelopment analysis and balanced scorecard approach. Appl. Math. Model. **38**(21–22), 5092–5112 (2014)
72. Sharif, A.M., Irani, Z., Lloyd, D.: Information technology and performance management for build-to-order supply chains. Int. J. Oper. Prod. Manag. **27**(11), 1235–1253 (2007)
73. Shepherd, C., Günter, H.: Measuring supply chain performance: current research and future directions. In: Behavioral Operations in Planning and Scheduling, pp. 105–121 (2010)
74. Sillanpää, I.: Empirical study of measuring supply chain performance. Benchmark. Int. J. **22**(2), 290–308 (2015)
75. Stephens, S.: Supply chain operations reference model version 5.0: a new tool to improve supply chain efficiency and achieve best practice. Inf. Syst. Front. **3**(4), 471–476 (2001)
76. Stern, J., Stewart, B., Chew, D.: The EVA financial management system (1998). Available at SSRN 6704
77. Tan, K.C., Kannan, V.R., Handfield, R.B., Ghosh, S.: Supply chain management: an empirical study of its impact on performance. Int. J. Oper. Prod. Manag. **19**(10), 1034–1052 (1999)
78. Tangen, S.: Improving the performance of a performance measure. Meas. Bus. Excell. **9**(2), 4–11 (2005)
79. Tangen, S.: Performance measurement: from philosophy to practice. Int. J. Prod. Perform. Manage.
80. Thakkar, J., Kanda, A., Deshmukh, S.G.: Supply chain performance measurement framework for small and medium scale enterprises. Benchmark. Int. J. **16**(5), 702–723 (2009)
81. Uur, B., Turan, E.E.: A model to evaluate supply chain performance and flexibility. Afr. J. Bus. Manage. **5**(11), 4263–4271 (2011)

82. Van Hoek, R.I.: "Measuring the unmeasurable"-measuring and improving performance in the supply chain. Supply Chain Manag. Int. J. (1998)
83. Varma, S., Wadhwa, S., Deshmukh, S. G. (2008). Evaluating petroleum supply chain performance: application of analytical hierarchy process to balanced scorecard. Asia Pacific J. Market. Logist.
84. Van Der Vorst, J.G.: Performance measurement in agrifood supply chain networks: an overview. Quantifying the Agri-food Supply Chain **15**, 13–24 (2005)
85. Wickramatillake, C.D., Koh, S.L., Gunasekaran, A., Arunachalam, S.: Measuring performance within the supply chain of a large scale project. Supply Chain Manage. Int. J. **12**(1), 52–59 (2007)

Chapter 3
Main Models and Approaches in Supply Chain Evaluation

3.1 Introduction

In today's global markets, companies are not units with unique brand names that can operate independently. The complexity of goods and services in today's world is such that it rarely happens that an organization or institution can produce a product or provide a service on its own without the help and cooperation of other organizations [116]. These challenges in the business world have led to the emergence of the supply chain approach. It has overcome the boundaries between companies and includes all interactions that include the supply of raw materials to the delivery of manufactured goods to the customer. Therefore, the supply chain has become an essential and vital factor in global markets, so the main competition occurs among their supply chains more than between organizations [10]. The importance of the concept of supply chain management and its performance is one of the paradigms of the twenty-first century to improve competitiveness that organizations have paid more attention to [21].

On the other hand, in recent decades, the supply chain has taken a significant share of the global economy and is the driving force of growth for many developing countries [43]. Also, globalization forces organizations to work together to provide services. To survive in dynamic and changing markets, they need tools to overcome environmental challenges and improve their industry performance [78]. In such a challenging and highly competitive environment, supply chains must constantly improve their performance to ensure survival. This can only be achieved by designing a performance evaluation system [67]. The correct design and evaluation of the supply chain are essential issues for managers and researchers, doubling the process's complexity due to the internal relationships of the units involved. Therefore, in the design of the performance evaluation system, all communications, interactions, priorities, influences, and limitations should be considered as much as possible so that the evaluation result of a supply chain provides correct feedback on the performance for improvement. Demand management, the optimal level of product access, and reduction of inventory costs (ordering, maintenance, and shortage) along the chain

© The Author(s), under exclusive license to Springer Nature Switzerland AG 2023
F. Hosseinzadeh Lotfi et al., *Supply Chain Performance Evaluation*,
Studies in Big Data 122, https://doi.org/10.1007/978-3-031-28247-8_3

are one of the critical goals of the supply chain, which requires the design of a correct performance evaluation system [116].

Therefore, the question always arises, what is the best way to evaluate the performance of the supply chain? On the surface, it may seem like a simple question, but the answer can be pretty complicated because; The supply chain can be examined from many perspectives such as financial, informational, strategic, operational, suppliers, customers, shareholders, and social [122]. Therefore, researchers in this field are trying to integrate the Supply Chain Performance Evaluation System. Different management approaches were created in this field: Lean, agile, Resilience and green, LARGe (Lean, Agile, Resilience, Green), and SCALE (Supply Chain Advisor Level Evaluation). For example, a lean factory is considered to have minimal inventory (close to zero).

In contrast, a stable (flexible) factory needs inventory to continue production when problems occur [24]. Lean approaches and Resilience appear contradictory [34]; however, in the best case, factories would like to have both the minimum level of inventory and not face production stoppages in the face of problems. The agile supply chain aims to respond immediately to the customer and the market. Finally, the green approach protects nature and the environment against direct and indirect waste.

Extensive supply chain management also aims to create integration in the supply chain in terms of lean, agile, sustainable, and green [22]. As mentioned, the supply chain has different approaches, each of which has advantages and disadvantages. Also, many models have been brought to evaluate the supply chain mentioned in the second chapter. Therefore, the complexity of dynamic systems can be best understood through system analysis and modeling. This way, it can gain insight into each subsystem's interactions. Therefore, performance evaluation cannot be done without modeling or an approach that identifies the values created for the entire supply chain [122]. Therefore, this chapter aims to explain the basic models in supply chain evaluation and the different approaches and network models in supply chain evaluation so that managers can model and evaluate their organization's supply chain more effectively.

3.2 Basic Models in Supply Chain Evaluation and Its Different Approaches

The second chapter thoroughly explained the reasons for the emergence of different performance evaluation models and their evolution and classification. In that chapter, several of the basic models of the supply chain are fully described. In this way, it is a guide for choosing and implementing the mentioned models in organizations. Supply chain performance measurement (SCPM) has become an important issue for organizations to pave the way to gain and maintain competitiveness. Different performance measurement and analysis tools have been developed in recent decades to evaluate supply chain performance from different perspectives. Therefore, supply chains today are trying to find a way to eradicate the intractable challenges associated with their supply chains by implementing innovative ideas, policies, and strategies. Therefore, an effective performance measurement system is in great demand to help organizations achieve their business goals by monitoring the effectiveness of implementing new strategies [59, 102]. Managers today have concluded that improving the performance of the supply chain requires the creation of a performance management system, and the ability to measure performance is considered one of the most critical elements of improvement and development at different levels of the supply chain [37]. In other words, supply chain management and improvement in it, like any other management system, requires performance evaluation to identify weaknesses and strengths, growth and success rates, determine the extent to which customers' needs are met, help organizations understand processes, and improve the program. Traditionally, the performance of companies is measured based on financial criteria with an emphasis on increasing revenues and efficiency and reducing costs. However, in parallel with the evolutionary process of organizations and moving towards a network approach and supply chain, performance measurement systems have also evolved [13] during this period, non-financial criteria such as customer satisfaction and product quality have also been included. Therefore, in response to the shortcomings of traditional methods, a wide range of different frameworks and models have been introduced and developed to evaluate and manage the performance of the supply chain, which were fully mentioned in the second chapter. In summary, it can be said that the most important of them are the following models:

- Activity-Based Costing
- Framework for Logistics Research
- Balanced Score Card
- Supply Chain Operation Reference Model (SCOR)
- Strategic Audit Supply Chain
- World Class Logistics Model
- Efficient customer response model
- Model of Excellence
- Strategic Profit Model
- Data Envelopment Analysis.

In Table 3.1, the above models are briefly explained:

Therefore, it can be said that various research has been conducted regarding the application of each of the mentioned methods to evaluate and manage supply chain performance [12, 13, 37, 63, 69, 74, 99, 117]. While many quality measures such as customer satisfaction measures, flexibility, the integrity of information and material flow, and effective supplier risk management are unquantifiable, data on other measures such as inventory turnover, cash conversion period, return on assets, value asset markets, profit margins, cost of goods sold, stock returns, etc. are readily available. As a result, quantitative measures are often preferred to qualitative measures due to their availability and use in various research. Also, there is no theoretical consensus regarding the most appropriate performance management method. However, among

Table 3.1 Supply chain performance evaluation models [37]

Supply chain performance evaluation models	
Activity-based costing	It was introduced in 1980, and its purpose is to analyze the cost and profit margin. This model requires in-depth knowledge of the company and group's activities through a logistic process using accounting information
Framework for logistics research	It was proposed in 1990 and discussed the description and theory of dependence between the levels of performance achieved, logistics organization, and competitive strategy. This model is used at the organizational and strategic level; It also divides logistics performance into several dimensions, including centralization, formalization, integration, and span of control
Balanced scorecard	It was proposed in 1990, and its purpose is a balanced evaluation based on the company's strategy. This model considers four dimensions: customers, financial, internal process and learning and growth, and the human dimension to evaluate performance
Supply chain operation reference model (SCOR)	It was proposed by the Supply Chain Association in 1996. It analyzes four dimensions: business performance reliability, flexibility/responsiveness, supply chain cost, and committed capital flow. This model can be used in manufacturing and service companies at the technical and operational levels to implement decisions related to performance management and strategic planning of the company
Strategic audit supply chain	It was proposed in 1999 and dealt with analyzing the value chain in the form of processes, information technology, and organization at the organizational level. The rules of this model divide the logistics chain into six competency areas; Customer-oriented, distribution, sales planning, lean production, partnership with suppliers, integrated chain management, and linking competencies to information technology and organizational chain
World class logistics model	Introduced in 1999 in automotive industry groups, it evaluates standard processes and functions and seeks continuous improvement. This model comprises six areas: strategy and progress, organized work, production planning, customer relations, process control, and supplier relations

(continued)

Table 3.1   (continued)

Supply chain performance evaluation models	
The efficient customer response model	In 1994, it was proposed by the association for effective response to customers, manufacturers, and retailers. This model evaluates appropriate inter-organizational practices and applies 45 criteria in four areas: customer demand management, supply chain management, technology database, and integration
Model of excellence	It was proposed in 1992 and had a questionnaire with 50 questions. This model is related to the efficiency of processes, including continual improvement in services and products, including employee management and operational improvement, and is suitable for all types of companies. This model is based on the first eight pillars, which describe customer focus, leadership, goal definition, process-based management, employee participation, continuous innovation, participatory improvement, and civic responsibility
Strategic profit mode	It was derived in 2002 that describes the interaction between strategic and operational levels of the "du Pont system" through financial ratios. This model processes strategic and financial performance based on cost drivers using return on assets
Data envelopment analysis	This technique was first created in 1957 and then expanded in 1978 and 1984 from these dates to now, various models of this model have been made to evaluate performance. In general, the data envelopment analysis model is based on linear programming, which compares decision-making units and measures the efficiency of decision-making units, including supply chains

the multiple approaches available in the field of performance management and improvement, it seems that the supply chain reference model (SCOR) is a more appropriate method due to its greater compliance with existing criteria in improving supply chain performance and information flow of supply chain [37], which will be explained below.

3.2.1   Supply Chain Operation Reference Model (SCORE)

SCOR, short for Supply Chain Operations Reference, is a process reference model that combines the concepts of business process, reengineering, modeling, and best practice. It is a model that links process elements, criteria, best practices, and characteristics related to supply chain implementation in a unique format [56] and is the most promising method for strategic decision-making and the most accurate method for evaluating chain performance, supply is seen [93]. This model was started in 1996 by the Supply Chain Council (SSC). This council is a non-profit consortium with many members around the world. The main activity of this association is to organize meetings to discuss and exchange opinions about supply chain problems and users on the standard framework and supply chain reference model.

The supply chain association, to support communication between supply chain components and increase effectiveness, a reference model named SCOR has provided an opportunity and a framework to help managers to measure the performance of supply chains in the industry, business processes, and how a company performs with market demand [73]. Company managers and researchers favored the SCOR model as a reference model for various supply chains. This model is the first general framework for evaluating and improving supply chain management and performance. It is the first model that can be used to configure the supply chain based on business strategy. Unlike other models, this model has standard definitions, vocabulary, and specific measurement criteria for measuring supply chain performance instead of mathematical formulas. Also, this model provides a standard and inclusive model, and its main advantage over previous models is its process-oriented perspective. As a result of this process-oriented view, a hierarchical and structured body of evaluations and criteria is created, which gives a general idea of the supply chain to all supply chain managers. Therefore, the SCOR model is a diagnostic tool for supply chain management (SCM) that enables users to understand the processes involved in a business organization and identify critical features that lead to customer satisfaction [93]. Also, the SCOR model maximizes measurable and actionable results when the supply chain vision strategy is aligned with the SCOR model [85].

SCOR model with five features; suggests managing the performance of operational processes, which include [79]:

• Reliability

Ability to perform tasks as expected. Reliability focuses on the predictability of the outcome of a process.

• Responsiveness

Speed of tasks. The speed at which a supply chain delivers products to the customer.

• Agility

The ability to respond to external influences is the ability to market changes to gain or maintain a competitive advantage.

• Asset Management Efficiency (Assets)

It is the ability to use assets optimally. Asset management strategies in a supply chain include inventory reduction and insourcing versus outsourcing.

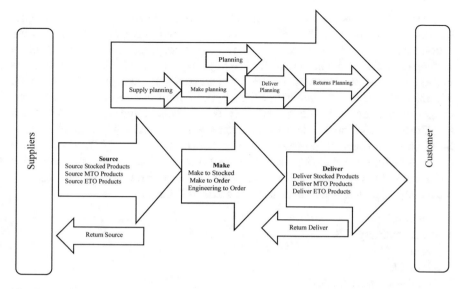

Fig. 3.1 SCORE model substructures [56]

- Costs

It includes the costs of operating supply chain processes such as labor, material, handling, and shipping.

Also, the main substructure of the score model based on opinion [56] is given in Fig. 3.1.

3.2.1.1 Supply Chain Operation Reference Model Versions (SCOR)

The Supply Chain Operations Reference (SCOR) model has been a product of APICS since the merger between the Supply Chain Council (SSC) and The Association for Operations Management (APICS) in 2014. Moreover, since its inception, it has been regularly updated to adapt to changes in supply chain business practices. Due to the adjustment of reality or industry conditions, the variables and features it already had have changed. APICS database members represent various industries, including manufacturers, distributors, and retailers. The extended APICS network also consists of technology providers and implementers, academics, and government agencies that participate in APICS activities and the development and maintenance of the SCOR model. For this reason, APICS is interested in providing the broadest possible release of the SCOR model, as the widespread use of this model will:

1. Enables communication using standard definitions and measurements.
2. It leads to better customer and supplier relationships.
3. Better support of software systems [1].

Therefore, its versions were constantly being changed and updated. For example, the first attempt of the SCOR model to consider environmental concerns was the introduction of the return process in its initial version. With its recent expansion to include GreenSCOR, it has gradually been applied to the field of natural resource management and the environmental performance of supply chain processes [7, 8, 108, 93, 55, 130]. Therefore, the Green SCOR extension of this model was introduced in its fifth version [93]. Cheng et al. [28] and Schoeman and Sanchez [113] note that Green SCOR is an amendment that integrates environmental considerations into SCM processes through processes, criteria, and best practices [93]. Also, criteria coding was introduced in the ninth version of this model to ensure that companies can adopt SCOR criteria without having to rename their existing criteria. For example, the coding of performance characteristics starts as follows:

- Reliability: RL
- Accountability: RS
- Agility: AG
- Cost: CO
- Asset management: AM.

Also, the latest model of SCOR is SCOR 12.0, which has many changes compared to previous versions. The reason for these changes is that the SCOR model is developed and maintained by the voluntary efforts of APICS members and supply chain industry subject matter experts. APICS depends on the participation of its members to actively advance the state of knowledge in the supply chain by identifying required model changes, researching and validating those changes, and building consensus on proposed changes. Similar to the Job Task Analysis (JTA) process used to update APICS certification content, a research survey was distributed to approximately 60,000 supply chain professionals worldwide to gauge industry-wide adoption and/or acceptance of new, activity-related business process methodologies. Evaluate them and vote for all updates to the SCOR framework, and many of the variables and features it previously had have changed. Also, among other significant changes and updates of the particular program of this version, the Green SCOR suffix is replaced by Sustainable SCOR.

Sustainable SCOR is based on the Global Reporting Standards (GRI), which is within the scope of the SCOR model. The GRI standards were chosen as a reference because the GRI has created a common language for organizations and stakeholders to communicate and understand the economic, environmental, and social impacts of organizations. The GRI standards are designed to increase the global comparability and quality of information about these impacts, thereby increasing the transparency and accountability of organizations. GRI standards are free to use and publicly available at www.globalreporting.org/standards. Sustainable SCOR uses GRI definitions and criteria when dealing with environmental sustainability issues (GRI Topic Standards 300 Series). This approach is used to help supply chain professionals see the environmental issues in their supply chain and value chain network and enable them to model and manage these impacts. A value chain covers the full range of upstream and downstream activities of an organization, including the complete life cycle of a product or service, from conception to end-use. Only GRI criteria in the supply chain management, sourcing, and risk management related to supply chain operations are included in the scope of the SCOR model.

The specific GRI disclosure number is cross-referenced when the SCOR model uses an element that aligns with a GRI disclosure. Please note that the GRI reporting guidelines must be followed when organizations make reporting claims. Care must be taken when transitioning from Green SCOR to Sustainable SCOR. Total air emissions are the only metric directly linked from one framework to another framework. However, the Green SCOR definition was not as precise as the Sustainable SCOR definition for total air emissions.

Other metrics to pay close attention to are as follows: The carbon footprint metric from the Green SCOR model is similar to greenhouse gas emissions but not quite the same. Sustainable SCOR has scope 1, 2, and 3 greenhouse gas emissions following GRI standards. In addition, there is also the emission of ozone-depleting substances (ODS). Green SCOR uses a % Recycled metric. However, the recycled and recycled terms used in Sustainable SCOR are very different from Green SCOR and follow the GRI standards. Green SCOR has a liquid release. According to Sustainable SCOR, following the GRI, liquid releases can be in the form of water discharges or hazardous or non-hazardous releases. Green SCOR measures solid emissions, and Sustainable SCOR measures hazardous or non-hazardous emissions following GRI standards.

In the past and previous versions, the SCORE model consisted of five main supply chain processes, which included (Plan, Source, Make, Deliver, and Return). Still, today this model consists of six main supply chain processes (Plan, Source, Make, Deliver, Return, Enable) is formed, which is shown in Fig. 3.2 and will be discussed further [1].

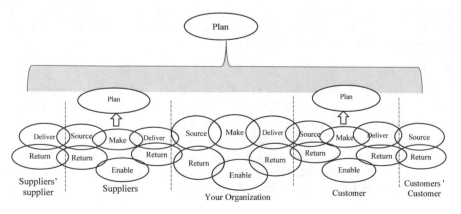

Fig. 3.2 Schematic display of SCOR model processes [1]

3.2.1.2   Score Performance Management Processes Based on Its Latest Version (Version 12)

The SCOR model includes the business processes of a company, which consists of (Fig. 3.2) [37, 56, 73]:

• Plan

The plan is the main activity in a supply chain, which includes production, material requirements, finances, distribution, and planning to provide value to the customer. The planning process includes processes that balance aggregated supply and demand by developing a set of actions that best achieve business goals. Planning processes deal with supply and demand planning, which includes activities for evaluating supply sources and integrating and prioritizing demand requirements, inventory planning, distribution requirements, production, materials, and capacity planning for all products and channels.

• Source

The source relates to preparing raw materials and materials needed for business processes. So it will be significantly associated with suppliers. The sourcing process includes methods that provide goods and services to meet actual and planned demand. Sourcing and supply include obtaining, receiving, inspecting, maintaining, and issuing materials. Sourcing infrastructure management has vendor approval and feedback, sourcing quality, internal transportation, materials engineering, sales contracts, and sales payments.

• Make

It is the primary step in providing value-added processes for products offered to customers. This stage includes the production process, work in progress until it becomes a semi-finished or finished product.

• Delivery

Delivery is related to the process of distributing products and services to customers. This stage also plays a vital role in measuring supply chain performance because of the relationship with the customer, which is the core or central point of the product being made or offered. Also, the delivery process includes processes that provide final goods and services to meet actual or planned demand. This process includes order management, shipping, and distribution activities. The delivery process has business rules, ordering rules, inventory, and quality.

• Return

The return is the process of returning products. Either in the condition that the customer has rejected or to improve the product. These conditions occur at a specific moment, such as a mismatch with market demand or other conditions. Returns also handle the reverse flow of materials and information about defective and excess products. This process includes approving, scheduling, receiving, arranging, and replacing or returning credit for the listed items.

• Enable

It creates, maintains, and monitors the information, relationships, resources, assets, business rules, compliance, and contracts required to perform supply chain processes. This process is related to top management, finance, human resources, information technology, facilities management, product management, sales, and support.

Therefore, every primary supply chain is a chain of planning, sourcing, Making, delivery, return and Enable. Any interaction between these processes is called a relationship in the supply chain. In this situation, planning is above these and manages them. All these cases are shown in Fig. 3.2.

The SCOR model was developed to describe business activities related to all stages of satisfying customer demand. The model includes multilingual sections and is organized around these six primary management processes Plan, Source, Make, Deliver, Return and Enable (shown in Fig. 3.2). By describing supply chains using these process building blocks, the model can describe supply chains that are very simple or very complex using a set of standard definitions. As a result, different industries can be linked to describe the depth and breadth of almost any supply chain. This model has successfully represented and provided a basis for supply chain improvement for global and site-specific projects.

3.2.1.3   The Structural Framework of the Score Model Based on Its Latest Version (12)

SCOR is a process reference model. A process reference model or business process framework aims to define the process architecture in a way that aligns with essential business functions and goals. The architecture here refers to how processes interact and operate, how these processes are configured, and the requirements (skills) of the employees who perform the procedures. Therefore, the SCOR reference model consists of 4 main parts [1]:

• Performance

It includes describing standard criteria for describing process performance and defining strategic goals.

• Processes

Includes standard descriptions of management processes and process relationships

• Procedures

It includes management practices that create better performance in the SCORE model process.

Table 3.2 SCOR performance attributes [1]

Performance attributes	Definition
Reliability	The ability to perform tasks as expected. Reliability focuses on the predictability of the outcome of a process. Typical metrics for the reliability attribute include On-time and the right quality
Responsiveness	The speed at which tasks are performed. The speed at which a supply chain provides products to the customer. Examples include cycle-time metrics
Agility	The ability to respond to external influences and marketplace changes to gain or maintain a competitive advantage. SCOR agility metrics include adaptability and overall value at risk
Costs	The cost of operating the supply chain processes. This includes labor costs, material costs, and management and transportation costs. A typical cost metric is the cost of goods sold
Asset management efficiency (assets)	The ability to efficiently utilize assets. Asset management strategies in a supply chain include inventory reduction and in-sourcing vs. outsourcing. Metrics include Inventory days of supply and capacity utilization

- People

Includes standard definitions for skills required to perform supply chain processes.

The SCOR model also includes a section for specific applications. Section 5, Special Applications, is used for proposed SCOR add-ons that have not yet been thoroughly tested for integration into the model but that APICS believes will be helpful to SCOR users.

Performance Section of the Score Model Based on the Latest Version (12)

SCOR's performance section focuses on measuring and evaluating the supply chain process implementation results. A comprehensive approach to understanding, evaluating, and diagnosing supply chain performance consists of three elements: Performance Attributes, Metrics, and Process/Practice Maturity, each of which describes different aspects or dimensions of performance [1]:

- **Performance Attributes**

The strategic characteristics of supply chain performance are used to prioritize and align supply chain performance with business strategy. Table 3.2 defines the Performance Attributes of the supply chain:

- **Metrics**

Discrete performance metrics that are themselves composed of related hierarchy levels. The coding of criteria is specified in this section. For example, every metric starts with this two-letter code; Reliability—RL, Responsiveness—RS, Agility—AG, Cost—CO, and Asset Management—AM, followed by a number to indicate the level, followed by a unique identifier. For example, complete order fulfillment is indicated by RL.1.1, which is a level 1 criterion in the Reliability feature.

- **Process/Practice Maturity**

Objective, specific descriptions use a reference tool to assess how supply chain processes and practices incorporate and implement best-accepted process models and leading practices. Process/procedure maturity provides a qualitative comparison of supply chain processes and methods with descriptive representations of different process acceptance and implementation levels. This assessment of supply chain process and operational effectiveness typically follows widely used models for maturity/operation (sometimes known as capability maturity models). Several maturity models for supply chain management usually follow a scale of "stages of maturity" where "high maturity" processes employ and often extend the best practice and are executed with a high degree of consistency. In contrast, those "low maturity" processes are characterized by outdated rules and/or a lack of discipline and consistency. SCOR does not currently embed the framework and content of the prescribed maturity model directly into the SCOR model document. The performance section provides an overview of this essential supply chain performance element. The SCOR user is encouraged to use existing maturity models to develop and tailor the content to their industry and company.

The Process and Levels Section of the Score Model is Based on the Latest Version (12)

The SCOR model has four levels [1, 37, 56]:

- **Level 1. Types of processes**

At this level, a broad definition of supply chain processes is defined as six integrated processes: planning, sourcing, manufacturing, delivery, return, and activation. Also, at this level of analysis, the scope of work, general components of the supply chain, and six types of mentioned processes and their performance are specified.

- **Level 2. Supply chain configuration and process classification**

The configuration of the supply chain is determined in the previous step by defining several processes. In fact, at this level, the structure of the supply chain is determined based on the procedures described in level 1. Also, at this level, processes are aligned

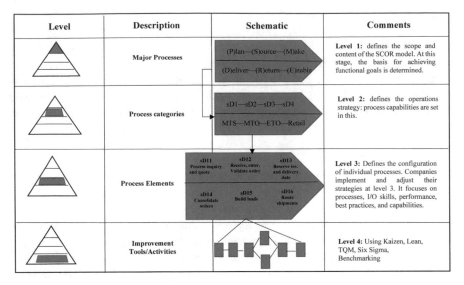

Level	Description	Schematic	Comments
	Major Processes	(P)lan—(S)ource—(M)ake (D)eliver—(R)eturn—(E)nable	**Level 1:** defines the scope and content of the SCOR model. At this stage, the basis for achieving functional goals is determined.
	Process categories	sD1—sD2—sD3—sD4 MTS—MTO—ETO—Retail	**Level 2:** defines the operations strategy: process capabilities are set in this.
	Process Elements	sD11 Process inquiry and quote sD12 Receive, enter, Validate order sD13 Reserve inv. and delivery date sD14 Consolidate orders sD15 Build loads sD16 Route shipments	**Level 3:** Defines the configuration of individual processes. Companies implement and adjust their strategies at level 3. It focuses on processes, I/O skills, performance, best practices, and capabilities.
	Improvement Tools/Activities		**Level 4:** Using Kaizen, Lean, TQM, Six Sigma, Benchmarking

Fig. 3.3 SCOR model process and level

with operational strategies. In the "Supply Chain Operations Reference" model, there are different types of supply chains with varying combinations of six main processes by default, which is necessary when analyzing, according to each chain's specific characteristics and conditions, one of these configurations should be selected by the analyst. For example, the manufacturing process can be in the form of production based on order, storage, and production to develop a new product. There can also be different situations for other processes.

- **Level 3. Analysis of processes**

It determines the necessary information for the design, and by detailing the information of each of the level 2 methods, it sets the chain improvement goals more precisely. Also, at this level, the components of the processes are defined. The information related to them and the inputs and outputs are specified. Measurement criteria are defined. The best experiences and similar techniques in successful companies are examined, and the system requirements necessary to reach the best performance level are discussed. At this level, companies make improvements in their operational processes and strategies.

- **Level 4. Implementation and analysis of process components**

This level also focuses on execution. This level exists when the company has implemented its supply chain improvement projects, and these projects are determined depending on the needs of each company. This level is not part of the primary levels of analysis in the "Supply Chain Operations Reference" model, and it only deals with providing suggestions to achieve a competitive advantage. Figure 3.3 gives more explanations of these levels.

Modeling is also a method recommended by the "Supply Chain Operations Reference" model to improve process indicators. A benchmark is a reference point against which a company compares itself and determines its current performance level. The Supply Chain Council provides members with a list of successful experiences in each process measure. What successful companies have done can be an enabler for other companies.

As shown in Fig. 3.3, this model is designed to support supply chain analysis at multiple levels. SCC focuses on the top three process levels, which are industry neutral. Therefore, SCOR does not attempt to explain how a particular organization should conduct its business or regulate its systems/information flow. However, any organization implementing supply chain improvements using the SCOR model should extend the model to at least level 4, using industry, organization, and/or location-specific processes, systems, and practices. It is also important to note that this model describes procedures, not functions. In other words, the model focuses on the activity involved, not the person or organizational element that acts.

Practices Section, Model Score Based on the Latest Version (12)

The Practices section, formerly known as "Best Practices," provides a set of industry-neutral practices companies have recognized for their value. An action is a unique way to configure a process or set of procedures. Uniqueness can be related to process automation, the technology used in the process, special skills applied to the method, the unique sequence for performing the process, or the unique way of distributing and connecting processes between organizations.

SCOR recognizes that there are several different method competencies within each organization (SCOR ID):

- Emerging practices (BP.E)
- Best practices (BP.B)
- Standard practices (BP.S).

SCOR recognizes that a practice's eligibility may vary by industry or geography. For some industries, a practice may be standard, while the same practice may be considered an emerging practice or best practice in another industry. Therefore, the SCOR classification of practices has been created based on the data of specialists and experts in various industries. All SCOR practices are mapped to one or more categories. SCOR12 recognizes 21 classifications. Taxonomies help identify practices by area of focus, for example, inventory management or new product introduction.

People Section of the Score Model Based on the Latest Version (12)

The People section of SCOR was introduced in SCOR 10 and provided a standard for describing the skills required to perform tasks and manage processes. In general, these skills are specific to the supply chain, and some of the skills identified may be applicable outside the scope of the supply chain process. Skills are described by a standard definition and related to other aspects of people: experiences, training, and level of competence. The competency level is not included in the framework description.

SCOR recognizes five accepted competency levels:

- Novice: Untrained beginner, no experience, requires and follows detailed documentation.
- Beginner: Performs the work with limited situational perception.
- Competent: Understands the work and can determine priorities to reach goals.
- Proficient: Oversees all aspects of the work and can prioritize based on situational aspects.
- Expert: Intuitive understanding. Experts can apply experience patterns to new situations.

These competency levels are similarly used as procedural or operational maturity levels. The profile of the person or job is evaluated based on the level of competence found (person) or desired (job profile). Coding in the people section includes coding skills, experiences, and training defining abilities. All People elements begin with a capital H followed by a capital letter representing the element: S for Skills, E for Experiences, and T for Training. A dot and a four-digit number follow them. For example, HS.0046 is the "ERP Systems" skill code, and HT.0007 is the APICS CSCP training code.

3.2.1.4   Advantages of the SCOR Model

The global era requires companies to change their business supply chain performance measurement system. For this reason, a proper supply chain performance measurement system is required. One of the appropriate supply chain performance measurement systems is the supply chain operations reference model (SCOR). Why so; one of the reasons for the various failures of supply chain improvement programs is the inability to get an overview of the company's current state, including its cultural aspects. Using the SCOR model in building the concept of measuring supply chain performance based on the process allows the company to evaluate the supply chain performance in general to achieve the following [98]:

- Monitor and control.
- Communicate organizational goals to functions in the supply chain.
- Find out where the organization is compared to its competitors and take action to improve its position and create a competitive advantage.

SCOR is a form of supply chain operation model that is renewed to become three intact entities, which are [97]:

1. Business processes
2. Criteria

3. Cross-functional frameworks in the supply chain.

Also, the SCORE model pursues three main goals:

- Defining the current state of the processes and specifying the desired state,
- Defining the target values for the indicators of the defined processes using benchmarking,
- Examining the best practices and management tools and software used by them to achieve the best results.

Therefore, the SCOR model is one of the appropriate methods for modeling and evaluating the supply chain [65]. This standard is a tool to check the supply chain configuration and identify and measure the criteria. In addition, the scoring method helps to adapt and use the best practices when they are similar to the supply chain under investigation. Therefore, the SCORE model has created a common language for communication in the supply chain [85]. Also, [37] described the strengths and weaknesses of the model as follows:

Strengths of the SCOR model

- Considers all company-customer communications and interactions and describes standard supply chain processes.
- Provides a framework of relationships among standard processes.
- It has standard definitions, vocabulary, and criteria for measuring supply chain performance.

Weaknesses of the SCOR model

- Marketing and sales processes, research and technology development, product development, and after-sales service do not explain or describe it.
- It includes education, quality, and information technology, but it is not explicitly mentioned in the model.

Also, the general advantages of the SCOR model widely published by the Supply Chain Council with illustrative case histories can be listed as follows [16, 56]:

- Cost reduction and improved customer service, delivering an average 3% increase in total operating income.
- Approximately 2–6% improvement in return on investment (ROI) within 12 months of project implementation.
- Due to informed investment decisions, there is a significant improvement in return on assets (ROA).
- Standard supply chain definitions and interpretations facilitate using standard features of information technology systems and drastically reduce operational costs.
- 1–3% increase in profit through continuous improvement in supply chain management.

As a practical example of the advantage of the SCOR model, we can mention the Intel Company; in particular, Intel is one of the beneficiaries of SCOR worldwide, which is forced to improve supply chain networks due to the emergence of business-to-business (B2Bi) integration. Virtual supply became complex and dynamic. Intel Corporation has appropriately adapted and integrated the SCOR methodology into its supply chain improvement efforts [58]. SCOR initiatives at Intel are intended to achieve the following goals:

- Documentation of current supply chain and process improvement efforts.
- Identify short-term improvements.
- Identify owners for long-term improvements.
- Most importantly, learn and incorporate the SCOR methodology into improvement efforts.

It also facilitates a mechanism for Intel to benchmark using industry best-in-class practices. Performance indicators are aligned and customized at different operational and strategic levels [56].

3.2.2 New Supply Chain Paradigms or Approaches

The supply chain network consists of different layers such as suppliers, manufacturers, distributors, and customers; in addition to the physical flow (flow of goods and materials), financial and informational flows also flow. What has turned supply chain management into a new management philosophy is dealing with the supply chain as an integrated whole and creating an intelligent alliance between the chain members to provide a product with high quality and low cost, or in other words, making the highest value to gain satisfaction. The customer is in a competitive environment. Therefore, in this direction, new SCM strategies and approaches were created with the ultimate goal of the competition, product quality and service level to customers, and optimal performance in terms of operational, economic, and environmental compatibility [39], which are described below Paid.

3.2.2.1 Lean Approach

Various definitions of the lean model can be found in the literature, but they all share the same principle: cost minimization and waste elimination. The basic concept of lean is to do more with fewer resources (e.g., less human effort, less equipment, less time, and less space) while being closer to the customer's ongoing needs. The term "lean" refers to a series of activities or solutions to eliminate waste, reduce non-value-added operations, and improve value-added processes [128]. Rossini and Staudacher [109] believe that lean supply refers to the spread of lean principles throughout the supply chain, both downstream and upstream. The lean supply chain is not exactly a new field for lean philosophy. The lean supply chain has appeared in the literature under different names: lean procurement, lean distribution, lean supply, and lean organization [109]. *Lean supply chain management* is a set of organizations directly related to the upstream and downstream flows of products, services, information, and financial resources. It works collaboratively to reduce costs and waste [125].

3.2.2.2 Agile Approach

Today, due to the speed of demand changes, short product life cycles, and the existing competitive environment, companies and operations managers need to adopt agile policies to meet customer demand in the short term. The origin of the agile supply chain goes back to flexible production systems in the 1960s [104]; the production system seeks to create flexibility and responsiveness by using automation. Also, an agile supply chain integrates business partners to enable new companies to respond quickly and effectively to market change, resulting in customized products and services [19]. The essential characteristic of agility that can be seen in most definitions is the ability to quickly respond to market changes, which is a critical component in the success and survival of companies in the market [27].

Christopher [29] defines an *agile supply chain* as a broad business capability that includes organizational structure, information systems, logistics, processes, and mindsets. In general, an agile supply chain is fast and flexible. Van Hoek and Mitchell [54] also believe that an agile supply chain should be considered at all company levels, from the strategic to the operational level. Companies must quickly respond to short-term changes in demand or supply and be able to react to sudden increases [104].

Following [29, 54] stated the following four essential characteristics of an agile supply chain:

1. Always sensitive to the market and able to identify and respond to actual demand.
2. Information is shared between suppliers and buyers to create a virtual supply chain, especially when inventory has been replaced with e-commerce information.
3. Process integration takes place between partners for collaborative work methods, product development, and shared systems between suppliers and buyers.
4. It is a network based on shared goals [104].

Also, [77] stated the following steps to implement an agile supply chain:

1. Promoting the flow of information with customers and suppliers,
2. Development of cooperative relations with suppliers,
3. Design for postponement,
4. Building inventory buffers by keeping a stock of inexpensive but essential components,
5. Having a reliable procurement system.
6. Developing contingency plans and crisis management groups [104].

Actions related to the agile supply chain approach can be classified as follows:

- De-risking measures include increasing excess capacity, supply chain risk management plans, and de-risking supply chain strategy.

- Flexibility and responsiveness activities include flexible resources, flexible transportation, Flexibility in product design, and Flexibility in production.
- Using supply chain knowledge for innovation, such as customer participation, mechanisms to promote supply chain innovation, dynamic alliances, and virtual networks for product development [31].

3.2.2.3 Resilience Approach

A sustainable supply chain refers to programs and activities of the company that have integrated environmental and social issues with supply chain management to improve social and environmental performance; To improve the social and environmental performance of the company and its customers and suppliers without jeopardizing the economic performance [44]. Sustainable supply chain management is the management of materials, information, and capital flows, as well as the coordination between the components in the supply chain that define their goals to meet the needs of customers and stakeholders based on the three dimensions of sustainability (Economic, Environmental, Social) [114]. Sustainable supply chain management seeks to integrate economic, environmental, and social aspects into a supply chain [100]. The main goal of a sustainable supply chain is to provide customers with better products and services at low costs and in a healthy environment [111]. Companies can improve internal production processes, economic benefits, and environmental and social impacts [44]. In the Resilience model, which is also referred to as the flexible model, unlike the lean model, which focuses on cost minimization, it is necessary to have the capacity to overcome problems and respond effectively to unexpected disturbances. The sustainable supply chain cannot be the lowest cost chain, instead, compared to other approaches, it is more able to cope with the uncertain business environment [25].

3.2.2.4 Green Approach

Michigan State University Industrial Research Association introduced green supply chain management in 1996, a new management model for environmental protection. The company can reduce negative environmental impacts and achieve optimal use of resources and energy by using supply chain management and green technology. Green supply chain management can reduce the environmental impact of industrial activities without sacrificing quality, cost, reliability, and performance or reducing the efficient use of energy. In contrast, environmental regulations minimize ecological damage and lead to economic profit [120]. A green supply chain management approach has been developed to balance economic and environmental performance [91]. Therefore, green supply chain management is the result of integrating environmental thinking with supply chain management, which includes product design, material selection and collection, production processes, final product delivery to customers, and end-of-life management of products after their useful life is defined

[120]. In other words, due to the increasing trend of industrialization, increasing environmental pollution, the pressure of government regulations to obtain environmental standards, increasing customer awareness, and as a result, the growing demand of customers for the supply of green products (products that do not harm the environment), environmental factors must be considered in all stages of the supply chain, from product design to product recycling. However, these stages are carried out associated with pollution and environmental damage. The green supply chain measures taken by the organization are known as internal measures and are intended to improve the organization's performance. The management of internal factors leads to the improvement of the environmental performance of the organization. External factors such as customer and supplier involvement were also studied as influential factors on organizational performance. Developing communication with external factors such as the government, suppliers, customers, and other competitors leads to better environmental supply chain performance. In addition to internal and external factors, the product must be designed to reduce both waste and the waste created can be recycled.

Also, the green supply chain must have six components in the entire process, from design to recycling, which are:

- Green design
- Green marketing
- Green production
- Green shopping
- Green packaging and transportation
- Green recycling.

Therefore, policies such as recycling raw materials, reuse of materials, clean production, and waste management can help achieve environmental goals. Implementing green supply chain management can help an organization gain a competitive advantage over its competitors [91].

3.2.2.5 LARGe Approach

The word LARGe is composed of the first letter of the Latin word for the four supply chain approaches (lean, agile, Resilience, green) that we explained earlier. The idea of effective supply chain management was developed in the Mechanical and Industrial Engineering Research Unit of the Faculty of Science and Technology of the Universidade Nova de Lisboa. Currently, this research unit is the primary reference in this field. Each of the four supply chain approaches has advantages and disadvantages. Exploiting the benefits of these approaches and planning to eliminate the weaknesses of each one increases the potential of creating value in the supply chain. Extensive supply chain management covers a variety of topics, including:

- Specifications and methodology [22]
- Organizational structure and performance indicators [36]
- Human factors [32]

- Informatics and integrative model [83].

Therefore, extensive supply chain management tries to combine lean, agile, Resilient, and green approaches in the space of supply chain management to benefit from each one of them and cover their shortcomings simultaneously [6].

Also, this model examines supply chain integration from five perspectives, which are [83]:

- Integration of supply of materials and services
- Integration of relationships
- Integration of planning and technology
- Internal integrity
- Customer integrity.

3.2.2.6   SCALE (Supply Chain Advisor Level Evaluation)

This model identifies and improves the processes that need to be enhanced to create value. The skill model monitors the levels (operational, tactical, strategic decision-making) and is classified into the following seven components:

1. Definition of supply chain strategy
2. Definition of chain goals
3. Creating procedures
4. Resource planning
5. Coordination of relationships between chain channels
6. Evaluating and monitoring the performance of each partner involved in the supply chain
7. Optimizing the entire supply chain.

Also, the focus of this model is on three processes of value creation, considering their indicators, which are:

1. Creating value for the company (profitability, innovation, creation of knowledge and their sharing, innovation, growth)
2. Creating value for the customer (delivery time, price, quality, performance, brand image, honesty, responsiveness)
3. Creating value for actors in the chain (sharing information, aligning decisions, sharing resources, sharing profit and risk, creating knowledge and sharing it, aligning innovation, speed of the chain, overall profitability, honesty, and sustainable development).

In the whole skill model, 57 indicators based on value creation are proposed from the seven levels of the process [122].

3.3   Network Models in Supply Chain Evaluation

The Supply chain network design is the most important strategic decision in supply chain management, which plays a vital role in the environmental and economical implementation of the supply chain. Supply chain network design generally includes determining network facilities' location, number, and capacity and creating material flow between them [87].

Two important features that are considered in the design of the supply chain network are Reliability and Robustness. According to Bundschuh et al. in 2003, Reliability was defined as follows:

The probability that a system or its components will perform their routine tasks within a given time horizon and environment.

He also defined Robustness as follows:

The system can perform expected tasks relatively well despite defects in components or subsystems. Therefore, a supply chain is resilient when it plays a role correctly, considering the uncertainty in future conditions such as demand, procurement time, supply, Etc. For example, when a distribution center is unavailable due to weather conditions, it should play its role properly.

There are different definitions of supply chain network design in the literature, some of which are mentioned below:

- The design of the supply chain network includes issues such as which facilities should be in the supply chain network (factories and warehouses), their size and location, and the establishment of transportation relationships between the members of the supply chain as to how materials flow between them. Also, according to him, appropriate network design decisions lead to a cost reduction between 5 and 60%, despite the 10% expected output from the supply chain network design project. Supply chain network design is a complex undertaking and mission. He also believes that supply chain network design is the core of strategic planning in supply chain management, whether creating a new network configuration or redesigning an existing network [50].
- A standard supply chain network design issue includes how to configure the network and its location mission. In this design, some open or may close. Each selected facility is assigned to one or more product, assembly, or distribution activities based on available capacity at each location. The mission of each facility should also be defined. Suppliers of vital raw materials should be selected. Each product market must choose a marketing policy, inventory levels, and minimum and maximum sales levels. The common goal in this chain is to maximize net profit over the planning horizon. Standard costs include fixed costs of location and facility configuration, fixed vendor costs, market policy choices, and variable costs, including production, handling, shortages, inventory, and transportation [5].
- Also, designing and establishing the supply chain network is a strategic decision. Its effects will last for several years; some parameters of the business environment, such as customer demand, may change during these years. Therefore, some basic parameters, such as customer demand, are entirely uncertain. In other

words, building or not a facility is time-consuming and expensive. It is impossible to change the facility's location considering the fluctuation of parameters in a short time. Therefore, the supply chain should be designed considering the non-deterministic and resistant parameters. Otherwise, the effect of parameter fluctuations over time will be huge [86].

- In 2006, Noorul Haq and Kannan concluded that the supply chain is a chain that connects every input of the production and supply process, from raw materials to the final consumer. A supply chain includes various systems such as supply. Different suppliers are manufacturing, warehousing, transportation, and retail systems, and state that the term supply chain network design is sometimes used as a synonym for supply chain strategic planning. They consider it essential that in today's competitive world, a network supply chain can survive for a significant period, during which several parameters can change. It may be necessary to consider making future modifications to the network configuration to accommodate gradual changes in the structure of the supply chain or facility capacity.

In general, in supply chain management, three levels of planning are usually considered according to the time horizon [9, 92]:

1. Strategic level
2. Tactical level
3. Operational level.

At the strategic level, decisions that have long-term effects on the organization are reviewed, like the; number, location, and capacity of warehouses and production factories or the flow of materials within the logistics network. Since tactical and operational activities are implemented after making strategic decisions, the arrangement of the logistics network will become a limitation for the level of tactical and strategic decisions. Since establishing or non-establishment of the supply chain network is time-consuming and costly, changing the network arrangement is impossible. Therefore, the supply chain arrangement is a critical strategic issue that affects tactical and operational activities and needs to be optimized for a long time and effective action in the entire supply chain [101].

3.3.1 Definitions of Supply Chain Network (SCN)

The supply chain is a chain that includes all related actions to achieve the production process and material changes, from the raw material supply/procurement stage to the final delivery stage of the product to the consumer. Here, apart from the goods process, there are two other processes, one of which is the information process and the other is financial resources and debt. So:

- The supply chain is a network of topological organizations comprising independent and semi-autonomous institutions. In each supply chain, there is a primary

entity responsible for the supply chain's structure based on demand-related information, financial process materials, and knowledge to achieve value in the entire
chain [115].

- The supply chain is a complex network of facilities designed to provide, produce
  and distribute goods to customers in the right quantities, places, and at the right
  time [66].
- The term "network" in the context of SCM represents an attempt to provide a
  broader and more strategic perspective by using the various potential resources
  of network actors in a more effective manner [62].
- Shen [84] defined the supply chain as "a supply chain network that includes
  all steps that directly or indirectly play a role in meeting customer demand."
  He also emphasized that the supply chain network includes manufacturing units
  and suppliers, warehouses, retailers' distribution centers, goods carriers, and
  customers. [121] believe that a supply chain includes several organizations;
  although these organizations are legally separate, they are connected through
  material, information, and financial flows. In other words, a supply chain network
  does not focus only on the flow within a chain. Still, it works on complex convergent and divergent flow networks that involve many disruptive customer orders
  that must be fulfilled in parallel.

Therefore, the network perspective questions the concept of using a linear and one-
dimensional approach in SC by arguing the issues of relational aspects of a particular
(fixed) situation in SC [41]. Moreover, it reflects the pattern of focal companies'
relationships with their partners and third parties in the business network context
[129].

Also, the supply chain network is an open and dynamic complex network. Supply
chain complexity is an integral part of supply chain management, mainly represented
by structural complexity, operational complexity, and decision-making complexity
[80, 133].

Therefore, with the expansion of the supply chain scale and the increase in network
complexity, the overall structure of the supply chain has received more and more
attention. The supply chain is an upstream and downstream organizational model
that usually consists of three or more levels. A multi-level supply chain has become
one of the focuses of research. Different companies from different supply chain
structures play a role in upstream and downstream connectivity, and each level is
not only a provider of information but also a demand for information. Defining
the role of node companies helps supply chain decision makers to understand and
identify the way and the strength of the relationship between node companies in
the entire supply chain, identify the position of each node company in the supply
chain and make reasonable adjustments and effective allocation of resources in the
supply chain. The multi-layer network model can accurately reflect the properties
of multiple node connectivity features [71]. Moreover, the topological connection
between levels also demonstrates the position of node companies in the supply chain
to a certain extent [133]. In the following, the difference between supply chain (SC)
and supply chain network (SCN) is given to clarify this issue.

3.3.2 Difference Between Supply Chain Network and Supply Chain

Van Der Zee and Van Der Vorst [131] presents the difference between supply chain network and supply chain in the form of Fig. 3.4.

According to Fig. 3.4, it can be said that each actor in SCN belongs to at least one SC (shadowy figure). However, each layer has different actors that can affect the shadow SC. An SCN is more like a rooted tree than a pipeline or chain; its branches and roots are a vast network of customers and suppliers [131]. In the following, citing [105], other differences between a supply chain network and a supply chain are given:

- SC has focused on examining traditional dyadic buyer–seller relationships. Instead, SCN has concentrated on examining relationships beyond dyadic [17] and includes related actors involved in the procurement, production, and delivery of goods or services that are applied to the final customers [70, 103].
- Pan and Nagi [95] define SC as "a set of mostly collaborative activities and relationships that link firms together in the value creation process to deliver the appropriate value mix of products and/or services to the end customer." Moreover, they define an SCN as "a set of active actors in an organization's SCs, as well as passive actors with whom an organization is associated, who can actively contribute to an SC when needed." Therefore, based on these definitions, some SCN actors are active, and some are inactive. Passive actors are not directly involved in the production process of final goods. Still, they play an essential role in increasing SC flexibility, especially in times of supply crisis, by providing support resources [95].
- Another significant distinction between SCN and SC is that the issues addressed in SC are usually focused on operational areas as well as improving efficiency through the development of better systems throughout the SC, including sourcing, product design, manufacturing, Refers to delivery, and recycling [33, 70, 90]. However, firms develop appropriate relationships with different SCN actors to access their valuable resources and effectively implement their strategies [70].

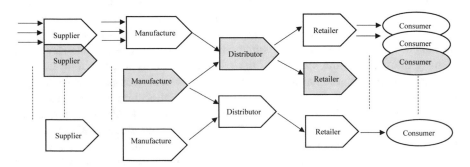

Fig. 3.4 A generic supply chain (shaded) within a supply chain network [131]

- In addition, in the case of overall SCM modeling, traditional approaches have usually focused on technical issues and have not paid enough attention to measuring the various complexities in the structural and behavioral aspects of SCM systems [14]. However, companies must examine the interrelationships and influences between SCN actors to find appropriate strategies to meet stakeholders' expectations [45].
- Mizgier et al. [88] provide the main distinctions between SCs and SCNs. They argue that SCs operate structured, while SCNs are more dynamic and complex.

So, Fig. 3.5 shows the difference in the position of actors in the supply chain (SC) and the supply chain network (SCN).

Figure 3.5 shows that SC actors are vertically connected. Also, SCN actors are actors that exist in each layer. SCN may also include non-corporate actors [33, 105, 123]. Therefore, SCN actors are SC actors and actors who have relationships with them in each layer [76]. In general, SCN actors can be placed on three levels [26]:

1. The upstream network level is related to supply-side interactions.
2. The level of the focal company is related to the interaction on the customer side.
3. The downstream network level is related to interaction on the customer side.

Also, the focal company is a relative perspective, which means that any company can be a focal company because it can make strategic decisions [26]. A focal firm represents the researchers' entry point and the focal firm's upstream and downstream business partners from the entire supply chain [119]. Therefore, actors (SCN) can be

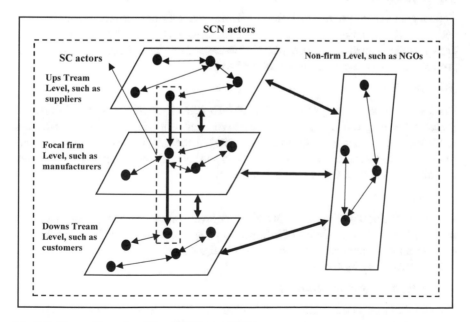

Fig. 3.5 The position of an SC and SCN actor [105]

identified based on their focal knowledge and knowledge of their vast network [38]. Also, the focal point can be placed at any level [105].

3.3.3 Supply Chain Network Structure

An essential element of analyzing the relationships in the supply chain network is understanding the configuration of the supply chain network structure [70]. The structure of the supply chain network shows how different companies are configured by their relationships with each other to create some values. A deep understanding of the structure of the supply chain network is vital for focal companies because the formation of connections between different actors in the supply chain network can affect behaviors, strategies, and the implementation of supply chain management practices [126]. The structure of SCN can be examined by referring to the horizontal and vertical dimensions of SC that different companies may use specific relationships to achieve their goals [94]. Some researchers present the SCN structure as a directed graph network $G = (N, A)$, where 'N' refers to the nodes representing SCN actors such as suppliers, manufacturers, and customers. Moreover, 'A' refers to the set of arcs. The arcs represent the relationship between actors, such as purchase interactions between buyers and suppliers [88, 95, 105]. Also, Rezaei Vandchali et al. [106] described the factors affecting the network structure as follows:

(A) Supplier and buyer dependency

Dependency is used to analyze the level of influence at the node level of a supply chain network, which can be considered one of the critical aspects of supply chain relationship management [53]. Also, [42] define dependency as indicating that suppliers' products and services are essential to the buyer.

(B) Distance

Distance is used to analyze the level of information available at the node level. Because the length of the path between two agents of the supply chain network can affect the exchange of information between them [96], also, three categories can be used to measure the distance between supply chain actors: physical distance, organizational distance, and cultural distance.

(C) Transparency

The level of information available at a network level is analyzed through transparency, which refers to the availability of information about one supply chain network actor to other supply chain network actors. Also, the different topologies of the supply chain network can be seen below.

A. Series supply chain network

The simplest case is related to the point where there is a company at each level of the chain, and these companies are connected in series. This critical manner is shown in Fig. 3.6.

Fig. 3.6 Serial supply chain network

The supply chain starts with procuring raw materials from suppliers; then, the production process is carried out on the raw materials. After that, the semi-finished materials are sent to the next stage, assembly. After the assembly, the desired products are ready, sent to the distribution centers and retailers, and finally reached the final customers [121].

B. Divergent chain network

Several researchers have also studied more advanced systems. In this case, each company supplies its product to several companies. For example, a central warehouse supplies the inventory of several regional distributors. This situation is shown in Fig. 3.7.

1. Converged supply chain network

This mode is often related to assembling parts to produce the final product. For example, several suppliers are associated with a factory, and this is shown in Fig. 3.8.

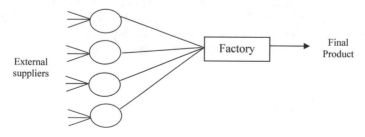

Fig. 3.7 Divergent supply chain network

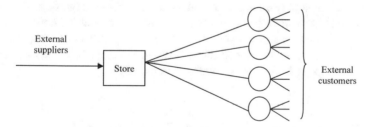

Fig. 3.8 converged supply chain network

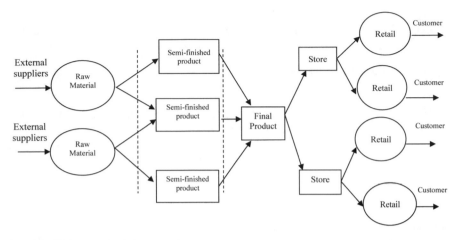

Fig. 3.9 Mixed supply chain network

2. **Mixed supply chain network**

This network is a general mode of the supply chain and is a combination of the above networks, where several companies are at each level of the supply chain. This situation is shown in Fig. 3.9.

The set of companies connected from the beginning to the end of the chain form the domain of the supply chain. In the mixed supply chain network of Fig. 3.9, the range is from the raw material company to the retailer company. Similar companies in the supply chain are called a supply chain level. In Fig. 3.9, one level of the supply chain is defined between the dotted lines. A two-level supply chain means a supply chain whose scope is limited to two types of companies.

3.3.4 How to Analyze the Supply Chain Network?

Companies have limited knowledge about the boundary of the network in which they are involved [52, 105] this limited knowledge is due to the invisible increase of network relationships and interactions because it expands without limits through connected relationships [38, 105]. By accepting the arbitrary nature of the network boundary, network analysis can be seen at three levels [55, 132]:

1. Network context

Network context refers to the parts of the network that the focal firms are typically associated with and includes all actors and links that can be relevant to the focal firms' business.

2. Network horizon

The middle level is called the "network horizon," which refers to the parts of the network that the focal firms are aware of.

3. Network environment

The network environment refers to the parts of the network that the focal firms are unaware of, so where the network horizon ends, the network environment begins [55].

Also, as mentioned, an SCN consists of nodes (actors) and links (flows) that connect these nodes [46, 70, 107]. Therefore, on this basis, the structure of SCN can be analyzed based on three distinct levels [14, 107] :

(A)   Knot

At the node level, the analysis is based on how the actors are placed in the network. Actors have characteristics that distinguish them from other actors [18], such as the number of connections an actor has with other actors [49].

(B)   Link

At the nexus level, the analysis concerns the types of flows among actors and their power. Flows between different actors also have characteristics such as the dollar volume of trade between two actors [18].

(C)   Network

The analysis refers to the overall network structure at the network level [70]. The overall network also has characteristics such as how well connected the network is with the number of links between actors [18]. Considering these three levels together can help companies conceptualize their SCN structure. Analyzing the structure of the SCN is essential because it can affect the behavior of each actor in the SCN. In fact, by using the model of interaction between SCN actors as the unit of analysis, companies can see themselves as part of an interconnected network. "This, in turn, means that there is a broader focus on relationship management" [108]. As a result, companies must create and maintain different types of relationships with SCN actors based on each actor's position in the SCN [20, 26].

3.3.5 Common Types of the Supply Chain Network

Due to the internal and external motivations for the re-collection of sold goods that have tangible benefits both economically and environmentally, the management of supply chain networks faced new concepts, which are generally divided into the following groups:

- Forward supply chain

In Forward mode, the supply chain network is designed so that all stages of the flow, from the extraction of raw materials to the delivery of the final product to the customer, are examined. It is tried to ensure that the customer's demand is always met in the right amount and at the right time. Also, the Forward supply chain includes sourcing, production, distribution, and delivery of new products/parts to the customer [15]. By doing this series of activities, a flow of new products is created. According to [35], the direct supply chain is a network of suppliers, manufacturers, distribution centers, and customers.

- Reverse supply chain

In the opposite case, the supply chain network is designed in such a way that it should be planned to collect consumed goods, improve economic conditions, comply with government laws, satisfy customers and improve environmental conditions. The reverse supply chain includes collection activities, obsolete customer products, reverse logistics, inspections, sorting, recovery operations, and remarketing [75]. Alegoz and Kaya [2] believe that the reverse supply chain focuses on the return of products from customers to create added value through the recovery of all products or some parts and components.

According to [15], the reverse supply chain includes the activities of use/collection, return of the used product to remanufacturers, reprocessing, and delivery of products with new parts to the same customers or new customers. After the waste products have been collected, one of the following processes takes place:

- Reuse: It is the process of reusing the product after partial parts repair during cleaning and inspection.
- Recycling: It is the process of recovering raw materials; during this process, the shape of raw materials changes, and finally, the recovered raw materials are used in the new product production cycle.
- Dismantling: This process recovers suitable parts and components of worn-out products to be used again in the production cycle.
- Repair: When the product specification is returned to a condition where the product is operable, not a new product's condition.
- Remanufacturing or reassembly of the product: It is a process during which the used parts are returned to parts equal to new parts [15].
- Closed-loop supply chain (CLSC).

The simultaneous consideration of direct and reverse supply chain networks together and in an integrated manner creates a new concept known as a Closed-loop Supply Chain (CLSC). The reason for the creation of this supply chain network was that in the early twentieth century, Henry Ford introduced his Fordism theory; this theory led to mass consumption and, as a result, mass production of paper machines [72] and products were designed in such a way that the costs of raw materials, assembly and distribution were minimized. As a result, manufacturers ignored things such as repairing, reusing, and disposing of the product. Although they believed that it would be helpful to consider these items, they were not considered due to customers' reluctance to pay more for a green product, and most purchase decisions were made to minimize purchase costs rather than optimizing the life cycle through maintenance and repair, reuse, and disposal. Therefore, most products were buried or burned in developed countries, and much environmental damage was done [124]. Over time, this process caused environmental concerns among consumers, which required more responsibility from producers to eliminate these concerns. To meet this customer's need, the Procurement Management Association initiated the study of reverse procurement in the early nineties. Since then, much research has been done on this topic. Since most of the issues related to reverse procurement are also found in the literature on direct procurement, it is essential to consider reverse procurement in addition to the direct procurement system. Combining these two concepts created a new concept called a closed-loop supply chain [72]. In 1995, Thierry et al. depicted one of the early models of the combination of direct and reverse supply chains according to Fig. 3.10. This figure is an integrated supply chain network model that includes servicing, product recovery, and waste management. Returned products and parts can be directly sold, recycled, or disposed of (including incineration or burial). According to this model, there are five ways to recover the product, which are: repair, reassembly of the product, restoration, Disassembly, and recycling. Each recovery method includes collecting end-of-life products and parts, reprocessing, and redistribution. The main difference between these impaired pathways is reprocessing [124].

In 1999, Jayaraman et al. used the term closed-loop logistics. They proposed a zero–one mixed-integer programming model in Fig. 3.11 to locate optimal amounts of refurbishing/distribution, transportation, production, and storage facilities simultaneously. Identify refurbished products.

Then in 2000, Fleischmann et al. presented a closed loop recovery chain network according to Fig. 3.12.

According to [64], with an extensive study in the field of the closed loop supply chain network, Fleischmann et al. proposed the characteristics of the supply chain, production planning, distribution planning, and inventory control of the reverse supply chain [64]. Next, [68] presented a closed-loop supply chain network model according to Fig. 3.13.

Moreover, [61] designed a closed-loop supply chain network with multiple recovery methods:

The closed-loop supply chain network has two main tasks [47]:

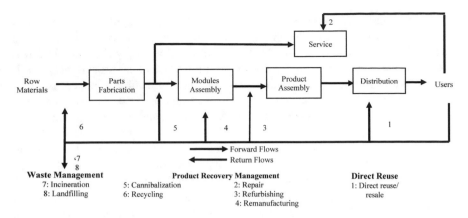

Fig. 3.10 Integrated closed loop supply chain [124]

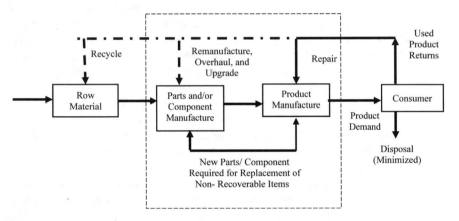

Fig. 3.11 A recoverable production system [60]

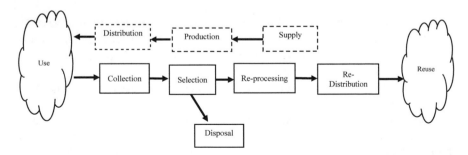

Fig. 3.12 Closed loop recovery supply chain network [40]

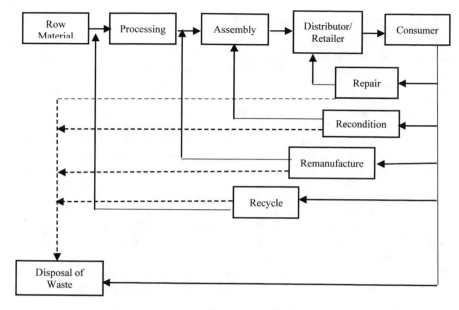

Fig. 3.13 Ways of expanding the product in the closed loop supply chain network process [68]

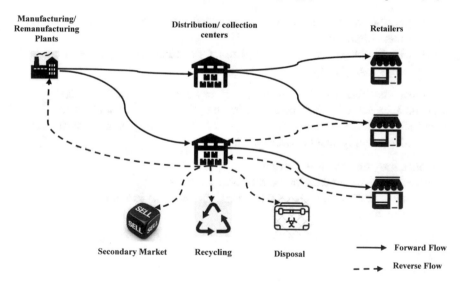

Fig. 3.14 Closed loop supply chain network structure [61]

Table 3.3 Different definitions of the closed supply chain network

Researchers	Definition
[17]	A closed-loop supply chain network maintains direct flows and demand for new products, while the reverse flows indicate the collection, restoration, or recycling of returned products
[118]	A classic or direct supply chain consists of a network of suppliers, manufacturers and distributors formed to produce and deliver a specific product or service. Reverse logistics includes all issues related to their collection, reprocessing, reconstruction, and destruction. If direct and reverse supply chains are considered simultaneously, the resulting network is a closed-loop supply chain
[110]	They believe that the closed-loop supply chain network includes direct and reverse logical flows at the same time. The direct flow includes value-adding processes to produce the final product and provide it to the customer. In contrast, the reverse flow involves collecting worn-out products from the customer to reduce environmental pollution and obtain reasonable economic profit
[89]	A closed-loop supply chain network is a supply chain in which product recovery activities are integrated with the direct supply chain
[51]	A closed-loop supply chain network extends the traditional direct supply chain by using reverse supply chain channels for product returns, recycling/recovery, refurbishing, and resale

1. Responsible for processes that create added value; these processes can cover customers' demands.
2. He tries to collect used (returned products) from customers and determine the best ways to use them.

Kumar and Yamaoka [72] believe that in an ideal closed-loop supply chain network, access to a landfill should not be an optional option, but all materials used in end-of-life products should be reused in the direct supply chain. Return flows in a closed-loop supply chain network include:

- Collecting products from consumers
- Reverse logistics to reclaim the collected products
- Screening, sorting, and disposal to determine the most economical reuse options
- Reconstruction.

Below and in Table 3.3, some other definitions of a closed supply chain network are mentioned.

3.4 Conclusion

This chapter describes the main models and approaches and supplies chain networks in three essential parts. In the first part, the primary supply chain models were explained. In the second part, different supply chain approaches were described. Moreover, in the third part, the network models were described. So that the readers are aware of how to organize and model the supply chain and are also mindful of the value creation and characteristics of the supply chain; in this way, they get to know the concept of supply chain performance evaluation more effectively.

References

1. APICS: Quick Reference Guide SCOR (Supply Chain Operations Reference) 12.0. APICS, Chicago (2017)
2. Alegoz, M., Kaya, O.: Coordinated dispatching and acquisition fee decisions for a collection center in a reverse supply chain. Comput. Ind. Eng. **113**, 475–486 (2017)
3. Allahviranloo, T., Ezadi, S.: Z-Advanced Numbers Processes, vol. 480, pp. 130–143 (2019)
4. Allahviranloo, T.: Uncertain Information and Linear Systems, vol. 254 (2020). https://doi.org/10.1007/978-3-030-31324-1
5. Amrani, H., Martel, A., Zufferey, N., Makeeva, P.: A variable neighborhood search heuristic for the design of multicommodity production–distribution networks with alternative facility configurations. OR Spectrum **33**(4), 989–1007 (2011)
6. Azevedo, S.G., Carvalho, H., Cruz-Machado, V., Grilo, F.: The influence of agile and resilient practices on supply chain performance: an innovative conceptual model proposal. In: Hamburg International Conference of Logistics, pp. 273–281 (2010)
7. Bai, C., Sarkis, J., Wei, X.: Addressing key sustainable supply chain management issues using rough set methodology. Manag. Res. Rev. **33**(12), 1113–1127 (2010)
8. Bai, C., Sarkis, J., Wei, X., Koh, L.: Evaluating ecological sustainable performance measures for supply chain management. Supp. Chain Manage. Int. J. **17**(1), 78–92 (2012)
9. Ballou, R.H.: Business logistics: importance and some research opportunities. Gestão Produção **4**, 117–129 (1997)
10. Baltacioglu, T., Ada, E., Kaplan, M.D., Yurtand, O., Cem Kaplan, Y.: A new framework for service supply chains. Serv. Ind. J. **27**(2), 105–124 (2007)
11. Banker, R.D., Chang, H.: The super-efficiency procedure for outlier identification, not for ranking efficient units. Eur. J. Oper. Res. **175**(2), 1311–1320 (2006)
12. Banker, R.D., Chang, H., Janakiraman, S.N., Konstans, C.: A balanced scorecard analysis of performance metrics. Eur. J. Oper. Res. **154**(2), 423–436 (2004)
13. Beamon, B.M.: Measuring supply chain performance. Int. J. Oper. Prod. Manag. **19**(3), 275–292 (1999)
14. Bellamy, M.A., Basole, R.C.: Network analysis of supply chain systems: a systematic review and future research. Syst. Eng. **16**(2), 235–249 (2013)
15. Bhattacharya, R., Kaur, A., Amit, R.K.: Price optimization of multi-stage remanufacturing in a closed loop supply chain. J. Clean. Prod. **186**, 943–962 (2018)
16. Bolstorff, P., Rosenbaum, R.: Supply chain excellence: a handbook for dramatic improvement using the SCOR model. J. Supply Chain Manag. **39**(4), 38 (2003)
17. Borgatti, S.P., Everett, M.G., Johnson, J.C.: Analyzing Social Networks. Sage (2018)
18. Borgatti, S.P., Li, X.: On social network analysis in a supply chain context. J. Supp. Chain Manag. **45**(2), 5–22 (2009)
19. Bottani, E.: A fuzzy QFD approach to achieve agility. Int. J. Prod. Econ. **119**(2), 380–391 (2009)
20. Byrne, R., Power, D.: Exploring agency, knowledge and power in an Australian bulk cereal supply chain: a case study. Supp. Chain Manage. Int. J. **19**(4), 431–444 (2014)
21. Campuzano, F., Mula, J., Peidro, D.: Fuzzy estimations and system dynamics for improving supply chains. Fuzzy Sets Syst. **161**(11), 1530–1542 (2010)
22. Carvalho, H., Azevedo, S.G., Cruz-Machado, V.: Supply chain performance management: lean and green paradigms. Int. J. Bus. Perform. Supp. Chain Modell. **2**(3–4), 304–333 (2010)

23. Carvalho, H., Azevedo, S.G., Cruz-Machado, V.: Agile and resilient approaches to supply chain management: influence on performance and competitiveness. Logist. Res. 4(1), 49–62 (2012)
24. Carvalho, H., Barroso, A.P., Machado, V.H., Azevedo, S., Cruz-Machado, V.: Supply chain redesign for resilience using simulation. Comput. Ind. Eng. 62(1), 329–341 (2012)
25. Carvalho, H., Machado, V.C.: Lean, agile, resilient and green supply chain: a review. In: Proceedings of the Third International Conference on Management Science and Engineering Management, pp. 66–76. International Society of Management Science and Engineering Management, Chengdu (2009)
26. Chan, H.L., Shen, B., Cai, Y.: Quick response strategy with cleaner technology in a supply chain: coordination and win-win situation analysis. Int. J. Prod. Res. 56(10), 3397–3408 (2018)
27. Charles, A., Lauras, M., Van Wassenhove, L.: A model to define and assess the agility of supply chains: building on humanitarian experience. Int. J. Phys. Distrib. Logist. Manag. 40(8/9), 722–741 (2010)
28. Cheng, J.C., Law, K.H., Bjornsson, H., Jones, A., Sriram, R.D.: Modeling and monitoring of construction supply chains. Adv. Eng. Inform. 24(4), 435–455 (2010)
29. Christopher, M.: The agile supply chain: competing in volatile markets. Ind. Mark. Manage. 29(1), 37–44 (2000)
30. Ciccullo, F., Pero, M., Caridi, M., Gosling, J., Purvis, L.: Integrating the environmental and social sustainability pillars into the lean and agile supply chain management paradigms: a literature review and future research directions. J. Clean. Prod. 172, 2336–2350 (2018)
31. Ciccullo, F., Pero, M., Caridi, M., Gosling, J., & Purvis, L.: Integrating the environmental and social sustainability pillars into the lean and agile supply chain management paradigms: A literature review and future research directions. J. Clea. Prod. 172, 2336-2350 (2018)
32. Correia, N., Machado, V.C., Nunes, I.: Strategy in Human Performance Management in Lean Environment's (2010)
33. Crespin-Mazet, F., Dontenwill, E.: Sustainable procurement: building legitimacy in the supply network. J. Purch. Supp. Manag. 18(4), 207–217 (2012)
34. Cruz-Machado, V., Duarte, S.: Tradeoffs among paradigms in Supply Chain Management. Proceedings of the 2010 International Conference on Industrial Engineering and Operations Management, Dhaka, Bangladesh, January 9–10 (2010)
35. Dai, Z.: Multi-objective fuzzy design of closed-loop supply chain network considering risks and environmental impact. Hum. Ecol. Risk Assess. Int. J. 22(4), 845–873 (2016)
36. Duarte, S., Carvalho, H., Cruz-Machado, V.: Exploring relationships between supply chain performance measures. In: The Fourth International Conference on Management Science and Engineering Management, pp. 3–7 (2010)
37. Ellinger, A.E., Natarajarathinam, M., Adams, F.G., Gray, J.B., Hofman, D., O'Marah, K.: Supply chain management competency and firm financial success. J. Bus. Logist. 32(3), 214–226 (2011)
38. Eng, T.Y.: Customer portfolio planning in a business network context. J. Mark. Manag. 24(5–6), 567–587 (2008)
39. Espadinha-Cruz, P., Grilo, A., Puga-Leal, R., Cruz-Machado, V.: A model for evaluating lean, agile, resilient and green practices interoperability in supply chains. In: 2011 IEEE International Conference on Industrial Engineering and Engineering Management, pp. 1209–1213. IEEE (2011)
40. Fleischmann, M., Krikke, H.R., Dekker, R., Flapper, S.D.P.: A characterisation of logistics networks for product recovery. Omega 28(6), 653–666 (2000)
41. Frostenson, M., Prenkert, F.: Sustainable supply chain management when focal firms are complex: a network perspective. J. Clean. Prod. 107, 85–94 (2015)
42. Gao, T., Sirgy, M.J., Bird, M.M.: Reducing buyer decision-making uncertainty in organizational purchasing: can supplier trust, commitment, and dependence help? J. Bus. Res. 58(4), 397–405 (2005)

43. Giannakis, M.: Management of service supply chains with a service-oriented reference model: the case of management consulting. Supp. Chain Manage. Int. J. **16**(5), 346–361 (2011)
44. Gimenez, C., Sierra, V., Rodon, J.: Sustainable operations: their impact on the triple bottom line. Int. J. Prod. Econ. **140**(1), 149–159 (2012)
45. Gimenez, C., Tachizawa, E.M.: Extending sustainability to suppliers: a systematic literature review. Supp. Chain Manage. Int. J. **17**(5), 531–543 (2012)
46. Gold, S., Seuring, S., Beske, P.: Sustainable supply chain management and inter-organizational resources: a literature review. Corp. Soc. Responsib. Environ. Manag. **17**(4), 230–245 (2010)
47. Govindan, K., Soleimani, H.: A review of reverse logistics and closed-loop supply chains. J. Clean. Prod. Focus **142**, 371–384 (2017)
48. Haddadsisakht, A., Ryan, S.M.: Closed-loop supply chain network design with multiple transportation modes under stochastic demand and uncertain carbon tax. Int. J. Prod. Econ. **195**, 118–131 (2018)
49. Hanneman, R.A., Riddle, M.: Concepts and measures for basic network analysis. In: The SAGE Handbook of Social Network Analysis, pp. 340–369 (2011)
50. Harrison, T.P.: Principles for the strategic design of supply chains. In: The Practice of Supply Chain Management: Where Theory and Application Converge, pp. 3–12. Springer, Boston, MA (2004)
51. He, Y.: Acquisition pricing and remanufacturing decisions in a closed-loop supply chain. Int. J. Prod. Econ. **163**, 48–60 (2015)
52. Hearnshaw, E.J., Wilson, M.M.: A complex network approach to supply chain network theory. Int. J. Oper. Prod. Manag. **33**(4), 442–469 (2013)
53. Hoejmose, S.U., Grosvold, J., Millington, A.: Socially responsible supply chains: power asymmetries and joint dependence. Supp. Chain Manage. Int. J. **18**(3), 277–291 (2013)
54. Van Hoek, R.I., Mitchell, A.J.: The challenge of internal misalignment. Int. J. Logist. **9**(3), 269–281 (2006)
55. Holmen, E., Pedersen, A.C.: Strategizing through analyzing and influencing the network horizon. Ind. Mark. Manage. **32**(5), 409–418 (2003)
56. Huang, S.H., Sheoran, S.K., Keskar, H.: Computer-assisted supply chain configuration based on supply chain operations reference (SCOR) model. Comput. Ind. Eng. **48**(2), 377–394 (2005)
57. Hwang, Y.D., Wen, Y.F., Chen, M.C.: A study on the relationship between the PDSA cycle of green purchasing and the performance of the SCOR model. Total Qual. Manag. **21**(12), 1261–1278 (2010)
58. Intel. (2002). SCOR Experience at Intel: Methods and Tools for Supply Chain Management. Intel Information Technology White Paper.
59. Jayaram, J., Dixit, M., & Motwani, J.: Supply chain management capability of small and medium sized family businesses in India: A multiple case study approach. Int. J. Prod. Econ. **147**, 472–485 (2014)
60. Jayaraman, V., Guide, V.D.R., Jr., Srivastava, R.: A closed-loop logistics model for remanufacturing. J. Oper. Res. Soc. **50**(5), 497–508 (1999)
61. Jerbia, R., Boujelben, M.K., Sehli, M.A., Jemai, Z.: A stochastic closed-loop supply chain network design problem with multiple recovery options. Comput. Ind. Eng. **118**, 23–32 (2018)
62. Jin, Y., Edmunds, P.: Achieving a competitive supply chain network for a manufacturer: a resource-based approach. J. Manuf. Technol. Manag. **26**(5), 744–762 (2015)
63. Johnson, M., Templar, S.: The relationships between supply chain and firm performance: the development and testing of a unified proxy. Int. J. Phys. Distrib. Logist. Manag. **41**(2), 88–103 (2011)
64. Kadambala, D.K., Subramanian, N., Tiwari, M.K., Abdulrahman, M., Liu, C.: Closed loop supply chain networks: designs for energy and time value efficiency. Int. J. Prod. Econ. **183**, 382–393 (2017)
65. Kasi, V.: Systemic assessment of SCOR for modeling supply chains. In: Proceedings of the 38th Annual Hawaii International Conference on System Sciences, pp. 87b-87b. IEEE (2005)

66. Kavilal, E.G., Venkatesan, S.P., Kumar, K.H.: An integrated fuzzy approach for prioritizing supply chain complexity drivers of an Indian mining equipment manufacturer. Resour. Policy **51**, 204–218 (2017)
67. Khan, S.A., Chaabane, A., Dweiri, F.: Supply chain performance measurement systems: a qualitative review and proposed conceptual framework. Int. J. Ind. Syst. Eng. **34**(1), 43–64 (2020)
68. Khor, K.S., Udin, Z.M.: Impact of reverse logistics product disposition towards business performance in Malaysian E&E companies. J. Supp. Chain Custom. Relation. Manage. **2012**, 1 (2012)
69. Kim, S.W.: An investigation on the direct and indirect effect of supply chain integration on firm performance. Int. J. Prod. Econ. **119**(2), 328–346 (2009)
70. Kim, Y., Choi, T.Y., Yan, T., Dooley, K.: Structural investigation of supply networks: a social network analysis approach. J. Oper. Manag. **29**(3), 194–211 (2011)
71. Kivelä, M., Arenas, A., Barthelemy, M., Gleeson, J.P., Moreno, Y., & Porter, M.A.: Multilayer networks. J. Compl. Netw. **2**(3), 203–271
72. Kumar, S., Yamaoka, T.: System dynamics study of the Japanese automotive industry closed loop supply chain. J. Manuf. Technol. Manag. **18**(2), 115–138 (2007)
73. Kusrini, E., Caneca, V.I., Helia, V.N., Miranda, S.: Supply chain performance measurement using supply chain operation reference (SCOR) 12.0 model: a case study in AA leather SME in Indonesia. In: IOP Conference Series: Materials Science and Engineering, vol. 697, no. 1, p. 012023. IOP Publishing (2019)
74. Lanier, D., Jr., Wempe, W.F., Zacharia, Z.G.: Concentrated supply chain membership and financial performance: Chain-and firm-level perspectives. J. Oper. Manag. **28**(1), 1–16 (2010)
75. Larsen, S.B., Masi, D., Jacobsen, P., Godsell, J.: How the reverse supply chain contributes to a firm's competitive strategy: a strategic alignment perspective. Prod. Plann. Contr. **29**(6), 452–463 (2018)
76. Lazzarini, S., Chaddad, F., Cook, M.: Integrating supply chain and network analyses: the study of netchains. J. Chain Netw. Sci. **1**(1), 7–22 (2001)
77. Lee, H.L.: The triple-A supply chain. Harv. Bus. Rev. **82**(10), 102–113 (2004)
78. Lee, H.K., Fernando, Y.: The antecedents and outcomes of the medical tourism supply chain. Tour. Manage. **46**, 148–157 (2015)
79. Lima-Junior, F.R., Carpinetti, L.C.R.: Predicting supply chain performance based on SCOR® metrics and multilayer perceptron neural networks. Int. J. Prod. Econ. **212**, 19–38 (2019)
80. Lin, Y.H., Wang, Y., Lee, L.H., Chew, E.P.: Consistency matters: revisiting the structural complexity for supply chain networks. Phys. A **572**, 125862 (2021)
81. Machado, V.C., Duarte, S.: Tradeoffs among paradigms in supply chain management. In: International Conference on Industrial Engineering and Operations Management, pp 9–10 (2010)
82. Mahmoodirad, A., Allahviranloo, T., Niroomand, S.: A New Effective Solution Method for Fully Intuitionistic Fuzzy Transportation Problem, vol. 23, no. 12, pp. 4521–4530 (2019)
83. Maleki, M., da Cruz, P.E., Valente, R.P., Machado, V.C.: Supply chain integration methodology: large supply chain. Encontro Nacional de Engenharia e Gestão Industrial **57** (2011)
84. Max Shen, Z.J.: Integrated supply chain design models: a survey and future research directions. J. Indus. Manage. Optim. **3**(1), 1–27 (2007)
85. McCormack, K., Ladeira, M.B., de Oliveira, M.P.V.: Supply chain maturity and performance in Brazil. Supp. Chain Manage. Int. J. **13**(4), 272–282 (2008)
86. Meepetchdee, Y., Shah, N.: Logistical network design with robustness and complexity considerations. Int. J. Phys. Distrib. Logist. Manag. **37**(3), 201–222 (2007)
87. Melo, M.T., Nickel, S., Saldanha-Da-Gama, F.: Facility location and supply chain management—a review. Eur. J. Oper. Res. **196**(2), 401–412 (2009)
88. Mizgier, K.J., Jüttner, M.P., Wagner, S.M.: Bottleneck identification in supply chain networks. Int. J. Prod. Res. **51**(5), 1477–1490 (2013)

89. Mohammed, F., Selim, S.Z., Hassan, A., Syed, M.N.: Multi-period planning of closed-loop supply chain with carbon policies under uncertainty. Transp. Res. Part D: Transp. Environ. **51**, 146–172 (2017)
90. Morgan, C.: Supply network performance measurement: future challenges? Int. J. Logist. Manage. **18**(2), 255–273 (2007)
91. Mumtaz, U., Ali, Y., Petrillo, A.: A linear regression approach to evaluate the green supply chain management impact on industrial organizational performance. Sci. Total Environ. **624**, 162–169 (2018)
92. Noorul Haq, A., Kannan, G.: Design of an integrated supplier selection and multi-echelon distribution inventory model in a built-to-order supply chain environment. Int. J. Prod. Res. **44**(10), 1963–1985 (2006)
93. Ntabe, E.N., LeBel, L., Munson, A.D., Santa-Eulalia, L.A.: A systematic literature review of the supply chain operations reference (SCOR) model application with special attention to environmental issues. Int. J. Prod. Econ. **169**, 310–332 (2015)
94. Otto, A.: Supply chain event management: three perspectives. Int. J. Logist. Manage. **14**(2), 1–13 (2003)
95. Pan, F., Nagi, R.: Multi-echelon supply chain network design in agile manufacturing. Omega **41**(6), 969–983 (2013)
96. Park, J., Shin, K., Chang, T.W., Park, J.: An integrative framework for supplier relationship management. Ind. Manag. Data Syst. **110**(4), 495–515 (2010)
97. Paul, J.: Panduan Penerapan Transformasi Rantai Suplai Dengan Model SCOR 15 Tahun Aplikasi Praktis Lintas Industri. PPM Manajemen ISBN, pp. 979–442 (2014)
98. Putri, Y.D., Huda, L.N., Sinulingga, S.: The concept of supply chain management performance measurement with the supply chain operation reference model (Journal review). In: IOP Conference Series: Materials Science and Engineering, vol. 505, no. 1, p. 012011. IOP Publishing (2019)
99. Qi, Y., Zhao, X., Sheu, C.: The impact of competitive strategy and supply chain strategy on business performance: the role of environmental uncertainty. Decis. Sci. **42**(2), 371–389 (2011)
100. Raj, A., Biswas, I., Srivastava, S.K.: Designing supply contracts for the sustainable supply chain using game theory. J. Clean. Prod. **185**, 275–284 (2018)
101. Ramezani, M., Bashiri, M., Tavakkoli-Moghaddam, R.: A new multi-objective stochastic model for a forward/reverse logistic network design with responsiveness and quality level. Appl. Math. Model. **37**(1–2), 328–344 (2013)
102. Ramezankhani, M.J., Torabi, S.A., Vahidi, F.: Supply chain performance measurement and evaluation: a mixed sustainability and resilience approach. Comput. Ind. Eng. **126**, 531–548 (2018)
103. Razavi, S.M., Safari, H., Shafie, H., Rezaei Vandchali, H.: How customer satisfaction, corporate image and customer loyalty are related? Eur. J. Sci. Res. **78**(4), 588–596 (2012)
104. Reza Abdi, M., Edalat, F.D., Abumusa, S.: Lean and agile supply chain management: a case of IT distribution industry in the Middle East. Green Lean Manage. 37–69 (2017)
105. Rezaei Vandchali, H., Cahoon, S., Chen, S.L.: Developing relationship management strategies in a network context. J. Adminis. Bus. Stud. JABS **5**(3), 179–192 (2019)
106. Rezaei Vandchali, H., Cahoon, S., Chen, S.L.: Creating a sustainable supply chain network by adopting relationship management strategies. J. Bus. Bus. Mark. **27**(2), 125–149 (2020)
107. Rezaei Vandchali, H., Cahoon, S., & Chen, S. L.: Developing relationship management strategies in a network context. J. Adm. Bus. Stud. JABS, **5**(3), 179-192 (2019)
108. Roscoe, S., Cousins, P.D., Lamming, R.C.: Developing eco-innovations: a three-stage typology of supply networks. J. Clean. Prod. **112**, 1948–1959 (2016)
109. Rossini, M., Staudacher, A.P.: Lean supply chain planning: a performance evaluation through simulation. In: MATEC Web of Conferences, vol. 81, p. 06002. EDP Sciences (2016)
110. Safaei, A.S., Roozbeh, A., Paydar, M.M.: A robust optimization model for the design of a cardboard closed-loop supply chain. J. Clean. Prod. **166**, 1154–1168 (2017)

111. Sarkar, B., Ahmed, W., Kim, N.: Joint effects of variable carbon emission cost and multi-delay-in-payments under single-setup-multiple-delivery policy in a global sustainable supply chain. J. Clean. Prod. **185**, 421–445 (2018)
112. Schnetzler, M.J., Lemm, R., Bonfils, P., Thees, O.: The supply chain operations reference (SCOR)-model to describe the value-added chain in forestry. Allgemeine Forst-und Jagdzeitung **180**(1/2), 1–14 (2009)
113. Schoeman, C., Sanchez, V.R.: Green supply chain overview and a South African case study. In: Southern African Transport Conference (SATC) (2009)
114. Seuring, S., Müller, M.: From a literature review to a conceptual framework for sustainable supply chain management. J. Clean. Prod. **16**(15), 1699–1710 (2008)
115. Shafiee, M., Ghotbi, M.: The performance measurement of supply chain with network DEA. Int. J. Data Envelop. Anal. **8**(3), 65–74 (2020)
116. Shafiee, M., Saleh, H., Ghaderi, M.: Benchmarking in the supply chain using data envelopment analysis and system dynamics simulations. Iranian J. Supp. Chain Manage. **23**(70), 55–70 (2021)
117. Shi, M., Yu, W.: Supply chain management and financial performance: literature review and future directions. Int. J. Oper. Prod. Manag. **33**, 1283–1317 (2013)
118. Soleimani, H., Govindan, K., Saghafi, H., Jafari, H.: Fuzzy multi-objective sustainable and green closed-loop supply chain network design. Comput. Ind. Eng. **109**, 191–203 (2017)
119. Spekman, R.E., Kamauff, J.W., Myhr, N.: An empirical investigation into supply chain management: a perspective on partnerships. Supp. Chain Manage. Int. J. **3**(2), 53–67 (1998)
120. Srivastava, S. K.: Green supply-chain management: a stateof- the-art literature review. Int. J. Manag. Rev. 9(1), 53–80 (2007)
121. Stadtler, H., & Kilger, C.: Supply chain management and advanced planning overview and challenges. Eur. J. Oper. Res. **163**(3), 575-588 (2005)
122. Stampe, D.: Supply Chain Performance and Evaluation Models. John Wiley & Sons, Inc (2014)
123. Tanskanen, K.: Who wins in a complex buyer-supplier relationship? A social exchange theory based dyadic study. Int. J. Oper. Prod. Manag. **35**(4), 577–603 (2015)
124. Thierry, M., Salomon, M., Van Nunen, J., Van Wassenhove, L.: Strategic issues in product recovery management. Calif. Manage. Rev. **37**(2), 114–136 (1995)
125. Tortorella, G.L., Miorando, R., Marodin, G.: Lean supply chain management: empirical research on practices, contexts and performance. Int. J. Prod. Econ. **193**, 98–112 (2017)
126. Wilhelm, M., Blome, C., Wieck, E., Xiao, C.Y.: Implementing sustainability in multi-tier supply chains: strategies and contingencies in managing sub-suppliers. Int. J. Prod. Econ. **182**, 196–212 (2016)
127. Wu, S., Wee, H.M.: How Lean supply chain effects product cost and quality—a case study of the Ford Motor Company. In: 2009 6th International Conference on Service Systems and Service Management (2009)
128. Wu, S., & Wee, H. M.: How Lean supply chain effects product cost and quality-A case study of the Ford Motor Company. In 2009 6th International Conference on Service Systems and Service Management (2009, June)
129. Wuyts, S., Geyskens, I.: The formation of buyer–supplier relationships: detailed contract drafting and close partner selection. J. Mark. **69**(4), 103–117 (2005)
130. Xiao, R., Cai, Z., Zhang, X.: An optimization approach to risk decision-making of closed-loop logistics based on SCOR model. Optimization **61**(10), 1221–1251 (2012)
131. Van Der Zee, D.J., Van Der Vorst, J.G.: A modeling framework for supply chain simulation: opportunities for improved decision making. Decis. Sci. **36**(1), 65–95 (2005)
132. Zhang, X., Wang, H., Nan, J., Luo, Y., Yi, Y.: Modeling and numerical methods of supply chain trust network with the complex network. Symmetry **14**(2), 235 (2022)
133. Zhang, X., Wang, H., Nan, J., Luo, Y., & Yi, Y.: Modeling and Numerical Methods of Supply Chain Trust Network with the Complex Network. Symmetry, **14**(2), 235 (2022)

Chapter 4
Supplier Performance Evaluation Models

4.1 Introduction

Supplier selection is one of the critical issues that purchase and operations managers make to help maintain the competitive position of organizations. The wrong choice of a supplier can turn the financial and operational situation of a company upside down, and on the other hand, choosing the right supplier or determining the optimal amount of the order can reduce purchase costs, improve competitiveness in the market, and improve the satisfaction of the final consumer [35].

Since the performance of an organization in the supply chain depends on the performance of its suppliers, the selection of suitable suppliers in the supply chain requires evaluation criteria that include information such as customer focus, competitive advantage, strategic purchasing, information technology, and the support of a company's top management. A company must have outstanding competitive priorities to manage its production, such as producing high-quality products with excellent logistics arrangements; a competitive priority close to top management's support is the default for evaluation [124].

Among suppliers with shared resources, manufacturers should look for those with whom they can develop long-term relationships [42]. Currently, the increase in demand, rapid changes in the global arena, the existence of great uncertainty, the presence of domestic competitors, and the increase in foreign competitors have caused companies to increasingly rely on their suppliers and try to cooperate with suppliers who have the necessary ability to meet the ever-increasing and new needs of customers.

Therefore, Fig. 4.1 shows the position of supplier selection in a supply chain [67].

Therefore, one of the critical components of supply chain management is evaluating, ranking, and selecting suppliers. Due to the importance of this issue, this chapter describes supplier performance and evaluation models.

© The Author(s), under exclusive license to Springer Nature Switzerland AG 2023
F. Hosseinzadeh Lotfi et al., *Supply Chain Performance Evaluation*,
Studies in Big Data 122, https://doi.org/10.1007/978-3-031-28247-8_4

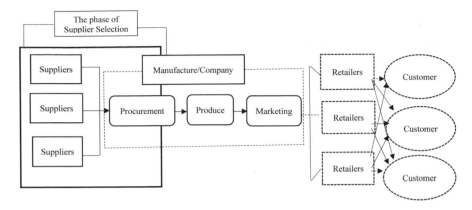

Fig. 4.1   The position of supplier selection in the supply chain [67]

4.2   Existing Models of Supplier Selection and Evaluation

4.2.1   Total Cost of Ownership (TCO) Models

Traditional methods are generally focused on the price element. Still, most supplier evaluation and selection strategies often ignore many costs [45]. As a result, providing a comprehensive approach that considers all the costs related to a service or production seems necessary in evaluating and managing the supply chain. For this purpose, the concept of "total cost of ownership" was created as a suitable indicator for performance evaluation and, therefore, a good solution for improving productivity. This indicator is a purchasing tool and a philosophy that seeks to understand the actual costs of purchasing a product or providing a particular service from specific suppliers. This model reduces the possible costs of an investment or purchase, and as a result, it will effectively improve productivity. Therefore, TCO-based models try to consider all purchase costs in the cycle. Why so, the costs before, after, and during the purchase differ [26].

This method seeks a correct understanding of the costs of working with a specific supplier. It is based on the idea that the decision to purchase significantly impacts different departments of a company (purchasing, logistical constraints, quality, accounting, etc.), which causes the spending of resources and creates costs.

Therefore, a basic need is a proper understanding of the costs caused by a supplier and paying attention to the purchase price. Although the total cost of ownership often measures the performance of suppliers, it does not consider only the direct costs that occur from choosing an inappropriate supplier; Rather, according to many researchers, the total cost of ownership refers to all the costs that are created in connection with a supplier for the company, which requires attention to other activities and other aspects of communication. These activities include order management, delivery planning, lack of packaging, materials and business management,

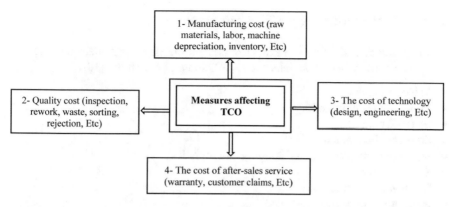

Fig. 4.2 Criteria (direct and indirect costs) affecting TCO

accounting matters, placement, logistics facilities, competencies of different information systems, etc. [126]. Also, many researchers have used this method in supplier selection, such as Mandolini et al. [76], Ellram and Siferd [27] and Ellram [28].

4.2.1.1 Criteria (Direct and Indirect Costs) Affecting the Total Cost of Ownership (TCO)

The concept of the total cost of ownership as a criterion for evaluating and selecting the performance of suppliers represents the criteria presented by Bhutta and Huq in 2002, which are given in Fig. 4.2.

These criteria include direct and indirect costs that affect production.

4.2.1.2 Advantages of the TCO Technique

The TCO technique has many advantages, which are described below [12, 35]:

- TCO is a technique that looks beyond the purchase price and includes many other costs associated with the purchase.
- It focuses on the actual costs associated with the entire purchase cycle, thus taking into account all charges related to the purchase, use, maintenance, and tracking of purchased goods or services and the purchase price.
- Improves buyer understanding of supplier performance and cost structure issues and provides excellent data for negotiation and improvement.
- Justifies higher upfront prices based on better quality/lower total costs in the long run.

120 4 Supplier Performance Evaluation Models

4.2.1.3 Factors Affecting the TCO Technique

The importance and value of each TCO component in supply chain management depend on certain factors and characteristics, some of which are [18]:

- Value of items
- Minimum order quantities
- Material content
- Physical characteristics
- Delivery method/lead time/shipping and logistics
- Source of supply
- Fluctuations in demand
- Product life cycle/obsolescence
- Order processing fees
- Application
- Program management costs
- Opportunity costs.

Item value, minimum order quantities (order requirements), material content, physical properties, delivery method and lead time, supply source, demand fluctuations, and product life cycle affect ordering and stocking decisions in a well-managed supply chain. Understanding these components and their impact on investment requirements and risk is critical to understanding TCO. These components affect processing and inventory holding costs, such as warehouse labor, employment, and financing costs. For example, high-value items result in higher financing costs, while smaller, lightweight objects (physical properties) may require little or no facilities or labor.

These components can also affect investment risk, which is often an underestimated component of TCO. Time to maturity, demand fluctuations, and the product life cycle all risk investing in inactive or obsolete inventory. Inactive inventory is inventory without activity for an extended period, and obsolete inventory has reached the end of its product life cycle. Detailed analysis is necessary to optimize the inventory considering these components.

Order processing costs may seem like a prominent part of TCO. Still, its impact on TCO can be more significant based on the previously mentioned components and even from one organization or industry to another. For example, a capital equipment manufacturer may have different systems and efficiencies in its order processing methods than an electronics assembly house with systems designed to handle more significant numbers of parts and purchasing transactions. Supply chain management systems drive order processing methodologies that focus on timely and efficient communication with suppliers that optimize order processing costs and relieve the burden on manufacturers. The order processing time should be included in the delivery time and is directly related to the order processing cost. Optimized systems with proper part profiles complete orders processed instantly.

The demand for an item directly affects the total cost of ownership. It defines the program's understanding of engineering, quality, and technical requirements. The best supply chain management companies use programs to add specialized expertise to outsourced product classes. For example, products used in high-temperature or corrosive environments may require certification of suppliers and materials along with specialized plating, all with little or no variance in manufacturing processes.

Vendor-managed inventory costs are the most significant variable in understanding the total cost of ownership in supply chain management after determining product characteristics. Developing programs that meet customer needs and expectations requires paying attention to details and understanding the cost of those services. Some customers may experience capacity issues that limit available space, leading to off-site storage and preprogram-type changes required to facilitate the smoothest flow of goods to assembly lines. Others may have excess capacity and want a higher volume of on-site inventory that requires lower replenishment costs. Supply chain experts analyze the processes and create a plan to optimize the situation's total cost of ownership.

Opportunity cost is the most overlooked component of the total cost of ownership. Where do organizations generate the most value? What is the lost opportunity if organizations spend time on their core competencies? The extra work and attention to detail to accurately measure and optimize the total cost of ownership attracts the focus and attention of a partner who recognizes this process as part of their competency.

4.2.2 Multiple Attribute Decision Making (MADM)

In another division, the suppliers are selected by Multiple Attribute Decision Making (MADM). Multiple Attribute Decision Making (MADM) refers to the decision-making problem of choosing the best alternative or ranking alternatives with various attributes and is a critical component of modern decision-making science [128]. MADM has enabled the development of many tools and solutions for problems, including selection, sorting, ranking, description, elimination, and design [134]. Multiple Attribute Decision Making is based on the assumption that decision-makers first specify several indicators to express their preferences regarding different decision-making options. Second, indicators are evaluated for each option. And third, the results are compared with each other to choose the best option according to the decision maker's priorities [21].

The goal of MADM is to select the optimal solution from among all alternatives, which plays an increasing role in decision theories. Also, an essential advantage of most MADM techniques is that they can analyze quantitative and qualitative evaluation criteria together. Also, the decision maker may express the ranking for the attributes as importance/weight. MADM aims to obtain the optimal alternative with the highest degree of satisfaction for all relevant attributes [92].

The use of new approaches to choose the right supplier in the supply chain, stable in guiding and managing supply chains, has proven their success in guaranteeing organizational goals. In recent years, choosing the right supplier in the supply chain has become an important strategic issue. Therefore, the nature of these decisions is usually complex. On the other hand, the selection process of the right supplier who can provide the buyer's needs in terms of quality products, time and volume is one of the most necessary processes to create a supply chain. Therefore, since the proper functioning of the supply chain plays a vital role in the success of an organization, identifying the new approaches and criteria of the supply chain can effectively help in the production, timely and cheap delivery of an organization. Therefore, the use of Multiple Attribute Decision-Making methods and the closeness of the conditions to real-world conditions help decision-makers in the supply chain to have an innovative aspect in providing management solutions to the managers of this field. Basically, the quality of supply chain management depends on the quality of decision-making because the quality of plans and programs, the effectiveness and efficiency of strategies and the quality of the results obtained from the implementation of supply chain management depend on manager decisions. In most decision-making problems, it is desirable and satisfactory to the decision-maker that the decision-making has been examined based on several criteria. Criteria may be quantitative or qualitative. Therefore, in the Multiple Attribute Decision-Making methods that have attracted the attention of researchers in recent decades, several measurement indicators are used instead of one optimality measurement criterion. For this reason, MADM is widely used in supplier selection problems [52, 68, 70, 72, 83, 86, 135, 143]. In the following, the most crucial Multiple Attribute Decision-Making models are described.

4.2.2.1 Analytical Hierarchy Process Model (AHP)

The Analytical Hierarchy process was first introduced by Tomas L. Satty in 1980 for the allocation of scarce resources and also for the planning needs of the military. This method is one of the Multiple Attribute Decision-Making (MADM) that enables decision-makers to determine the mutual and simultaneous effects of many complex and uncertain situations. Also, many researchers have used this method in supplier selection, such as [24, 25, 33, 39, 84, 129].

To solve decision-making problems through AHP, the problem must be defined and explained carefully and with all the details, and its elements must be drawn in the form of a hierarchical structure [95, 106]. The algorithm of the AHP method is as follows [87]:

1. Making pairwise comparisons between indicators.

Create a list of all the options you want to compare. Write your options as both titles of rows and columns.

Then compare each option in each row with each option in each column. In other words, express the preference of each option over another option in the form of a quantitative or qualitative criterion.

2. Normalizing the matrix of pairwise comparisons.

Divide each number of the matrix of pairwise comparisons by the sum of the column corresponding to that number.

3. Calculate the relative weights by calculating the arithmetic mean of each row.
4. Multiplying the relative weights of the indicators in the arithmetic mean of the options.
5. Ranking the options.

4.2.2.2 Analytic Network Process Model (ANP)

The network analysis process method was introduced in 1996 by Saati. This is used to solve problems whose indicators are not independent. So, this technique overcomes the assumption of independence between indicators, which is one of the limitations of the AHP method [64]. In other words, the ANP method provides a better understanding of the complex relationships between the evaluation criteria, decreasing reliability and trust in decision-making [53]. And a rotating arrow indicates internal dependencies. The relationship between nodes in a decision-making network is divided into two categories [130].

– **External dependency**: This dependency exists when the elements in one node affect another node.
– **Internal dependency**: This dependency exists when the elements in a node affect each other.

It should be noted that solving problems with the help of the network depends a lot on the art of the modeler, and the formation of the network does not follow a specific rule. Therefore, each problem has its complexity, and a general rule or formula cannot be assigned to solve the network problem [107]. Also, many researchers have used this method in supplier selection, such as Giannakis et al. [38], Asadabadi [5], Galankashi and Chegeni [34] and Vinodh et al. [125]. Figure 4.3 shows the steps of ANP.

Despite the similarities between ANP and AHP, these two methods also have differences. The similarities and differences between the two techniques can be explained in the application and the way of expressing preferences:

1. Both techniques are used to determine the priority of elements.
2. The priority determination in both techniques is based on pairwise comparisons.
3. The AHP model has a clear and regular structure of the target sequence, criteria, and sub-criteria, but the ANP model does not have any specific and predictable structure.
4. AHP model compares each element based on its immediate upstream element. But in ANP no predetermined rule. The model's features and the problem designer's perspective determine which elements should be compared.

Fig. 4.3 Network analysis
steps (ANP)

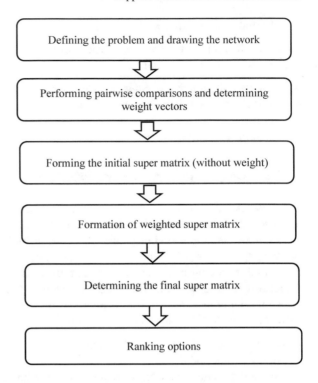

5. In the AHP model, the final weight is obtained based on the simple multiplication
    of the importance of each element in its upper cluster. In the ANP, the elements'
    weights will be obtained by calculating the supermatrix.

4.2.2.3   Taxonomy Analysis

One of the essential Multiple Attribute Decision-Making methods is the Taxonomy
Analysis method. This method was proposed for the first time in 1763 by Adenson
and was extended by a group of mathematicians in 1950. In 1968, it was offered
by UNESCO as an essential tool in classifying the degree of development between
different nations, and today it uses in various fields of science. This method can
divide a set into almost homogeneous subsets and grade them by recognizing the
extent to which the options have the same indicators [107]. Also, many researchers
have used this method in supplier selection, such as Mckone-sweet et al. [78], Xiao
et al. [140] and He et al. [44]. This method includes the following steps.

- Specifying options and determining different indicators.
- Forming the decision matrix ($X = [x_{ij}]$).

The decision matrix of this method includes a series of criteria and options where
the criteria are placed in the columns, and the options are in the rows. In this matrix,
each element of the matrix is a score. This score is assigned to each option based on the

desired criteria. So the decision matrix should be completed with experts' opinions. Usually, this process is done using the Likert scale. In cases where the criterion is quantitative, such as cost or production rate, etc., for which we have a real number, we put the number for each option. In cases where the criterion is qualitative, and the quantitative value has no meaning, we use the range of 1–9 or 1–5. In other words, in this step, a group of experts evaluates m options (A_1–A_m) according to n indicators (C_1–C_n) and forming the decision matrix, and then calculates the index's average ($\bar{x}_j$) and standard deviation (σ_j). Also, at this stage, positive and negative indicators should be identified.

- Forming the normalized (standard) matrix (Z): $z_{ij} = \frac{x_{ij} - \bar{x}_j}{\sigma_j}$.
- Determining   the   compound   distance   between   options: $d_{ab} =$ $\sqrt{\sum_{j=1}^{m} (z_{aj} - z_{bj})^2}$.

Such that "a" and "b" are the two options to be evaluated. Finally, a composite distance matrix is formed between the options and $D = [d_{ab}]$.

- Determining the shortest distance.

In this step, the minimum distance of each row of the matrix is determined (d_r). Then the average of each of the distances of the options and their standard deviation are obtained. The upper bound and lower bound are used as follows:

$$O_r(+) = \bar{d}_r + 2\,\sigma_{dr} \tag{4.1}$$

$$O_r(-) = \bar{d}_r - 2\,\sigma_{dr} \tag{4.2}$$

If d_r between the upper and lower limits, then it is compatible. So the options outside this range should be removed. Again, the decision matrix is formed without the deleted options, and the steps are repeated.

- Determining the pattern of options.

In this step, the distance of each option from the ideal value is obtained (C_{io}). So

$$C_{io} = \sqrt{\sum_{j=1}^{m} (z_{ij} - \bar{z}_j)^2}.$$

- Calculate the degree of development of an option: $F_i = \frac{C_{io}}{C_o}$ so that $C_o = \bar{C}_{io} + 2\sigma_{C_{io}}$. F_i is between zero and one, and the closer it is to zero, it indicates the development of choice (ranking higher). Moreover, the closer it gets to one, shows its lack of development. So options are ranked.

4.2.2.4   The Elimination Et Choice Translating Reality (ELECTRE)

Another supplier selection model is the ELECTRE method. The ELECTRE method or approximate mastery is one of the MADM methods. It is derived from the first letters of the sentence Elimination Et Choice Translating Reality, which means the method of elimination and choice consistent with reality. This method was introduced by Benayoun in 1966 and then developed by Bernard Roy, Van Delf, and Nijkamp. The basis of the superior choice in this method is pairwise comparisons. This method compares the options two by two.

In this method, the best choice is based on the maximum advantage and minimum conflict based on different criteria. The ELECTRE method is used to select the best action from among a set of actions and was later named the ELECTREI method. Various versions of the ELECTRE method have been presented, among which ELEC-TREI, II, III, IV, and TRI methods can be mentioned. All these methods are based on the same basic concepts, but they have differences in operationalization and the type of decision-making problem.

Many researchers have used this method in the field of supplier selection, such as Fei et al. [31], Shojaie et al. [115], Zhong and Yao [144], Kumar et al. [61], Azadnia et al. [7], Liu and Zhang [69], Sevkli [105], and Birgün and Cihan [11]. In general, to use this method, the following assumptions are required:

- The criteria should be quantitative or can be converted into quantitative.
- Criteria should be completely heterogeneous.
- Basically, the goal of the ELECTRE method is to separate the options that are preferred in the evaluation based on most of the criteria.

In the following, the steps of the algorithm for solving decision problems through the ELECTREI method are described:

Step 1: Formation of the decision matrix ($X = [x_{ij}]$).

Step 2: Normalized the decision matrix.

$$r_{ij} = \frac{x_{ij}}{\sqrt{\sum_i x_{ij}}} \text{ and } R = [r_{ij}] \tag{4.3}$$

Step 3: Determining the criteria's weight matrix. $V = [v_{kj}]$ such that $v_{kj} = w_j \times r_{ij}$, $j = 1, ..., n; i = 1,..., m$.

Step 4: Forming a set of criteria for and against

For each pair of options k and e, the set of criteria is divided into two subsets for and against. The agreement set (S_{ke}) is a set of criteria in which option k is preferred to option e, and the complementary set is the opposite (D_{ke}). The set of agreement criteria for positive and negative criteria are defined as relations (4.4) and (4.5), respectively.

$$S_{ke} = \{j | v_{kj} \geq v_{ej}\} \tag{4.4}$$

$$S_{ke} = \{j | v_{kj} \leq v_{ej}\} \tag{4.5}$$

Also, the set of opposite criteria for positive and negative are respectively defined as (4.6) and (4.7).

$$D_{ke} = \{j|v_{kj} < v_{ej}\} = j - S_{ke} \tag{4.6}$$

$$D_{ke}\{j|v_{kj} > v_{ej}\} = j - S_{ke} \tag{4.7}$$

Step 5: Formation of agreement matrix (coordination matrix)

The agreement matrix is a square. Its dimension equals the number of options. It is indicated by $C = [C_{ke}]$ and $C_{ke} = \sum_{j \in S_{ke}} w_j$.

Step 6: Formation of the opposite matrix

The opposite matrix is a square. The elements of this matrix are indicated by d_{ke} and $d_{ke} = \frac{\max|V_{kj}-V_{ej}, \ j \in D_{ke}|}{\max|V_{kj}-V_{ej}, \ j \in \text{All indicators}|}$.

Step 7: Formation of the matrix of agreement dominance (effective coordination matrix)

The agreement dominance matrix (F) is a Boolean matrix. It is formed according to relation (4.8).

$$f_{ke} = \begin{cases} 1 & C_{ke} \geq \overline{C} \\ 0 & C_{ke} < \overline{C} \end{cases} \tag{4.8}$$

Such that: $\overline{C} = \sum_{k=1}^{m} \sum_{e=1}^{m} \frac{C_{ke}}{m(m-1)}$.

Step 8: Formation of the opposing dominance matrix (effective inconsistency matrix)

Suppose that $\overline{d} = \sum_{k=1}^{m} \sum_{e=1}^{m} \frac{d_{ke}}{m(m-1)}$. Similar to the agreement dominance matrix, the opposite dominance matrix (G) is formed based on the relation (4.9).

$$G_{ke} = \begin{cases} 1 & d_{ke} \leq \overline{d} \\ 0 & d_{ke} > \overline{d} \end{cases} \tag{4.9}$$

Step 9: Formation of the final dominance matrix (effective final matrix)

The final dominance matrix (H) is obtained by multiplying each row of the favorable dominance matrix (F) by the negative dominance matrix (G): $H_{ke} = G_{ke} \times f_{ke}$.

Step 10: Choosing the best option. The final dominance matrix (H) expresses the partial preferences of the options. For example, suppose the value of H_{ke} is equal to one. It means that the superiority of option k over option e is acceptable in both cases of agreement and opposition. However, option k may still be dominated by other options. In this way, all options are ranked.

4.2.2.5   PROMETHEE Decision-Making Method

Another method that can be used to rank and select suppliers is the PROMETHEE decision-making method. This method was first developed in 1982 by professor Brans and was widely used in the early years. A few years later, Professor Branes, with the help of his colleagues, presented newer versions of this method [13]. This method has several versions, which are mentioned below.

- PROMETHEE 1: This method ranks the alternatives partially.
- PROMETHEE 2: This method ranks discrete alternatives ultimately.
- PROMETHEE 3: This method defines preference and non-preference relationships based on the mean and standard deviation of preference indicators.
- PROMETHEE 4: This method is used continuous items, and, in this version, the ranking is done based on a continuous scale and is not practical for discrete alternatives.
- PROMETHEE 5: This method defines a multi-criteria method for choosing options and considering constraints.
- PROMETHEE 6: This method works like the human brain.

The PROMETHEE method is a modified, simple, and understandable form of the ELECTRE method [14]. In using the PROMETHEE technique, there is a limitation to compensating for one criterion's weakness or another criterion's strength. Therefore, an ideal option must obtain the minimum of all criteria. In addition, the PROMETHEE method can easily use criteria with different measurement scales (without the need to equate the criteria scale). This issue is done through six separate functions according to the information and the criteria scale. Since the criteria usually have different scales in multi-criteria decision-making, this issue is considered a strong point for decision-makers.

PROMETHEE's method begins by expanding the scale of criteria to measure the intensity of preference of one option over another by converting the resulting levels for the options to a scale of 0–1 (where 0 represents the worst and 1 represents the best).

This method is widely applied in supplier selection problems and supply chain management by researchers such as [2, 60, 94, 104, 123, 127].

In general, to use this method, the following steps are required:

- Effective criteria for evaluation.
- Specifying of weight or relative importance of criteria (The weight of criteria can be obtained through a questionnaire or some methods such as AHP).
- Specifying the type of each criterion (positive and negative).
- Formation of the decision matrix.
- The preference function of each criterion: In this method, the preference function of each criterion is determined according to the criteria's nature or the decision maker's opinion. In this method, six types of standard preference functions are used, that experience has shown that these functions are satisfactory for most real-world problems. However, there is no compulsion to use this type of six preference functions, and the decision maker can consider another function to make the decision criteria. In the preference functions, q and p are the indifference and superiority thresholds. Figure 4.4 shows these preference functions.

In the following, the steps of PROMETHEE I are described.

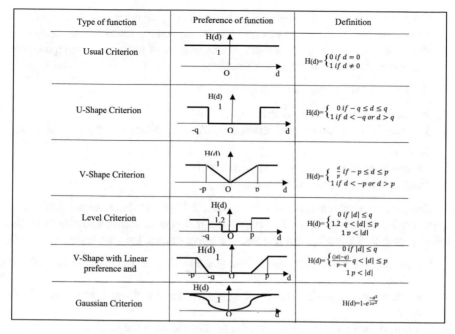

Fig. 4.4 Preference functions in PROMETHEE model

- Compute $d_j(a, b) = g_j(a) - g_j(b)$. Such that $g_j(a)$ and $g_j(b)$ are the evaluation score of options a and b relative to the j-th criteria.
- Compute $p_j(a, b) = H_j(d(a, b))$.
- $\pi(a, b) = \sum_{j=1}^{k} p_j(a, b)w_j$. The $\pi(a, b)$ expresses to what degree option "a" is superior to option "b" concerning all criteria. Also, w_j is the weight of criteria j.
- Calculation of the flow of preference ranking: $\varnothing(a) = \varnothing^+(a) - \varnothing^-(a)$. So that: $\varnothing^+(a) = \frac{\sum \pi(a,x)}{n-1}$ and $\varnothing^-(a) = \frac{\sum \pi(a,x)}{n-1}$.
- Ranking of options using the $\varnothing$.

4.2.2.6 Technique for Order of Preference by Similarity to Ideal Solution (TOPSIS)

Another supplier selection model is the TOPSIS method. The TOPSIS method was presented in 1981 by Hwang and Yoon. This method considers each option's distance from the positive and negative ideal solutions. This way, the option with the smallest distance from the positive ideal solution and the most considerable distance from the negative ideal solution is selected.

Also, many researchers have used this method in the field of supplier selection, such as [1, 54, 59, 62, 66, 79, 91, 118, 142].

The basic assumptions of this method are:

- The desirability of each criterion must be uniformly increasing or decreasing. So the best available value can be considered ideal, and the worst is anti-ideal.
- The criteria should be designed to be independent of each other (independence means the absence of internal relationships).

The steps of this method are as follows:

1. Formation of the decision matrix
2. Descale the decision matrix (decision matrix normalization): $r_{ij} = \dfrac{x_{ij}}{\sqrt{\sum_i x_{ij}}}$ and $R = [r_{ij}]$
3. Determining the normalized weighted matrix.
4. Finding the ideal and anti-ideal solution.

Here, the type of criteria should be specified. The criteria are either positive or negative. Positive criteria are criteria whose increase causes improvement in the system, such as the quality of a product. Negative criteria are vice versa.

- For criteria with a positive nature, the positive ideal is the largest value of that criterion.
- For criteria with a positive nature, the negative ideal is the smallest value of that criterion.
- For criteria with a negative nature, the positive ideal is the smallest value of that criterion.
- For criteria with a negative nature, the negative ideal is the largest value of that criterion.

5. The distance of the options from the ideal and anti-ideal solution is calculated as the Euclidean distance. And they are shown by (d_i^+) and (d_i^-).
6. Obtain cl_i^* so that: $cl_i^* = \dfrac{d_i^-}{d_i^- + d_i^+}$. The closer this index is to the number 1, it suggests the superiority of that option.

4.2.2.7   VIse Kriterijumska Optimizacija I Kompromisno Resenje (VIKOR)

VIKOR method was presented in 1988 by Opricovic and Tzeng. This method is derived from a Serbian expression with multi-criteria optimization and compromise solution. VIKOR method is a multi-criteria decision making for solving a decision problem with different and conflicting measurement criteria.

Many researchers have used this method for supplier selection, such as Fei et al. [31], Wu et al. [136], Wu et al. [137], Sahu et al. [96], and You et al. [141].

The steps for solving the problem by the VIKOR method are as follows:

- Formation of the decision matrix.
- Normalization of the decision matrix (the de-scaling process of the decision matrix).
- Determining the weight vector of the criteria and forming the weighted matrix.

- Determining the ideal positive point and the negative anti-ideal point.
- Calculation of usefulness (S) and regret (R) values for each index, using relations (4.10) and (4.11).

The value of usefulness (S) indicates the relative distance of the option from the ideal point, and the value of regret (R) indicates the maximum discomfort of the option from the ideal point.

$$S_i = \sum_{j=1}^{n} w_j \times \frac{f_j^* - f_{ij}}{f_j^* - \overline{f_j}} \tag{4.10}$$

$$R_i = \max\left[w_j \times \frac{f_j^* - f_{ij}}{f_j^* - \overline{f_j}} \right] \tag{4.11}$$

- The VIKOR index for each option is calculated from the (4.12).

$$Q_I = v\left[\frac{S_i - S^*}{\overline{S} - S^*}\right] + (1 - v)\left[\frac{R_i - R^*}{\overline{R} - R^*}\right] \tag{4.12}$$

Such that: $S^* = \text{Min}\{S_i\}$, $\overline{S} = \text{Max}\{S_i\}$, $R^* = \text{Min}\{R_i\}$, and $\overline{R} = \text{Max}\{R_i\}$.

- Sort options by S, Q, and R

In the final step of the VIKOR technique, the options are sorted into three groups, from minor to large, based on the values of Q, R, and S. The best option is the one with the highest rank in all three values. Otherwise, the best option is the one with the smallest Q, provided that the following two conditions hold:

- **Condition 1**: If options A_1 and A_2 rank first and second among m options, the relation (4.13) must be established:

$$Q(A_1) - Q(A_2) \geq \frac{1}{m - 1} \tag{4.13}$$

- **Condition 2**: Option A_1 must be recognized as the first rank in at least one of the R and S groups.

If the first condition is not fulfilled, both options will be the best, and also, if the second condition is not fulfilled, options A_1 and A_2 are selected as the best.

The Difference Between the Two Methods, TOPSIS and VIKOR

As seen, the TOPSIS technique, like the VIKOR method, aims to rank options, but despite the similarities between the two methods, there are also differences between the two methods. In the TOPSIS technique, the selected option should have the smallest distance from the ideal solution and the farthest distance from the anti-ideal solution. But the relative importance of the distances from these two points must be considered. Also, these two methods each use a different normalization method. TOPSIS method uses vector normalization, but the VIKOR method uses linear normalization.

4.2.2.8   Decision-Making Trial and Evaluation (DMATEL)

DMATEL's method is one of the decision-making methods that determine cause and effect relationships between complex factors, which was used for the first time in the Swiss Battelle Memorial Institute (BMI) in 1972 and in the research center of Geneva in 1976. This method takes advantage of the principles of graph theory to extract the influencing relationships and the mutual influence of the elements in the graph under study so that the intensity of the effect of the mentioned relationships are determined in the form of a numerical score. One of the advantages of this method is the use of feedback relationships; that is, each element affects all the elements above and below it and, in turn, is also affected by each of them. In other words, the DMATEL is used to identify and examine the relationship between the criteria and build the mapping of network relationships [110].

Also, many researchers have used this method in supplier selection, such as Li et al. [65], Daniel [20], Mirmousa and Dehnavi [81], Hsu et al. [47], Mavi et al. [77], Gharakhani [36], and Chang et al. [15].

This method includes the following steps:

- Making pairwise comparisons between indicators. ($A = [a_{ij}]$).

When several people's points of view are used, we use the average of opinions and form the matrix.

- Normalize the direct correlation matrix (N):

$$N = k*A \text{ so that } k = \max \left\{ \frac{1}{\max \sum_{j=1}^{n} a_{ij}}, \frac{1}{\max \sum_{i=1}^{n} a_{ij}} \right\} \qquad (4.14)$$

- Calculation of the complete correlation matrix:

$$T = N*(1 - N)^{-1} \qquad (4.15)$$

- Cause and effect diagram:
- The sum of each line (D) for each factor indicates its influence on the other factors of the system.

- The sum of the column elements (R) for each factor indicates the influence on the other factors of the system.
- The horizontal vector (D + R) is the degree of influence of the desired factor in the system. In other words, the higher the value of D + R, the more interaction that factor has with other system factors.
- The vertical vector (R − D) shows the influence of each factor. In general, if R − D is positive, the variable is considered a causal variable, and if it is negative, it is regarded as an effect.
- Finally, a Cartesian coordinate system is drawn.

In this device, the horizontal axis is D + R, and the vertical axis is R − D. each point is illustrated by the coordinates $(R_i + D_j, \ R_i - D_j)$. In this way, a graphic diagram will also be obtained.

4.2.3 Mathematical Programming Models

Although the MADM methods explained in sub-section () are suitable for comparing suppliers and ranking and, finally, for choosing the best supplier, these methods require using a pairwise comparison matrix or a decision matrix. As mentioned, the pairwise comparison matrix or the decision matrix is made based on an expert's opinion, so in some cases, some scores may be applied based on personal taste. This will affect the choice of the final option. Also, these methods are only suitable for weighting criteria, ranking criteria and options (suppliers and supply chains). In short, we do the following steps in these methods.

1. Definition of criteria.
2. Determining the importance (value) of criteria.
3. Determining the impact of each criterion in each option (intended supply chain).
4. Use the desired technique to compare alternatives (suppliers or supply chains).

While evaluating the performance of supply chains and comparing them with each other, we need to consider some limitations and special conditions in different industries, times or societies with varying goals in mind. Therefore, researchers suggested using mathematical programming models in this field.

Mathematical programming is a mathematical method used to optimize (maximize or minimize if necessary) some functions whose variables are under constraints, provided that the function and the constraints dependent on the variables. Therefore, mathematical programming is a mathematical technique to best use the organization's limited resources.

The resources available in the organization, such as raw materials, labor, capital, time, machinery capacity, space, Etc. are limited. In general, a mathematical programming problem is displayed as (4.16).

$$\min(\max)z = f(x)$$
$$s.t \qquad\qquad\qquad\qquad (4.16)$$
$$g_i(x) \leq = \geq b_i, i = 1, ..., m.$$

Therefore, the necessary elements to create a linear programming model suitable for a practical situation are:

- Objective function
- Decision-making variables
- Several constraints.

Mathematical programming models in the selection of suppliers are divided into the following categories.

4.2.3.1   Linear Programming

Linear programming was developed independently by American physicist and mathematician George Dantzig (1914–2005) and Russian mathematician and economist Leonid Kantorovich (1986–1912) immediately after World War II. Linear programming has a wide range of applications, which are mentioned below:

- Different types of planning
- Scheduling
- Project control
- Management
- Allocation of resources
- Industrial systems design
- Transportation issues
- Repairs and maintenance
- Theory of networks
- Economy and military affairs.

Each linear programming model has been formed from a linear objective function and several linear constraints. In this technique, limited resources can be allocated to different activities so that the optimal solution, for example, achieves maximum profit or minimum cost. There are different types of linear programming models in supplier selection, which we explain below.

Simple Linear Programming

According to the items mentioned above, the classical form of linear programming is as follows [101, 146]:

$$\min(\max)z = \sum_{j=1}^{n} c_j x_j$$

$$s.t$$

$$\sum_{j=1}^{n} a_{ij} x_j \leq = \geq b_i, i = 1, ..., m \tag{4.17}$$

$$x_j \geq 0, j = 1, ..., n.$$

In model (4.17), z is the objective function, c_j is the coefficient of the objective function, a_{ij} is the technology coefficient, and b_i is the right-hand side value. A great deal of research has been focused on this method for supplier selection, such as [37, 80], and [102].

Fuzzy Linear Programming

In the real world, in some systems under evaluation, some information is expressed in an imprecise and ambiguous manner. One of the assumptions of mathematical programming is that the model's parameters are definite. In some systems faced with this information, to simplify the modeling and solve this problem, this information is approximated with real numbers. Although this method solves the problem to some extent, it causes the accuracy of the obtained results to decrease. One of the solutions to this problem is the fuzzy linear programming technique and the fuzzy sets theory. The fuzzy set theory was first introduced by Professor Lotfi Aliasker Zadeh (1965). Therefore, fuzzy linear programming is more applicable and flexible than classical linear programming models for optimization problems. The results are more reliable because they allow decision-makers to include qualitative and ambiguous data in model parameters. In a fuzzy environment, at least one of the decision-maker's goals, resource constraints, or decision variables is expressed in a fuzzy form.

The concept of fuzzy linear programming was first proposed by Tanaka et al. in 1974, in line with the fuzzy decision-making framework presented by Bellman and Zadeh [9]. Zimmerman proposed the first formulation of the fuzzy linear programming problem in 1978. Since then, various types of fuzzy programming and their solution methods have been presented.

Also, many researchers have used this method in supplier selection, such as Guneri et al. [40], and Kumar et al. [63].

A general fuzzy linear programming model is shown in the model (4.18) [101, 146].

$$\min(\max) \sum_{j=1}^{n} \tilde{c}_j \tilde{x}_j$$

$$s.t$$

$$\sum_{j=1}^{n} \tilde{a}_{ij} \tilde{x}_j \leq \approx \geq \tilde{b}_i, i = 1, ..., m \tag{4.18}$$

$$\tilde{x}_j \geq 0, j = 1, ..., n.$$

Where $\tilde{x}_j$ is the fuzzy decision variable, $\tilde{c}_j$, $\tilde{a}_{ij}$, $\tilde{b}_i$ are fuzzy parameters. And $\preccurlyeq$, $\approx$ and $\succcurlyeq$ show the fuzziness of the operators. The following chapters will explain the fuzzy set theory and its use in detail.

Stochastic Programming

Stochastic programming discusses situations where some or all parameters or variables of the optimization problem are expressed by random variables instead of definite quantities. The basic idea in all stochastic optimization methods is to transform the problem into an equivalent deterministic problem. Therefore, stochastic programming is a framework for modeling nondeterministic probabilistic optimization problems in which probability distributions are known or can be estimated despite unknown parameters. This technique is valuable for addressing supplier selection problems in the real world. In Stochastic Programming, each uncertain parameter is considered a random variable. These random variables are usually expressed by a limited set of scenarios [10]. Also, many researchers have used this method in supplier selection, such as Zhou et al. [145], Sun et al. [117] and Rabbani et al. [89].

A stochastic linear programming problem can be formulated as the model (4.19), but $\tilde{c}_j$, $\tilde{b}_i$, and $\tilde{a}_{ij}$ are random variables with known probability distributions, and x_j are assumed to be definite decision variables.

$$\min(\max) \sum_{j=1}^{n} \tilde{c}_j x_j$$

$$s.t$$
$$\sum_{j=1}^{n} \tilde{a}_{ij} x_j \leq \tilde{b}_i, \ i = 1, ..., m \tag{4.19}$$
$$x_j \geq 0, \ j = 1, ..., n.$$

In the stochastic linear programming model, there are several methods, including:

1. A two-step programming method transforms a stochastic LP problem into an equivalent deterministic problem. This is done by increasing the problem size.
2. The method of programming with random constraints, as its name suggests, is a method that can be used to solve problems involving random adverbs, that is, adverbs that have a certain probability of happening. Programming with random constraints allows constraints to be violated with a certain (small) probability, while two-stage programming allows no constraints to be violated. The random constraint programming method was initially developed by Charnes and Cooper [16].

Integer Linear Programming (ILP)

One of the features of the linear programming model is divisibility; that is, the variables can be any number (integer, fractional and so on). But, in some practical problems, decimal values are not acceptable. For example, if the variables of the problem are the number of employees or the number of machines, a number like 4.5 will not make sense. Therefore, in some real cases, the variables are integers. Therefore, Integer Linear Programming (ILP) is a type of linear programming in which one or more variables can be integers. In fact, like linear optimization, the purpose of ILP is to find the minimum or maximum value of a linear function in a space with linear constraints. Due to discrete variables, this space is not continuous and convex. These models have many applications in real problems because many variables in the real world are integers.

In ILP, the objective function is linear, the constraints are linear, and some variables are integers. These integer variables can be only zero and one or other integer variables. The types of ILP are:

1. Pure integer programming: All variables are integers in this integer programming.
2. Mixed integer programming: In this integer programming, some variables are integers, and others are real numbers.
3. Zero and one integer programming (binary programming): In this integer programming, all variables are zero and one.

There are techniques for solving integer linear programming problems, which are mentioned below:

(A)   Gommory's fractional method or the method of cutting plane

This method solves the problem without considering the integer condition on the decision variables. If the obtained answer applies to the condition of integer variables, we stop at this stage. Otherwise, adding some constraints creates the necessary conditions for the solution. The additional constraint cuts a part of the feasible region so that the separated area does not include the integer solution. So this method is called the cutting plane.

(B)   Branch and bound method or branching and limitation

This method starts with the solve a problem without regard to integer conditions. It systematically divides the problem into more minor issues so that the part of the answer space that does not contain integers is removed. Among the researchers who studied supplier selection through integer programming are Ahmmed et al. [3], Anna and Fhiliantie [4], and Kaur et al. [56].

4.2.3.2   Non-linear Programming

Several studies on supplier selection have also used nonlinear programming models. For example, if the profit or cost of a product unit is not the same at different production levels, nonlinear programming should be used. In the nonlinear programming model, the goal is to find the values of $x = (x_1, x_2, ..., x_n)$ so that:

$$\min(\max) f(x)$$
$$s.t \qquad\qquad\qquad\qquad (4.20)$$
$$g_i(x) \leq 0, \quad i = 1, ..., m$$

The condition of non-negativity of variables should be considered as $-x_j \leq 0$ in the set of constraints of that problem, if necessary.

Unlike linear programming, nonlinear programming allows some objective functions and constraints to be nonlinear. Also, in nonlinear programming, the optimal solution is not necessarily located on the frontier and is not an extreme point. Also, a relative maximum solution (local maximum) is not necessarily an absolute maximum (global maximum), or a relative minimum solution (local minimum) is not necessarily an absolute minimum solution (global minimum).

Therefore, it is essential to know under what conditions a relative optimal solution (local optimum) for model () is also the absolute optimum. For example: If there are no constraints, the concavity of the objective function guarantees that every relative maximum solution has an absolute maximum solution, or the convexity of the objective function guarantees that every relative minimum solution is an absolute minimum solution. Suppose there are some constraints and the feasible region is a convex set. Hence, the relative maximum solution is also an absolute maximum solution (if the constraints are convex, the feasible region will also be a convex set).

Since we are not necessarily dealing with a convex space in the nonlinear discussion. So, no algorithm can generally solve all types of nonlinear programming problems. Therefore, nonlinear problems are classified as follows:

1. Constrained nonlinear programming. Among the solution methods for this category of NLP problems are: quadratic programming, convex programming, programming with linear constraints
2. Nonlinear programming without constraints. These problems are divided into two categories: multivariable and univariate functions. There are various methods for solving these problems, including Lagrange, Fibonacci, Hook, and Jeeves.

Among the available research in supplier selection using NLP are [93, 132], and [43].

4.2.3.3 Multi-objective Programming

Standard mathematical programming problems involve an objective function and some constraints. However, many issues in the real world, including supply chains, involve different objectives and limitations. Sometimes, these objective functions conflict. In such conditions, multiple objectives should be optimized on the constraints. Multi-objective programming is a subset of Multiple Objective Decision Making (MODM). MOPl is widely used in various sciences, including performance evaluation and supply chain management [81, 120, 133, 138].

A general multi-objective programming problem is as model (4.21):

$$\min(\max)z_1$$
$$\min(\max)z_2$$
$$\vdots$$
$$\min(\max)z_k \tag{4.21}$$
$$s.t$$
$$g_i(x) \leq 0, i = 1, ..., m.$$

In this model, which is known as Vector Maximization (or Minimization) Problem (VMP) or Vector Optimization Problem (VOP), Z_1, Z_2, ..., Z_k represent the objective functions and $g_i(x) \leq 0, i = 1, ..., m$ define the problem's constraints. Multi-objective models seek to simultaneously optimize several objectives, for example, minimizing time and environmental pollution. Also, goals may conflict in multi-objective models, and their simultaneous optimization is impossible. Therefore, efforts are made to achieve a satisfactory level of multiple purposes according to the possibilities.

In a particular case, multi-objective programming becomes multi-objective linear programming. In such a problem, the objective functions and constraints are linear. But unlike classical linear programming, there are several objective functions for optimization. MOLP has been shown in model (4.22).

$$\min(\max)c_1 x$$
$$\min(\max)c_2 x$$
$$\vdots$$
$$\min(\max)c_k x \tag{4.22}$$
$$s.t$$
$$Ax \leq => b,$$
$$x \geq 0.$$

There are many methods to solve multi-objective planning problems, including goal programming, objective function weighting techniques, etc., and we choose the appropriate method according to the conditions of the problem. For more details, please refer to [119].

4.2.3.4 Data Envelopment Analysis Models

As seen so far, in the described methods, the goal is to evaluate the performance of supply chains based on different criteria, restrictions, and multiple objectives. Although different criteria are considered to assess the opposite components of supply chains and, finally, the entire supply chain in different industries, in many cases, these criteria can be divided into two general categories: input (what enters the system) and output (what leaves the system). Traditionally, managers in different systems usually seek to reduce input and increase output. Therefore, in evaluating the performance of different systems, we are faced with the problem of multi-objective planning. In this way, a new branch of mathematical programming called Data Envelopment Analysis was introduced by researchers.

Data envelopment analysis is a mathematical programming method for evaluating the efficiency of decision-making units with multiple inputs and outputs. Efficiency measurement is always the attention of researchers because of its importance in assessing the performance of a company or organization. In 1957, Farel measured the efficiency of a production unit using the same method of measuring efficiency in engineering. The performance evaluation problem that Farrell focused on included only one input and one output. Charnes, Cooper, and Rhodes developed Farrell's view and presented a model that could measure efficiency with multiple inputs and outputs. This method was named data envelopment analysis (DEA). It was used for the first time in Edward Rhodes' doctoral thesis under the guidance of Cooper under the title of evaluating the educational progress of American national school students in 1976 at Carnegie University. Since Charnes, Cooper, and Rhodes presented this model, it became known as the CCR model, which consists of the first letters of the names of the three mentioned individuals, and was introduced in 1978 in an article entitled measuring the efficiency of decision-making units [17, 109]. Also, many researchers have used this method in the field of supplier selection, such as [22, 23, 30, 71, 75, 88, 113, 121].

The CCR model is shown in the model (4.23).

$$
\begin{aligned}
&\max \sum_{r=1}^{s} u_r y_{ro} \\
&s.t \\
&\sum_{i=1}^{m} v_i x_{io} = 1 \\
&\sum_{r=1}^{s} u_r y_{rj} - \sum_{i=1}^{m} v_i x_{ij} \leq 0, \ j = 1, ..., n, \\
&u_r, v_i \geq 0, \ i = 1, ..., m, \ r = 1, ..., s.
\end{aligned}
\tag{4.23}
$$

In the model (4.1), x_{ij} (i = 1, 2, ... m) and y_{ro} (r = 1,2,...,s) are the i-th and r-th of the input and output of the unit under evaluation, respectively. n is the number of under-evaluation units.

In 1984, Banker, Charnes, and Cooper introduced a new model by changing the CCR model, which became known as the BCC model due to the first letter of their name [8]. The BCC model is shown in the model (4.24).

$$\max \sum_{r=1}^{s} u_r y_{ro} + u_o$$

$$s.t$$

$$\sum_{i=1}^{m} v_i x_{io} = 1 \qquad\qquad (4.24)$$

$$\sum_{r=1}^{s} u_r y_{rj} - \sum_{i=1}^{m} v_i x_{ij} + u_o \leq 0, \ j = 1, ..., n,$$

$$u_r, v_i \geq 0, \ i = 1, ..., m, \ r = 1, ..., s.$$

With the development of efficiency evaluation in different industries, other models were introduced for performance evaluation. The types of DEA models and how to use them will be explained in detail in the following chapters. Also, unlike the methods mentioned earlier, by using DEA models, in addition to the performance evaluation of supply chains, it is possible to analyze them based on different criteria and perspectives. For instance, cost efficiency models consider how to combine inputs to achieve better performance, and profit efficiency models were applied to evaluate units to achieve more profit [48, 97, 99]. Also, for considering internal exchanges, network DEA models were introduced by Färe and Grosskopf and then developed by [29], Kao and Hwang [55], Tone and Tsutsui [122], Cook et al. [19], Wu [139], Shafiee et al. [106], and Sahoo et al. [98].

It is necessary to say that because of imprecise data in some practical cases, the DEA model can be applied to systems with fuzzy input and output [41, 49, 90, 103, 111], integer data (Lozano and Villa 2007) [108, 138], stochastic indicators [46, 73, 114], and also interval criteria [49, 116, 131].

Another benefit of using DEA models is that this technique can be determined returns to scale and congestion in the under-evaluation units [51, 57, 58, 74, 85, 100]. In addition, the DEA technique can introduce a benchmark for inefficient units [6, 112].

Considering the mentioned features, many researchers suggested using the DEA technique to analyze the performance of the supply chains under evaluation. So, in the following, we'll describe the use of DEA models for analyzing supply chain performance.

4.3 Conclusion

Evaluation and proper selection of suppliers is an issue that has received special attention in recent years in academic and industrial environments, because choosing the right set of suppliers to work with is very important and vital for the success of a company. Therefore, due to the importance of this issue, in this chapter, the importance and position of suppliers was first described, then the various models of supplier performance evaluation were described, which this section includes three sub-sets; Models of total cost of ownership (TCO) models, models based on Multiple Attribute Decision Making (MADM) and mathematical programming models. In this way, the readers will become familiar with the various models of supplier performance evaluation and take steps to optimize their work space and choose the right suppliers.

References

1. Abdel-Basset, M., Saleh, M., Gamal, A., Smarandache, F.: An approach of TOPSIS technique for developing supplier selection with group decision making under type-2 neutrosophic number. Appl. Soft Comput. **77**, 438–452 (2019)
2. Abdullah, L., Chan, W., Afshari, A.: Application of PROMETHEE method for green supplier selection: a comparative result based on preference functions. J. Indus. Eng. Int. **15**(2), 271–285 (2019)
3. Ahmmed, M.S., Ghosh, S.K., Zoha, N., Chowdhury, T.Z.: Supplier selection using integer linear programming model. Glob. J. Res. Eng. **18**(J4), 27–30 (2018)
4. Anna, I.D., Fhiliantie, P.R.: Supplier selection and order quantity allocation of raw material using integer linear programming. J. ASRO **9**(1), 98–105 (2018)
5. Asadabadi, M.R.: A customer based supplier selection process that combines quality function deployment, the analytic network process and a Markov chain. Eur. J. Oper. Res. **263**(3), 1049–1062 (2017)
6. Avkiran, N.K., Shafiee, M., Saleh, H., Ghaderi, M.: Benchmarking in the supply chain using data envelopment analysis. Theor. Econom. Lett. **8**(14), 2987 (2018)
7. Azadnia, A.H., Ghadimi, P., Mat Saman, M.Z., Wong, K.Y., Sharif, S.: Supplier selection: a hybrid approach using ELECTRE and fuzzy clustering. In: International Conference on Informatics Engineering and Information Science, pp. 663–676. Springer, Berlin, Heidelberg (2011)
8. Banker, R.D., Charnes, A., Cooper, W.W.: Some models for estimating technical and scale inefficiencies in data envelopment analysis. Manage. Sci. **30**(9), 1078–1092 (1984)
9. Bellman, R.E., Zadeh, L.A.: Decision-making in a fuzzy environment. Manage. Sci. **17**(4), B-141 (1970)
10. Birge, J.R., Louveaux, F.: Introduction to stochastic programming. Springer Science & Business Media (2011)
11. Birgün, S., & Cihan, E.: Supplier selection process using ELECTRE method. In: 2010 IEEE International Conference on Intelligent Systems and Knowledge Engineering, pp. 634–639. IEEE (2010)
12. Bhutta, K.S., Huq, F.: Supplier selection problem: a comparison of the total cost of ownership and analytic hierarchy process approaches. Supp. Chain Manage. Int. J. **7**(3), 126–135 (2002)
13. Brans, J.P., Mareschal, B.: PROMETHEE V: MCDM problems with segmentation constraints. INFOR: Inform. Syst. Operation. Res. **30**(2), 85–96 (1992)
14. Bouyssou, D., Marchant, T., Pirlot, M., Perny, P., Tsoukias, A., Vincke, P.: Evaluation and Decision Models: A Critical Perspective, vol. 32. Springer Science & Business Media (2000)
15. Chang, B., Chang, C.W., Wu, C.H.: Fuzzy DEMATEL method for developing supplier selection criteria. Expert Syst. Appl. **38**(3), 1850–1858 (2011)
16. Charnes, A., Cooper, W.W.: Chance-constrained programming. Manage. Sci. **6**(1), 73–79 (1959)
17. Charnes, A., Cooper, W.W., Rhodes, E.: Measuring the efficiency of decision-making units. Eur. J. Oper. Res. **3**(4), 339–338 (1979)
18. Chrzanowski, J.: 12 Key Elements of Total Cost of Ownership. Taken from the link https://www-supplytechnologies-com.translate.goog/blog/12-key-elements-of-total-cost-of-ownership?_x_tr_sl=en&_x_tr_tl=fa&_x_tr_hl=fa&_x_tr_pto=op,sc (2022)
19. Cook, W.D., Zhu, J., Bi, G., Yang, F.: Network DEA: additive efficiency decomposition. Eur. J. Oper. Res. **207**(2), 1122–1129 (2010)

20. Daniel, J.: Developing a conceptual model to evaluate green suppliers: decision making method using DEMATEL (2016)

21. De Tré, G., De Mol, R., Bronselaer, A.: Handling veracity in multi-criteria decision-making: a multi-dimensional approach. Inf. Sci. **460**, 541–554 (2018)

22. Dobos, I., Vörösmarty, G.: Inventory-related costs in green supplier selection problems with data envelopment analysis (DEA). Int. J. Prod. Econ. **209**, 374–380 (2019)

23. Dutta, P., Jaikumar, B., Arora, M.S.: Applications of data envelopment analysis in supplier selection between 2000 and 2020: a literature review. Ann. Oper. Res., 1–56 (2021)

24. Dweiri, F., Kumar, S., Khan, S.A., Jain, V.: Designing an integrated AHP based decision support system for supplier selection in automotive industry. Expert Syst. Appl. **62**, 273–283 (2016)

25. Dweiri, F., Kumar, S., Khan, S.A., Jain, V.: Corrigendum to "designing an integrated AHP based decision support system for supplier selection in automotive industry". Expert Syst. Appl. **62**, 273–283 (2016). Expert Syst. Appl. **100**(72), 467–468 (2017)

26. Ellram, L.M.: The supplier selection decision in strategic partnerships. J. Purchas. Mater. Manage. **26**(4), 8–14 (1990)

27. Ellram, L.M., Siferd, S.P.: Total cost of ownership: a key concept in strategic cost management decisions. Mater. Eng. **19**(1), 55–84 (1998)

28. Ellram, L.: Total cost of ownership: elements and implementation. Int. J. Purch. Mater. Manag. **29**(3), 2–11 (1993)

29. Färe, R., Grosskopf, S.: Theory and application of directional distance functions. J. Prod. Anal. **13**(2), 93–103 (2000)

30. Farzipoor Saen, R.: Developing a new data envelopment analysis methodology for supplier selection in the presence of both undesirable outputs and imprecise data. Int. J. Adv. Manuf. Technol. **51**(9), 1243–1250 (2010)

31. Fei, L., Xia, J., Feng, Y., Liu, L.: An ELECTRE-based multiple criteria decision making method for supplier selection using Dempster-Shafer theory. IEEE Access **7**, 84701–84716 (2019)

32. Fei, L., Deng, Y., Hu, Y.: DS-VIKOR: a new multi-criteria decision-making method for supplier selection. Int. J. Fuzzy Syst. **21**(1), 157–175 (2019)

33. Fu, Y.K.: An integrated approach to catering supplier selection using AHP-ARAS-MCGP methodology. J. Air Transp. Manag. **75**, 164–169 (2019)

34. Galankashi, M.R., Chegeni, A., Soleimanynanadegany, A., Memari, A., Anjomshoae, A., Helmi, S.A., Dargi, A.: Prioritizing green supplier selection criteria using fuzzy analytical network process. Procedia Cirp **26**, 689–694 (2015)

35. Garfamy, R.M.: A data envelopment analysis approach based on total cost of ownership for supplier selection. J. Enterp. Inf. Manag. **19**(6), 662–678 (2006)

36. Gharakhani, D.: The evaluation of supplier selection criteria by fuzzy DEMATEL method. J. Basic Appl. Sci. Res. **2**(4), 3215–3224 (2012)

37. Ghodsypour, S.H., O'Brien, C.: A decision support system for supplier selection using an integrated analytic hierarchy process and linear programming. Int. J. Prod. Econ. **56**, 199–212 (1998)

38. Giannakis, M., Dubey, R., Vlachos, I., Ju, Y.: Supplier sustainability performance evaluation using the analytic network process. J. Clean. Prod. **247**, 119439 (2020)

39. Gold, S., Awasthi, A.: Sustainable global supplier selection extended towards sustainability risks from (1+n) th tier suppliers using fuzzy AHP based approach. Ifac-Papersonline **48**(3), 966–971 (2015)

40. Guneri, A.F., Yucel, A., Ayyildiz, G.: An integrated fuzzy-lp approach for a supplier selection problem in supply chain management. Expert Syst. Appl. **36**(5), 9223–9228 (2009)

41. Guo, P., Tanaka, H.: Fuzzy DEA: a perceptual evaluation method. Fuzzy Sets Syst. **119**(1), 149–160 (2001)

42. Ha, S.H., Krishnan, R.: A hybrid approach to supplier selection for the maintenance of a competitive supply chain. Expert Syst. Appl. **34**(2), 1303–1311 (2008)

43. Hao-dong, C.H.E.N., Zhi-ping, W.A.N.G., Yan, C.: Mixed-integer non-linear program model of dynamic supplier selection under fuzzy environment. Oper. Res. Manage. Sci. **24**(4), 128 (2015)
44. He, T., Wei, G., Lu, J., Wei, C., Lin, R.: Pythagorean 2-tuple linguistic taxonomy method for supplier selection in medical instrument industries. Int. J. Environ. Res. Public Health **16**(23), 4875 (2019)
45. Ho, W., Dey, P.K., Lockström, M.: Strategic sourcing: a combined QFD and AHP approach in manufacturing. Supp. Chain Manage. Int. J. **16**(6), 446–461 (2011)
46. Hosseinzadeh Lotfi, F., Nematollahi, N., Behzadi, M.H., Mirbolouki, M., Moghaddas, Z.: Centralized resource allocation with stochastic data. J. Comput. Appl. Math. **236**(7), 1783–1788 (2012)
47. Hsu, C.W., Kuo, T.C., Chen, S.H., Hu, A.H.: Using DEMATEL to develop a carbon management model of supplier selection in green supply chain management. J. Clean. Prod. **56**, 164–172 (2013)
48. Jahanshahloo, G.R., Soleimani-Damaneh, M., Mostafaee, A.: A simplified version of the DEA cost efficiency model. Eur. J. Oper. Res. **184**(2), 814–815 (2008)
49. Jahanshahloo, G.R., Lotfi, F.H., Shahverdi, R., Adabitabar, M., Rostamy-Malkhalifeh, M., Sohraiee, S.: Ranking DMUs by l1-norm with fuzzy data in DEA. Chaos, Solitons Fractals **39**(5), 2294–2302 (2009)
50. Jahanshahloo, G.R., Lotfi, F.H., Malkhalifeh, M.R., Namin, M.A.: A generalized model for data envelopment analysis with interval data. Appl. Math. Model. **33**(7), 3237–3244 (2009)
51. Jahanshahloo, G.R., Lotfi, F.H., Zohrehbandian, M.: Finding the efficiency score and RTS characteristic of DMUs by means of identifying the efficient frontier in DEA. Appl. Math. Comput. **170**(2), 985–993 (2005)
52. Jia, F., Liu, Y., Wang, X.: An extended MABAC method for multi-criteria group decision making based on intuitionistic fuzzy rough numbers. Expert Syst. Appl. **127**, 241–255 (2019)
53. Jharkharia, S., Shankar, R.: Selection of logistics service provider: an analytic network process (ANP) approach. Omega **35**(3), 274–289 (2007)
54. Kamalakannan, R., Ramesh, C., Shunmugasundaram, M., Sivakumar, P., Mohamed, A.: Evaluvation and selection of suppliers using TOPSIS. Mater. Today Proc. **33**, 2771–2773 (2020)
55. Kao, C., Hwang, S.N.: Efficiency decomposition in two-stage data envelopment analysis: an application to non-life insurance companies in Taiwan. Eur. J. Oper. Res. **185**(1), 418–429 (2008)
56. Kaur, H., Singh, S.P., Glardon, R.: An integer linear program for integrated supplier selection: a sustainable flexible framework. Glob. J. Flex. Syst. Manag. **17**(2), 113–134 (2016)
57. Khezri, S., Dehnokhalaji, A., Lotfi, F.H.: A full investigation of the directional congestion in data envelopment analysis. RAIRO-Oper. Res. **55**, S571–S591 (2021)
58. Khoveyni, M., Eslami, R., Khodabakhshi, M., Jahanshahloo, G.R., Lotfi, F.H.: Recognizing strong and weak congestion slack based in data envelopment analysis. Comput. Ind. Eng. **64**(2), 731–738 (2013)
59. Kilic, H.S., Yalcin, A.S.: Modified two-phase fuzzy goal programming integrated with IF-TOPSIS for green supplier selection. Appl. Soft Comput. **93**, 106371 (2020)
60. Krishankumar, R., Ravichandran, K.S., Saeid, A.B.: A new extension to PROMETHEE under intuitionistic fuzzy environment for solving supplier selection problem with linguistic preferences. Appl. Soft Comput. **60**, 564–576 (2017)
61. Kumar, P., Singh, R.K., Vaish, A.: Suppliers' green performance evaluation using fuzzy extended ELECTRE approach. Clean Technol. Environ. Policy **19**(3), 809–821 (2017)
62. Kumar, S., Kumar, S., Barman, A.G.: Supplier selection using fuzzy TOPSIS multi criteria model for a small scale steel manufacturing unit. Proc. Comp. Sci. **133**, 905–912 (2018)
63. Kumar, P., Shankar, R., Yadav, S.S.: An integrated approach of analytic hierarchy process and fuzzy linear programming for supplier selection. Int. J. Oper. Res. **3**(6), 614–631 (2008)
64. Lee, Y., Wu, W.: Development strategies for competency models. International Trade Department, Ta Hwa Institute of Technology, Taiwan (2005)

65. Li, Y., Diabat, A., Lu, C.C.: Leagile supplier selection in Chinese textile industries: a DEMATEL approach. Ann. Oper. Res. **287**(1), 303–322 (2020)
66. Li, J., Fang, H., Song, W.: Sustainable supplier selection based on SSCM practices: a rough cloud TOPSIS approach. J. Clean. Prod. **222**, 606–621 (2019)
67. Liao, C.N., Kao, H.P.: An integrated fuzzy TOPSIS and MCGP approach to supplier selection in supply chain management. Expert Syst. Appl. **38**(9), 10803–10811 (2011)
68. Liou, J.J., Chuang, Y.C., Zavadskas, E.K., Tzeng, G.H.: Data-driven hybrid multiple attribute decision-making model for green supplier evaluation and performance improvement. J. Clean. Prod. **241**, 118321 (2019)
69. Liu, P., Zhang, X.: Research on the supplier selection of a supply chain based on entropy weight and improved ELECTRE-III method. Int. J. Prod. Res. **49**(3), 637–646 (2011)
70. Liu, P., Wang, Y., Jia, F., Fujita, H.: A multiple attribute decision making three-way model for intuitionistic fuzzy numbers. Int. J. Approximate Reasoning **119**, 177–203 (2020)
71. Liu, J., Ding, F.Y., Lall, V.: Using data envelopment analysis to compare suppliers for supplier selection and performance improvement. Supp Chain Manage Int J (2000)
72. Lo, H.W., Liaw, C.F., Gul, M., Lin, K.Y.: Sustainable supplier evaluation and transportation planning in multi-level supply chain networks using multi-attribute-and multi-objective decision making. Comput. Ind. Eng. **162**, 107756 (2021)
73. Lotfi, F.H., Nematollahi, N., Behzadi, M.H., Mirbolouki, M.: Ranking decision making units with stochastic data by using coefficient of variation. Math. Comput. Appl. **15**(1), 148–155 (2010)
74. Lotfi, F.H., Jahanshahloo, G.R., Esmaeili, M.: An alternative approach in the estimation of returns to scale under weight restrictions. Appl. Math. Comput. **189**(1), 719–724 (2007)
75. Mahdiloo, M., Saen, R.F., Lee, K.H.: Technical, environmental and eco-efficiency measurement for supplier selection: an extension and application of data envelopment analysis. Int. J. Prod. Econ. **168**, 279–289 (2015)
76. Mandolini, M., Marilungo, E., Germani, M.: A TCO model for supporting the configuration of industrial plants. Proc. Manuf. **11**, 1940–1949 (2017)
77. Mavi, R.K., Kazemi, S., Najafabadi, A.F., Mousaabadi, H.B.: Identification and assessment of logistical factors to evaluate a green supplier using the fuzzy logic DEMATEL method. Polish J. Environ. Stud. 22(2) (2013)
78. Mckone-Sweet, K.A.T.H.L.E.E.N., LEE, Y.T.: Development and analysis of a supply chain strategy taxonomy. J. Supp. Chain Manage. **45**(3), 3–24
79. Memari, A., Dargi, A., Jokar, M.R.A., Ahmad, R., Rahim, A.R.A.: Sustainable supplier selection: a multi-criteria intuitionistic fuzzy TOPSIS method. J. Manuf. Syst. **50**, 9–24 (2019)
80. Mendoza, A., Ventura, J.A.: An effective method to supplier selection and order quantity allocation. Int. J. Bus. Syst. Res. **2**(1), 1–15 (2008)
81. Mirmousa, S., Dehnavi, H.D.: Development of criteria of selecting the supplier by using the fuzzy DEMATEL method. Proc. Soc. Behav. Sci. **230**, 281–289 (2016)
82. Moheb-Alizadeh, H., Handfield, R.: Sustainable supplier selection and order allocation: a novel multi-objective programming model with a hybrid solution approach. Comput. Ind. Eng. **129**, 192–209 (2019)
83. Ning, B., Wei, G., Lin, R., Guo, Y.: A novel MADM technique based on extended power generalized Maclaurin symmetric mean operators under probabilistic dual hesitant fuzzy setting and its application to sustainable suppliers selection. Expert Syst. Appl. **204**, 117419 (2022)
84. Nirmala, G., Uthra, G.: AHP based on triangular intuitionistic fuzzy number and its application to supplier selection problem. Mater. Today Proc. **16**, 987–993 (2019)
85. Noura, A.A., Lotfi, F.H., Jahanshahloo, G.R., Rashidi, S.F., Parker, B.R.: A new method for measuring congestion in data envelopment analysis. Socioecon. Plann. Sci. **44**(4), 240–246 (2010)
86. Ocampo, L.A., Labrador, J.J.T., Jumao-as, A.M.B., Rama, A.M.O.: Integrated multi-phase sustainable product design with a hybrid quality function deployment–multi-attribute decision-making (QFD-MADM) framework. Sustain. Prod. Consum. **24**, 62–78 (2020)

87. Peters, M., Zelewski, S.: Pitfalls in the application of analytic hierarchy process to performance measurement. Manage. Decis. (2008)
88. Pratap, S., Daultani, Y., Dwivedi, A., Zhou, F.: Supplier selection and evaluation in e-commerce enterprises: a data envelopment analysis approach. Benchmark. Int. J. 29(1), 325–341 (2021)
89. Rabbani, M., Molana, S.M.H., Sajadi, S.M., Davoodi, M.H.: Sustainable fertilizer supply chain network design using evolutionary-based resilient robust stochastic programming. Comp. Indus. Eng. 108770 (2022)
90. Rahmani, A., Hosseinzadeh Lotfi, F., Rostamy-Malkhalifeh, M., Allahviranloo, T.: A new method for defuzzification and ranking of fuzzy numbers based on the statistical beta distribution. Adv. Fuzzy Syst. (2016)
91. Ramakrishnan, K.R., Chakraborty, S.: A cloud TOPSIS model for green supplier selection. Facta Universitatis. Ser. Mech. Eng. 18(3), 375–397 (2020)
92. Razmi, J., Seifoory, M., Pishvaee, M.S.: A fuzzy multi-attribute decision making model for selecting the best supply chain strategy: Lean, agile or leagile. Adv. Indus. Eng. 45(Special Issue), 127–142 (2011)
93. Razmi, J., Rafiei, H.: An integrated analytic network process with mixed-integer non-linear programming to supplier selection and order allocation. Int. J. Adv. Manuf. Technol. 49(9), 1195–1208 (2010)
94. Safari, H., Fagheyi, M.S., Ahangari, S.S., Fathi, M.R.: Applying PROMETHEE method based on entropy weight for supplier selection. Bus. Manage. Strat. 3(1), 97–106 (2012)
95. Saaty, T.L.: How to make a decision: the analytic hierarchy process. Eur. J. Oper. Res. 48(1), 9–26 (1990)
96. Sahu, A.K., Datta, S., Mahapatra, S.S.: Evaluation and selection of resilient suppliers in fuzzy environment: exploration of fuzzy-VIKOR. Benchmark. Int. J. (2016)
97. Sahoo, B.K., Mehdiloozad, M., Tone, K.: Cost, revenue and profit efficiency measurement in DEA: a directional distance function approach. Eur. J. Oper. Res. 237(3), 921–931 (2014)
98. Sahoo, B.K., Saleh, H., Shafiee, M., Tone, K., Zhu, J.: An alternative approach to dealing with the composition approach for series network production processes. Asia-Pacific J. Oper. Res. 38(06), 2150004 (2021)
99. Saleh, H.I.L.D.A., Hosseinzadeh Lotfi, F., Rostmay-Malkhalifeh, M., Shafiee, M.: Provide a mathematical model for selecting suppliers in the supply chain based on profit efficiency calculations. J. New Res. Mathe. 7(32), 177–186 (2021)
100. Saleh, H., Hosseinzadeh, F., Rostamy, M., Shafiee, M.: Performance evaluation and specifying of return to scale in network DEA. J. Adv. Mathe. Model. 10(2), 309–340 (2020)
101. Salski, A., Noell, C.: Fuzzy linear programming for the optimization of land use scenarios. In: N. Mastorakis, V. Mladenov, B. Suter & L. J. Wang (eds.) Advances in Scientific Computing, Computational Intelligence and Applications, Mathematics and Computers in Science and Engineering, pp. 355–360 (2001)
102. Sanayei, A., Mousavi, S.F., Abdi, M.R., Mohaghar, A.: An integrated group decision-making process for supplier selection and order allocation using multi-attribute utility theory and linear programming. J. Franklin Inst. 345(7), 731–747 (2008)
103. Sanei, M., Rostami-Malkhalifeh, M., Saleh, H.: A new method for solving fuzzy DEA models. Int. J. Indus. Mathe. 1(4), 307–313 (2009)
104. Senvar, O., Tuzkaya, G., Kahraman, C.: Multi criteria supplier selection using fuzzy PROMETHEE method. In: Supply Chain Management Under Fuzziness, pp. 21–34. Springer, Berlin, Heidelberg (2014)
105. Sevkli, M.: An application of the fuzzy ELECTRE method for supplier selection. Int. J. Prod. Res. 48(12), 3393–3405 (2010)
106. Shafiee, M., Afifian, E.: Ranking of software units of Shiraz Rivers Software Engineering Company using VIKORE-AHP-DEA technique. In: 9th Conference of Iranian Association of Operations Research (2016)
107. Shafiee, M., Honarvar, A.: Assessing of supply chain risks via ANP technique (Case Study: Zagros Petrochemical Company Located in South Pars Special Economic Zone). Sanandaj Indus. Manage. Quart. 11(35), 85–102 (2016)

108. Shafiee, M., Lotfi, F.H., Saleh, H., Ghaderi, M.: A mixed integer bi-level DEA model for bank branch performance evaluation by Stackelberg approach. J. Indus. Eng. Int. **12**(1), 81–91 (2016)
109. Shafiee, M.: Designing a multi-level data envelopment analysis model to evaluate the efficiency of financial organizations. Jor **14**(2), 41–66 (2017)
110. Shafiee, M., Saleh, H., Akbarpour, S.: Presenting the combination model of EFQM excellence model, balanced scorecard, and network data envelopment analysis to compile teamwork performance evaluation at Shiraz emergency bases. Jorar **10**(2), 58–73 (2018)
111. Shafiee, M., Saleh, H.: Evaluation of strategic performance with fuzzy data envelopment analysis. Int. J. Data Envelop. Anal. **7**(4), 1–20 (2019)
112. Shafiee, M., Hosseinzade Lotfi, F., Saleh, H.: Benchmark forecasting in data envelopment analysis for decision making units. Int. J. Indus. Mathe. **13**(1), 29–42 (2021)
113. Shafiee, M., Akbarpoor, S., Akhlaghi Nik, A.: The performance evaluation of the instrumentation equipment suppliers of the Borzouyeh petrochemical company using the data envelopment analysis and the nash game approach. Iranian J. Supp. Chain Manage. **23**(72), 41–53 (2022)
114. Shang, J.K., Wang, F.C., Hung, W.T.: A stochastic DEA study of hotel efficiency. Appl. Econ. **42**(19), 2505–2518 (2010)
115. Shojaie, A.A., Babaie, S., Sayah, E., Mohammaditabar, D.: Analysis and prioritization of green health suppliers using Fuzzy ELECTRE method with a case study. Glob. J. Flex. Syst. Manag. **19**(1), 39–52 (2018)
116. Smirlis, Y.G., Maragos, E.K., Despotis, D.K.: Data envelopment analysis with missing values: an interval DEA approach. Appl. Math. Comput. **177**(1), 1–10 (2006)
117. Sun, J., Ozawa, M., Zhang, W., & Takahashi, K. (2022). Electricity supply chain management considering environmental evaluation: A multi-period optimization stochastic programming model. Clean. Respons. Consum. 100086
118. Sureeyatanapas, P., Sriwattananusart, K., Niyamosoth, T., Sessomboon, W., Arunyanart, S.: Supplier selection towards uncertain and unavailable information: an extension of TOPSIS method. Oper. Res. Perspect. **5**, 69–79 (2018)
119. Tamiz, M. (Ed.). (2012). Multi-objective Programming and Goal Programming: Theories and Applications, vol. 432. Springer Science & Business Media.
120. Tirkolaee, E.B., Mardani, A., Dashtian, Z., Soltani, M., Weber, G.W.: A novel hybrid method using fuzzy decision making and multi-objective programming for sustainable-reliable supplier selection in two-echelon supply chain design. J. Clean. Prod. **250**, 119517 (2020)
121. Toloo, M., Nalchigar, S.: A new DEA method for supplier selection in presence of both cardinal and ordinal data. Expert Syst. Appl. **38**(12), 14726–14731 (2011)
122. Tone, K., Tsutsui, M.: Network DEA: a slacks-based measure approach. Eur. J. Oper. Res. **197**(1), 243–252 (2009)
123. Tong, L.Z., Wang, J., Pu, Z.: Sustainable supplier selection for SMEs based on an extended PROMETHEE II approach. J. Clean. Prod. **330**, 129830 (2022)
124. Tseng, M.L., Chiang, J.H., Lan, L.W.: Selection of optimal supplier in supply chain management strategy with analytic network process and choquet integral. Comput. Ind. Eng. **57**(1), 330–340 (2009)
125. Vinodh, S., Ramiya, R.A., Gautham, S.G.: Application of fuzzy analytic network process for supplier selection in a manufacturing organisation. Expert Syst. Appl. **38**(1), 272–280 (2011)
126. Visani, F., Barbieri, P., Di Lascio, F.M.L., Raffoni, A., Vigo, D.: Supplier's total cost of ownership evaluation: a data envelopment analysis approach. Omega **61**, 141–154 (2016)
127. Wan, S.P., Zou, W.C., Zhong, L.G., Dong, J.Y.: Some new information measures for hesitant fuzzy PROMETHEE method and application to green supplier selection. Soft. Comput. **24**(12), 9179–9203 (2020)
128. Wang, W., Zhan, J., Zhang, C.: Three-way decisions based multi-attribute decision making with probabilistic dominance relations. Inf. Sci. **559**, 75–96 (2021)

129. Wang, Y.C., Chen, T.: A Bi-objective AHP-MINLP-GA approach for flexible alternative supplier selection amid the COVID-19 pandemic. Soft Comput. Lett. **3**, 100016 (2021)
130. Wang, J., Xing, R.: Decision making with the analytic network process: economic, political, social and technological applications with benefits, opportunities, costs and risks (2007)
131. Wang, Y.M., Greatbanks, R., Yang, J.B.: Interval efficiency assessment using data envelopment analysis. Fuzzy Sets Syst. **153**(3), 347–370 (2005)
132. Ware, N.R., Singh, S.P., Banwet, D.K.: A mixed-integer non-linear program to model dynamic supplier selection problem. Expert Syst. Appl. **41**(2), 671–678 (2014)
133. Weber, C.A., Ellram, L.M.: Supplier selection using multi-objective programming: a decision support system approach. Int. J. Phys. Distrib. Logis. Manage. (1993)
134. Wu, J.Z., Tiao, P.J.: A validation scheme for intelligent and effective multiple criteria decision-making. Appl. Soft Comput. **68**, 866–872 (2018)
135. Wu, X., Liao, H.: Geometric linguistic scale and its application in multi-attribute decision-making for green agricultural product supplier selection. Fuzzy Sets Syst. (2022)
136. Wu, Q., Zhou, L., Chen, Y., Chen, H.: An integrated approach to green supplier selection based on the interval type-2 fuzzy best-worst and extended VIKOR methods. Inf. Sci. **502**, 394–417 (2019)
137. Wu, Y., Chen, K., Zeng, B., Xu, H., Yang, Y.: Supplier selection in nuclear power industry with extended VIKOR method under linguistic information. Appl. Soft Comput. **48**, 444–457 (2016)
138. Wu, D.D., Zhang, Y., Wu, D., Olson, D.L.: Fuzzy multi-objective programming for supplier selection and risk modeling: a possibility approach. Eur. J. Oper. Res. **200**(3), 774–787 (2010)
139. Wu, D.D.: Bilevel programming data envelopment analysis with constrained resource. Eur. J. Oper. Res. **207**(2), 856–864 (2010)
140. Xiao, L., Zhang, S., Wei, G., Wu, J., Wei, C., Guo, Y., Wei, Y.: Green supplier selection in steel industry with intuitionistic fuzzy Taxonomy method. J. Intell. Fuzzy Syst. **39**(5), 7247–7258 (2020)
141. You, X.Y., You, J.X., Liu, H.C., Zhen, L.: Group multi-criteria supplier selection using an extended VIKOR method with interval 2-tuple linguistic information. Expert Syst. Appl. **42**(4), 1906–1916 (2015)
142. Yu, C., Shao, Y., Wang, K., Zhang, L.: A group decision making sustainable supplier selection approach using extended TOPSIS under interval-valued Pythagorean fuzzy environment. Expert Syst. Appl. **121**, 1–17 (2019)
143. Zhang, H., Wei, G., Chen, X.: SF-GRA method based on cumulative prospect theory for multiple attribute group decision making and its application to emergency supplies supplier selection. Eng. Appl. Artif. Intell. **110**, 104679 (2022)
144. Zhong, L., Yao, L.: An ELECTRE I-based multi-criteria group decision making method with interval type-2 fuzzy numbers and its application to supplier selection. Appl. Soft Comput. **57**, 556–576 (2017)
145. Zhou, R., Bhuiyan, T.H., Medal, H.R., Sherwin, M.D., Yang, D.: A stochastic programming model with endogenous uncertainty for selecting supplier development programs to proactively mitigate supplier risk. Omega **107**, 102542 (2022)
146. Zimmerman, H.J.: Fuzzy programming and linear programming with several objective function. Fuzzy Set Syst. **1**, 45–55.

Chapter 5
Examining Supply Chain Crises and Disruptions

5.1 Introduction

During the last few years, various types of unpredictable events, including intentional non-terrorist acts, terrorist acts, accidents, natural disasters, Etc., have occurred, showing that our world is increasingly uncertain and vulnerable. In addition, today's supply chains seem more vulnerable than in the past due to the multiplicity of industries and work activities, decentralized production, increasing outsourcing, reducing the number of suppliers, and focusing on reducing inventories. Although these various industries and work activities have lowered the costs of the supply chain, they have exposed the supply chain to more risk and disruption and have made the supply chain more complex [41].

Supply chain crises and breakdowns are unplanned and unpredictable events that disrupt the normal flow of goods and materials in the supply chain and, as a result, expose companies within the supply chain to financial and operational risks. Therefore, it is necessary to have a coherent and scientific crisis management system in the supply chain that can prevent crises by predicting and identifying them. Also, in the event of a crisis, it can be done by; Prioritizing, planning, organizing, guiding, leading, and controlling the necessary activities to intervene, guide, and manage the crisis and complete the rehabilitation after the crisis. Therefore, we can safely say that the most important influencing factor in the success of the crisis management process is crisis logistics; it plays an essential and decisive role in the entire crisis supply chain. Crisis logistics includes all the techniques of estimation, supply, transportation, storage, and distribution of goods, equipment, services, and all the needs of the victims and relief teams, which should be given in the shortest possible time (appropriate time) and in designated places (appropriate place) to the required amount (appropriate amount) to individuals and teams (certain people) and with a scientific and accurate method and with the minor problems for the needy (proper method) reach them.

© The Author(s), under exclusive license to Springer Nature Switzerland AG 2023
F. Hosseinzadeh Lotfi et al., *Supply Chain Performance Evaluation*,
Studies in Big Data 122, https://doi.org/10.1007/978-3-031-28247-8_5

Therefore, considering the importance of this issue, the purpose of this chapter is to identify supply chain crises and risks and ways to deal with them, and finally, the concepts of reliability in the supply chain, which are discussed further.

5.2 Definitions of Supply Chain Disruptions

Disruption in the supply chain is defined as follows:

- Any event interrupts the flow of materials in the supply chain and leads to a sudden stop of the product flow [56].
- *Disturbance* can be defined as a sudden quantitative or qualitative deviation from the normal situation [5].
- Factors that disrupt the flow of materials in the supply chain and cause a sudden stop in the movement of the product to reach the final customer [18, 45].
- Disruption equals any stoppage in production [17].
- Supply chain disruptions are planned and unanticipated events that disrupt the flow of materials or products within the chain) [16, 23].

Rice and Caniato [40] and Xanthopoulos et al. [57] have identified six essential disruptions in the supply chain, which are mentioned below:

- Supply disruption
- Disruption in transportation
- Facility disruption
- Communication disorder
- Demand disruption
- Disruption of transfer agreements.

Also, [7] point to sixteen events that disrupt the supply chain and categorize these events into four main categories:

1. Natural events (any phenomenon such as volcanoes, storms, torrential rains);
2. Operational events (events that are caused by companies' operations, such as supplier delays, product defects, Etc.);
3. Intentional fake accidents and events (events that are caused by intentional accidents, such as factory explosions, supplier strikes, terrorist attacks, fires, Etc.);
4. Financial events (events caused by financial problems such as financial crises or supplier bankruptcy).

The following examples of these events and crises in the supply chain are described.

5.3 Examples of Crises (Disruptions) in the Supply Chain

- The 1995 earthquake in Kobe, Japan, destroyed a large area of transmission links in this city and one of the most critical global transportation centers [44].
- When the 1999 earthquake in Taiwan disrupted Dell's suppliers, they could keep their customers satisfied by considering policies such as dynamic pricing and low-cost system upgrades [43].
- A fire in March 2000 at the Philips semiconductor manufacturing plant in Albuquerque, New Mexico, caused one of the company's main customers, Ericsson, to lose $400 million in potential revenue. At the same time, another customer of this company (Nokia) was able to reduce the effect of this disruption by allocating reserve suppliers [12].
- The terrorist incident of September 11 caused the cancellation of flights in a few days. These delays caused the idleness of several production lines of the Ford Company due to the lack of components from suppliers [45, 28].
- Apple lost many customers due to the lack of DRAM chips due to the 1999 earthquake in Taiwan [30].
- The Motorola mobile phone company in Singapore was closed due to the SARS outbreak [30].
- North American blackouts on August 14, 2003, negatively affected many businesses, and a fire at one of Ericsson's second-tier suppliers caused severe problems for the company [24].
- Reference [16] examined the disruptions in the Wall Street Journal during the 1990s. They showed disruptions in these organizations' capital and stock performance, as well as their operational performance (cost, sales, and profit), a significant negative impact.
- Winter storms in China in 2008 and 2010 disrupted vegetable and coal supply chains [27].
- The 2011 earthquake in Japan disrupted Toyota's twelve assembly lines, which reduced the production of 14,000 vehicles, and many factories in this country faced a shortage of raw materials, fuel, and energy [4].
- The COVID-19 pandemic.

In recent decades, some extraordinary disease outbreaks have confused business enterprises and created significant challenges for business processes. The scope of such challenges largely depends on the severity of the epidemic in question. In general, any pandemic, reducing efficiency and performance quality and disrupting supply chains (known as ripple effects), hurts flexibility and sustainability and significantly on businesses and supply chains [58].

Statistics published by the World Health Organization (WHO) show that 1,438 epidemics occurred during seven years, from 2011 to 2018 [19]. However, the current COVID-19 pandemic is an exceptional case with far more severe, diverse, and dynamic effects compared to the SARS epidemic (in 2003) or the H1N1 epidemic (in 2009) [25]. Corona disease, abbreviated as COVID-19. It appeared in December

2019 and was declared a pandemic by the Director-General of WHO on March 11, 2020 [20].

The COVID-19 crisis is having a devastating impact on the energy sector globally. Production and supply chains have been disrupted by response measures such as widespread quarantines, reduced demand for goods and services, reduced commodity prices, and a significant economic contraction worldwide. In addition to the health crisis, the epidemic caused several people to lose their jobs and threatened their livelihoods [58].

Furthermore, it was only the COVID-19 pandemic that could affect all nodes (supply chain members) and edges (links) in a supply chain simultaneously [14]. The virus led to a significant disruption in the flow of the supply chain. For example, the pandemic increased demand for essential items such as dry and canned foods, personal protective equipment, and ventilators. In the meantime, the capacity of vital sectors such as production, supply, and transportation decreased drastically. This situation was due to the closing of the borders, the closure of the supply market, the lack of labor, the interruption in the traffic of cars and international trade, the requirement to maintain physical distance in production facilities, Etc. [1].

Therefore, due to such multi-dimensional effects on supply chains and many other financial/economic challenges, Covid-19 has severely affected international trade worldwide. For example, the World Trade Organization (WTO) reported a 13–32% decrease in global trade in 2020 due to the pandemic [58]. The crisis has also shocked the renewable energy sector, shutting down many factories in China, accounting for 50 percent of the global wind energy supply chain. The problem has led to a worldwide "ripple effect" that has slowed the pace of renewable energy deployment in several parts of the world (International Renewable Energy Agency and Post-COVID19, 2020). The massive impact of COVID-19 on various sectors of the renewable energy value chain was reported by the International Renewable Energy Agency (IRENA) in 2020. This report notes that the pandemic has significantly impacted manufacturing, installation, procurement, transportation, and logistics.

On the other hand, only a little impact on project planning, operation, and maintenance were reported. In particular, concerning the solar energy sector, the materials and components used to make solar arrays and panels have slowed down. This is because most manufacturing companies operate in countries seriously affected by the epidemic, such as China, Vietnam, South Korea, Malaysia, Singapore, and Thailand [53]. China was under an almost complete blockade, prohibiting importing and exporting goods. In addition, India's renewable energy sector has also been seriously affected as it imports about 80% of PV modules from China [37]. The pandemic also disrupted the wind industry in India. (For example, lack of financing and supply chain barriers) caused significant companies such as Siemens Gamesa, Vestas, and LM Wind Power (the three main competitors in the wind energy market) to stop production. This caused a delay in producing 600 MW of wind power until 2022 [37].

Today, supply chains are undergoing many changes that have led to their greater complexity, which can be mentioned as factors such as the Globalization of businesses, acceptance of some business philosophies such as lean, effective response to customers, and quick response programs. Implementing these philosophies and practices may create new problems and issues and make the supply chain more vulnerable to disruptions [25, 34 48].

Also, manufacturing institutions must manage their supply chains efficiently to increase efficiency and flexibility in a business environment with complex characteristics and uncertainties. One of the goals of chain management is to reduce the adverse effects of external disturbances and manage specific risks within the chain [49]. According to Hishamuddin et al. in 2013, the nature and complexity of today's supply chain have made them more vulnerable to various risks. These risks may be included in different terms, including disruption, uncertainty, and Disturbance [18].

Disruption in supply chains has undoubtedly attracted more attention after the September 11, 2001, incident, while supply chains have always faced issues related to identifying their vulnerabilities and managing them. Disruptions in the supply chain have different types and may be caused by internal or external factors of the supply chain. Disruptions in the supply chain are costly, so it is necessary to understand how disruption affects the supply chain so that we can adopt an appropriate strategy to deal with it. Reference [8] believe that since today's organizations operate in a global environment, academic experts and practitioners understand that supply chain disruption is one of the most critical risks in the supply chain. Therefore, the disruption issue and its vulnerability to the supply chain will be addressed in the following.

5.4 Vulnerability of the Supply Chain and Its Aggravating Factors Against Disruption

Disruption in the supply chain causes much damage. Still, it seems that the supply chain's readiness to be damaged by such a situation is also essential, called the "vulnerability of the supply chain." There are many definitions and approaches in the field of supply chain vulnerability, which are discussed below:

- Reference [9] have defined *supply chain vulnerability* as "the emergence of a severe disruption." They have also mentioned that in the literature related to crisis management and natural hazards, *vulnerability* is defined as the capacity of an individual or group to predict, cope with and sustain after the effect of a natural threat. Some supply chain researchers have defined *vulnerability* as readiness and the ability to suffer losses due to organizational conditions. The vulnerability of the supply chain is a function of the specific characteristics of the supply chain. In other words, the supply chain's vulnerability is a function of the particular characteristics of the supply chain that causes damage and loss to the supply chains and companies.
- The conceptual basis of vulnerability evaluates the suddenness of events [36].
- *Supply chain vulnerability* can be defined as "facing a serious disruption caused by supply chain risks" that affect the ability of the supply chain to estimate the needs

of the final customer. According to this statement, to reduce these vulnerabilities, organizations must determine and manage their own internal risk and the risk arising from their collaborations and links with other companies [21].

• The inability of the supply chain, at one moment, to react to disturbances and thus achieve its goals can be defined as the supply chain's vulnerability [3].

Many researchers also pointed out how the specific characteristics of a supply chain affect the increase or decrease of its vulnerability. Some hypotheses increase the supply chain's vulnerability due to customer dependence, dependence on suppliers, focus on the supplier, having a single source, and a global source [36].

The following are the most critical factors that increase the probability of disruption and thus increase the vulnerability of the supply chain [16, 48, 36, 44]:

• **Competitive environment**

Businesses today are more competitive than ever. Factors such as runaway demands, increased demand for personalization, increased variety of products, and short product life cycles intensify competition. These conditions make it very challenging to coordinate supply and demand. So, organizations face increasing problems with forecasting.

• **Increasing complexity**

The supply chain complexity is becoming increasingly complex due to the existence of globalized resources and the management of many partners. This complexity makes it more challenging to coordinate supply and demand, increasing the risk of disruption. This risk is more pronounced when each member seeks Local Optimization, and the cooperation between partners is low, and on the other hand, supply chain flexibility is lacking.

• **Outsourcing and participation**

Increasing outsourcing and participation increases internal dependencies, and disorders created in one member are easily transferred to another. Despite the benefits of this partnership and outsourcing, to achieve these benefits, the members must cooperate, share information and programs, and the members' operations are visible to each other. Such changes require significant investments in communication systems, changes in performance versions, commitment to sharing findings, and building trust among chain members, which are not easy to achieve.

• **Limited Buffers**

Focusing on reducing inventory, increasing capacity, and reducing supply chain redundancies are strongly related to the possibility of error. On-time deliveries and zero inventory are among the goals that are often included among organizations' goals, even though these strategies can make the supply chain fragile (inflexible) and ultimately more vulnerable.

• **Focus on efficiency**

Supply chains focus a lot on increasing efficiency (reducing cost), and the cost of this increased efficiency is increasing the risk of disruption. Many organizations do not pay attention to the relationship between efficiency and risk.

- **Over aggregation of operations**

Organizations tend to over-concentrate their operations in one location to gain advantages from economical production quantities, volume discounts, and fewer interactions concerning customers and suppliers. These actions reduce the flexibility of the supply chain to respond to changes in the environment and lead to the supply chain being susceptible to disruptions and more vulnerable to the supply chain.

- **Dependence on customer and supplier**

Dependency is a characteristic of buyer–supplier relationships. Supplier dependence is seen when an organization's resources come from one (or more) suppliers when there are only a few alternative sources. In such a situation, the buyer's company is vulnerable because it has few possibilities to maneuver, while the supplier dominates the relationship, and the power is in his hands. In the event of a supply disruption, the purchasing company will experience significant difficulties in replacing the supply with an unexpected source. The importance of the purchased items significantly intensifies the disruption. Customer dependency is similar to supplier dependency transferred to downstream relationships. The primary company is dependent on some customers, which may be due to their high sales volume. In the event of a disruption, the company may have to bear a more significant share of the effects due to customer dominance.

- **Being a centralized or single-source supplier**

A concentrated supplier is a situation where the buying organization has only a small number of suppliers. Reducing the number of suppliers has many benefits, such as improving product quality or communication. However, the reliance on external sources and the reduced number of suppliers affects the supply chain's vulnerability. However, when an organization focuses on a small number of suppliers, it loses the ability to replace potential future suppliers in the event of a disruption.

Single-source strategies reduce the purchase price and handling costs but make the supply chain highly vulnerable if that source is unable to deliver on time.

A common strategy to protect against the results of a sudden shortage in supply is to vary the number of supplies. Therefore, the rational way to reduce vulnerability to supply failures is to develop possible sources. Having several competing suppliers, the buying company can change the amount and number of orders when a particular supplier dies.

- **Relying on global resources**

The benefits of global sourcing are highly dependent on the parameters of the geographic location of the supplier, the product purchased, or the type of transportation. Compared to local resources, global resources are usually associated with increased uncertainty and reduced transparency. Also, the more complicated factors of longer delivery times due to long transportation routes, reliance on yard infrastructures (ports and communication systems), taxes, customs duties, and exchange rate fluctuations should be investigated. In general, having global resources causes more complexity in the structure of the supply chain.

5.5 Identification of Supply Chain Risks

In most cases, risk and uncertainty are assumed to be the same, but the risk results from uncertainty. Risk in a supply chain is the possibility of potential changes in interactions that affect the reduced added value of each member of a chain. Therefore, to identify and control it, we must know the meaning of risk and risk management, the types of risk, and the sources from which the risk originates, which are discussed below.

5.5.1 Risk Definition and Supply Chain Risk Management

There are different definitions of risk in the literature. According to Merriam-Webster's culture, *the risk* is the possibility of loss or damage or the chance that an investment will lose its value [11]. According to the definition of the Association of Project Managers, a risk is an uncertain event or series of events that, if they occur, will affect the achievement of one or several goals [52]. Waters also believes that risk arises due to uncertainty about the future [54]. Also, according to the definition of [59], supply chain risk is the risk that is included in the supply chain and affects customers.

Therefore, risk management should evaluate the risk in organizational communication and within the supply chain to reduce the vulnerability of the supply chain by managing it effectively [48]. Therefore, risk management can be defined as follows:

Supply chain risk management means the management of external risks and supply chain risks through a coordinated approach between the members of the supply chain to reduce the vulnerability of the supply chain as a whole. From this definition, it is concluded that the risks should not necessarily be the risks between the members of the supply chain but can be the management of supply chain risks within a company. Supply chain risk management deals with the identification of types of risks that can lead to interruptions and disruptions in the supply chain, and its purpose is to prevent interruptions inside or outside the supply chain that can lead to adverse effects on the entire supply chain [11].

Also, *supply chain risk management* can be defined as a management activity to identify and manage risks for the supply chain through a collaborative approach among supply chain members to reduce the vulnerability of the supply chain. Therefore, supply chain risk management (SCRM) includes the same multiplicity of business processes and tasks as supply chain management [21].

Also, supply chain risk management can be considered from two operational and strategic perspectives. From an operating point of view, management Risk assumes the risks associated with traditional purchasing tasks expressed as production support. Moreover, from a strategic point of view, risk management requires the development of strategies, approaches, and methods to identify the benefits of strategic risk management and opportunities resulting from timely responses to risks and hazards.

Therefore, risk management is a vital part of supply chain management. According to the different objectives of the supply chain, if the risk is a multifaceted phenomenon, it can be used. Factors such as uncertainty in supply and demand, globalization of markets, shortening of technology and product life cycles, and increasing use of outsourcing have been listed as reasons for the importance of SCRM [10]. The supply chain is considered a problematic issue:

- Identifying risks is complicated because there are two-way interactions.
- Risks can occur in any part of the supply chain.
- There are a few well-defined techniques and tools for SCRM.

5.5.2 Different Forms of Supply Chain Risks

Risk in the supply chain has been presented in different forms in the literature. Below is a comprehensive reference to the categories made in the field of Supply Chain Risk:

1. **Reference [24] divided supply chain risk types into two categories**:
   (a) The risk of coordination between demand and supply.
   (b) Risk of disruption.

2. **Reference [8] classified supply chain risk into nine categories**:
   (a) The risk of disorders includes Natural disasters, employee strikes, supplier bankruptcy, war, and terrorism.
   (b) The risk of delays includes Inflexibility of supply sources, inflexibility, and low quality of the supplier.
   (c) Systems risk includes Infrastructure problems, product variety, seasonality, and short life cycle.
   (d) Predictive risk includes wrong predictions of the leather whip effect.
   (e) The characteristic intellectual risk or virtual assets, including Vertical integration of supply chain, outsourcing, and global trade.
   (f) Procurement risks, including currency exchange rate risk.
   (g) Acquired and receivable risk includes the Number of customers.
   (h) Inventory risk includes Inventory holding cost and uncertainty in supply and demand.
   (i) Capacity risk includes Capacity cost.

3. **Reference [48] divided supply chain risks into two categories**:

    (a) Operational risks: These risks include things that have inherent uncertainty, such as customer demand, supply, and prices.
    (b) Disruption risks: includes natural disasters such as floods and earthquakes and factors such as war, terrorism, economic crises, Fluctuations in the national currency, Etc.

4. **Reference [55] divided the sources of risk in the supply chain into two general categories**:

    (a) Risks within the chain (which can be controlled) include delays in delivery of customer orders, excess inventory, poor forecasts, human errors, and Malfunctions in IT systems.
    (b) Risks caused by external chain factors (which cannot be controlled) include Earthquake, war, storm, increase in prices, and …

5. **Also, Waters introduced risks in three general categories in another category**:

    (a) Environmental risk: this includes the uncertainties caused by the interactions of the supply chain with the environment, such as terrorist attacks, wars, earthquakes, etc.
    (b) Organizational risk: It is caused by factors within the supply chain, such as workers' strikes, production uncertainties, such as machinery breakdowns, and uncertainty in information technology systems.

6. **Reference [6] divided supply chain risks into 14 general categories, and each class has internal and external risks**:

    (a) Disorders/diseases
    (b) Logistics
    (c) Dependence on the supplier
    (d) Quality
    (e) Information systems
    (f) Prediction
    (g) Mental assets
    (h) Procurement
    (i) Law
    (j) Receivables
    (k) Capacity
    (l) Management
    (m) Security
    (n) Inventory.

7. **Reference [50] divided supply chain risks into two categories**:

    (a) Internal risk includes:

- Market turbulence, such as new products, price sensitivity, level of competition, and fluctuating demand.
- Technological turbulence, such as rapid changes, is considered a high-level technological impact.

(b) External risk also includes:

- Continuous risks such as interest rate, commodity price
- Discrete risks such as terrorism, disasters, strikes.

8. **Reference [35] divide supply chain risks into external and internal categories**:

(a) External risks include:

- Natural disasters such as floods, earthquakes, fires,
- A political system such as war, terrorism, traditions, and laws.
- Competitors and the market, such as price fluctuations, economic recession, volatility of consumer demand, the substitution of options.

(b) Internal risks include:

- Available capacity such as capacity cost, structural capacity.
- Internal operations such as error prediction, safety, whiplash effect, just-in-time delivery.
- Information systems such as information system failure, distorted information, hackers, and viruses.

9. **Reference [33] divided supply chain risks into two categories**:

(a) The natural environment includes Flood, earthquakes, fire, lightning, storm, blizzard, and volcanic explosion.
(b) The risks of the social environment include the risk of a single supplier (the existence of a supplier), the risk of information technology technologies, and the risk of the institution's culture.

10. **Reference [29] also divided supply chain risks into external and internal categories**:

(a) External risks: Includes policy, supply, and shipping.
(b) Internal risks include Program, information, organization structure, production, and research and development.

11. **Reference [46] divided the risk in the supply chain into four categories**:

(a) Supply risks include:

- Working with inappropriate suppliers
- Density and transport capacity
- Customs clearance in ports.

(b) Operational risks include:

- High cost of transportation
- Service quality, including responsiveness and delivery performance
- Supplier obligations and inventory maintenance cost
- Poor product quality
- Human resources
- Failure of information technology.

(c) Demand risk includes:

- Excessive customer demand
- The volatility of customer demand.

(d) Environmental risks include:

- Economic risk
- Natural disasters
- Law and government risk
- Identify the risk of social uncertainty.

12. **Also, the risk associated with the supply chain network can be divided as follows: These risks, caused by the interaction of organizations in a supply chain, can be divided into two groups** [22]:

(a) Systematic risks: These risks are related to unavoidable environmental factors such as disruptions in supply and demand, disciplinary, legal, bureaucratic changes, catastrophic events, and infrastructural disruptions.
(b) Unsystematic risks are factors that can be controlled mainly by the company, such as disruptions in production system facilities.

13. **Also, based on thematic literature and according to the study of sources [23, 34, 8], risks can be divided into two general categories:**

(a) Risks with high probability and low impact (inherent and frequent risks)
(b) Risks with low probability and high impact (destructive and rare risks).

14. **Also, based on the study of other sources (Chopra and Meindel 2007), [13, 32], the types of supply chain risks can be divided as follows:**

(a) Supplier Risks:

- Delivery errors
- Return of materials
- Timely supply of raw materials
- Lack of direct communication between suppliers and customers
- Quality of raw materials
- Increase in the price of raw materials
- Inability to fulfill demands
- Due to wrong forecasting and seasonality, and the short life period of the product
- Lagging behind rapid changes in technology

- Conditions of competitors
- Inadequate transportation of materials
- Information technology problems
- Insufficient inventory in the warehouse
- Supplier's bankruptcy
- Environmental factors (sanctions and labor strikes, war and terrorism)
- Producer Risks:
- Quality of raw materials
- Technology transfer
- Changing product design and engineering
- Changing the life wheel of the product
- Mistakes in production planning
- Improper production control
- Inventory (obsolescence and maintenance and going out of fashion)
- Environmental risk (such as legal laws and government policies and taxes and economic developments and sanctions)
- Dependence on a supplier
- The inflexibility of the supplier
- Environmental problems
- Financial ability of customers

(b)   Distributor Risks:

- Incorrect forecasting of demand (change in demand over time, from one market to another and from one product to another)
- Market share
- Expected product quality
- International supply rules and regulations
- Product price changes
- Receiving outstanding claims
- Product return by the customer
- Transportation risk
- Environmental factors (war, disease, environmental problems, Etc.)

(c)   End customer Risks:

- Reasonable price
- After-sales service
- Quality assurance
- Timely delivery.

These items are shown in Fig. (5.1):

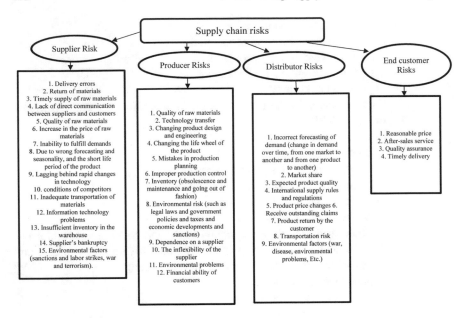

Fig. 5.1 Supply chain risks

5.5.3 Supply Chain Risk Management Processes

Tummala and Schoenherr [51] presented a comprehensive and interconnected approach to risk management in the supply chain. They concluded that risks could be effectively managed by using supply chain risk management processes. They consider three phases for risk management:

1. The first phase includes the steps of risk identification, measurement, and evaluation: In the identification phase, it deals with the complete and structured determination of the potential risks of the supply chain. In the risk measurement stage, it determines the consequences of risks, and in the evaluation stage, it determines the probability of each risk. Methods such as the Delphi, parameter estimation, point estimation, probabilistic coding, and Monte Carlo simulation are used at this stage.
2. The second phase includes risk assessment, mitigation, and contingency plans: The measurement phase deals with risk ranking and acceptance, and mitigation provides solutions to risk.
3. Finally, the third phase includes risk control and monitoring: The third phase deals with corrective measures in coping with deviations in achieving the optimal performance of the supply chain and preparing guidelines for future improvements.

Therefore, supply chain risk management processes are as shown in Fig. (5.2):

Fig. 5.2 Supply chain risk
management processes

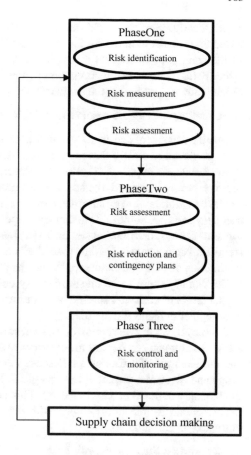

5.5.4   Disruption Management Solutions in the Supply Chain

One of the critical areas of literature is specific to actions related to dealing with disorders. Supply chain issues can be divided into two main groups [48]:

1. Supply management: Supply management issues include supplier selection, supplier relationships, supply planning, transportation, logistics, Etc.
2. Demand management: Demand management issues include new product introduction, product line management, demand planning, product pricing, promotion planning, Etc.

In this regard, [48] has introduced nine solid strategies to deal with supply chain disruptions, which are mentioned below:

1. **Postponement strategy**

A Postponement strategy uses the product or process design concepts such as standardization, commonality, modular design, and operation reversal to delay the point of product differentiation. This strategy enables a firm to produce a generic product based on the total demand for all products and later customize the generic product. In the context of disruption recovery, the postponement strategy provides a cost-effective and time-efficient contingency plan that allows the supply chain to quickly reconfigure the product in the event of a supply disruption.

2. Creating a strategic reserve (strategic stock)

Before the concept of just-in-time manufacturing (JIT), excess inventories of products were produced to ensure that the supply chain could continue to operate smoothly in the event of a supply disruption. However, as product life cycles shortened and product variety increased, the costs of inventory holding, and obsolescence of these excess inventories were prohibitive. However, instead of carrying more safety stock, a company can store some of its inventory in specific "strategic" locations (warehouses, logistics centers, distribution centers) and share it with several supply chain partners (retailers, repair centers, Etc.) be placed. For example, Toyota and Sears keep specific inventories of cars and home appliances in particular locations so that all retailers in the vicinity share these inventories. By doing so, Toyota and Sears can achieve higher levels of customer service without incurring high inventory costs when faced with regular demand fluctuations.

Moreover, when a disruption occurs, these shared inventories in strategic locations allow the company to deploy these strategic stocks to the affected area quickly. As another example, the Center for Disease Control (CDC) stores large quantities of drugs and medical equipment in strategic locations in the United States, known as the strategic national stockpile (SNS). This strategic stockpile is helpful to protect the American people in a public health emergency (such as a terrorist attack, flu outbreak, or Earthquake).

3. Creating a Flexible supply base

Although sourcing from a single supplier enables the company to reduce costs (lower supply management costs, lower unit costs due to quantity discounts, Etc.), it can create problems for managing inherent demand fluctuations or significant disruptions. For example, to reduce the risks associated with sole sourcing, HP explained how they use their factories in Washington and Singapore as supply bases to produce inkjet printers. HP used the Singapore plant for base volume production to handle regular demand fluctuations and the Washington plant for excess base volume production. A flexible supply base enables the company to manage changes in demand regularly. Still, it can also be used to maintain a continuous supply of materials in the event of significant disruption. As another example, Li and Fungs have established a network of 4000 suppliers. This Li and Fung's 4000-supplier network offers Li and Fung great flexibility to quickly change production among suppliers in different countries, mainly when a disruption occurs in a particular country.

4. **The strategy of combining production and outsourcing (Make-and-buy strategy) (such as product classification)**

In case of possible supply disruptions, the supply chain is more flexible to outsource other automotive products to different suppliers if it produces certain products internally. For example, HP manufactured part of its DeskJet printers in its Singapore factory and outsourced the rest of its production to a contract manufacturer in Malaysia. In addition, Brooks Brothers and Zara manufacture their fashion items in their domestic factories and outsource other essential things to their suppliers in China. This make-and-buy strategy offers flexibility that allows companies to change production during supply disruptions quickly.

5. **Proposing economic incentives for suppliers to increase the number of suppliers (Economic supply incentives)**

In many cases, the buyer cannot switch production among different suppliers due to the limited number of suppliers in the market. To gain the flexibility to shift production among suppliers, the buyer can provide specific economic incentives to cultivate additional suppliers. For example, due to the uncertainty of producing a particular flu vaccine formulation each year, uncertain market demand, and price pressure from the US government, many flu vaccine manufacturers, including Wyeth Pharmaceuticals, have exited the market. A decline in the number of flu vaccine manufacturers has put many Americans at risk. In October 2004, Chiron, one of two remaining vaccine manufacturers for the US market, was suspended due to bacterial contamination at Chiron's Liverpool plant. With a shortage of 48 million flu shots from Chiron, the US government was initially able to provide flu shots only to people in high-risk groups. To prevent this type of failure in the future, the US government has put in place special economic incentives to encourage more suppliers to re-enter the flu vaccine market. For example, the government shared some financial risks with suppliers, buying a certain amount of flu vaccine at a specific price and buying back the unsold stock at a lower price at the end of the flu season. So, with more potential suppliers, the US government will have the flexibility to quickly change its orders from different suppliers in the event of significant disruptions. Even without major disruptions, economical supply incentives can be beneficial. As another example, when Intercon Japan became more concerned about the "monopoly" mindset of its leading supplier, it offered economic incentives to attract a new supplier, Nagoya Steel, to develop a new steel process technology to produce different types of cable joints. These incentives included minimum order quantities, technical advice on new steel process technology, and information on market demand for this new process technology to make Nagoya Steel competitive. By establishing additional suppliers that used different process technologies, Intercon Japan maintained pressure on both suppliers to keep costs down.

6. **Strategy of using Flexible transportation**

In supply chain management, transportation can be the Achilles heel that disrupts the supply chain. As such, increasing flexibility should be considered proactively. Here are three basic approaches to doing this:

(a) **Multi-modal transportation**

To prevent supply chain operations from being interrupted when disruptions occur in the ocean, air, road, Etc., some companies use a flexible logistics strategy that relies on multiple modes of transportation. For example, Japan's Seven-Eleven requires its logistics partner to diversify its mode of transportation, which includes trucks, motorcycles, bicycles, ships, and helicopters. This flexible logistics strategy won the hearts of many Japanese people, as Seven-Eleven Japan was able to use 125 motorcycles and seven helicopters to quickly deliver 64,000 rice balls to earthquake victims in Kobe shortly after an earthquake that destroyed many roads in the late 1980s helped.

(b) **Multi-carrier transportation**

To ensure the continuous flow of materials in case of political disturbances (landing rights, labor strikes, Etc.), various air transport companies such as Aeroméxico Cargo, KLM Cargo, Delta Air Logistics, Air France Cargo, CSA Czech Airline Cargo, Korean Air Cargo Etc. have formed an alliance called SkyTeam Cargo. It allows them to change carriers in the event of political disruption quickly. In addition, the coalition has enabled SkyTeam Cargo to offer low-cost global delivery to 500 destinations in 110 countries.

(c) **Multiple routes**

Companies are considering alternative ways to ensure the smooth flow of materials along the supply chain to avoid a complete shutdown. For example, due to long delays at West Coast ports and heavy traffic along the various West Coast freeways, some East Coast shipping companies are encouraging shipping companies to develop new routes in addition to the traditional way (i.e., ocean shipping from Asia to the west coast and then rail transport from the west coast to the east coast). In particular, after the closure of West Coast ports for two weeks in 2002, some shippers considered shipping various manufactured goods from Asia to East Coast ports through the Panama Canal.

7. **Revenue management via dynamic pricing and promotion**

Dynamic pricing is a common mechanism for selling perishable products/services. For example, when selling limited seats on an airplane with uncertain demand, airlines constantly adjust their ticket prices to meet uncertain demand with limited supply. This revenue management has generated "approximately $1 billion in incremental annual revenue" at American Airlines through dynamic pricing. Revenue management through dynamic pricing and promotions can also be an effective way to manage demand when the supply of a particular product is disrupted. In fact, a retailer can use a pricing mechanism to induce customers to choose widely available products. For example, when Dell faced supply disruptions from its Taiwanese suppliers after the 1999 earthquake, it immediately responded by offering customers special "low-cost upgrade" options if they chose the same computers with parts from other suppliers. Dell implemented the emergency strategy. This dynamic pricing and promotion strategy enabled Dell to satisfy its customers during the supply crisis.

8. Assortment planning

Brick-and-mortar retailers have used assortment planning (the assortment of products displayed, the location of each product on the shelves, and the number of finishes per product) to influence consumer product choice and customer demand. Studies conducted in five supermarkets in the United States showed that a store manager could manipulate product selection and customer demand by rearranging the assortment of products displayed, the location of each product on the shelves, and the number of covers for each product. Their findings suggest that assortment planning can be used to induce customers to buy products that are widely available when certain products face supply disruptions.

9. Silent product rollover

Under the silent product movement strategy, new products are slowly "leaked" into the market without any formal announcement. As such, customers are not fully aware of the unique features of any special product and are more likely to choose existing products over those that are out of stock or being phased out. For example, since Swatch only produces each watch model once, Swatch uses a silent product rotation strategy to launch new watches so that its customers see all existing Swatch watches as collectibles. Using the same approach as Swatch, Zara is quietly launching its latest fashion collection. Since Zara usually does not repeat the production run for the same design of clothes, many of Zara's fashion-conscious customers buy the clothes in their stores immediately. Finally, all products are essentially "interchangeable" at Swatch and Zara. Interchangeable products are very desirable to handle fluctuations in demand under normal conditions and when there is a disruption in supply or demand.

While these nine powerful strategies are beneficial under normal circumstances and during a significant disruption, they also present the following challenges:

- **Costs versus benefits**

Some companies may express concern about the necessary costs associated with these robust strategies, while others recognize the additional benefits. At a conceptual level, these powerful strategies improve a firm's competitive position, especially when other firms' supply chains are more vulnerable to disruptions. However, the value of competition is difficult to quantify. Theoretically, the costs of implementing these proactive strategies can be considered as an "insurance premium" that protects the supply chain from significant disruptions [44]. However, it is difficult to assess the return of these premiums, especially in the absence of reliable data (probability of disruption, potential loss due to disruption, etc.).

- **Strategic fit**

Although these robust strategies enhance the company's ability to manage supply and demand better, they may not align with its overall business strategy. For example, suppose a company chooses to reduce product variety to rationalize its product lines, then the value of the postponement strategy decreases. Second, if the retailer has positioned itself as an "everyday low price" store, its dynamic pricing and promotion strategy is inconsistent with its strategic market position.

- **Proactive execution**

A strong strategy is useless unless a company can playfully implement the plan. For example, when Longshoreman's contract was up for renewal in 2002, NUMMI (a joint venture between Toyota and General Motors), Ralph Lauren, and Tommy Hilfiger developed various alternative transportation programs. As the Coastal Workers Union and the Port Authority disagreed over the labor contract, Ralph Lauren and Tommy Hilfiger proactively implemented contingency plans by rerouting their shipments to the East Coast. However, NUMMI used six days of extra inventory instead of an alternative route. Unfortunately, when NUMMI ran out of stock, truckers could still not unload parts from the West Coast port and deliver them to the NUMMI plant. By then, it was too late for NUMMI to reroute its shipments. As a result, NUMMI had to shut down for a few days.

Therefore, managers should be able to analyze destructive events and their adverse effects on the supply chain to determine the most effective way to deal with disruptions. Supply chain disruptions are costly, and managers must take appropriate measures to reduce their adverse effects. Disturbance management is one of the fields of study in this field, which has recently received much attention from researchers. The most crucial goal of disruption management is to choose and implement the most appropriate strategy to restore the supply chain to a normal state by minimizing related costs [39]. Therefore, due to the specific characteristics of disorders, their management in the supply chain requires special conditions and patterns. To be more precise, supply chain disruption management strategies can be divided into two groups:

- Preventive Strategies
- Restorative Strategies.

In the group of preventive strategies, the discussed strategies are usually introduced with keywords such as increasing chain security and inventory management. In some cases, the expressions of resilience and agility have also been used. In summary, preventive strategies have been divided into four categories by researchers [45, 16, 40, 48, 8, 44, 50–42]:

(A) **A strategy to promote stability to deal with disruption**

These strategies have two characteristics:

- Attention to efficiency: this strategy must be able to manage the chain efficiently in any situation.
- Attention to resilience: this strategy should be able to keep the chain stable during disruption and shorten the recovery time as much as possible.

When a strategy is stable, the organization can perform analyzes such as cost, profit, rate of return on investment, etc. to improve efficiency under risk.

(B) Supply chain resilience strategy to deal with disruption

There are boosters for supply chain resilience. The concept of resilience often refers to a feature of the organization that gives the organization the ability to react quickly to disruption. In addition to maintaining the standard conditions of operations, elasticity is also a competitive ability for the organization. Especially in situations where competitors do not have this characteristic, it provides the possibility of increasing market share and attracting customers to competing organizations. Even when organizations are equally affected by the disruption, they will compete for resilience.

Among the various methods of improving elasticity, flexibility and doubling are the best solutions; each of these two methods has different characteristics and cost burdens in supply chain design. Resilience means creating capabilities to respond in the supply chain, developed before disruptions occur through investment in infrastructure and resources. Among the flexibility requirements, we can mention a multi-skilled workforce, production system design with the ability to produce various products and adaptability to changes (flexible production systems), and use of resources with the ability to adapt. If there is flexibility, the organization can use part of the capacity of one department to produce the product of another department. Of course, this issue imposes costs for the design of processes, products, labor, etc. to the chain.

Redundancy means maintaining the extra capacity for disruption conditions, often achieved by investing in excess capacity. Duplication of work involves inventory management, repairs and maintenance of production lines or facilities, material supply contracts, and transportation systems. In its simplest form, doubling means an increase. Within the organization, this increase can take the form of people, capacity, process steps, inventory, Etc. In general, there are two types of double work:

- Rework: Rework is when more than two units do a specific task.
- Overlap: Overlap also means that more than one unit does part of the same work.

From an organization's perspective, redundancy provides a buffer to protect against disruptions through resources preserved somewhere. These resources include safety stock, reserve suppliers, information technology systems, or excess capacity. Despite the potential benefits, the cost of these can be high. At the same time, increasing resources increases the potential for complexity. Regarding costs, flexibility has advantages over redundancy because it uses predetermined capacities.

Therefore, one of the main differences between flexibility and redundancy is that redundancy has a capacity that may not be used. On the other hand, flexibility means reapplying previously committed capacity; In other words, this issue creates a balance between producing different products or serving different customers, which may not be considered double work. Of course, in the end, every organization often uses a combination of the two.

(C) The strategy of improving security in the supply chain to deal with disruption

Organizations seek to increase security in different ways, but these activities can be divided into three main groups: physical security, information security, and fleet security. All organizations do not need to use all the activities related to each group. They perform a group of these activities according to their characteristics and needs.

Organizations without security assurance may be unable to implement timely and lean inventory conditions. Some supply chains are forced to source their needs from local sources at a higher cost. One of the efficient strategies is to apply lessons from successful quality improvement programs. Therefore, it should be possible to use the principles of comprehensive quality management to design and implement supply chain security assurance processes. The significant point in total quality management is that it claims higher quality with lower cost, which can be extended to supply chain security. Using the proper management approaches, new technology, and open-engineered processes, we can achieve higher security at a lower cost.

Nowadays, disruption in the supply chain is not a strange problem. So many chains try to increase their security against these cases by increasing the amount of safety stock, using alternative sources or at a higher cost, and considering additional capacity.

Six Sigma was primarily developed for quality control. Therefore, it can be used for security management. The nature of the supply chain is such that it is vulnerable to disruptions; today, researchers are looking for the design of secure and resilient supply networks, which means that while looking for the security of processes and procedures, they want sufficient resilience to respond to disruptions and return quickly to the usual conditions before the disorder. Considering the existing risks for supply chains, the proper relationship between organizations and the government and coordination with customers and suppliers can become the basis for creating a secure and resilient network. Therefore, the first step in creating a safe and resilient supply network is to understand the difference between the two, and many activities only improve one of these things discussed.

(D) Agility strategy to deal with disruption

The purity of the supply chain can be considered a factor in strengthening the risk of the supply chain. Moreover, the chain's agility has been identified as one of the influential factors in dealing with the risk of disruption. Therefore, with the timely release of information, other members will be informed about the known disorder, and appropriate measures can be taken to mitigate the effects of the disorder. Timely transmission of information increases the supply chain's agility while improving the chain's stability and overall performance.

The second group of discussed strategies includes repair strategies based on the supply chain's flexibility. Among these solutions, we can mention rerouting the demand from the disrupted supplier to another supplier. In this way, an alternative supplier is considered for specific conditions. So that the parts are supplied from them during the leading supplier disruption. Supply rerouting is a viable tactic if the trusted supplier can increase capacity; it means having flexibility in capacity. Supply rerouting is often part of an optimal disruption management strategy and significantly reduces costs for the organization. Among the nine strategies mentioned by [48], such as flexible conditions in suppliers and flexible transportation, are included in this group. Also, Hendricks and Singhal [16] have suggested the following solutions to reduce the possibility of disruption in the supply chain:

(A) Improving and increasing the accuracy of demand forecasts

One of the main reasons for not matching supply and demand is inaccurate forecasts. Organizations in their planning should pay attention to the error of this forecast in addition to the demand forecast. Organizations should be aware that long-term forecasts are more imprecise than short-term forecasts, just as the forecasts of each component individually are more imprecise than the overall forecast. Predictions often go astray if the organization does not dynamically adjust forecasts, does not consider background issues, and ignores factors outside the organization; these issues significantly impact forecasts. Most organizations assume fixed supply and shipping time, capacity, and distribution and shipping routes. These assumptions should be carefully questioned and checked, and changed if necessary. Fixed long-time horizons also make accurate forecasting difficult.

(B) Coordination and integration of planning and implementation

Plans are often separate from operational realities. Programs often change during execution. Managers are responsible for implementing adjustments to respond to environmental changes in programs. Such adjustments are usually made. However, the planners are not informed, which results in the lack of integration of the plan and implementation. With the integration of planning and implementation, many mismatches between supply and demand are resolved.

(C) Reducing the average and variance of supply time

Inaccurate forecasting and a lack of communication between planning and execution can be devastating when supply times are long and variable. Reducing the mean and variance of the supply time can lead to reducing uncertainty in the supply chain. Among the activities that can be used in this field are:

- Elimination of activities and steps without added value.
- Increasing the reliability and stability of production, management, and logistics processes.
- Special attention to critical processes, resources, and materials.
- Attention to dynamic supply time in planning delivery times.
- Cooperation and collaboration with chain members.

Accessing this issue is not an easy task. A leader in this collaboration creates trust among chain members, increases agreement on the benefits of information sharing, and shows a willingness to change from the previous situation. In such a situation, chain members collectively make decisions and solve problems, and information about strategies, plans, and performance is shared among chain members. Such activities reduce the deviation in information and lack of coordination, which are factors for the occurrence of disorders.

(D) **Invest in accessibility**

Organizations must be aware of chain events to reduce the risk of disruptions. These events include internal operations, customers, suppliers, inventory locations, capacity, and critical assets. The following are required for access:

- Identify and select foresight indicators or supply chain performance leadership (suppliers, internal operations, and customers).
- Data collection and analysis of these indicators.
- Determining modeling levels in these indicators.
- Monitoring these indicators compared to modeling.
- Discussing deviations from expected performance with relevant managers.
- Providing and implementing processes to deal with these deviations.
- Creating flexibility in the supply chain.

Organizations must make the right decisions to create flexibility and, as a result, the ability to respond. Depending on environmental conditions, there are different indicators for the flexibility an organization requires.

(E) **Creating flexibility in product design**

Standardization and modularization and the use of routine parts provide the ability to react quickly to disruptions in the delivery of parts.

(F) **Creating flexibility in sourcing**

This issue can be resolved through flexible contracts and the use of local markets for purchase and supply. Local markets can supply parts to meet unpredictable increases in demand and prevent inventory build-up (if demand is lower than anticipated).

(G) **Creating flexibility in production**

This issue can arise through flexibility in capacity, which can focus on other products depending on the demand. Also, they can divide their capacity into two parts, primary and reactive. The essential capacity belongs to products whose demand can be accurately predicted, and the reactive capacity belongs to things whose forecasting is complicated. This category can include goods with a short life cycle or fluctuating demand. Delaying product differentiation can be used as another strategy for production flexibility.

(H) **Delay strategy**

Delayed differentiation is a strategy in which product differentiation is delayed until closer to product demand. This strategy requires the design and production of standardized products quickly and cost-effectively configured and customized as soon as demand is received. Therefore, supply and demand mismatches are minimized. The critical factors for implementing this strategy are:

- Multi-skilled teams are providing design and manufacturing practices.
- Product and process reengineering to increase standardization.
- Modular production.

- Multipurpose parts.
- Performance versions and goals that resolve conflicts and ensure accountability.

(I) Invest in technology

Investing in the right technologies can significantly reduce the chance of disruption. Today, web-based technologies are available to link the supply chain partners' database, enabling tracking inventory, capacity, facility conditions, and demands of different parts of the supply chain, preventing passive (reactive) action. RFID Radio Frequency Identification technology guarantees that the accuracy of inventory accounts will increase, and accurate Information will be provided on time in transportation, Etc.

5.6 The Concept of Reliable Supply Chain

Supply chain management is the field of interest of many researchers in different fields. The supply chain emerged in the 1990s when issues related to material circulation were developed. This category occupied a wide range of articles in magazines and publications; besides, it was a topic of interest to many professors and pioneers. A supply chain generally consists of various activities, including procurement, inventory, sourcing and purchasing, production planning, inter-, intra-organizational relations, and performance measurement [2].

The success of many private, government, and military organizations depends on their ability to deliver acceptable outputs. Providing better products in a wide range and at a low cost and doing it quickly, the optimal presentation of these outputs (cost, quality, performance, delivery, flexibility, and innovation) depends on the organization's ability to manage the flow of materials, information, and money inside and outside the organization. This flow is known as the supply chain. Because supply chains can be long and complex and involve many business partners, problems arise. If these problems are delayed, they lead to customer dissatisfaction and loss of sales, and the organization incurs high costs to solve them. In recent years, with the increasing complexity and expansion of supply chains, the management has faced many uncertainties (due to uncertainty in the chain links) and has been forced to take risks. Finally, achieving successful and reliable supply chain management is impossible without increasing confidence and trust between the members [38].

Therefore, the reliability of a system or supply chain is the probability of satisfactory performance of that system under specific working conditions and for a certain period. Reliability includes four main parts of probability, good performance, time, and certain working conditions, whose probability is expressed with a number, which is the reliability evaluation index. In many cases, this index is the most important.

It can also be said that uncertainty in the supply chain affects the performance of the supply chain. Supply chain uncertainty comes from three sources [47]:

(1) Uncertainty of the supplier is caused by the supplier's inability to meet the needs of the production unit.
(2) Uncertainty of the process that occurs due to the unreliability of the production process. It is due to the failure of the machines.
(3) Uncertainty of demand comes from the inability to predict demand accurately.

In the following, the description of reliability in each of the parts of the supply chain is discussed [38].

5.6.1 Reliability of the Supplier

Suppliers are one of the influencing factors in the organization. One of the ways to produce products with a lower cost and cost is to supply raw materials and required parts cheaper than suppliers. On the other hand, the quality of the product is related to the quality of the raw materials. It is unbreakable, so suppliers' role in the product's quality is significant and thoughtful; finally, it can be said that without timely receipt of raw materials and parts, the product cannot be delivered to the customer on time.

It will be the art of management that can increase the reliability of the supply chain by communicating with more suppliers. However, organizations should also consider the essential point that improving reliability always requires more investment. The general trend of the ratio between cost and reliability with increasing reliability is an incremental process. In other words, less improvement in reliability is obtained by spending a particular investment on higher reliability levels. In any case, achieving high reliability is costly.

5.6.2 Reliability of Producer

The reliability of the production sector is strongly dependent on factors such as flexibility of production lines, design capacity, amount of investment, Etc. In other words, the reliability of a manufacturer with a flexible production line will be higher. At the same time, flexibility within the organization is related to the type of arrangement and planning of the organization. Depending on the serial or parallel structure, the level of flexibility will change, and the flexible organization can adapt to different changes in demand levels. Also, the reliability of this sector is highly dependent on the reliability of suppliers as long as the production department receives the materials it needs in a specific time. With the desired quantity and quality, it can deliver the final product to the distribution department with higher reliability. As a result, the higher the reliability of the suppliers, the higher the manufacturer's reliability will be.

5.6.3 Reliability of the Distributor

The distributor section forms the last link in the chain. Due to its direct relationship with the customer, this department conveys the customer's demands, expectations, and opinions to other chain links (producer and supplier). The more complete and faster this information is provided to other links, the more dynamic the chain will be with higher reliability and flexibility. The reliability of the distributor is the same as that of the supplier. By increasing the number of distributors, customers' demands can be met faster and better, but this increase will lead to increased costs. Finally, like the supplier sector, an optimal situation should be achieved between the number of distributors and the investment cost.

5.6.4 Reliability of the Entire Supply Chain

In the supply chain, reliability depends on the reliability of each part and considering that the links of the supply chain are usually series. Therefore, the reliability of the whole supply chain is minor than each link, so the chain management should try to improve the reliability of each member as much as possible. Increase and increase the reliability of the entire chain by choosing the best suppliers and distributors in the best arrangement and the best manufacturer with high flexibility.

Also, Fault Tree Analysis (FTA) is a valuable tool for reliability evaluation. This technique can identify and control the risk or the possibility of failure in different parts of the system. Fault tree analysis is a logical method to determine the cause contributing to the occurrence of a dangerous and unpleasant event by using logical graphic symbols. The stages of designing a fault tree in summary are:

- Identification of a final adverse event or peak event
- Determining the subsets that can cause the peak event
- Determining the relationship between subsets by drawing a fault tree
- Quantitative analysis of the drawn fault tree
- Estimating the probability of the peak event under the influence of the risks of each of the subcategories.

To quantitatively analyze the drawn fault tree and estimate the probability of occurrence of the vertex event, it is possible to convert the fault tree into a Reliability Block Diagram (RBD) or convert the fault tree into Bayesian networks.

5.7 Conclusion

This chapter has been done to identify supply chain crises and risks. Five primary sections about supply chain crises and disruptions were given in this chapter. In the first part, the definitions of supply chain disorders were discussed. The second part presents examples of problems (disruptions) in the supply chain. In the third part, the vulnerability of the supply chain and its aggravating factors against disruption were discussed. In the fourth part, the supply chain risks were identified, and in the last part, the concepts of reliability in the supply chain were given. So that while familiarizing the readers with the concepts of disruption and supply chain risks, they can also take measures to manage and control risks and reduce the supply chain's vulnerability.

References

1. Amankwah-Amoah, J.: Note: mayday, mayday, mayday! responding to environmental shocks: Insights on global airlines' responses to COVID-19. Transp. Res. Part E: Logistics Transp. Rev. **143**, 102098 (2020)
2. Arshinder, K., Kanda, A., Deshmukh, S.G.: A review on supply chain coordination: coordination mechanisms, managing uncertainty and research directions. Supply Chain Coord. Under Uncertainty 39–82 (2011)
3. Azevedo, S.G., Machado, V.H., Barroso, A.P., Cruz-Machado, V.: Supply chain vulnerability: environment changes and dependencies. Int. J. Logist. Transp. **2**(1), 41–55 (2008)
4. Baghalian, A., Rezapour, S., Farahani, R.Z.: Robust supply chain network design with service level against disruptions and demand uncertainties: a real-life case. Eur. J. Oper. Res. **227**(1), 199–215 (2013)
5. Blackhurst, J., Dunn, K.S., Craighead, C.W.: An empirically derived framework of global supply resiliency. J. Bus. Logist. **32**(4), 374–391 (2011)
6. Blackhurst, J.V., Scheibe, K.P., Johnson, D.J.: Supplier risk assessment and monitoring for the automotive industry. Int. J. Phys. Distrib. Logist. Manag. **38**(2), 143–165 (2008)
7. Carvalho, H., Barroso, A.P., Machado, V.H., Azevedo, S., Cruz-Machado, V.: Supply chain redesign for resilience using simulation. Comput. Ind. Eng. **62**(1), 329–341 (2012)
8. Chopra, S., Sodhi, M.S.: Supply-chain breakdown. MIT Sloan Manag. Rev. **46**(1), 53–61 (2004)
9. Christopher, M., Lee, H.: Mitigating supply chain risk through improved confidence. Int. J. Phys. Distrib. Logist. Manag. **34**(5), 388–396 (2004)
10. Christopher, M., Peck, H.: Building the resilient supply chain. Int. J. Logist. Manag. **15**(2), 1–14 (2004)
11. Dan, W., Zan, Y.: Risk management of global supply chain. In: 2007 IEEE International Conference on Automation and Logistics, pp. 1150–1155. IEEE (August 2007).
12. Friesz, T.L., Kim, T., Kwon, C., Rigdon, M.A.: Approximate network loading and dual-time-scale dynamic user equilibrium. Transp. Res. Part B: Methodol. **45**(1), 176–207 (2011)
13. Goh, M., Lim, J.Y., Meng, F.: A stochastic model for risk management in global supply chain networks. Eur. J. Oper. Res. **182**(1), 164–173 (2007)
14. Gunessee, S., Subramanian, N.: Ambiguity and its coping mechanisms in supply chains lessons from the Covid-19 pandemic and natural disasters. Int. J. Oper. Prod. Manag. (2020)
15. Hale, T., Moberg, C.R.: Improving supply chain disaster preparedness: a decision process for secure site location. Int. J. Phys. Distrib. Logist. Manag. **35**(3), 195–207 (2005)
16. Hendricks, K.B., Singhal, V.R.: The effect of supply chain disruptions on shareholder value. Total Qual. Manag. **19**(7–8), 777–791 (2008)
17. Hishamuddin, H., Sarker, R.A., Essam, D.: A disruption recovery model for a single stage production-inventory system. Eur. J. Oper. Res. **222**(3), 464–473 (2012)
18. Hishamuddin, H., Sarker, R.A., Essam, D.: A recovery model for a two-echelon serial supply chain with consideration of transportation disruption. Comput. Ind. Eng. **64**(2), 552–561 (2013)
19. Hudecheck, M., Sirén, C., Grichnik, D., Wincent, J.: How companies can respond to the coronavirus. In: MIT Sloan Management Review (2020)

20. IRENA.: The post-COVID recovery: an agenda for resilience, development and equality (2020)
21. Jüttner, U.: Supply chain risk management: Understanding the business requirements from a practitioner perspective. Int. J. Logist. Manag. **16**(1), 120–141 (2005)
22. Kar, A.K.: Risk in supply chain management (2010)
23. Kleindorfer, P.R., Saad, G.H.: Managing disruption risks in supply chains. Prod. Oper. Manag. **14**(1), 53–68 (2005)
24. Kleindorfer, P.R., Wu, D.J.: Integrating long-and short-term contracting via business-to-business exchanges for capital-intensive industries. Manage. Sci. **49**(11), 1597–1615 (2003)
25. Koonin, L.M.: Novel coronavirus disease (COVID-19) outbreak: now is the time to refresh pandemic plans. J. Bus. Contin. Emer. Plan. **13**(4), 298–312 (2020)
26. Lee, H.L., Whang, S.: Higher supply chain security with lower cost: lessons from total quality management. Int. J. Prod. Econ. **96**(3), 289–300 (2005)
27. Li, J., Wang, S., Cheng, T.E.: Competition and cooperation in a single-retailer two-supplier supply chain with supply disruption. Int. J. Prod. Econ. **124**(1), 137–150 (2010)
28. Li, Q., Zeng, B., Savachkin, A.: Reliable facility location design under disruptions. Comput. Oper. Res. **40**(4), 901–909 (2013)
29. Lin, Y., Zhou, L.: The impacts of product design changes on supply chain risk: a case study. Int. J. Phys. Distrib. Logist. Manag. **41**(2), 162–186 (2011)
30. Martha, J., Subbakrishna, S.: Targeting a just-in-case supply chain for the inevitable next disaster. Supply Chain Manag. Rev. **6**(5), 18–23 (Sept/Oct 2002). ILL
31. Martin, C., Towill, D.R.: Supply chain migration from lean and functional to agile and customised. Supply Chain Manag.: Int. J. **5**(4), 206–213 (2000)
32. Micheli, G.J., Cagno, E., Zorzini, M.: Supply risk management vs supplier selection to manage the supply risk in the EPC supply chain. Manag. Res. News **31**(11), 846–866 (2008)
33. Mu, J., Wan, Z.: A fuzzy approach for supply chain risk assessment. In: 2010 Seventh International Conference on Fuzzy Systems and Knowledge Discovery, vol. 1, pp. 429–431. IEEE (Aug 2010)
34. Norrman, A., Jansson, U.: Ericsson's proactive supply chain risk management approach after a serious sub-supplier accident. Int. J. Phys. Distrib. Logist. Manag. **34**(5), 434–456 (2004)
35. Olson, D.L., Wu, D.D.: A review of enterprise risk management in supply chain. Kybernetes **39**(5), 694–706 (2010)
36. Peck, H.: Drivers of supply chain vulnerability: an integrated framework. Int. J. Phys. Distrib. Logist. Manag. **35**(4), 210–232 (2005)
37. Pradhan, S., Ghose, D., Shabbiruddin.: Present and future impact of COVID-19 in the renewable energy sector: a case study on India. In: Energy Sources, Part A: Recovery, Utilization, and Environmental Effects, pp. 1–11 (2020)
38. Pryke, S. (ed.).: Construction Supply Chain Management: Concepts and Case Studies, vol. 3. Wiley (2009)
39. Qi, X., Bard, J.F., Yu, G.: Supply chain coordination with demand disruptions. Omega **32**(4), 301–312 (2004)
40. Rice, J.B., Caniato, F.: Building a secure and resilient supply network. Supply Chain Manage. Rev. **7**(5), 22–30 (Sept/Oct 2003). ILL
41. Saghafi, M.M.: Optimal pricing to maximize profits and achieve market-share targets for single-product and multiproduct companies. Issues Pricing: Theory Res. 239–253 (1988)
42. Schmitt, A.J., Singh, M.: A quantitative analysis of disruption risk in a multi-echelon supply chain. Int. J. Prod. Econ. **139**(1), 22–32 (2012)
43. Shao, X.F.: Demand-side reactive strategies for supply disruptions in a multiple-product system. Int. J. Prod. Econ. **136**(1), 241–252 (2012)
44. Sheffi, Y.: Supply chain management under the threat of international terrorism. Int. J. Logist. Manage. **12**(2), 1–11 (2001)
45. Sheffi, Y.: The resilient enterprise: overcoming vulnerability for competitive advantage. Pearson Education India (2007)
46. Sofyalıoğlu, Ç., Kartal, B.: The selection of global supply chain risk management strategies by using fuzzy analytical hierarchy process—a case from Turkey. Procedia Soc. Behav. Sci. **58**, 1448–1457 (2012)

47. Stadtler, H.: Supply chain management and advanced planning—basics, overview and challenges. Eur. J. Oper. Res. **163**(3), 575–588 (2005)
48. Tang, C.S.: Robust strategies for mitigating supply chain disruptions. Int. J. Log. Res. Appl. **9**(1), 33–45 (2006)
49. Thun, J.H., Hoenig, D.: An empirical analysis of supply chain risk management in the German automotive industry. Int. J. Prod. Econ. **131**(1), 242–249 (2011)
50. Trkman, P., McCormack, K.: Supply chain risk in turbulent environments—a conceptual model for managing supply chain network risk. Int. J. Prod. Econ. **119**(2), 247–258 (2009)
51. Tummala, R., Schoenherr, T.: Assessing and managing risks using the supply chain risk management process (SCRMP). Supply Chain Manage.: Int. J. **16**(6), 474–483 (2011)
52. Tuncel, G., Alpan, G.: Risk assessment and management for supply chain networks: a case study. Comput. Ind. **61**(3), 250–259 (2010)
53. Vaka, M., Walvekar, R., Rasheed, A.K., Khalid, M.: A review on Malaysia's solar energy pathway towards carbon-neutral Malaysia beyond Covid'19 pandemic. J. Clean. Prod. **273**, 122834 (2020)
54. Vilko, J.P., Hallikas, J.M.: Risk assessment in multimodal supply chains. Int. J. Prod. Econ. **140**(2), 586–595 (2012)
55. Waters, D.: Supply Chain Risk Management: Vulnerability and Resilience in Logistics. Kogan Page Publishers (2011)
56. Wu, T., Blackhurst, J., O'grady, P.: Methodology for supply chain disruption analysis. Int. J. Prod. Res.**45**(7), 1665–1682 (2007)
57. Xanthopoulos, A., Vlachos, D., Iakovou, E.: Optimal newsvendor policies for dual-sourcing supply chains: a disruption risk management framework. Comput. Oper. Res. **39**(2), 350–357 (2012)
58. Zahraee, S.M., Shiwakoti, N., Stasinopoulos, P.: Agricultural biomass supply chain resilience: COVID-19 outbreak versus sustainability compliance, technological change, uncertainties, and policies. Cleaner Logist. Supply Chain **4**, 100049 (2022)
59. Zsidisin, G.A.: A grounded definition of supply risk. J. Purch. Supply Manag. **9**(5–6), 217–224 (2003)

Chapter 6
Data Envelopment Analysis

6.1 Introduction

Throughout history, humanity has tried to make maximum use of the available facilities and resources, considering the limitations on the way. In this regard, performance evaluation is considered one of the managers' most essential issues. In fact, for a manager, knowing the performance of supervised units is the most critical task in making a decision and adopting a suitable strategy. The complexity of information, the massive amount of data, and the influence of other factors make managers unable to get informed about the performance of the units under their supervision without a scientific approach. One of the concepts in performance evaluation is calculating the efficiency of the units under the assessment. Efficiency is one of the essential management concepts obtained from the output ratio to input. But due to the number of inputs and outputs and their different scales, it is not easy to compute this ratio. Therefore, it is felt necessary to use scientific methods to evaluate performance. One of the appropriate and efficient tools for performance evaluation and efficiency measurement is data envelopment analysis (DEA), which is used as a non-parametric method to calculate the efficiency of decision-making units. Farrell presented the initial idea of this method in 1957. Today, the DEA technique is expanding rapidly and is used in evaluating various organizations and industries such as the banking industry [62, 69, 67, 59], hospitals [46], building materials including the cement industry [60], educational centers [29], power plants [24], refineries [8], supply chain [61], evaluation of projects [70] and so on.

Among the advantages of using DEA models are that considering several inputs and outputs for units is possible. Also, with the development of the theoretical aspects of DEA models, in contrast to classical models, in the recent DEA models, there are certain limitations for the types of inputs and outputs in the evaluation of the units under investigation is not considered. In addition to measuring relative efficiency, DEA models specify the organization's weak points in various indicators. The sources affecting organizations' inefficiency are identified using DEA models and presenting

© The Author(s), under exclusive license to Springer Nature Switzerland AG 2023
F. Hosseinzadeh Lotfi et al., *Supply Chain Performance Evaluation*,
Studies in Big Data 122, https://doi.org/10.1007/978-3-031-28247-8_6

their optimal level. Thus, the organization's policy toward improving efficiency specifies, and the efficient benchmark is introduced to the inefficient units. Also, in using the DEA technique, there is no need to provide an initial function concerning the relationship between inputs and outputs of the units under evaluation. This causes DEA to be classified as a non-parametric method. These reasons have caused this technique to grow increasingly from the theoretical and practical aspects and become one of the popular branches in the science of operations research. Of course, it is essential to pay attention to the fact that although DEA has many strengths, it should be noted that the results of DEA models are highly dependent on the selection of inputs and outputs. As a result, the wrong choice of each of them affects the final results. And to ensure the results, you must pay attention to this issue.

In recent years, many theoretical and practical developments have happened in DEA models, making it indispensable to know its various aspects for a more precise application of DEA models for the performance evaluation of a supply chain. Thus, in the rest of this chapter, we will explain the DEA definitions and models needed in the following chapters. Thus, in the rest of this chapter, we will explain the DEA definitions and models required for the following chapters.

6.2   Basic Concepts and Definitions

Before entering into the main discussion on data envelopment analysis, we will review some of the essential definitions in the topic of data envelopment analysis.

Definition 6.1
Everything that enters and consumes in an under-evaluation system is input.

Definition 6.2
Everything externalized due to the input consumption from an under-evaluation system is called output.

Definition 6.3
The unit that produces the output y by receiving the input x is called a decision-making unit (DMU). Data envelopment analysis assumes that the decision-making units are homogeneous; in other words, DMUs produce the same outputs using the same inputs.

In decision-making issues, "doing things righ" is called efficiency, which is obtained based on internal organizational indicators. In contrast, effectiveness, which means "doing the right things", is obtained from comparing external corporate indicators [48]. Finally, productivity is a function of efficiency and effectiveness, which is displayed as the relation (6.1). It is necessary to explain that the nature of f is uncertain.

$$Productity = f \, (efficiency, \, effectivness) \tag{6.1}$$

Therefore, to evaluate each unit's efficiency, it is enough to compare the indicators of that unit with the standards. Since the standard criteria are from outside or inside the society, several classifications and definitions have been provided for measuring efficiency. One of the most common classifications of efficiency is absolute and relative efficiency. We will explain these concepts in the following.

Definition 6.4

Suppose each DMU has one input and one output. If y^* and $\bar{y}$ are the maximum output and obtained output from the consumption of the one unit of the input, respectively, then the absolute efficiency is equal to: $\frac{\bar{y}}{y^*}$.

Definition 6.5

If e_j is the absolute efficiency of $DMU_j, j = 1, \ldots, n$, then the relative efficiency of DMU_o is calculated by relation (6.2).

$$Re_o = \frac{e_o}{\max\{e_j | j = 1, \ldots, n\}} \tag{6.2}$$

Therefore DMU_o is relatively efficient if only if $Re_o = 1$.

Now let's assume that we are faced with an evaluation problem of three furniture factories that produce two outputs of tables and chairs by consumption of three inputs wood, paint, and fabric. So in such cases, we will have problems calculating the relative efficiency using the formula (6.2) because the fraction (6.2) can only be used for the unit that produces one output using one input. As a result, to solve this problem, we need to provide a new definition of efficiency, which is called economic efficiency.

In general, suppose that DMU_j by consuming the input vector $x = (x_1, x_2, \ldots, x_m), x \geq 0, x \neq 0$, produces the output vector $y = (y_1, y_2, \ldots, y_s), y \geq 0, y \neq 0$. If u_r is the price (weight) of rth output and v_i is the cost (weight) of the ith input, then the economic efficiency is defined using the relation (6.3).

$$\frac{u_1 y_{1o} + \cdots + u_s y_{so}}{v_1 x_{1o} + \cdots + v_m x_{mo}} \tag{6.3}$$

And so, to calculate the relative efficiency of DMUs with multiple inputs and outputs, relation (6.4) is used.

$$Re_o = \frac{\frac{\sum_{r=1}^{s} u_r y_{ro}}{\sum_{i=1}^{m} v_i x_{io}}}{\max\left\{\frac{\sum_{r=1}^{s} u_r y_{rj}}{\sum_{i=1}^{m} v_i x_{ij}} | j = 1, \ldots, n\right\}} \tag{6.4}$$

Now suppose that we are dealing with the problem of performance evaluation of hospitals. In such a case, the number of nurses, the number of doctors, the number of operating rooms, and the number of intensive care units are inputs, and the number of recovered patients is considered output. To calculate the relative efficiency of these

hospitals using (6.4), we must first determine the output value and the cost of the
inputs. But as seen, we are in trouble at this stage because the pricing on the number
of doctors or nurses or determining the values of patients recovered is not easily
possible. Therefore, we face a new problem using the relation (6.4).

Because determining the weight (value) of each of the inputs and outputs is a
vital point in using (6.4) to calculate the relative efficiency, in most cases, it is not
possible to quickly determine the cost of the input or the value of the output of each
unit. Therefore, the crucial question in this situation is how to choose the weights.

One of the main goals of the DEA technique is to answer this question and find
the optimal weights u and v to calculate the relative efficiency of the units under
evaluation.

Therefore, in the rest of this chapter, we focus on a suitable solution to find the
answer to this question.

6.3 Multiplier Models

As explained in the previous section, the main challenge in calculating the relative
efficiency of DMUs is to find a solution to obtain weights corresponding to the input
and output of the under-evaluation units. Therefore, [10] presented model (6.5) to
calculate the relative efficiency of DMUs, which was known as CCR.

$$\max \frac{\frac{\sum_{r=1}^{s} u_r y_{ro}}{\sum_{i=1}^{m} v_i x_{io}}}{\max\left\{\frac{\sum_{r=1}^{s} u_r y_{rj}}{\sum_{i=1}^{m} v_i x_{ij}} \middle| j = 1, \ldots, n\right\}}$$

$$u_r, v_i \geq \varepsilon, i = 1, \ldots, m, r = 1, \ldots, s. \tag{6.5}$$

It should be explained that model (6.5) is also known as the T.D.T model. In this
way, the relative efficiency of DMU_o is achieved by solving the model (6.5) and
obtaining the optimal weights (v^*, u^*). Since model (6.5) is a non-linear programming problem, solving model (6.5) and getting its optimal solution is complex.
Therefore, the following Change of the variable is used to linearize the model (6.5),
proposed by [11].

$$t = \frac{1}{\max\left\{\frac{\sum_{r=1}^{s} u_r y_{rj}}{\sum_{i=1}^{m} v_i x_{ij}} \middle| j = 1, \ldots, n\right\}} \tag{6.6}$$

As a result, by using variable change (6.6) and the definition of the new variable
$u_r = t u_r$, model (6.5) becomes (6.7).

$$\max \frac{\sum_{r=1}^{s} u_r y_{ro}}{\sum_{i=1}^{m} v_i x_{io}}$$

$$s.t$$

$$\frac{\sum_{r=1}^{s} u_r y_{rj}}{\sum_{i=1}^{m} v_i x_{ij}} \leq 1, j = 1, \ldots, n,$$

$$u_r, v_i \geq \varepsilon, \ i = 1, \ldots, m, \ r = 1, \ldots, s. \tag{6.7}$$

Model (6.7) is a fractional programming problem, so (6.7) is also called the fractional form of the CCR model. So, using variable change $v_i = t v_i$, $u_r = t u_r$ and $t = \frac{1}{\sum_{i=1}^{m} v_i x_{io}}$, model (6.7) becomes the linear programming problem (6.8).

$$\max \sum_{r=1}^{s} u_r y_{ro}$$

$$s.t$$

$$\sum_{i=1}^{m} v_i x_{io} = 1,$$

$$\sum_{r=1}^{s} u_r y_{rj} - \sum_{i=1}^{m} v_i x_{ij} \leq 0, j = 1, \ldots, n,$$

$$u_r, v_i \geq \varepsilon, \ i = 1, \ldots, m, \ r = 1, \ldots, s. \tag{6.8}$$

Since model (6.8) has been created based on optimal multipliers for the inputs and outputs of DMUs, model (6.8) is called the multiplier form of the CCR model. Now, suppose (v^*, u^*) is the optimal solution of model (6.8) when DMU_o is under evaluation, so it is evident that DMU_o is CCR-efficient if only if $\sum_{r=1}^{s} u_r^* y_{ro} = 1$.

The vector form of model (6.8) is displayed as the model (6.9).

$$\max \sum_{r=1}^{s} u_r y_{ro}$$

$$s.t$$

$$\sum_{i=1}^{m} v_i x_{io} = 1,$$

$$\sum_{r=1}^{s} u_r y_{rj} - \sum_{i=1}^{m} v_i x_{ij} \leq 0, j = 1, \ldots, n,$$

$$u_r, v_i \geq \varepsilon, \ i = 1, \ldots, m, \ r = 1, \ldots, s. \tag{6.9}$$

Now Consider the constraint $\sum_{r=1}^{s} u_r y_{rj} - \sum_{i=1}^{m} v_i x_{ij} \leq 0, j = 1, \ldots, n$, in the model (6.8) again. If DMU_o is under evaluation, it is easy to conclude that: $\sum_{r=1}^{s} u_r y_{ro} - \sum_{i=1}^{m} v_i x_{io} + s_o = 0$, as a result: $s_o = 1 - \sum_{r=1}^{s} u_r y_{ro}$. Therefore, the minimum s_o can be obtained instead of maximizing the objective function (6.8).

Consequently, model (6.8) and model (6.10) are equivalents.

$$\min s_o$$

$$s.t$$

$$\sum_{i=1}^{m} v_i x_{io} = 1,$$

$$\sum_{r=1}^{s} u_r y_{ro} - \sum_{i=1}^{m} v_i x_{io} + s_o = 0,$$

$$\sum_{r=1}^{s} u_r y_{rj} - \sum_{i=1}^{m} v_i x_{ij} \leq 0, \ j = 1, \ldots, n, \ j \neq o,$$

$$u_r, v_i \geq \varepsilon, \ i = 1, \ldots, m, \ r = 1, \ldots, s. \qquad (6.10)$$

Unlike model (6.8), the optimal value of model (6.10) represents the inefficiency score of DMU_o.

As seen in the modeling process, determining the weights of inputs and outputs of under-evaluation DMU is independent of other DMUs. In other words, in the evaluation of DMU_o, the DEA model searches to find the best possible weight for indicators of DMU_o, without considering the importance of these indicators in measuring the performance of other DMUs. Therefore, the efficiency score of DEA models is an optimistic value.

Another point that should consider in the modeling process is that the model (6.5)–(6.9) is all based on the definition of relative efficiency presented in (6.4). Now, we consider the relation (6.11) to compute the efficiency of DMUs in output oriented.

$$\frac{v_1 x_{1o} + \cdots + v_m x_{mo}}{u_1 y_{1o} + \cdots + u_s y_{so}} \qquad (6.11)$$

As explained earlier, to improve efficiency, the decision maker (DM) seeks solutions to have the most output from the lowest input, so the fraction (6.11) must be reduced. As a result, the definition of relative efficiency needs to change in the form of relation (6.12).

$$Re_o = \frac{\frac{\sum_{i=1}^{m} v_i x_{ij}}{\sum_{r=1}^{s} u_r y_{rj}}}{\min\left\{ \frac{\sum_{i=1}^{m} v_i x_{ij}}{\sum_{r=1}^{s} u_r y_{rj}} \middle| j = 1, \ldots, n \right\}} \qquad (6.12)$$

Therefore, model (6.13) is obtained to calculate the relative efficiency of DMU_o, based on (6.12).

$$\min \sum_{i=1}^{m} v_i x_{io}$$

s.t

$$\sum_{r=1}^{s} u_r y_{ro} = 1,$$

$$\sum_{i=1}^{m} v_i x_{ij} - \sum_{r=1}^{s} u_r y_{rj} \geq 0, \; j = 1, \ldots, n,$$

$$u_r, v_i \geq \varepsilon, \; i = 1, \ldots, m, \; r = 1, \ldots, s. \tag{6.13}$$

Model (6.8) and (6.13) are called the multiplier form of the CCR model in the input-oriented and output-oriented, respectively.

Consider again the problem of evaluating the three hospitals described earlier. Suppose that, in addition to the mentioned inputs and outputs, the fixed cost of building or buying a hospital is also considered. In this case, we cannot apply definition (6.4) to calculate the relative efficiency directly. So relation (6.14) is introduced, such that $\sum_{r=1}^{s} u_r y_{ro} - u_o$ represents the profit of DMUo.

$$Re_o = \cfrac{\frac{\sum_{r=1}^{s} u_r y_{ro} - u_o}{\sum_{i=1}^{m} v_i x_{io}}}{\max \left\{ \frac{\sum_{r=1}^{s} u_r y_{rj} - u_o}{\sum_{i=1}^{m} v_i x_{ij}} \middle| j = 1, \ldots, n \right\}} \tag{6.14}$$

Therefore, to improve efficiency, e_o should be maximized. As a result, we obtain model (6.15) with a process similar to model (6.8).

$$\max \sum_{r=1}^{s} u_r y_{ro} - u_o$$

s.t

$$\sum_{i=1}^{m} v_i x_{io} = 1,$$

$$\sum_{r=1}^{s} u_r y_{rj} - \sum_{i=1}^{m} v_i x_{ij} - u_o \leq 0, \; j = 1, \ldots, n,$$

$$u_r, v_i \geq \varepsilon, \; i = 1, \ldots, m, \; r = 1, \ldots, s. \tag{6.15}$$

The model (6.15) was presented for the first time by Banker and Cooper [6], so this model is known as the BCC model. Similar to model (6.8), model (6.15) is called the multiplier form of the BCC model in the input-oriented.

6.4   Envelopment Models

In the last section, to evaluate the performance of DMUs, we explained different definitions of relative efficiency and presented the modeling process based on each one.

In this section, we discuss evaluating the units' performance from another point of view. The production function is the function that provides the maximum output for each combination of inputs. However, this function is usually unavailable due to the complexity of the production process, changes in the production technology, and the multi-valued production function. Therefore, managers use an approximation of the production function in practical cases. Production function approximation methods are divided into two general categories:

- Parametric methods.
- Non-parametric methods.

For a long time, parametric methods such as the Cobb–Douglas method were one of the most popular methods for estimating the production function. In this method, a specific form of the function is first considered, then the function's parameters are estimated using mathematical techniques such as Least Absolute Deviations. Nevertheless, having an initial guess regarding the shape of the production function makes the use of parametric methods difficult. Therefore, to solve this problem, parametric methods were proposed by Farrell in [28]. This method uses a concept called production possibility set (PPS) to estimate the production function, and part of the frontier created by PPS is considered as production function. Suppose that production possibilities set denote with the symbol T, so:

$$T = \{(x, y) | x \ can \ produce \ y\} \tag{6.16}$$

Definition 6.6
$DMUo$ is efficient in T, if there does not exist a DMU in T that dominates $DMUo$.

According to the conditions of each society, production technologies have some of the following five principles. So different PPS be created according to the production technologies and their properties.

1. Non-empty
   All observations belong to the production possibility set. In other words: $(x_j, y_j) \in T, j = 1, \ldots, n$.
2. Constant return to scale
   Suppose x produces y and $\lambda \geq 0$ then λx produces λy. In other words, for each $\lambda \geq 0$ efficiency of (x, y) is equal to $(\lambda x, \lambda y)$.
3. Convexitytiy
   If x_1, x_2 produces y_1 and y_2 respectively, then every convex combination of observations belongs to PPS, in other words, for every $0 \leq \lambda \leq 1$: $((\lambda x_1 + (1 - \lambda)x_1), (\lambda y_1 + (1 - \lambda)y_1)) \in T$.

Fig. 6.1 Geometric
interpretation of T_c

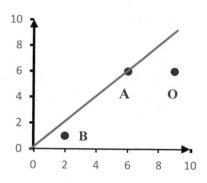

4. Plausilibility (Free disposal)

   If x produces y, in other words, $(x, y) \in T$ so for each $(\bar{x}, \bar{y})$ such that $\bar{x} \geq x$ and $\bar{y} \leq y$ in this case $(\bar{x}, \bar{y}) \in T$.

5. Minimality

   T is the smallest set that includes selected principles.

Suppose each DMU_j produces an output $y_j = (y_{1j}, y_{2j}, \ldots, y_{sj})$ using input $x_j = (x_{1j}, x_{2j}, \ldots, x_{mj})$. In this case, it can be easily shown that T_c is the smallest set that includes principles 1–4.

$$T_c = \left\{ (x, y) \,\middle|\, \sum_{j=1}^{n} \lambda_j x_j \leq x, \sum_{j=1}^{n} \lambda_j y_j \geq y, \lambda_j \geq 0, j = 1, \ldots, n \right\} \quad (6.17)$$

Example 6.1 If A $= (6, 6)$ B $= (2, 1)$, and O $= (9, 6)$, then T_c is displayed geometrically in Fig. 6.1.

Now suppose that the principle of constant returns to scale is not established in the desired production technology, so by removing this principle from the T_c, T_v is presented as follows:

$$T_v = \left\{ (x, y) \,\middle|\, \sum_{j=1}^{n} \lambda_j x_j \leq x, \sum_{j=1}^{n} \lambda_j y_j \geq y, \sum_{j=1}^{n} \lambda_j = 1, \lambda_j \geq 0, j = 1, \ldots, n \right\} \quad (6.18)$$

In the same way, other PPS can be obtained considering different combinations of stated principles.

Example 6.2 If A $= (7, 6)$, B $= (2, 1)$, C $= (4, 3)$ O $= (7, 5)$ then T_v is displayed geometrically in Fig. 6.2.

Fig. 6.2 Geometric
interpretation of T_v

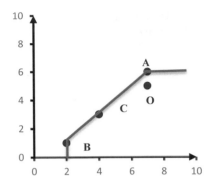

6.4.1   Radial Models

Consider Fig. 6.1 again. Suppose that goal is to evaluate the performance of DMU_o. In this case, to check the performance of DMU_o, we must answer this question, does DMU_o have inputs waste or output deficiency? In other words, is it possible to improve the input (output) without changing the output (input)? In such a case, we say that is there a production possibility in T_c that dominates DMU_o? There are three ways to answer this question, and we will explain these ways below.

6.4.1.1   Input-Oriented of CCR Models

Is there a DMU that produces output 6 (or even more) with input less than 9? In other words, is $0 \le \theta < 1$ that $(\theta x_o, y_o) \in T_c$? Also, it is evident that if $\theta \ge 1$, DMU_j is not dominant DMU_o. Therefore, it is enough to solve the model (6.19).

$$\min \theta$$
$$s.t$$
$$(\theta x_o, y_o) \in T_c \qquad (6.19)$$

Therefore:

$$\min \theta$$
$$s.t$$
$$6\lambda_1 + 2\lambda_2 + 9\lambda_3 \le 9\theta$$
$$6\lambda_1 + 1\lambda_2 + 6\lambda_3 \ge 6$$
$$\lambda_1, \lambda_2, \lambda_3 \ge 0 \qquad (6.20)$$

Using the simplex method, $\theta^* = 0.6667$ and it is concluded that DMU_o is inefficient, and as seen in Fig. 6.1, this issue was not far from the mind. Because DMU_A,
unlike DMU_o, produces an output equal to 6 by consuming only six units.

Therefore, to evaluate the performance of DMU_o, it is enough to solve the model
(6.21).

$$\min \theta$$
$$s.t$$
$$\sum_{j=1}^{n} \lambda_j x_{ij} \leq \theta x_{io}, \ i = 1, \ldots, m,$$
$$\sum_{j=1}^{n} \lambda_j y_{rj} \geq y_{ro}, \ r = 1, \ldots, s,$$
$$\lambda_j \geq 0, j = 1, \ldots, n. \tag{6.21}$$

The model (6.21) presented in 1978 by Charnes, Cooper, and Rhodes is called
the envelopment form of the CCR model. Since in model (6.21), the priority is to
reduce the input, this model is called the envelopment form of the CCR model in the
input-oriented. Also, DMU_o is CCR-efficient if only if $\theta^* = 1$.

Theorem 6.1
If (θ^*, λ^*) is an optimal solution of model (6.21), then:

(A) $\lambda^* \neq 0$.
(B) $0 < \theta^* \leq 1$.

Proof of Part A (proof by contradiction) suppose $\lambda^* = 0$ at a result $\sum_{j=1}^{n} \lambda_j^* y_{rj} = 0 \geq y_{ro} \geq 0$ so, $y_{ro} = 0, r = 1, \ldots, s$ and this is a contradiction; Because in DEA
models, it is assumed that $y_o \geq 0, y_o \neq 0$.

Proof of Part B It is evident that $\theta = 1, \lambda_j = 0, j \neq o, \lambda_o = 1$ is a feasible solution
to the model (6.21) so $\theta^* \leq 1$. We know that $x_{ij}, \lambda_j \geq 0$ and $\sum_{j=1}^{n} \lambda_j x_{ij} \leq \theta x_{io}$
therefore, it follows that $0 \leq \theta^* \leq 1$.

Now it is enough to prove: $\theta^* \neq 0$. Suppose that $(\bar{\lambda}, \bar{\theta})$ is a feasible solution and
$\bar{\theta} = 0$ so, $0 \leq \sum_{j=1}^{n} \lambda_j x_{ij} \leq 0$ consequently, $\lambda_j x_{ij} = 0, j = 1, \ldots, n$. We know that
$x_o \geq 0, x_o \neq 0$, hence there is an index like t: $x_{tj} > 0$ and $\lambda_j x_{tj} = 0, j = 1, \ldots, n$,
according: $\lambda_j = 0, j = 1, \ldots, n$. Therefore, it is concluded that $0 = \sum_{j=1}^{n} \lambda_j y_{rj} \geq y_{ro}$
and $y_{ro} = 0, r = 1, \ldots, s$ and this is a contradiction, so $0 < \theta^* \leq 1$.

Example 6.3 The information related to the input and output of 4 DMUs is explained
in Table 6.1.

It is evident that $(x_{1j}, x_{2j}, y_{1j}) \in \mathbb{R}^3$, therefore $T_c \subset \mathbb{R}^3$. Because drawing in $\mathbb{R}^3$ is
somewhat complicated, we depict the PPS image in 2D. Due to the same output of

Table 6.1 Inputs and outputs

	x_{1j}	x_{1j}	y_{1j}
A	2	5	1
B	5	2	1
C	8	2	1
D	6	4	1

DMUs, it is enough to obtain the cross-section in 323D PPS with the plane $y = 1$. The frontier shown in Fig. 6.4 is also called the Farrell frontier.

We use model (6.21) to evaluate the performance of DMU_c. And it follows that $\theta^* = 1$ and $\lambda^* = (0, 1, 0, 0)$. Therefore, unit c is apparently efficient, but as shown in Fig. 6.4, b is dominant over c. Because consumption of input 1 in DMU_B, as much as $8-5 = 3$ units, is less than the consumption input of DMU_c. In other words, this value is equal to: $s_1^- = \theta^* x_{1c} - \sum_{j=1}^n \lambda_j^* x_{1j} = 1 \times 8 - (0 \times 2 + 1 \times 5 + 0 \times 8 + 0 \times 6)$. In such a situation, DMU_c is called weak efficient.

Therefore, $s^{-*} \neq 0$ and $s^{+*} \neq 0$ indicate waste in input and lack in output, respectively and so DMU_o is strong CCR-efficient if and only if, in each optimal solution of the model (6.21), $s^{-*} = s^{+*} = 0$.

As a result, the model (6.22) can be used to calculate input wastage or output deficiency of DMU_o.

$$\min \sum_{i=1}^m s_i^- + \sum_{r=1}^s s_r^-$$

s.t

$$\sum_{j=1}^n \lambda_j x_{ij} + s_i^- = \theta^* x_{io}, \ i = 1, \ldots, m,$$

$$\sum_{j=1}^n \lambda_j y_{rj} - s_r^+ = y_{ro}, \ r = 1, \ldots, s,$$

$$s_i^-, s_r^+ \geq 0, i = 1, \ldots, m, \ r = 1, \ldots, s,$$
$$\lambda_j \geq 0, j = 1, \ldots, n. \tag{6.22}$$

So that θ^* is the optimal solution of the model (6.21). Therefore, to check the strong efficiency in phase I, model (6.21), and in phase II, we solve the model (6.22). As a result, this method is called the two-phase method.

Definition 6.7
Projection of DMU_o or so-called improved activity for inefficient units is defined using relations (6.23) and (6.24).

$$\hat{x}_o = \theta^* x_o - s^{-*} \tag{6.23}$$

$$\hat{y}_o = y_o + s^{+*} \tag{6.24}$$

So that θ^* is the optimal solution of the model (6.21) and s^{-*}, s^{+*} is the optimal solution of the model (6.22).

Theorem 6.2
Projection of DMU_o is strongly efficient.

Proof To evaluate $(\hat{x}_o, \hat{y}_o)$, consider the model (6.25).

$$\min \theta$$

s.t

$$\sum_{j=1}^{n} \lambda_j x_{ij} + s_i^- = \theta \hat{x}_{io}, \ i = 1, \dots, m,$$

$$\sum_{j=1}^{n} \lambda_j y_{rj} - s_r^+ = \hat{y}_{ro}, \ r = 1, \dots, s,$$

$$s_i^-, s_r^+ \geq 0, i = 1, \dots, m, \ r = 1, \dots, s,$$

$$\lambda_j \geq 0, \ j = 1, \dots, n. \tag{6.25}$$

Assume that $(\hat{\theta}, \hat{\lambda}, \hat{s}^-, \hat{s}^+)$ is the optimal solution of the model (6.25); therefore, by placing relations (6.23) and (6.24) in the model (6.26), we have: It is necessary to say that

$$\sum_{j=1}^{n} \hat{\lambda}_j x_{ij} + \hat{s}_i^- = \hat{\theta}(\theta^* x_{io} - s_i^{-*}), \ i = 1, \dots, m, \tag{6.26}$$

$$\sum_{j=1}^{n} \hat{\lambda}_j y_{rj} - \hat{s}_r^+ = y_{ro} + s_r^{+*}, \ r = 1, \dots, s, \tag{6.27}$$

By changing the variable: $\tilde{\theta} = \hat{\theta}\theta^*$, $\tilde{s}_r^+ = \hat{s}_r^+ + s^{+*}$, $\tilde{s}_i^- = \hat{s}_i^- + \theta^* s_i^{-*}$ we have:

$$\sum_{j=1}^{n} \hat{\lambda}_j x_{ij} + \tilde{s}_i^- = \tilde{\theta} x_{io} \tag{6.28}$$

$$\sum_{j=1}^{n} \hat{\lambda}_j y_{rj} - \tilde{s}_r^+ = y_o \tag{6.29}$$

$(\tilde{\theta}, \hat{\lambda}, \tilde{s}^-, \tilde{s}^+)$ is a feasible solution of the model (6.21) when DMU_o is under evaluation. we know that $\hat{\theta} \leq 1$, assume that: $\hat{\theta} < 1$ so $\hat{\theta}\theta^* < \theta^*$ at result $\tilde{\theta} < \theta^*$ and in it is a contradiction thuse $\hat{\theta} = 1$.

Now, in phase II, we form the model (6.30).

$$\max \sum_{i=1}^{m} s_i^- + \sum_{r=1}^{s} s_r^+$$

$s.t$

$$\sum_{j=1}^{n} \lambda_j x_{ij} + s_i^- = \hat{\theta}\hat{x}_{io}, \quad i = 1, \ldots, m,$$

$$\sum_{j=1}^{n} \lambda_j y_{rj} - s_r^+ = \hat{y}_{ro}, \quad r = 1, \ldots, s,$$

$$s_i^-, s_r^+ \geq 0, i = 1, \ldots, m, \quad r = 1, \ldots, s,$$

$$\lambda_j \geq 0, \quad j = 1, \ldots, n. \tag{6.30}$$

consequently:

$$\sum_{j=1}^{n} \hat{\lambda}_j x_{ij} + \hat{s}_i^- = \hat{\theta}\hat{x}_{io}, \quad i = 1, \ldots, m, \tag{6.31}$$

$$\sum_{j=1}^{n} \hat{\lambda}_j y_{rj} - \hat{s}_r^+ = \hat{y}_{ro}, \quad r = 1, \ldots, s, \tag{6.32}$$

By placing (6.23) and (6.24) in (6.31) and (6.32), it is easy to conclude that $(\tilde{\theta}, \hat{\lambda}, \tilde{s}^-, \tilde{s}^+)$ is a feasible solution for phase II while DMU_o is under evaluation. so:

$$(\tilde{s}^- + \tilde{s}^+) \leq (s^{-*} + s^{+*}) \Rightarrow [(\hat{s}^- + \theta^* s^{-*}) + (\hat{s}^- + s^{+*})] \leq (s^{-*} + s^{+*}) \tag{6.33}$$

According to $\hat{s}^- \geq 0$ and $\hat{s}^- \geq 0$, the relation (6.33) is established in a state where $\hat{s}^+ = \hat{s}^- = 0$. □.

Consider again Fig. 6.4. the projection of C is a real DMU, DMU_B, because B belongs to the set of under-evaluation DMUs, but the projection of D is D'. Although D' dominant over D, D' is not a real DMU, so D' is called a virtual DMU. Since the point D' is on the line segment between A and B, the coordinates D' are obtained using the convex combination of points A and B.

Therefore, another advantage of DEA is that, in addition to comparing the unit under evaluation with actual units, the DMU under evaluation is also compared with all the units that can be produced (virtual DMUs).

Now suppose that $(v_1, \ldots, v_m, u_1, \ldots, u_s)$ are the dual variables correspond to the constraints of the model (6.21); therefore, the dual of the model (6.21) is obtained as the model (6.34).

$$\max \sum_{r=1}^{s} u_r y_{ro}$$

$$s.t$$

$$\sum_{i=1}^{m} v_i x_{io} = 1,$$

$$\sum_{r=1}^{s} u_r y_{rj} - \sum_{i=1}^{m} v_i x_{ij} \leq 0, \ j = 1, \ldots, n,$$

$$u_r, v_i \geq 0, \ i = 1, \ldots, m, \ r = 1, \ldots, s. \tag{6.34}$$

The model (6.34) is the multiplier form of the CCR model in the input-oriented. Therefore, the multiplier form and the envelopment form of the CCR model are duals of each other; thus, the optimal solution of both models is the same. Hence, according to the desired approach, any of the two multiplier or envelopment forms can choose for performance evaluation.

In particular circumstances, we may want to change the unit of measurement of the DMUs indicators; for example, vary the unit of measure from millimeters to meters, so all input and output values must be multiplied by 1000. This will be allowed when the efficiency score of the units under evaluation with the new data does not change compared to the efficiency value with the current data. In such a case, we say that the model is unit invariant. DEA models are unit invariant if Changes (6.35) and (6.36) do not vary the efficiency of DMUs.

$$x_{ij} \rightarrow \alpha_i x_{ij}, \alpha_i > 0, i = 1, \ldots, m, j = 1, \ldots, n \tag{6.35}$$

$$y_{rj} \rightarrow \beta_r x_{rj}, \beta_r > 0, r = 1, \ldots, m, j = 1, \ldots, n \tag{6.36}$$

Theorem 6.3
CCR model is unit invariant.

Proof You can easily see that by placing the relations (6.35) and (6.36) in the first and second constraints of the model (6.21), the coefficients α_i and β_r are removed from the sides. Therefore, the CCR model is unit invariant. $\square$

As it has been explained so far, the projection obtained from DEA is a benchmark for an inefficient unit, but the goals and preferences of the decision maker, considered external criteria for DMUs, may be beyond the current image. Also, nessercerilly, benchmarking is not limited to inefficient DMUs because, in some cases, the inputs and outputs of an efficient DMU also may the manager' goals and standards are far away and need to be modified.

Since the ideals are not necessarily accessible, we seek to obtain the points that have the slightest deviation f from the ideal issues. As a result:

$$x_{io}^* - \delta_{io}^I \leq g_{io} \tag{6.37}$$

$$y_{ro}^* + \delta_{ro}^O \geq h_{ro} \tag{6.38}$$

Of course, it should be noted that the criteria should be realistic and not too far from the current position of DMU. Therefore, we are looking for the points between the current position of DMU and the ideal points. In other words, the new input projection of i-th input is located on the line segment between x_{io} and g_{io} or the convex combination of x_{io} and the idea point also must belong to PPS. As a result:

$$\sum_{j=1}^{n} \lambda_j x_{ij} - \delta_{io}^I \leq \alpha g_{io} + (1 - \alpha) x_{io}, \tag{6.39}$$

In the same way, a new projection for the outputs can also be defined:

$$\sum_{j=1}^{n} \lambda_j y_{rj} + \delta_{ro}^O \geq \alpha h_{ro} + (1 - \alpha) y_{ro}, \tag{6.40}$$

Finally, the model (6.41) was introduced by [71] as follows:

$$\min\{\delta_{1o}^I, \ldots, \delta_{mo}^I, \delta_{1o}^O, \ldots, \delta_{so}^O\}$$

$$s.t$$

$$\sum_{j=1}^{n} \lambda_j x_{ij} - \delta_{io}^I \leq \alpha g_{io} + (1 - \alpha) x_{io}, \quad i = 1, \ldots, m,$$

$$\sum_{j=1}^{n} \lambda_j y_{rj} + \delta_{ro}^O \geq \alpha h_{ro} + (1 - \alpha) y_{ro}, \quad r = 1, \ldots, s,$$

$$\lambda_j \geq 0, \quad j = 1, \ldots, n. \tag{6.41}$$

Model (6.41) is a multi-objective linear programming (MOLP) problem with m + s as the objective function. To solve this problem, we use the technique of scalarizing objective functions. Readers can refer to the reference [15] to learn more about the types of Scalarizing Functions methods in MOLP problems. In this section, we use the Augmented Chebyshev method for Scalarizing Functions. In other words, if $f_1, \ldots, f_n$ are objective functions and $z_1, \ldots, z_n$ are ideals correspond to each objective function, then the optimal value $\max[w_i|f_i - z_i||i = 1, \ldots, n] + \varepsilon[\sum_{i=1}^{n} w_i|f_i - z_i|]$ can be achieved instead of optimization of $f_1, \ldots, f_n$. Therefore, model (6.41) becomes model (6.42).

$$\min\left[\max\left[\max\{w_{io}^I\delta_{io}^I|i=1,\ldots,m\},\max\{w_{ro}^O\delta_{ro}^O|r=1,\ldots,s|\}\right]\right]$$
$$+\varepsilon\left[\sum_{i=1}^m w_{io}^I\delta_{io}^I+\sum_{r=1}^s w_{ro}^O\delta_{ro}^O\right]$$

s.t

$$\sum_{j=1}^n \lambda_j x_{ij}-\delta_{io}^I\le \alpha g_{io}+(1-\alpha)x_{io},\ i=1,\ldots,m$$

$$\sum_{j=1}^n \lambda_j y_{rj}+\delta_{ro}^O\ge \alpha h_{ro}+(1-\alpha)y_{ro},\ r=1,\ldots,s$$

$$\lambda_j\ge 0,\ j=1,\ldots,n. \tag{6.42}$$

Using changing the variable $\Delta=\max\left[\max\{w_{io}^I\delta_{io}^I|i=1,\ldots,m\},\right.$ $\left.\max\{w_{ro}^O\delta_{ro}^O|r=1,\ldots,s|\}\right]$, we rewrite model (6.42) as the model (6.43).

$$\min \Delta+\varepsilon\left[\sum_{i=1}^m w_{io}^I\delta_{io}^I+\sum_{r=1}^s w_{ro}^O\delta_{ro}^O\right]$$

s.t

$$\sum_{j=1}^n \lambda_j x_{ij}-\delta_{io}^I\le \alpha g_{io}+(1-\alpha)x_{io},\ i=1,\ldots,m,$$

$$\sum_{j=1}^n \lambda_j y_{rj}+\delta_{ro}^O\ge \alpha h_{ro}+(1-\alpha)y_{ro},\ r=1,\ldots,s,$$

$$\Delta-w_{io}^I\delta_{io}^I\ge 0,\ i=1,\ldots,m,$$

$$\Delta-w_{ro}^O\delta_{ro}^O\ge 0,\ r=1,\ldots,s,$$

$$\lambda_j\ge 0,\ j=1,\ldots,n. \tag{6.43}$$

6.4.1.2 Output-Oriented of CCR Model

In the second case, as shown in Fig. 6.5, we try to answer the question, is there a DMU (virtual or real) that produces an output greater than 6 consuming an input equal to 9? In other words, is there $\varphi>1$ such that: $(x_o,\varphi y_o)\in T_c$? Finding the answer to this question is enough to solve the model (6.44).

$$\max \varphi$$

$$s.t$$

$$\sum_{j=1}^{n} \lambda_j x_{ij} \leq x_{io}, \ i = 1, \ldots, m,$$

$$\sum_{j=1}^{n} \lambda_j y_{rj} \geq \varphi y_{ro}, \ r = 1, \ldots, s,$$

$$\lambda_j \geq 0, \ j = 1, \ldots, n. \tag{6.44}$$

Model (6.44), the envelopment form of the CCR model in output-oriented is called. Similar to the input-oriented, other cases can also be stated.

Theorem 6.4
If θ^* and φ^* are the optimal solution of model (6.21) and (6.44) respectively, then $\varphi^* = \frac{1}{\theta^*}$.

Proof According to $0 < \theta \leq 1$, model (6.21) can be rewritten as the model (6.45).

$$\min \theta$$

$$s.t$$

$$\sum_{j=1}^{n} \frac{\lambda_j}{\theta} x_{ij} \leq x_{io}, \ i = 1, \ldots, m,$$

$$\sum_{j=1}^{n} \frac{\lambda_j}{\theta} y_{rj} \geq \frac{1}{\theta} y_{ro}, \ r = 1, \ldots, s,$$

$$\lambda_j \geq 0, \ j = 1, \ldots, n. \tag{6.45}$$

we define: $\varphi = \frac{1}{\theta}$ and $\mu_j = \frac{\lambda_j}{\theta}$ and it can be easily shown that: $\max \varphi = \max \frac{1}{\theta} = \min \theta$. As a result, the model (6.45) which is equivalent to (6.46) is obtained. Thuse: $\varphi^* = \frac{1}{\theta^*}$.

$$\max \varphi$$

$$s.t$$

$$\sum_{j=1}^{n} \mu_j x_{ij} \leq x_{io}, \ i = 1, \ldots, m,$$

$$\sum_{j=1}^{n} \mu_j y_{rj} \geq \varphi y_{ro}, \ r = 1, \ldots, s,$$

$$\mu_j \geq 0, \ j = 1, \ldots, n. \tag{6.46}$$

6.4.1.3   Directional CCR Models

In this case, a new question arises, is there a DMU (virtual or real) that produces output more than 6 using input less than 9? To understand the issue better, consider the Fig. 6.6. As can be seen, there is a possibility of improvement along each of the vectors d that exist in the cone $(Ao\acute{o})$.

Therefore, in this case, we have to answer this question, is there any direction such as d and any positive scaler like $\theta \geq 0$, such that: $(x_o - \theta d_1, y_o + \theta d_2) \in T_c$. Therefore, to answer this question, it is enough to obtain the optimal answer of the model (6.47).

$$\max \theta$$
$$s.t$$
$$(x_o - \theta d_1, y_o + \theta d_2) \in T_c \tag{6.47}$$

consequent that:

$$\max \theta$$
$$s.t$$
$$\sum_{j=1}^{n} \lambda_j x_j \leq x_o - \theta d_1,$$
$$\sum_{j=1}^{n} \lambda_j y_j \geq y_o + \theta d_2,$$
$$\lambda_j \geq 0, \ j = 1, \ldots, n. \tag{6.48}$$

Based on the model (6.48), different models can be obtained for different directions. [55] defined (d_1, d_2) as relations (6.49) and (6.50).

$$d_{1i} = x_{io} - \min\{x_{ij} | j = 1, \ldots, n\}, i = 1, \ldots, m \tag{6.49}$$

$$d_{2r} = \max\{y_{rj} | j = 1, \ldots, n\} - y_{ro}, r = 1, \ldots, s \tag{6.50}$$

The new model based on (6.49) and (6.50) is called the RDM model. Also, if $(d_1, d_2) = (x_o, y_o)$, then model (6.48) becomes model (6.51). The (6.51) model is called the hybrid-oriented of the CCR model.

$$\max \theta$$
$$s.t$$
$$\sum_{j=1}^{n} \lambda_j x_j \leq x_o(1 - \theta),$$

$$\sum_{j=1}^{n} \lambda_j y_j \geq y_o(1 + \theta),$$

$$\lambda_j \geq 0, \ j = 1, \ldots, n. \tag{6.51}$$

6.4.1.4   BCC Model

Now consider $DMU_j, j = 1, \ldots, 4$ is presented in the Example 6.2, again. Similar to the CCR model, this question is raised: is there a DMU (virtual or real) in T_v that prevails over DMU_o? in other words, is there $0 \leq \theta < 1$ so that $(\theta x_o, y_o) \in T_v$? According to the Fig. 6.7, o' dominates over o. Therefore DMU_o is inefficient.

As a result, to evaluate the performance of DMU_o in T_v, it is enough to solve the model (6.52). In other words:

$$\min \theta$$
$$s.t$$
$$(\theta x_o, y_o) \in T_v \tag{6.52}$$

Hence it follows that:

$$\min \theta$$
$$s.t$$
$$\sum_{j=1}^{n} \lambda_j x_{ij} \leq \theta x_{io}, \ i = 1, \ldots, m,$$
$$\sum_{j=1}^{n} \lambda_j y_{rj} \geq y_{ro}, \ r = 1, \ldots, s,$$
$$\sum_{j=1}^{n} \lambda_j = 1,$$
$$\lambda_j \geq 0, \ j = 1, \ldots, n. \tag{6.53}$$

The model (6.53), known as the BCC model, was introduced in 1985 by Banker et al. Since in model (6.53), the priority is on decreasing input, this model is called the envelopment form of the BCC model in the input-oriented. also we define Phase II and the definition of the projection in the same way the CCR model. Also, the multiplier form of the BCC model is obtained using the dual of the model (6.54).

$$\max \sum_{r=1}^{s} u_r y_{ro} + u_o$$

$$s.t$$

$$\sum_{i=1}^{m} v_i x_{io} = 1,$$

$$\sum_{r=1}^{s} u_r y_{rj} - \sum_{i=1}^{m} v_i x_{ij} + u_o \leq 0, j = 1, \ldots, n,$$

$$u_r, v_i \geq 0, i = 1, \ldots, m, r = 1, \ldots, s. \tag{6.54}$$

Now consider the changing variable $u'_o = -u_o$. In this case, model (6.54) becomes model (6.55):

$$\max \sum_{r=1}^{s} u_r y_{ro} - u'_o$$

$$s.t$$

$$\sum_{i=1}^{m} v_i x_{io} = 1,$$

$$\sum_{r=1}^{s} u_r y_{rj} - \sum_{i=1}^{m} v_i x_{ij} - u'_o \leq 0, \ j = 1, \ldots, n,$$

$$u_r, v_i \geq 0, \ i = 1, \ldots, m, \ r = 1, \ldots, s. \tag{6.55}$$

model (6.55) is the same as the model (6.15), so in some books, the multiplier form of the BCC model is displayed as the model (6.55). Also, the model (6.56) shows the output-oriented of the BCC model.

$$\min \varphi$$

$$s.t$$

$$\sum_{j=1}^{n} \lambda_j x_{ij} \leq x_{io}, \ i = 1, \ldots, m,$$

$$\sum_{j=1}^{n} \lambda_j y_{rj} \geq \varphi y_{ro}, \ r = 1, \ldots, s,$$

$$\sum_{j=1}^{n} \lambda_j = 1,$$

$$\lambda_j \geq 0, \ j = 1, \ldots, n. \tag{6.56}$$

Also, unlike the CCR model, θ^* is not necessarily equal to $\frac{1}{\varphi^*}$. Therefore, when using the BCC model, it is essential to be careful in choosing the input-oriented or output-oriented model. Also, like the CCR model, the BCC model is unit invariant. As seen so far, one of the basic assumptions in DEA models is that all input and output are non-negative. However, this assumption is very limiting in some practical issues, especially financial issues. Therefore, we face problems in evaluating units with negative dat with DEA models.

One of the other properties that exist in some DEA models is translation invariant. In other words, if a constant value is added to all inputs or all outputs, the efficiency value does not change.

Theorem 6.5
The BCC model in input-oriented is invariant concerning output transmission; the output-oriented BCC model is invariant in input transmission.

Proof Suppose that:

$$x_{ij} \rightarrow \alpha_i + x_{ij}, \alpha_i > 0, i = 1, \ldots, m, j = 1, \ldots, n \tag{6.57}$$

$$y_{rj} \rightarrow \beta_r + y_{rj}, \beta_r > 0, r = 1, \ldots, s, j = 1, \ldots, n \tag{6.58}$$

as a result:

$$\sum_{j=1}^{n} \lambda_j(\alpha_i + x_{ij}) = \alpha_i \sum_{j=1}^{n} \lambda_j + \sum_{j=1}^{n} \lambda_j x_{ij} \leq \alpha_i + x_{io}$$

$$\Rightarrow \alpha_i + \sum_{j=1}^{n} \lambda_j x_{ij} \leq \alpha_i + x_{io} \Rightarrow \sum_{j=1}^{n} \lambda_j x_{ij} \leq x_{io} \tag{6.59}$$

so the BCC model input-oriented is translation invariant concerning output transmission. In the same way, we can discuss the output-oriented.

Similarly, it can be proved for the output-oriented BCC model.

Based on Theorem 6.5, in the presence of negative data instead of using input and output directly, we use $q_r + y_{rj}$ and $p_i + x_{ij}$. So that q_r and p_i are arbitrary scalers that by adding p_i (q_r), all inputs(outputs) convert to numbers greater than or equal to zero, for example, $q_r = \max\{y_{rj} | r = 1, \ldots, m\}$. It should be noted that if the inputs and outputs contain negative data simultaneously, then the BCC model is unsuitable for evaluating the performance of DMUs.

Consider Figs. 6.1, 6.2, 6.3, 6.4, 6.5, 6.6 and 6.7, again. As seen, the inputs (outputs) are contracted (expanded) along a radius. In other words, all inputs (outputs) are reduced (increased) in the same ratio, so this category of models is called radial models. In the following, we will discuss the performance evaluation of DMUs using non-radial models and their advantages and disadvantages compared to radial models.

Fig. 6.3 Geometric
interpretation of a projection
of inefficient DMU in the
input-oriented in T_c

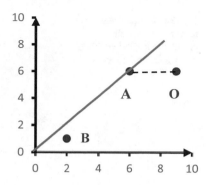

Fig. 6.4 Geometric
interpretation of Farrell
frontier

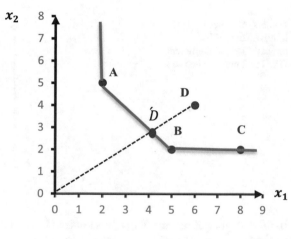

Fig. 6.5 Geometric
interpretation of a projection
of inefficient DMU in the
output-oriented in T_c

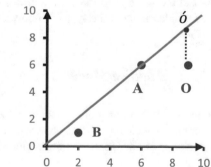

6.5 Non-radial Models

As stated, radial models such as CCR and BCC, to calculate the projection of under-
evaluation DMUs, all the inputs or all the outputs change with the same ratio. In this
section, we will get acquainted with other types of models called non-radial models.

Fig. 6.6 Geometrical
interpretation of the
projection inefficient DMU
along a desired direction

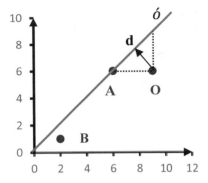

Fig. 6.7 Geometric
interpretation of the
projection of the inefficient
DMU in input-oriented in T_v

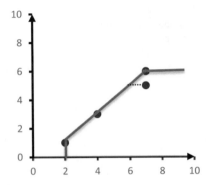

In this category of models, the inputs contraction and the expansion of the output are
not in the same proportion. Therefore, these types of models work non-radial, and
we will learn about the types of non-radial models in the following.

6.5.1 Additive Model

One of the first non-radial models is the additive model presented by [9]. Suppose
there are n decision-making units with input x_j and output y_j. In order to evaluate the
efficiency of DMUo, the additive model is shown as the model (6.60).

$$\max \sum_{i=1}^{m} s_i^- + \sum_{r=1}^{S} s_r^+$$

$$s.t$$

$$\sum_{j=1}^{n} \lambda_j x_{ij} + s_i^- = x_{io}, \quad i = 1, \ldots, m,$$

$$\sum_{j=1}^{n} \lambda_j y_{rj} - s_r^+ = y_{ro}, \ r = 1, \ldots, s,$$

$$s_i^-, s_r^+ \geq 0, i = 1, \ldots, m, \ r = 1, \ldots, s,$$

$$\lambda_j \geq 0, \ j = 1, \ldots, n. \tag{6.60}$$

It is necessery to explain that DMU_o is Additive-efficient if and only if and $s^{-*} = 0$ and $s^{+*} = 0$.

Although the model (6.60) is presented in the case of constant return to scale, it can easily be used for variable return to scale by adding $\sum_{j=1}^{n} \lambda_j = 1$.

Definition 6.8
Projection of DMU_o is defined by model (6.60) as follows:

$$\hat{y}_o = y_o + s^{+*} \tag{6.61}$$

$$\hat{x}_o = x_o - s^{-*} \tag{6.62}$$

such that s^{-*} and s^{-*} are gotten by (6.60).

Theorem 6.6
Projection of DMU_o is defined by additive model is efficient.

In other words, it has been presented in a hybrid nature. But unlike hybrid models, without considering an initial movement direction, the best direction of movement is obtained for imaging on the frontier. As seen in Fig. 6.8, D' and $\overline{D}$ are images correspond to DMU_D in radial and non-radial motion, respectively.

Another feature of the additive model is that it is stable to input and output transfers. Therefore, if the inputs and outputs include negative data simultaneously, it is recommended to use the additive model. Also, to obtain the multiplier form of the additive model, it is enough to obtain the dual of the model (6.60). So:

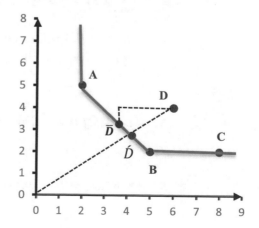

Fig. 6.8 Comparison of radial and non-radial approaches in evaluating DMU_o

$$\min \sum_{i=1}^{m} v_i x_{io} - \sum_{r=1}^{s} u_r y_{ro}$$

s.t

$$\sum_{i=1}^{m} v_i x_{ij} - \sum_{r=1}^{s} u_r y_{rj} \geq 0, \ j = 1, \dots, n,$$

$$u_r, v_i \geq 1, \ r = 1, \dots, s, \ i = 1, \dots, m. \tag{6.63}$$

Although the additive model, to some extent, solved the drawbacks of the radial models, the main shortcoming of the additive model is that it can only classify the DMUs into efficient and inefficient groups, and it is not possible to determine the efficiency score using the additive model.

6.5.2   RAM Model

The range-adjusted measure model (RAM) was presented by Cooper and Park [17]. The RAM model is a modification of the additive model and can measure the efficiency score of DMUs. The envelopment form of the RAM model is demonstrated in the model (6.64).

$$\max \frac{1}{m+s} \left(\sum_{i=1}^{m} \frac{s_i^-}{R_i^-} + \sum_{r=1}^{s} \frac{s_r^+}{R_r^+} \right)$$

s.t

$$\sum_{j=1}^{n} \lambda_j x_{ij} + s_i^- = x_{io}, \ i = 1, \dots, m,$$

$$\sum_{j=1}^{n} \lambda_j y_{rj} - s_r^+ = y_{ro}, \ r = 1, \dots, s,$$

$$s_r^+, s_i^- \geq 0, i = 1, \dots, m, \ r = 1, \dots, s,$$

$$\lambda_j \geq 0, \ j = 1, \dots, n. \tag{6.64}$$

so that:

$$R_i^- = \max\{x_{ij}, j = 1, \dots, n\} - \min\{x_{ij}, j = 1, \dots, n\} \tag{6.65}$$

and

$$R_r^+ = \max\{y_{rj}, j = 1, \dots, n\} - \min\{y_{rj}, j = 1, \dots, n\} \tag{6.66}$$

Thus: DMU_o, using the RAM model, is efficient if and only if $s^{-*} = 0$ and $s^{+*} = 0$.

Definition 6.9

Suppose that DMU_o is the under-evaluation unit by the RAM model, projection of DMU_o is achieved by (6.67) and (6.68).

$$\hat{x}_o = x_o - s^{-*} \tag{6.67}$$

$$\hat{y}_o = y_o + s^{+*} \tag{6.68}$$

So that s^{-*} and s^{-*} is the optimal solution of the model (6.64).

Theorem 6.7

Projection of DMU_o is gotten by the RAM model is efficient.

Proof Similar to Theorem 6.2.

The multiplier form of the RAM model is displayed as the model (6.69).

$$\min \sum_{i=1}^{m} v_i x_{io} - \sum_{r=1}^{s} u_r y_{ro} - w$$

$s.t$

$$\sum_{i=1}^{m} v_i x_{ij} - \sum_{r=1}^{s} \bar{u}_r y_{rj} - w \leq 0, \ j = 1, \ldots, n,$$

$$u_r \geq \frac{1}{m+s} R_r^+, \ r = 1, \ldots, s.$$

$$v_i \geq \frac{1}{m+s} R_i^-, \ i = 1, \ldots, m. \tag{6.69}$$

6.5.3 SBM Model

SBM model as a general modification for the additive model with various returns to scale was proposed for the first time by [73]. The envelopment of the SBM model is indicated in (6.70).

$$\rho^* = \min \frac{1 - \frac{1}{m} \sum_{i=1}^{m} \frac{s_i^-}{x_{io}}}{1 + \frac{1}{s} \sum_{r=1}^{s} \frac{s_r^+}{y_{ro}}}$$

$s.t$

$$\sum_{j=1}^{n} \lambda_j x_{ij} + s_i^- = x_{io}, i = 1, \ldots, m,$$

$$\sum_{j=1}^{n} \lambda_j y_{rj} - s_r^+ = y_{ro}, \ r = 1, \ldots, s,$$

$$\lambda_j, s_i^-, s_r^+ \geq 0, j = 1, \ldots, n, \ i = 1, \ldots, m, \ r = 1, \ldots, s. \tag{6.70}$$

some significant properties of the SBM model are described as follows:

1. $0 < \rho^* \leq 1$.
2. The objective function is stable against unit change.
3. Any increase in the auxiliary variables causes the objective function to decrease monotonously and any reduction of auxiliary variables causes the objective function to increase monotonically.

So DMU_o is SBM-efficient if only if $\rho^* = 1$ on the other hand DMU_o is SBM-efficient if only if $s_i^- = 0, i = 1, \ldots, m$ and $s_r^+ = 0, r = 1, \ldots, s$.

Definition 6.10
Projection of DMU_o, using the SBM model, is achieved by (6.71) and (6.72).

$$\hat{x}_o = x_o - s^{-*} \tag{6.71}$$

$$\hat{y}_o = y_o + s^{+*} \tag{6.72}$$

So that s^{-*} and s^{-*} is the optimal solution of the model (6.70).

Theorem 6.8
Projection of DMU_o is gotten by the SBM model is efficient.

Proof Similar to Theorem 6.2.

It is obvious that the SBM model is a non-linear programming problem. Therefore, using Channers-Cooper transformations, we define:

$$t = \frac{1}{1 + \frac{1}{s} \sum_{r=1}^{s} \frac{s_r^+}{y_{ro}}} \tag{6.73}$$

As a result, model (6.70) is rewritten as the model (6.74).

$$\min t - \frac{1}{m} \sum_{i=1}^{m} \frac{t s_i^-}{x_{io}}$$

$$s.t$$

$$t + \frac{1}{s} \sum_{r=1}^{s} \frac{ts_r^+}{y_{ro}} = 1,$$

$$\sum_{j=1}^{n} \lambda_j x_{ij} + s_i^- = x_{io}, \quad i = 1, \ldots, m,$$

$$\sum_{j=1}^{n} \lambda_j y_{rj} - s_r^+ = y_{ro}, \quad r = 1, \ldots, s,$$

$$\lambda_j, s_i^-, s_r^+ \geq 0, \, j = 1, \ldots, n, \, i = 1, \ldots, m, \, r = 1, \ldots, s. \qquad (6.74)$$

By changing the variable $S_i = ts_i^-$, $S_r = ts_r^+$ and $\gamma_j = t\lambda_j$ the model (6.75) is obtained.

$$\min t - \frac{1}{m} \sum_{i=1}^{m} \frac{S_i^-}{x_{io}}$$

s.t

$$t + \frac{1}{s} \sum_{r=1}^{s} \frac{S_r^+}{y_{ro}} = 1,$$

$$\sum_{j=1}^{n} \gamma_j x_{ij} + S_i^- = tx_{io}, \quad i = 1, \ldots, m,$$

$$\sum_{j=1}^{n} \gamma_j y_{rj} - S_r^+ = ty_{ro}, \quad r = 1, \ldots, s,$$

$$\gamma_j, S_i^-, S_r^+ \geq 0, \, j = 1, \ldots, n, \, i = 1, \ldots, m, \, r = 1, \ldots, s. \qquad (6.75)$$

Also, the dual of the SBM model is obtained as model (6.76).

$$\max \xi$$

s.t

$$\xi + \sum_{i=1}^{m} v_i x_{io} - \sum_{r=1}^{s} u_r y_{ro} = 1,$$

$$\sum_{i=1}^{m} v_i x_{ij} - \sum_{r=1}^{s} u_r y_{rj} \leq 0, j = 1, \ldots, n,$$

$$v_i \geq \frac{1}{m} \left[\frac{1}{x_{io}} \right], i = 1, \ldots, m,$$

$$u_r \geq \frac{1 - \sum_{r=1}^{s} u_r y_{ro} - \sum_{i=1}^{m} v_i x_{io}}{s} \left[\frac{1}{y_{ro}} \right], r = 1, \ldots, s. \qquad (6.76)$$

6.5.4   Enhanced Russell Model

Pastor et al. [54] presented another type of non-radial models. This model considers changes in each input and output independent of other inputs and outputs. In other words, the input x_i and output y_r changes by a factor of θ_i, and φ_r respectively. As a result, the enhanced Russell proposed as (6.77)

$$\text{Re} = \min \frac{\frac{1}{m}\sum_{i=1}^{m}\theta_i}{\frac{1}{s}\sum_{r=1}^{s}\varphi_r}$$

$s.t$

$$\sum_{j=1}^{n}\lambda_j x_{ij} \leq \theta_i x_{io}, i = 1, \ldots, m,$$

$$\sum_{j=1}^{n}\lambda_j y_{rj} \geq \varphi_r y_{ro}, r = 1, \ldots, s,$$

$$\theta_i \leq 1, i = 1, \ldots, m, \varphi_r \geq 1, r = 1, \ldots, s, \lambda_j \geq 0, j = 1, \ldots, n. \qquad (6.77)$$

It can be easily shown that the objective function is stable concerning unit change and $0 \leq \text{Re} \leq 1$. Therefore, the properties of the Russell model are utterly similar to the SBM model. Also, DMU_o is efficient using (6.77) if only if $\rho^* = 1$, in the other way DMU_o is efficient if only if $\theta_i = 1, i = 1, \ldots, m$ and $\varphi_r = 1, r = 1, \ldots, s$.

Definition 6.11
If DMU_o is under evaluation and $\theta_i^*, i = 1, \ldots, m$ and $\varphi_r^*, r = 1, \ldots, s$ are the optimal solutions of the model (6.77) then, the projection of DMU_o is $(\theta_1^* x_{1o}, \theta_2^* x_{2o}, \ldots, \theta_m^* x_{mo}, \varphi_1^* y_{1o}, \varphi_2^* y_{2o}, \ldots, \varphi_s^* y_{so})$.
Similar to the SBM model, using variable change $t = \frac{1}{\frac{1}{s}\sum_{r=1}^{s}\varphi_r}, \Theta_i = t\theta_i, \Phi_r = t\varphi_r$ and $\Phi_r = t\varphi_r$ the model (6.77) becomes the linear programming problem (6.78).

$$\min \frac{1}{m}\sum_{i=1}^{m}\Theta_i$$

$s.t$

$$\frac{1}{s}\sum_{r=1}^{s}\Phi_r = 1,$$

$$\sum_{j=1}^{n}\gamma_j x_{ij} \leq \Theta_i x_{io}, i = 1, \ldots, m,$$

$$\sum_{j=1}^{n}\gamma_j y_{rj} \geq \Phi_r y_{ro}, r = 1, \ldots, s,$$

$$\Theta_i \leq 1, i = 1, \ldots, m, \Phi_r \geq 1, r = 1, \ldots, n, \gamma_j \geq 0, j = 1, \ldots, n. \qquad (6.78)$$

Esmaeili [25] obtained the dual of the enhanced Russell as (6.79).

$$\max \alpha - \beta$$

$$s.t$$

$$\sum_{r=1}^{s} u_r y_{rj} - \sum_{i=1}^{m} v_i x_{ij} \leq 0, \, j = 1, \ldots, n,$$

$$v_i x_{io} - \mu_i \leq \frac{1}{m}, \, i = 1, \ldots, m,$$

$$\frac{\alpha}{s} - u_r y_{ro} + f_r \leq 0, \, r = 1, \ldots, s,$$

$$\sum_{i=1}^{m} \mu_i - \sum_{r=1}^{s} f_r - \beta \leq 0,$$

$$\alpha, \beta, \mu_i, f_r, u_r, v_i \geq 0, i = 1, \ldots, m, r = 1, \ldots, s, \qquad (6.79)$$

Using the idea of Enhanced Russell model, [32] introduced a new non-radial model in the form of the model (6.80).

$$\max \left(\frac{1}{m} \sum_{i=1}^{m} \frac{\theta_i}{\overline{\theta}_o} + \frac{1}{s} \sum_{r=1}^{s} \frac{\varphi_r}{\overline{\varphi}_o} \right)$$

$$s.t$$

$$\sum_{j=1}^{n} \lambda_j x_{ij} \leq x_{io} - \theta_i |x_{io}|, \, i = 1, \ldots, m,$$

$$\sum_{j=1}^{n} \lambda_j y_{rj} \geq y_{ro} + \varphi_r |y_{ro}|, \, r = 1, \ldots, s,$$

$$\sum_{j=1}^{n} \lambda_j = 1,$$

$$\lambda_j \geq 0, j = 1, \ldots, n. \qquad (6.80)$$

So that:

$$\overline{\theta}_o = \max \left\{ \frac{x_{io} - x_{il}}{|x_{io}|}, x_{io} \neq 0, i = 1, \ldots, m \right\} \text{ and } x_{il} = \min\{x_{ij}, j = 1, \ldots, n\}$$

$$(6.81)$$

$$\overline{\varphi}_o = \max \left\{ \frac{y_{rl} - y_{ro}}{|y_{ro}|}, y_{ro} \neq 0, r = 1, \ldots, s \right\} \text{ and } y_{rl} = \max\{y_{rj}, j = 1, \ldots, n\}$$

$$(6.82)$$

Definition 6.12

If DMU_o is under evaluation and $\theta_i^* = 1, i = 1, \ldots, m$ and $\varphi_r^*, r = 1, \ldots, s$ are the optimal solutions of the model (6.80) then, the projection of DMU_o is $(\theta_1^* x_{1o}, \theta_2^* x_{2o}, \ldots, \theta_m^* x_{mo}, \varphi_1^* y_{1o}, \varphi_2^* y_{2o}, \ldots, \varphi_s^* y_{so})$.

6.6 Hybrid Models

One of the drawbacks of radial models is ignoring non-radial slacks when evaluating the performance of DMUs. While these slacks provide helpful information in managerial interpretations, missing this information causes decision-makers to be misled. On the other hand, non-radial models such as SBM only focus on non-radial changes (slack) and ignore proportional and radial changes in inputs and outputs. Therefore, considering these problems, [75] introduced the epsilon-based measure (EBM) model in input-oriented as the model (6.83). In this model, both radial and non-radial changes are considered. It is also easy to show that the objective function of the EBM model is stable to unit change.

$$\gamma^* = \min \theta - \varepsilon_x \sum_{i=1}^{m} \frac{w_i^- s_i^-}{x_{io}}$$

$s.t$

$$\sum_{j=1}^{n} \lambda_j x_{ij} + s_i = \theta x_{io}, \ i = 1, \ldots, m,$$

$$\sum_{j=1}^{n} \lambda_j y_{rj} \geq y_{ro}, \ r = 1, \ldots, s,$$

$$\lambda_j, s_i \geq 0, \ j = 1, \ldots, n, \ i = 1, \ldots, m. \tag{6.83}$$

ε_x is an important parameter in the model (6.83) because it causes the combination of radial movement (θ) and non-radial movement (s_i^-). If $\varepsilon_x = 0$, then the EBM model becomes the CCR model. If $\varepsilon_x = 1$ and $\theta = 1$, then the EBM model becomes the SBM model. To determine ε_x, perform the following steps:

1. Obtain the projection of DMUs under evaluation, $(\hat{x}_o, \hat{y}_o)$, using the SBM or additive model in VRS. Thus:

$$\begin{bmatrix} \hat{x}_{11}, \ldots, \hat{x}_{1n} \\ \hat{x}_{21}, \ldots, \hat{x}_{2n} \\ \vdots \\ \hat{x}_{m1}, \ldots, \hat{x}_{mn} \end{bmatrix} = \begin{bmatrix} \bar{x}_1 \\ \bar{x}_2 \\ \vdots \\ \bar{x}_m \end{bmatrix} \tag{6.84}$$

2. Forming the diversity matrix: The diversity matrix is displayed as $D = [D_{i,k}]_{m \times m}$. If $\bar{x}_i = (\hat{x}_{i1}, \hat{x}_{i2}, \ldots, \hat{x}_{in})$ and $\bar{x}_k = (\hat{x}_{k1}, \hat{x}_{k2}, \ldots, \hat{x}_{kn})$, then D_{ik} is calculated using the relation (6.85):

$$D_{ik} = D(\bar{x}_i, \bar{x}_k) = \begin{cases} \frac{\sum_{j=1}^{n} |c_j - \bar{c}|}{n(c_{max} - c_{min})} & c_{max} > c_{min} \\ 0 & c_{max} = c_{min} \end{cases} \tag{6.85}$$

so that: $c_j = \ln \frac{\bar{x}_{ij}}{\bar{x}_{kj}}, j = 1, \ldots, n, \bar{c} = \sum_{j=1}^{n} \frac{c_j}{n}, c_{max} = \max_j \{c_j\}$ and $c_{min} = \min_j \{c_j\}$.

3. Form the affinity matrix represented by the symbol $S = [S_{ik}]_{m \times m}, i, k = 1, \ldots, m$. so that $S_{ik} = 1 - 2 \times D_{ik}, i, k = 1, \ldots, m$ indicate the degree of affinity between $\bar{x}_i$ and $\bar{x}_k$.

4. Calculate ε_x and w_i using relations (6.86) and (6.87).

$$\varepsilon_x = \begin{cases} \frac{m - \rho_x}{m - 1}, & m > 1 \\ 0, & m = 1 \end{cases} \tag{6.86}$$

and

$$w_i = \frac{w_{ix}}{\sum_{i=1}^{m} w_{ix}} \tag{6.87}$$

So that ρ_x is the largest eigenvalue of the matrix S and $W_x = (w_{1x}, w_{2x}, \ldots, w_{m_hx})$ is the eigenvector corresponding to ρ_x.

After calculating ε_x, based on the above steps, the EBM model will be easily used to evaluate the units under evaluation. It is evident that DMU_o is EBM-efficient if only if $\gamma^* = 1$.

Definition 6.13
If DMU_o is under evaluation and s^{-*}, s^{+*} and θ^* are the optimal solutions of the model (6.83), then the projection of DMU_o consists of:

$$\hat{x}_o = \theta^* x_o - s^{-*} \tag{6.88}$$

$$\hat{y}_o = y_o + s^{+*} \tag{6.89}$$

Finally, the multiplier form of the EBM model is defined as follows using the the dual of the model (6.90).

$$\max \sum_{r=1}^{s} u_r y_{ro}$$

$$s.t$$

$$\sum_{i=1}^{m} v_i x_{io} = 1,$$

$$\sum_{r=1}^{s} u_r y_{rj} - \sum_{i=1}^{m} v_i x_{ij} \leq 0, \ j = 1, \ldots, n,$$

$$v_i \geq \left[\frac{\varepsilon_x w_i}{x_{io}} \right], \ i = 1, \ldots, m,$$

$$u_r \geq 0, \ r = 1, \ldots, s. \tag{6.90}$$

Although the EBM model is expressed in an input-oriented and constant return scale, it can easily be presented as output-oriented by considering each of the CRS or VRS states.

6.7 Cost and Revenue Efficiency

Due to limited resources and budgets, managers of companies and factory owners always seek to maintain or improve the quality of their services by minimizing the overall cost. In other words, managers behind the scenes of using resources and allocating costs for resources are always looking for ways to reduce resource-cost consumption in competition with other competitors while maintaining efficient performance in the unit under assessment. So how to spend and use available resources has become one of their management challenges. Therefore, companies' management is constantly trying to identify and eliminate the inefficiency reasons for the group under their evaluation in competition with others.

Therefore, Fare et al. [27] proposed the first concept of cost efficiency in DEA in its current form as a model (6.91). Because the evaluation of cost efficiency according to the use of resources and the unit price of each resource provides new insights to managers to save money by providing a suitable alternative.

$$\min \sum_{i=1}^{m} c_i x_i$$

$$s.t \ \sum_{i=1}^{m} \lambda_j x_{ij} \leq x_i, \ i = 1, \ldots, m,$$

$$\sum_{r=1}^{s} \lambda_j y_{rj} \leq y_{ro}, \ r = 1, \ldots, s,$$

$$\lambda_j, x_i \geq 0, \ i = 1, \ldots, m, \ j = 1, \ldots, n. \tag{6.91}$$

So that c_i is the price of one unit of the i-th input (x_{io}), and x_i is a decision variable that indicates the optimal amount for consumption from the i-th input.

If (λ^*, x^*) is the optimal solution of the model (6.91), then DMU_o is cost-efficient, if only if: $CE_o = \frac{cx^*}{cx_o} = 1$.

In the same way, if p_r is equal to the income of production of one unit of output y_{ro}, then the model (6.92) is used to calculate the revenue efficiency of DMU_o.

$$\max \sum_{r=1}^{s} p_r y_r$$

$$s.t \ \sum_{i=1}^{m} \lambda_j x_{ij} \leq x_{io}, \ i = 1, \ldots, m,$$

$$\sum_{r=1}^{s} \lambda_j y_{rj} \leq y_r, \ r = 1, \ldots, s,$$

$$\lambda_j, y_r \geq 0, \ j = 1, \ldots, n, \ r = 1, \ldots, s. \tag{6.92}$$

If (λ^*, y^*) is the optimal solution of the model (6.92), then DMU_o is revenue-efficient, if only if: $RE_o = \frac{py^*}{py_o} = 1$.

Also, if p_r is equal to the income of production of one unit of output y_{ro}, c_i is the price of one unit of input x_i, in this case, the current profit of production is equal to $\sum_{r=1}^{s} p_r y_{ro} - \sum_{i=1}^{m} c_i x_{io}$.

$$\max \sum_{r=1}^{s} p_r y_r - \sum_{i=1}^{m} c_i x_i$$

$$s.t \ \sum_{i=1}^{m} \lambda_j x_{ij} \leq x_i, \ i = 1, \ldots, m,$$

$$\sum_{r=1}^{s} \lambda_j y_{rj} \leq y_r, \ r = 1, \ldots, s,$$

$$\lambda_j, x_i, y_r \geq 0, \ j = 1, \ldots, n, \ i = 1, \ldots, m, \ r = 1, \ldots, s. \tag{6.93}$$

If (λ^*, x^*, y^*) is the optimal solution of the model (6.93), then DMU_o is profit-efficient, if only if: $PE_o = \frac{cx^* - py^*}{cx_o - py_o} = 1$.

6.8 Ranking

Before starting the discussion, consider the Example 6.4 as follows.

Example 6.4 Suppose DMU_j produces output y_j by consuming input (x_{1j}, x_{2j}). The input and output indicators are stated in the second, third, and fourth columns of the

Table 6.2 Input and output values and ranking results using the AP model

	x_{1j}	x_{2j}	y_{1j}	Efficiency	Super efficiency	Ranking
A	1.5	5	1	1	1.3333	2
B	2	3	1	1	1.1935	3
C	4	1	1	1	3	1
D	6	4	1	0.5	0.5	5
E	5	3	1	0.62	0.625	4

Table 6.2. Also, the efficiency score of DMUs using the CCR model is reported in fifth column of the Table 6.2.

As previously stated, one of the main reasons for calculating the efficiency score is to provide a general ranking for comparing the units under evaluation. As seen, DMU_E and DMU_D are ranked 4 and 5 among five units. There is a problem with comparing other DMUs and commenting on their rank. Although these DMUs ranked higher than DMU_E and DMU_D, it is impossible to compare them with each other. In other words, the question that arises in this issue is, if there are several efficient DMUs, using DEA models, which one of the efficient units performs better than the others? Or how to rank efficient DMUs?

To overcome this problem, a new category of models has been presented to rank efficient DMUs. Each methods uses a specific criterion for ranking, so each technique has different advantages and disadvantages.

6.8.1 Anderson and Peterson Method

One of the first and most widely used methods for ranking efficient units is the method proposed by Anderson and Peterson in [4]. Therefore, this method became known as the AP method. This method is based on the idea that removing the under-evaluation DMU from a PPS how much will affect the efficiency of other DMUs. Therefore, to use this method, the first DMU_o is removed from the desired PPS and creates a new PPS, then DMU_o is evaluated in the new PPS. For instance, removing DMU_o from the T_c is shown in relation (6.94).

$$\overline{T}_c = \left\{ (x, y) \, \middle| \, \sum_{\substack{j=1 \\ j \neq o}}^{n} \lambda_j x_j \leq x, \sum_{\substack{j=1 \\ j \neq o}}^{n} \lambda_j y_j \geq y, \lambda_j \geq 0, j = 1, \ldots, n \right\} \quad (6.94)$$

Therefore, Anderson and Patterson's proposed model for ranking efficient units using $\overline{T}_c$ is as follows.

$$\min \theta$$

$$s.t$$

$$\sum_{j=1, j\neq o}^{n} \lambda_j x_{ij} \leq \theta x_{io}, \quad i = 1, \ldots, m,$$

$$\sum_{j=1, j\neq o}^{n} \lambda_j y_{rj} \geq y_{ro}, \quad r = 1, \ldots, s,$$

$$\lambda_j \geq 0, \quad j = 1, \ldots, n. \tag{6.95}$$

It is easy to generalize this idea to other PPS, including T_v. It is also necessary to explain that unlike DEA models in input-oriented, the value of objective function of (6.95) may be greater than one, so this value is called super efficiency.

Therefore, to rank DMUs, calculate the efficiency of $DMU_j, j = 1, \ldots, n$ using the Anderson and Patterson (AP) model and then rank DMUs using the obtained super-efficiency values.

6.8.2 Cross Efficiency

Another method for ranking of DMUs is cross efficiency method. As seen in the previous sections, using DEA models, the efficiency score of DMUs is calculated by assigning the best weights to the unit under evaluation units; in other words, DMUs are considered in the best conditions. Therefore, although the efficiency score obtained from DEA provides valuable information, it can be misleading in some cases. Now, if the company's efficiency is calculated with weights assigned to indicators of other DMUs, its performance in its best condition is compared with the best state of other competitors. This way, a matrix is obtained, known as the cross-efficiency matrix. Cross-efficiency was first presented by [66] and then expanded by [22]. We perform the following steps to compute the cross-efficiency of under-evaluation DMUs.

1. Obtain efficiency of DMU_d using multiplier models such as the CCR model.
2. Obtain the efficiency of $DMU_j, j = 1, \ldots, n, j \neq d$ using the relation (6.96).

$$E_{jd} = \frac{\sum_{r=1}^{s} u_{rd}^* y_{rj}}{\sum_{i=1}^{m} v_{id}^* x_{ij}}, j = 1, \ldots, n, j \neq d \tag{6.96}$$

in a way that $(v_{1d}^*, v_{2d}^*, \ldots, v_{md}^*, u_{1d}^*, u_{2d}^*, \ldots, u_{sd}^*)$ are optimal weights obtained from step 1.
3. Achieve the cross-efficiency matrix:

$$CEM = \begin{bmatrix} E_{11} & E_{12} & \cdots & E_{1n} \\ E_{21} & E_{22} & & E_{2n} \\ & & & \\ \vdots & \vdots & & \vdots \\ E_{n1} & E_{n2} & \cdots & E_{nn} \end{bmatrix} \tag{6.97}$$

Note that the element located in row i and column j is the efficiency of DMU_i which is obtained with the optimal weights corresponding to DMU_j. Therefore, the elements on the main diameter of the matrix are the efficiency values obtained from multiplier DEA models.

4. Obtain the cross efficiency using the relation (6.98).

$$\overline{E}_j = \frac{\sum_{d=1}^{n} E_{jd}}{n}, j = 1, \ldots, n \tag{6.98}$$

Therefore, the calculation of the optimal weights plays a prominent role in the calculation of the cross-efficiency of DMUs. The vital issue we face when calculating the efficiency of DMU_j, as a result of calculating the optimal weights, is the existence of alternative optimal solutions and different optimal weights. Hence, for each optimal weight, we will have a different value for E_{jd} and, as a result, different values for $\overline{E}_j$.

A great deal of research has been to this problem by introducing a secondary objective; we will explain some of them in the following chapters. Readers are also recommended to study [38, 35, 52].

To learn about other ranking methods, readers can refer to [39, 33, 36, 30].

6.9   Return to Scale

One of the necessary topics in DEA is the issue of the returns to scale. The concept of return to scale can be analyzed from two economic and mathematical perspectives. Mathematically, return-to-scale discusses the relationship between changes in the inputs and the outputs.

For example, suppose that the amount of use of all inputs in a company increases by a small equal proportion; therefore, one of the following three conditions will be given for this company's production (output).

- The output will increase exactly in the same proportion. (constant return to scale)
- The output increases with a ratio greater than that value. (Increasing return to scale)
- The output increases with a ratio less than that value. (decreasing return to scale).

From an economic point of view, one of the primary and essential issues in production is the investigation of economies of scale. Economies of scale will exist when an increase in output by one percent causes a need to increase costs by less than one percent. In other words, the economy of scale answers the question of whether more significant production and service units have an advantage in terms of cost and efficiency in terms of production compared to smaller units. In economics, determining a company's return to scale can show different types of economies of scale in different product ranges. In other words, returns to scale may be increasing, decreasing, or constant at various production levels. In microeconomics, the type of returns to scale a firm faces is primarily related to the technology used and is not affected by economic decisions or market conditions. Another ability of the DEA is to specify the type of return to the scale of units. But it is necessary to pay attention to the fact that determining the type of return of the scale of the production possibilities set differs from the sort of RTS of each unit under evaluation. The type of return to the scale of a PPS can usually be done through an expert's opinion. Still, this issue is somewhat complicated about the return to the scale of each DMUs. Pay attention that if the RTS of the production technology is constant, then the RTS of all units is also constant. Still, by removing this principle from the PPS, the RTS of the DMUs will be different. Therefore, in this subsection, to determine the type of RTS of units, we focus on the DMUs belonging to T_v.

Due to the importance of RTS in management decisions, various methods have been provided to specify the RTS. Some of the essential techniques in this field explain in this section. In one of the most basic RTS identifying methods provided by Banker, we first define $\alpha(\beta) = \max\{\alpha | (\beta x_o, \alpha y_o) \in T_v\}$. By taking the definition of $\alpha(\beta)$, if $\gamma^+ = \lim_{\beta \to 1^+} \frac{\alpha(\beta)-1}{\beta-1}$, $\gamma^- = \lim_{\beta \to 1^-} \frac{\alpha(\beta)-1}{\beta-1}$, then:

1. If $\gamma^+, \gamma^- > 1$, then DMU_o has increasing return to scale (IRS).
2. If $\gamma^+, \gamma^- < 1$, then DMU_o has decreasing return to scale (DRS).
3. Otherwise, DMU_o has constant return to scale (CRS).

As seen, using the definition of γ for practical applications is not easily possible, so another method provided by Banker and Thrall to determine the type of RTS is explained as follows.

Suppose DMU_o is on the boundary $((x_o, y_o) \in \partial T_V)$, if the optimal solution of the multiplier form of the BCC model (v^*, u^*, u_0^*) is unique, then:

1. If $u_0^* = 0$, then DMU_o has constant return to scale.
2. If $u_0^* < 0$, then DMU_o has decreasing return to scale.
3. If $u_0^* > 0$, then DMU_o has increasing return to scale.

If the optimal solution of the BCC model is not unique, we perform the following steps.

1. Obtain the optimal solution of the model (6.99)

$$u_0^{*+} = \max u_0$$

$$s.t$$

$$\sum_{i=1}^{m} v_i x_{io} = 1$$

$$\sum_{r=1}^{s} u_r y_{ro} - \sum_{i=1}^{m} v_i x_{io} + u_0 = 0,$$

$$\sum_{r=1}^{s} u_r y_{rj} - \sum_{i=1}^{m} v_i x_{ij} + u_0 \leq 0, j = 1, \ldots, n, j \neq o,$$

$$u_r, v_i \geq 0, i = 1, \ldots, m, r = 1, \ldots, s. \tag{6.99}$$

2. Achieve Obtain the optimal solution of the model (6.100)

$$u_0^{*-} = \min u_0$$

$$s.t$$

$$\sum_{i=1}^{m} v_i x_{io} = 1$$

$$\sum_{r=1}^{s} u_r y_{ro} - \sum_{i=1}^{m} v_i x_{io} + u_0 = 0,$$

$$\sum_{r=1}^{s} u_r y_{rj} - \sum_{i=1}^{m} v_i x_{ij} + u_0 \leq 0, j = 1, \ldots, n, j \neq o,$$

$$u_r, v_i \geq 0, i = 1, \ldots, m, r = 1, \ldots, s. \tag{6.100}$$

3. If $u_0^{*+} \geq u_0^{*-} > 0$, then DMU_o has increasing return to scale.
4. If $u_0^{*-} \leq u_0^{*+} < 0$, then DMU_o has decreasing return to scale.
5. If $u_0^{*-} \leq 0 \leq u_0^{*+}$, then DMU_o has constant return to scale.

It can be easily shown that to determine the type of RTS, the calculation of u_0 or γ has the same results. Those interested can refer to [7] for information on the details of the technique and the proof of theorems. Also to learn about other methods of determining returns to scale, readers can refer to the sources of [18, 40, 41, 45, 49, 56].

6.10   Network DEA

Although data envelopment analysis is a powerful tool for evaluating the efficiency of decision-making units with multiple inputs and multiple outputs, decision-making units can take different forms, such as hospitals, universities, and banks. Therefore, in many practical applications, such as the evaluation of supply chains, DMUs act as a multi-stage process, as shown in Fig. 6.10. The internal relationships cause

complex evaluation processes in each multi-stage DMU under assessment. Therefore, one of the critical problems of conventional DEA models is that these models
consider the systems under evaluation as a closed unit, and internal processes are
ignored. This view, known as the black box, ignores much beneficial information
concerning DMUs, and the performance analysis of DMUs is done just based on the
initial inputs and final outputs of DMUs. While to convert initial inputs into final
outputs, several internal processes are required. Thus, the evaluation results obtained
from conventional DEA models prevent getting influential management information.
Hence, classic DEA models get into trouble evaluating the performance of complex
and multi-stage systems. [16]. Therefore, a new branch of DEA called the Network
Data Envelopment Analysis (NDEA) model was introduced by [26]. Various classifications for NDEA have been introduced so far, but one of the most common
classifications is based on the formation of each network. Therefore, networks are
divided into three general categories: series, parallel and mix. In the following, we
will study each of the network types and models presented for the evaluation of
series, parallel and mix systems.

6.10.1 Series Network

One of the simplest networks is the two-stage decision-making unit shown in Fig. 6.9.
The first stage produces the output z by consuming the input x. These values are also
called intermediate production. Then, in the second stage, z is considered input, and
the final output y is produced by consuming the input z.

With the development of the two-stage network, the series network, as shown in
Fig. 6.9, is acquired. In the first stage, the outputs of this stage (z_j^1) are begotten using
the initial inputs of the system. These outputs are deemed inputs of the second stage.
Then, the second stage uses the received inputs to generate new outputs (z_j^2) used
as inputs in the third stage, and in the same way, this process will continue until the
final production of the system is produced in the kth stage.

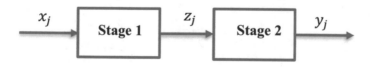

Fig. 6.9 A two-stage network

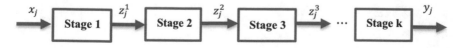

Fig. 6.10 A series multi-stage network

Chen and Zhou [14] proposed the model (6.101) to evaluate two-stage units. In this model, the efficiency of stages 1 and 2 is defined based on their production possibility, and then both stages are linked by intermediate variables.

$$\min(\alpha w_1 - \beta w_2)$$

$$\sum_{j=1}^{n} \lambda_j x_{ij} \leq \alpha x_{io}, \ i = 1, \ldots, m,$$

$$\sum_{j=1}^{n} \lambda_j z_{dj} \geq \tilde{z}_{do}, \ d = 1, \ldots, D,$$

$$\sum_{j=1}^{n} \lambda_j = 1,$$

$$\sum_{j=1}^{n} \mu_j z_{dj} \leq \tilde{z}_{do}, \ d = 1, \ldots, D,$$

$$\sum_{j=1}^{n} \mu_j y_{rj} \geq \beta y_{ro}, \ r = 1, \ldots, S,$$

$$\sum_{j=1}^{n} \mu_j = 1,$$

$$\mu_j, \lambda_j \geq 0, \ j = 1, \ldots, n. \tag{6.101}$$

w_1 and w_2 respectively symbolize the significance of steps 1 and 2, which the manager provides according to the importance of the first and second stages. It is also necessary to explain that in the model (6.101), it is always $\alpha^* \leq 1$ and $\beta^* \geq 1$ as a result $\alpha^* - \beta^* \leq 0$.

Kao and Hwang [44], to measure the total efficiency of two-stage processes and to be able to compare the performance of stages 1 and 2, expressed the total efficiency as the product of the efficiency of stages 1 and 2. So:

$$e_j^{Total} = e_j^1 \times e_j^2 = \frac{\sum_{d=1}^{D} w_d z_{dj}}{\sum_{i=1}^{m} v_i x_{ij}} \times \frac{\sum_{r=1}^{s} u_r y_{rj}}{\sum_{d=1}^{D} w_d z_{dj}} \tag{6.102}$$

By utilizing relation (6.102), model (6.101) is proposed to assess the efficiency of two-stage DMUs.

$$\max e_o^{Total}$$

$$s.t$$

$$e_j^1 \leq 1, \ e_j^2 \leq 1, j = 1, \ldots, n$$

$$v_i, w_d, u_r \geq 0, \ i = 1, \ldots, m, \ d = 1, \ldots, D, \ r = 1, \ldots, s. \tag{6.103}$$

Therefore, by substituting the relation (6.102) and then the transformations of Charnes and Cooper, the model (6.103) becomes model (6.104).

$$\max \sum_{r=1}^{s} u_r y_{ro}$$

$$s.t$$

$$\sum_{i=1}^{m} v_i x_{io} = 1,$$

$$\sum_{d=1}^{D} w_d z_{dj} - \sum_{i=1}^{m} v_i x_{ij} \le 1, \ j = 1, \dots, n,$$

$$\sum_{r=1}^{s} u_r y_{rj} - \sum_{d=1}^{D} w_d z_{dj} \le 1, \ j = 1, \dots, n,$$

$$v_i, w_d, u_r \ge 0, \ i = 1, \dots, m, \ d = 1, \dots, D, \ r = 1, \dots, s. \tag{6.104}$$

the dual of the model (6.104) is calculated as model (6.105).

$$\min \theta$$

$$s.t$$

$$\sum_{j=1}^{n} \lambda_j x_{ij} \le \theta x_{io}, \ i = 1, \dots, m,$$

$$\sum_{j=1}^{n} \mu_j y_{rj} \ge y_{ro}, \ r = 1, \dots, s,$$

$$\sum_{j=1}^{n} (\lambda_j - \mu_j) z_{dj} \ge 0, \ d = 1, \dots, D,$$

$$\mu_j, \lambda_j \ge 0, \ j = 1, \dots, n. \tag{6.105}$$

Suppose (v^*, w^*, u^*) is the optimal solution of the model (6.105). So the efficiency of stages 1, 2, and total efficiency is equal to $e_o^{*1} = \frac{\sum_{d=1}^{D} w_d^* z_{do}}{\sum_{i=1}^{m} v_i^* x_{io}}$, $e_o^{*2} = \frac{\sum_{r=1}^{s} u_r^* y_{ro}}{\sum_{d=1}^{D} w_d^* z_{do}}$, and $e_o^{*Total} = \sum_{r=1}^{s} u_r^* y_{ro}$, respectively. it is evivent that: $e_o^{*Total} = e_o^{*1} \times e_o^{*2}$. Due to the existence of alternative optimal solutions, the decomposition may not be unique. Therefore, we face problems in calculating e_j^1 and e_j^2. Thus, solving this problem, e_o^1 is first obtained using the model (6.106).

$$\max \sum_{d=1}^{D} w_d z_{do}$$

$$s.t$$

$$\sum_{i=1}^{m} v_i x_{io} = 1,$$

$$\sum_{r=1}^{s} u_r y_{rj} - \sum_{i=1}^{m} v_i x_{ij} \le 0, \ j = 1, \ldots, n, \ j \ne o$$

$$\sum_{r=1}^{s} u_r y_{ro} - e_o^{*Total} \times \sum_{i=1}^{m} v_i x_{io} = 0,$$

$$\sum_{d=1}^{D} w_d z_{dj} - \sum_{i=1}^{m} v_i x_{ij} \le 1, \ j = 1, \ldots, n,$$

$$\sum_{r=1}^{s} u_r y_{rj} - \sum_{d=1}^{D} w_d z_{dj} \le 1, \ j = 1, \ldots, n,$$

$$v_i, w_d, u_r \ge 0, \ i = 1, \ldots, m, \ d = 1, \ldots, D, \ r = 1, \ldots, s. \tag{6.106}$$

So that: e_o^{*Total} is the optimal solution of the model (6.104). In other words, the model (6.106), by keeping the total efficiency constant (e_o^{*Total}) and then searching among multiple weights while calculating the efficiency of step 1, compute the most considerable possible value for e_o^{*1}, and finally: $e_o^{*2} = \frac{e_o^{*1}}{e_o^{*Total}}$.

Now, we summarize the drawbacks of Kao and Hwang's model below:

- Does not guarantee relative efficiency.
- by removing constant return to scale becomes a non-linear model, and it is impossible to linearize it with usual methods.
- Failure to provide an image for the inefficient unit.
- It cannot be generalized for multi-stage structures.

To solve the mentioned problems in NDEA, [12] introduced the total efficiency as a convex combination of the efficiency of stages 1 and 2 as follows:

$$e_j^{Total} = w_1 e_j^1 \times w_2 e_j^2 = w_1 \frac{\sum_{d=1}^{D} w_d z_{dj}}{\sum_{i=1}^{m} v_i x_{ij}} \times w_2 \frac{\sum_{r=1}^{s} u_r y_{rj}}{\sum_{d=1}^{D} w_d z_{dj}}, \ w_1 + w_2 = 1, \quad (6.107)$$

Therefore, considering the relation (6.107), the model (6.108) for calculating the total efficiency in two-stage networks is presented as follows:

$$\max w_1 \frac{\sum_{d=1}^{D} w_d z_{do}}{\sum_{i=1}^{m} v_i x_{io}} \times w_2 \frac{\sum_{r=1}^{s} u_r y_{ro}}{\sum_{d=1}^{D} w_d z_{do}}$$

$$s.t$$

$$\frac{\sum_{r=1}^{s} u_r y_{rj}}{\sum_{d=1}^{D} w_d z_{dj}} \le 1, j = 1, \ldots, n,$$

$$\frac{\sum_{d=1}^{D} w_d z_{dj}}{\sum_{i=1}^{m} v_i x_{ij}} \leq 1, j = 1, \ldots, n,$$

$$v_i, w_d, u_r \geq 0, i = 1, \ldots, m, d = 1, \ldots, D, r = 1, \ldots, s. \qquad (6.108)$$

In a multi-stage system, the amount of consumption in each stage and the entire system are the influential factors in the performance of each stage. Thus [12] defined w1 and w2 as a ratio of stages 1 and 2 input to whole system input, which is shown in relations (6.109) and (6.110).

$$w_1 = \frac{\sum_{i=1}^{m} v_i x_{io}}{\sum_{i=1}^{m} v_i x_{io} + \sum_{d=1}^{D} w_d z_{do}} \qquad (6.109)$$

and

$$w_2 = \frac{\sum_{d=1}^{D} w_d z_{do}}{\sum_{i=1}^{m} v_i x_{io} + \sum_{d=1}^{D} w_d z_{do}} \qquad (6.110)$$

Therefore, using variable change $t = \frac{1}{\sum_{i=1}^{m} v_i x_{io} + \sum_{d=1}^{D} w_d z_{do}}$ and relation (6.109) and (6.110), model (6.108) becomes model (6.111).

$$\max \sum_{d=1}^{D} \pi_d z_{do} + \sum_{r=1}^{s} \mu_r y_{ro}$$

s.t

$$\sum_{i=1}^{m} \omega_i x_{io} + \sum_{d=1}^{D} \pi_d z_{do} = 1,$$

$$\sum_{r=1}^{s} \mu_r y_{rj} - \sum_{d=1}^{D} \pi_d z_{dj} \leq 0, j = 1, \ldots, n,$$

$$\sum_{d=1}^{D} \pi_d z_{dj} - \sum_{i=1}^{m} \omega_i x_{ij} \leq 0, j = 1, \ldots, n,$$

$$\omega_i, \pi_d, \mu_r \geq 0, \ i = 1, \ldots, n, \ d = 1, \ldots, D, \ r = 1, \ldots, S. \qquad (6.111)$$

Due to multiple optimal weights, similar to the model (6.106), the efficiency of stage 1 is calculated by the model (6.112).

$$\max \frac{\sum_{d=1}^{D} \pi_d z_{do}}{\sum_{i=1}^{m} \omega_i x_{io}}$$

s.t

$$\frac{\sum_{r=1}^{s} \mu_r y_{rj}}{\sum_{d=1}^{D} \pi_d z_{dj}} \leq 1, j = 1, \ldots, n, \qquad \frac{\sum_{d=1}^{D} \pi_d z_{dj}}{\sum_{i=1}^{m} \omega_i x_{ij}} \leq 1, j = 1, \ldots, n,$$

$$\frac{\sum_{r=1}^{s} \mu_r y_{ro} + \sum_{d=1}^{D} \pi_d z_{do}}{\sum_{d=1}^{D} \pi_d z_{do} + \sum_{i=1}^{m} \omega_i x_{io}} = e_o^{*Total},$$

$$\omega_i, \pi_d, \mu_r \geq 0, i = 1, \ldots, m, d = 1, \ldots, D, r = 1, \ldots, s. \qquad (6.112)$$

So, applying Charnes and Cooper's variable change, model (6.112) becomes model (6.113).

$$\max \sum_{d=1}^{D} \pi_d z_{do}$$

$$s.t$$

$$\sum_{i=1}^{m} \omega_i x_{io} = 1,$$

$$\sum_{r=1}^{s} \mu_r y_{rj} - \sum_{d=1}^{D} \pi_d z_{dj} \leq 0, \ j = 1, \ldots, n,$$

$$\sum_{d=1}^{D} \pi_d z_{dj} - \sum_{i=1}^{m} \omega_i x_{ij} \leq 0, \ j = 1, \ldots, n,$$

$$\frac{\sum_{r=1}^{s} \mu_r y_{ro} + \sum_{d=1}^{D} \pi_d z_{do}}{\sum_{d=1}^{D} \pi_d z_{do} + \sum_{i=1}^{m} \omega_i x_{io}} = e_o^{*Total},$$

$$\omega_i, \pi_d, \mu_r \geq 0, i = 1, \ldots, m, d = 1, \ldots, D, r = 1, \ldots, s. \qquad (6.113)$$

finally, $e_o^{1*} = \sum_{d=1}^{D} \pi_d^* z_{do}$ and $e_o^{2*} = \frac{e_o^{*Total} - w_1^* e_o^{1*}}{w_2^*}$, so that w_1^* and w_2^* are obtained from the optimal solution of model (6.113). Another advantage of the model (6.108) is that it is generalizable to the variable return to scale. Similarly, by defining $e_j^1 = \frac{\sum_{d=1}^{D} w_d z_{dj} + u^A}{\sum_{i=1}^{m} v_i x_{ij}}$ and $e_j^2 = \frac{\sum_{r=1}^{s} u_r y_{rj} + u^B}{\sum_{d=1}^{D} w_d z_{dj}}$, the model (6.114) is gotten to evaluate the performance of a two-stage without considering the constant return to scale.

$$\max \sum_{d=1}^{D} \pi_d z_{do} + u^A + \sum_{r=1}^{s} \mu_r y_{ro} + u^B$$

$$s.t$$

$$\sum_{i=1}^{m} \omega_i x_{io} + \sum_{d=1}^{D} \pi_d z_{do} = 1,$$

$$\sum_{d=1}^{D} \pi_d z_{dj} - \sum_{i=1}^{m} \omega_i x_{ij} + u^A \leq 0, \ j = 1, \ldots, n,$$

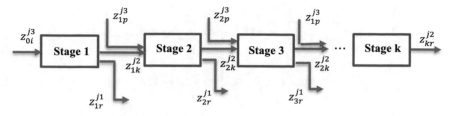

Fig. 6.11 A general series multi-stage network

$$\sum_{r=1}^{s} \mu_r y_{rj} - \sum_{d=1}^{D} \pi_d z_{dj} + u^B \leq 0, \; j = 1, \ldots, n,$$

$$\omega_i, \pi_d, \mu_r \geq 0, \; i = 1, \ldots, m, \; d = 1, \ldots, D, \; r = 1, \ldots, s. \quad (6.114)$$

Considering that the model presented by Chen et al. is only suitable for evaluating the performance of two-stage networks, [16] suggested a new model for assessing multi-stage DMUs in general. Regard the Fig. 6.11.

Suppose that:

z_{pr}^{j1} The rth component of the R_p-dimensional output vector from stage p of DMUj leaves the system.

z_{pk}^{j2} The kth component of the s_p-dimensional output vector from stage p of DMUj which enters as input to stage $p + 1$.

z_{pi}^{j3} The ith component of the I_p-dimensional input vector of DMUj which enters the system in stage $p + 1$.

In this case, the efficiency of stage 1 is:

$$\theta_1 = \frac{\sum_{r=1}^{R_1} u_{1r} z_{1r}^{j1} + \sum_{k=1}^{s_1} \eta_{1k} z_{1k}^{j2}}{\sum_{i=1}^{i_o} v_{0i} z_{0i}^{j}} \quad (6.115)$$

Also, the efficiency of other stages ($p = 2, \ldots, P$) is defined as follows:

$$\theta_p = \frac{\sum_{r=1}^{R_p} u_{pr} z_{pr}^{j1} + \sum_{k=1}^{s_p} \eta_{pk} z_{pk}^{j2}}{\sum_{k=1}^{s_{p-1}} \eta_{(p-1)k} z_{(p-1)k}^{j2} + \sum_{k=1}^{I_p} v_{(p-1)i} z_{(p-1)i}^{j3}} \quad (6.116)$$

So total efficiency of network is equal to $\theta = \sum_{p=1}^{P} w_p \theta_p$ such that $\sum_{p=1}^{P} w_p = 1$.

Similarly, we introduce model (6.117) to calculate the efficiency score of multi-stage networks with the mentioned process for the model (6.108).

$$\max \sum_{p=1}^{P} \left(\sum_{r=1}^{R_p} u_{pr} z_{pr}^{o1} + \sum_{k=1}^{s_p} \eta_{pk} z_{pk}^{o2} \right)$$

$s.t$

$$\sum_{i=1}^{i_o} v_{0i} z_{0i}^j + \sum_{p=2}^{P} \left(\sum_{k=1}^{s_{p-1}} \eta_{(p-1)k} z_{(p-1)k}^{o2} + \sum_{k=1}^{i_p} v_{(p-1)i} z_{(p-1)i}^{o3} \right) = 1,$$

$$\sum_{r=1}^{R_1} u_{1r} z_{1r}^{j1} + \sum_{k=1}^{s_1} \eta_{1k} z_{1k}^{j2} - \sum_{i=1}^{i_o} v_{0i} z_{0i}^j \leq 0, \; j = 1, \dots, n,$$

$$\sum_{r=1}^{R_p} u_{pr} z_{pr}^{j1} + \sum_{k=1}^{s_p} \eta_{pk} z_{pk}^{j2}$$

$$- \left(\sum_{k=1}^{s_{p-1}} \eta_{(p-1)k} z_{(p-1)k}^{j2} + \sum_{k=1}^{i_p} v_{(p-1)i} z_{(p-1)i}^{j3} \right) \leq 0, \quad j = 1, \dots, n,$$

$$p = 2, \dots, P, u_{pr}, \eta_{pk}, v_{pi}, v_{0i} \geq 0. \tag{6.117}$$

Although the described NDEA models solve some of the difficulties related to series multi-stage networks, one of the crucial disadvantages in utilizing NDEA models is the incapability of the models to introduce the projection of inefficient units. Chen et al. [13] proposed a new model equivalent to the (6.104) model to solve this problem. Consider the constraint $\sum_{j=1}^{n} (\lambda_j - \mu_j) z_{dj} \geq 0$ in the model (6.105) to describe Chen et al.'s method. By introducing a new set of intermediate measures in the form of $\tilde{z}_{do}$, we decompose the constraint $\sum_{j=1}^{n} (\lambda_j - \mu_j) z_{dj} \geq 0$ into two constraints, $\sum_{j=1}^{n} \mu_j z_{dj} \leq \tilde{z}_{do}$ and $\sum_{j=1}^{n} \lambda_j z_{dj} \geq \tilde{z}_{do}$. The first group of limitations is considered input, and the second group plays the output role. Thus, the model proposed by Chen et al. is expressed as follows:

$$\min \theta$$

$$\sum_{j=1}^{n} \lambda_j x_{ij} \leq \theta x_{io}, \; i = 1, \dots, m,$$

$$\sum_{j=1}^{n} \lambda_j z_{dj} \geq \tilde{z}_{do}, \; d = 1, \dots, D,$$

$$\sum_{j=1}^{n} \mu_j z_{dj} \leq \tilde{z}_{do}, \; d = 1, \dots, D,$$

$$\sum_{j=1}^{n} \mu_j y_{rj} \geq y_{ro}, \; r = 1, \dots, s,$$

$$\sum_{j=1}^{n} \mu_j = 1, \sum_{j=1}^{n} \lambda_j = 1,$$

$$\tilde{z}_{do}, \mu_j, \lambda_j \geq 0, \; j = 1, \dots, n, \; d = 1, \dots, D. \tag{6.118}$$

So network o is efficient if and only if $\theta^* = 1$.

To obtain the multiplier form of the model (6.118), it is enough to get the dual of the model (6.118) such as the model (6.119).

$$\max \sum_{r=1}^{s} u_r y_{ro}$$

$s.t$

$$\sum_{i=1}^{m} v_i x_{io} = 1,$$

$$\sum_{d=1}^{D} w_d^1 z_{dj} - \sum_{i=1}^{m} v_i x_{ij} \leq 1, \ j = 1, \ldots, n,$$

$$\sum_{r=1}^{s} u_r y_{rj} - \sum_{d=1}^{D} w_d^2 z_{dj} \leq 1, \ j = 1, \ldots, n,$$

$$w_d^2 - w_d^1 \leq 0,$$

$$v_i, w_d^1, w_d^2, u_r \geq 0, i = 1, \ldots, m, \ d = 1, \ldots, D, \ r = 1, \ldots, s. \quad (6.119)$$

Consider the model (6.118) again. According to the constraint $\sum_{j=1}^{n} \mu_j z_{dj} \leq \tilde{z}_{do}$, it is evident that $\tilde{z}_{do} \geq 0$, so by dismissing this condition from the model (6.120), the variable z is transformed into a free variable in the model (6.118) and so the constraint $w_d^2 - w_d^1 \leq 0$ in the model (6.119) becomes $w_d^2 - w_d^1 = 0$. At a result by placing the constraint $w_d^2 - w_d^1 = 0$ in the model (6.119), model (6.120) is gotten.

$$\max \sum_{r=1}^{s} u_r y_{ro}$$

$s.t$

$$\sum_{i=1}^{m} v_i x_{io} = 1,$$

$$\sum_{d=1}^{D} w_d z_{dj} - \sum_{i=1}^{m} v_i x_{ij} \leq 1, \ j = 1, \ldots, n,$$

$$\sum_{r=1}^{s} u_r y_{rj} - \sum_{d=1}^{D} w_d z_{dj} \leq 1, \ j = 1, \ldots, n,$$

$$v_i, w_d, u_r \geq 0, \ i = 1, \ldots, m, \ d = 1, \ldots, D, \ r = 1, \ldots, s. \quad (6.120)$$

Therefore, the model (6.120) is equivalent to the model presented by [44].

Definition 6.14
The projection of DMUo is equall to $(\theta^* x_o, \tilde{z}_o^*, y_o)$ so that θ^* and $\tilde{z}_o^*$ are optimal solution of the model (6.120).

Although model (6.120) solves the problem of not presenting the projection by NDEA models, the projection is not necessarily efficient. To solve this problem, [63] started their discussion by focusing on the PPS of a two-stage unit to present an efficient projection. Since the first stage consumes x and produces z, the second stage produces the output y by consumption z, thus, z plays both input and output roles; therefore, to build PPS, it is considered in both roles. So we display the network j as (x_j, z_j, z_j, y_j). The following steps are performed by considering the five principles described in Sect. 5.4 to create a PPS of the two-stage unit.

1. $(x_j, z_j, z_j, y_j), j = 1, ..., n$ belong to PPS.
2. Every convex combination of activities belonging to PPS belongs to the production possibility set. In other words, suppose $(x_i, z_i, z_i, y_i) \in T^{network}$ and $(x_k, z_k, z_k, y_k) \in T^{network}$ so:

$$\lambda(x_i, z_i, z_i, y_i) + (1 - \lambda)(x_k, z_k, z_k, y_k) \in T^{network} \text{ and } \lambda \geq 0$$

3. Suppose stage 1 produces the output z by consuming the input x, in this case, with any input $\bar{x}$ so that $x \leq \bar{x}$, it is able to produce $\bar{z}$ that $\bar{z} \leq z$. In the same way, suppose stage 2 produces the output y by consuming the input z, so with each input $\bar{z}$, this stage can create output $\bar{y}$ so that $\bar{y} \leq y$. Therefore, if $(x, z, z, y) \in T^{network}$ and $(x, -z, z, -y) \leq (\bar{x}, -\bar{z}, \bar{z}, -\bar{y})$, then $(\bar{x}, \bar{z}, \bar{z}, \bar{y}) \in T^{network}$.
4. If $(x, z, z, y) \in T^{network}$, then for all $\lambda \geq 0$: $\lambda(x, z, z, y) \in T^{network}$.

Therefore, the production possibility set a two-stage network is expressed as follows:

$$T^{network} = \left\{ (x_j, z_j, z_j, y_j) \middle| \sum_{j=1}^{n} \lambda_j x_j \leq x, \sum_{j=1}^{n} \lambda_j z_j \leq z, \right.$$
$$\left. \sum_{j=1}^{n} \lambda_j z_j \geq z, \sum_{j=1}^{n} \lambda_j y_j \geq y, \lambda \geq 0 \right\} \tag{6.121}$$

therefore:

$$T^{network} = \left\{ (x_j, z_j, z_j, y_j) \middle| \sum_{j=1}^{n} \lambda_j x_j \leq x, \sum_{j=1}^{n} \lambda_j z_j = z, \sum_{j=1}^{n} \lambda_j y_j \geq y, \lambda \geq 0 \right\}$$
$$\tag{6.122}$$

In this way, to evaluate the performance of the two-stage using the $T^{network}$, model (6.123) is applied:

$$\min$$
$$s.t (\theta x_o, z_o, y_o) \in T^{network} \tag{6.123}$$

Consequently:

$$\min \theta$$

$$s.t$$

$$\sum_{j=1}^{n} \lambda_j x_j \leq \theta x_o,$$

$$\sum_{j=1}^{n} \lambda_j z_j = z_o,$$

$$\sum_{j=1}^{n} \lambda_j y_j \geq y_o,$$

$$\lambda_j \geq 0, j = 1, \ldots, n. \qquad (6.124)$$

Similar to the black box models, to identify strong efficient DMUs, model (6.125) is used.

$$\min 1s^- + 1s^+$$

$$s.t$$

$$\sum_{j=1}^{n} \lambda_j x_j + s^- = \theta^* x_o,$$

$$\sum_{j=1}^{n} \lambda_j z_j = z_o,$$

$$\sum_{j=1}^{n} \lambda_j y_j - s^+ = y_o,$$

$$s^- \geq 0, s^+ \geq 0,$$

$$\lambda_j \geq 0, j = 1, \ldots, n. \qquad (6.125)$$

Definition 6.15
If θ^*, s^{+*} and s^{-*} are the optimal solutions of model (6.124) and (6.126), respectively, then the projection of DMUo is defined as follows:

$$\hat{x}_o = \theta^* x_o - s^{-*} \qquad (6.126)$$

$$\hat{z}_o = z_o \qquad (6.127)$$

$$\hat{y}_o = y_o + s^{+*} \qquad (6.128)$$

Theorem 6.9
The projection of DMUo using relations (6.126)–(6.128) is efficient.

Proof The proof is similar to Theorem 6.2.

In the explained methods, stages efficiency is obtained by decomposition of the total efficiency into the efficiency of stages 1 and 2 (and in the general case, stages 1 to k). Another group of methods that became famous as composition methods is the opposite of decomposition methods, and the efficiency of stages is used to calculate the total efficiency. Composition methods were first introduced by Despotis et al. [19]. Suppose that in a two-stage DMU, as in Fig. 6.9, the efficiency of stage 1 is regarded as $\frac{\sum_{i=1}^{m} v_i x_{io}}{\sum_{d=1}^{D} w_d z_{do}}$ (output-oriented), and the efficiency of stage 2 is considered as $\frac{\sum_{r=1}^{s} u_r y_{ro}}{\sum_{d=1}^{D} w_d z_{do}}$ (input-oriented). In this case, to improve the efficiency of stages 1 and 2, the model (6.129) is suggested.

$$\min \frac{\sum_{i=1}^{m} v_i x_{io}}{\sum_{d=1}^{D} w_d z_{do}}$$

$$\max \frac{\sum_{r=1}^{s} u_r y_{ro}}{\sum_{d=1}^{D} w_d z_{do}}$$

$$s.t$$

$$\frac{\sum_{r=1}^{s} u_r y_{rj}}{\sum_{d=1}^{D} w_d z_{dj}} \le 1, j = 1, \ldots, n, \quad \frac{\sum_{i=1}^{m} v_i x_{io}}{\sum_{d=1}^{D} w_d z_{do}} \ge 1, j = 1, \ldots, n,$$

$$v_i, w_d, u_r \ge 0, i = 1, \ldots, m, d = 1, \ldots, D, r = 1, \ldots, s. \quad (6.129)$$

The efficiency value should be minimized to improve the efficiency of stage 1, but for stage 2, it should be maximized to enhance the performance of the second stage. Therefore, the model (6.129) is obtained as a bi-objective fractional programming problem. In the following, utilizing Charnes and Cooper transformations, model (6.129) is converted into model (6.130).

$$\min \sum_{i=1}^{m} v_i x_{io}$$

$$\max \sum_{r=1}^{s} u_r y_{ro}$$

$$s.t$$

$$\sum_{d=1}^{D} w_d z_{do} = 1,$$

$$\sum_{r=1}^{s} u_r y_{rj} - \sum_{d=1}^{D} w_d z_{dj} \le 0, j = 1, \ldots, n,$$

$$\sum_{d=1}^{D} w_d z_{do} - \sum_{i=1}^{m} v_i x_{io} \leq 0, \ j = 1, \ldots, n,$$

$$v_i, w_d, u_r \geq 0, \ i = 1, \ldots, m, \ d = 1, \ldots, D, \ r = 1, \ldots, s. \tag{6.130}$$

Furthermore, to convert the problem (6.130) into a linear single-objective problem, the model (6.131) is attained by combining objective functions.

$$\min \sum_{i=1}^{m} v_i x_{io} - \sum_{r=1}^{s} u_r y_{ro}$$

s.t

$$\sum_{d=1}^{D} w_d z_{do} = 1,$$

$$\sum_{r=1}^{s} u_r y_{rj} - \sum_{d=1}^{D} w_d z_{dj} \leq 0, \ j = 1, \ldots, n,$$

$$\sum_{d=1}^{D} w_d z_{dj} - \sum_{i=1}^{m} v_i x_{ij} \leq 0, \ j = 1, \ldots, n,$$

$$v_i, w_d, u_r \geq 0, \ i = 1, \ldots, n, \ d = 1, \ldots, D, \ r = 1, \ldots, s. \tag{6.131}$$

If (v^*, u^*) is optimal solution of model (6.131), then $e_o^1 = \frac{\sum_{i=1}^{m} v_i^* x_{io}}{\sum_{d=1}^{D} w_d^* z_{do}}$ and $e_o^2 = \frac{\sum_{d=1}^{D} w_d^* z_{do}}{\sum_{r=1}^{s} u_r^* y_{rj}}$. Also, considering that: $\sum_{r=1}^{s} u_r y_{rj} \leq \sum_{d=1}^{D} w_d z_{dj} \leq \sum_{i=1}^{m} v_i x_{io} \Rightarrow \sum_{i=1}^{m} v_i x_{io} - \sum_{r=1}^{s} u_r y_{rj} \geq 0$, as a result, the value of the objective function is always non-negative.

It is evident that DMU_o is efficient if and only if $\sum_{i=1}^{m} v_i^* x_{io} - \sum_{r=1}^{s} u_r^* y_{rj} = 0$.

For more information about NDEA using composition method, references [20, 21, 59] are recommended.

6.10.2 Parallel Networks

In the real world, some systems under evaluation are made of independent and parallel subunits. Although subunits work independently, the performance of the entire system is affected by the performance of each of them. Consider the k-stage parallel network shown in Fig. 6.12.

So that x_j and y_j are the system's initial input and final output, respectively. Also, $x_j^p, y_j^p, p = 1, \ldots, P$ are each component's independent inputs and outputs. So it is evident that: $x_j = \sum_{p=1}^{P} x_j^p$ and $y_j = \sum_{p=1}^{P} y_j^p$. In parallel networks, three basic assumptions are considered:

Fig. 6.12 Parallel network

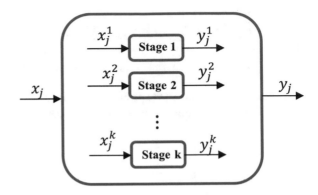

- None of the network subunits have shared inputs and shared outputs.
- There are no intermediate flows between network components.
- Each initial input or final output of the network belongs to at least one of the subunits.

Of course, some of the stated assumptions were ignored with the development of parallel network evaluation models. To compute the efficiency of parallel networks, consider the constraint $\sum_{r=1}^{s} u_r y_{ro} - \sum_{i=1}^{m} v_i x_{io} + s_o = 0$ in the model () again. By placing x_j and y_j values:

$$\sum_{r=1}^{s} u_r \left(\sum_{p=1}^{P} y_o^p \right) - \sum_{i=1}^{m} v_i \left(\sum_{t=1}^{k} x_o^p \right) + s_o = 0$$

$$\Rightarrow \sum_{p=1}^{P} \left(\sum_{r=1}^{s} u_r y_o^p - \sum_{i=1}^{m} v_i x_o^p \right) + s_o = 0 \qquad (6.132)$$

the relation: $\sum_{r=1}^{s} u_r y_o^t - \sum_{i=1}^{m} v_i x_o^t$ in the constraint (6.132), is related to stage p. Considering that the efficiency of each stage must be less than or equal to 1, so $\sum_{r=1}^{s} u_r y_o^p - \sum_{i=1}^{m} v_i x_o^p \le 0$. As a result: $\sum_{r=1}^{s} u_r y_o^p - \sum_{i=1}^{m} v_i x_o^p + s_o^p = 0$. Since each variable can be expressed as a sum of positive variables, therefor $s_o = \sum_{p=1}^{P} s_o^p$ and $s_o^p \ge 0$. Hence we have: $\sum_{p=1}^{P} \left(\sum_{r=1}^{s} u_r y_o^p - \sum_{i=1}^{m} v_i x_o^p \right) + \sum_{p=1}^{P} s_o^p = 0$. Based on the explained relations, model (6.133) was presented by [43] to assess the under-evaluation of parallel networks.

$$\min \sum_{p=1}^{P} s_o^p$$

$$\sum_{i=1}^{m} v_i x_o = 1,$$

$$\sum_{r=1}^{s} u_r y_o^p - \sum_{i=1}^{m} v_i x_o^p + s_o^p = 0, p = 1, \ldots, P,$$

$$\sum_{r=1}^{s} u_r y_j^p - \sum_{i=1}^{m} v_i x_j^p \leq 0, p = 1, \ldots, P, j = 1, \ldots, n, j \neq o,$$

$$\sum_{p=1}^{P} \left(\sum_{r=1}^{s} u_r y_o^p - \sum_{i=1}^{m} v_i x_o^p \right) + \sum_{p=1}^{P} s_o^p = 0,$$

$$u_r, v_i \geq 0, \ i = 1, \ldots, m, \ r = 1, \ldots, s$$

$$s_o^p \geq 0, \ p = 1, \ldots, P. \tag{6.133}$$

6.10.3   Mix Networks

The third configuration is the Mix network. These systems are composed of series and parallel structures at the same time. Therefore, this type of network has more complexities than the series and parallel networks but the research done on the Mix network is less than that of series and parallel networks.

Tone and Tsutsui introduced a new model [74] using the SBM model and its generalization to evaluate mix networks, as shown in Fig. 6.13. In this method, two points of view are considered. The first, intermediate products can change freely (free link), but this assumption is not considerd in the second point of view. According to the first point of view, the model (6.134) was presented as follows.

$$\min \frac{\sum_{k=1}^{K} w^k \left[1 - \frac{1}{m_k} \sum_{i=1}^{m_k} \frac{s_i^{k-}}{x_{io}^k} \right]}{\sum_{k=1}^{K} w^k \left[1 + \frac{1}{r_k} \sum_{r=1}^{r_k} \frac{s_r^{k+}}{y_{ro}^k} \right]}$$

s.t

$$\sum_{j=1}^{n} \lambda_j^k x_{ij}^k + s_i^{k-} = x_{io}^k, i = 1, \ldots, m_k, k = 1, \ldots, K,$$

$$\sum_{j=1}^{n} \lambda_j^k y_{rj}^k - s_r^{k+} = y_{ro}^k, r = 1, \ldots, r_k, k = 1, \ldots, K,$$

$$\sum_{j=1}^{n} \lambda_j^k z_{rj}^{(k,h)} = z^{(k,h)}, k = 1, \ldots, K, \ h \in H^k,$$

$$\sum_{j=1}^{n} \lambda_j^h z_{rj}^{(k,h)} = z^{(k,h)}, \ k = 1, \ldots, K, \ h \in H^k,$$

$$\sum_{j=1}^{n} \lambda_j^k = 1, \ k = 1, \ldots, K,$$

$$\lambda_j^k, s_i^{k-}, s_r^{k+} \geq 0, \ k = 1, \ldots, K, \ j = 1, \ldots, n, \ i = 1, \ldots, m_k, \ r = 1, \ldots, r_k.$$

$$(6.134)$$

w^k is the weight (degree of importance) of the kth division of the network, which the network manager specifies, and it should be noted that $\sum_{k=1}^{K} w^k = 1$. Also H^k is the set of indices related to the stages where the intermediate products left from stage k and are entered into those stages.

In model (6.134), $z^{(k,h)}$ is a variable. Therefore, intermediate products change freely. But if using the second point of view, relations (6.135) and (6.136) are added to the model (6.134).

$$\sum_{j=1}^{n} \lambda_j^k z_{rj}^{(k,h)} = z_o^{(k,h)}, k = 1, \ldots, K, h \in H^k \tag{6.135}$$

$$\sum_{j=1}^{n} \lambda_j^h z_{rj}^{(k,h)} = z_o^{(k,h)}, k = 1, \ldots, h \in H^k \tag{6.136}$$

If DMU_o is under evaluation and (s^{k-*}, s^{k+*}) is the optimal solution of the model (6.134), then the efficiency of the stage k is:

Fig. 6.13 A mix network

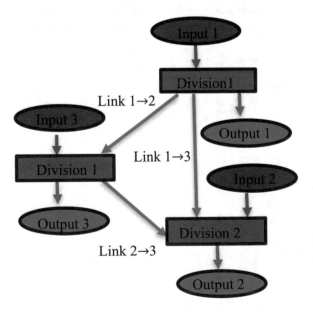

$$\rho^{k*} = \frac{1 - \frac{1}{m_k} \sum_{i=1}^{m_k} \frac{s_i^{k-*}}{x_{io}^k}}{1 + \frac{1}{r_k} \sum_{r=1}^{r_k} \frac{s_r^{k+*}}{y_{ro}^k}} \tag{6.137}$$

Also, to evaluate input or output-oriented, it is enough to replace the objective function of the model (6.134) using relations (6.138) and (6.139), respectively.

$$\min \sum_{k=1}^{K} w^k \left[1 - \frac{1}{m_k} \sum_{i=1}^{m_k} \frac{s_i^{k-}}{x_{io}^k} \right] \tag{6.138}$$

$$\max \sum_{k=1}^{K} w^k \left[1 + \frac{1}{r_k} \sum_{r=1}^{r_k} \frac{s_r^{k+}}{y_{ro}^k} \right] \tag{6.139}$$

In the same way, we use relations (6.140) and (6.141) to calculate the efficiency of stage k of network o in input or output-oriented.

$$\theta^{k*} = 1 - \frac{1}{m_k} \sum_{i=1}^{m_k} \frac{s_i^{k-*}}{x_{io}^k} \tag{6.140}$$

$$\tau^{k*} = \frac{1}{1 + \frac{1}{r_k} \sum_{r=1}^{r_k} \frac{s_r^{k+*}}{y_{ro}^k}} \tag{6.141}$$

So that s^{k-*} and s^{k+*} are the optimal solutions of model (6.134) considering the objective functions (6.138) and (6.139), respectively.

Definition 6.16
Suppose that $(\hat{s}^{k-}, \hat{s}^{k+}, \hat{\lambda}^k)$ is the optimal solution of the model (6.134). Therefore, the projection of the network o in the free link is defined as follows:

$$\hat{x}_o^k = x_o^k - \hat{s}^{k-}, k = 1, \dots, K \tag{6.142}$$

$$\hat{y}_o^k = y_o^k + \hat{s}^{k+}, k = 1, \dots, K \tag{6.143}$$

$$\hat{z}_o^k = \sum_{j=1}^{n} \hat{\lambda}_j^k z_j^{(k,j)}, k = 1, \dots, K \tag{6.144}$$

but the intermediate remains constant in the second point of view.

Theorem 6.10
The projection of the network o using the relation (6.142) to (6.144) is efficient.

Proof We prove the theorem in input-oriented. Assume that DMU_o is an under-evaluation network and $(\overline{s}^{k-}, \overline{s}^{k+}, \overline{\lambda}^k, \overline{z}_o^k)$ is the optimal solution for the model

(6.134). So:

$$\sum_{j=1}^{n} \bar{\lambda}_j^k x_{ij}^k + \bar{s}_i^{k-} = \hat{x}_{io}^k \Rightarrow \sum_{j=1}^{n} \bar{\lambda}_j^k x_{ij}^k + \bar{s}_i^{k-}$$

$$= x_o^k - \hat{s}^{k-} \Rightarrow \sum_{j=1}^{n} \bar{\lambda}_j^k x_{ij}^k + \bar{s}_i^{k-} + \hat{s}^{k-} = x_o^k$$

$$\sum_{j=1}^{n} \bar{\lambda}_j^k y_{rj}^k - \bar{s}_r^{k+} = \hat{y}_{ro}^k \Rightarrow \sum_{j=1}^{n} \bar{\lambda}_j^k y_{rj}^k - \bar{s}_r^{k+} = y_o^k + \hat{s}^{k+}$$

$$\Rightarrow \sum_{j=1}^{n} \bar{\lambda}_j^k y_{rj}^k - \bar{s}_r^{k+} - \hat{s}^{k+} = y_o^k$$

To prove the efficiency of image q, it is enough to show w in reverse, suppose w. It is enough to show $\bar{s}^{k-} = 0$. proof by contradiction, suppose that $\bar{s}^{k-} \neq 0$. It is easily proved that $(\bar{s}^{k-} + \hat{s}^{k-}, \bar{s}^{k+} + \hat{s}^{k+}, \bar{\lambda}^k, \bar{z}_o^k)$ is a feasible solution for model (), and the value of the objective function for this feasible solution is equal to: $\sum_{k=1}^{K} w^k \left[1 - \frac{1}{m_k} \sum_{i=1}^{m_k} \frac{\bar{s}^{k-} + \hat{s}^{k-}}{x_{io}^k} \right]$. It is evident that: $\sum_{k=1}^{K} w^k \left[1 - \frac{1}{m_k} \sum_{i=1}^{m_k} \frac{\bar{s}^{k-} + \hat{s}^{k-}}{x_{io}^k} \right] \leq \sum_{k=1}^{K} w^k \left[1 - \frac{1}{m_k} \sum_{i=1}^{m_k} \frac{\bar{s}^{k-}}{x_{io}^k} \right]$ and it is a contradiction. So $\bar{s}^{k-} = 0$.

6.11   Conclusion

At the end of this chapter, we mention some essential points that readers should consider. As definitely noticed, DEA models have different properties and approaches to measuring the efficiency of DMUs. Therefore, the proper choice of the model for performance evaluation is one of the crucial steps in evaluating decision-making units that should be considered. Because the efficiency value obtained from DEA models is different, the model selection affects the efficiency score and, in some cases, the type of classification of units in the set of efficient and inefficient units. Therefore, paying awareness to the following items is beneficial when choosing models.

- **PPS type**

Since the production possibility set is based on the properties of the under-evaluation society, the models based on each PPS also inherit these properties. Therefore, to choose the appropriate model, one must first know about the properties of the society under assessment. These properties are determined using experts' opinions or methods such as Angles Method Alirezaee et al. [1]. For instance, the use of the

CCR model, it is only for the conditions that the society under assessment has all five properties mentioned in Sect. 6.2, but by removing the principle of constant return to scale (taking into account the other four principles), the BCC model or VRS non-radial models should be used.

- **Selection of input or output—oriented models**

One of the advantages of using DEA models is providing an efficient image for inefficient units. According to the choice of the type of model, the obtained image will also be different.

For example, the projection of the input-oriented CCR model is $(\theta^* x_o - s^{-*}, y_o + s^{+*})$, while in the output-oriented CCR model, the projection is $(x_o - s^{-*}, \varphi^* y_o + s^{+*})$. Also, although using the CCR model, the efficiency obtained in the input and output-oriented are equal, this issue is not established concerning other models, for example, the BCC model. Therefore, the choice of the model type in the obtained projection and efficiency score is effective.

- **Stability of models**

As explained in the previous sections, some models are translation invariant, and some models are unit invariant. Therefore, suitable models can be used according to the available data and the desired goals. For example, stable model to translation invariant should be used if negative data exists among the inputs or outputs.

- **How to choose inputs and outputs and their number?**

Another important point that should be considered utilizing DEA models is the choice of inputs and outputs and their numbers. Suppose n, m, and s are the number of under-evaluation units' DMUs, inputs, and outputs. In that case, it is better to establish the relation () among the units under evaluation and their indicators:

$$n \geq 3(m + s) \tag{6.145}$$

Of course, it is emphasized that the relation () has no mathematical reasoning and is only an empirical relationship. Finally, experts believe that selecting inputs and outputs is an art, and reliable results can be obtained by choosing them artistically.

- **Using DEA or NDEA models**

DEA models are only suitable for evaluating one-component units. Because using DEA models, only input and primary outputs are considered, and no information about the subunits and internal relationships is provided. So when faced with multi-stage units and if need to be aware of information on the performance of each sub-unit, NDEA models should be used.

- **Typese of inputs and outputs**

DEA models assume that the input and output are always non-negative real numbers. However, this assumption is not necessarily valid in practical applications. For example, in some problems, numbers are expressed qualitatively; in such situations, the fuzzy DEA models are used, including [2, 3, 5, 23, 37, 42, 31, 47, 50, 51,50, 58, 57, 64, 6568].

Also, suppose that a company's profit is between 10 and 15 units. When the numbers are expressed as an interval, Interval DEA is used. To get knowledgeable about Interval DEA, readers can refer to [76], Azadeh et al. (2008), Jahanshahloo et al. [37, 42, 34], Tamaddon et al. [72].

Another critical indicator in company evaluation is the number of factory equipment. This indicator is an integer number, so the projection presented for this indicator must be an integer number. Therefore, in the presence of integer data, Integer DEA models are applied. Among the studies conducted in this field are [74, 78, 79].

References

1. Alirezaee, M., Hajinezhad, E., Paradi, J.C.: Objective identification of technological returns to scale for data envelopment analysis models. Eur. J. Oper. Res. **266**(2), 678–688 (2018)
2. Allahviranloo, T., Hosseinzadeh Lotfi, F., AdabitabarFirozja, M.: Efficiency in fuzzy production possibility set. Iranian J. Fuzzy Syst. **9**(4), 17–30 (2012)
3. Allahviranloo, T., Lotfi, F.H., Kiasari, M.K., Khezerloo, M.: On the fuzzy solution of LR fuzzy linear systems. Appl. Math. Model. **37**(3), 1170–1176 (2013)
4. Andersen, P., Petersen, N.C.: A procedure for ranking efficient units in data envelopment analysis. Manage. Sci. **39**(10), 1261–1264 (1993)
5. Bagheri, M., Ebrahimnejad, A., Razavyan, S., Hosseinzadeh Lotfi, F., Malekmohammadi, N.: Solving the fully fuzzy multi-objective transportation problem based on the common set of weights in DEA. J. Intell. Fuzzy Syst. **39**(3), 3099–3124 (2020)
6. Banker, R.D., Charnes, A., Cooper, W.W.: Some models for estimating technical and scale inefficiencies in data envelopment analysis. Manage. Sci. **30**(9), 1078–1092 (1984)
7. Banker, R.D., Cooper, W.W., Seiford, L.M., Thrall, R.M., Zhu, J.: Returns to scale in different DEA models. Eur. J. Oper. Res. **154**(2), 345–362 (2004)
8. Cavalheiro Francisco, C.A., Rodrigues de Almeida, M., Ribeiro da Silva, D.: Efficiency in Brazilian refineries under different DEA technologies. Int. J. Eng. Bus. Manage. **4**(Godište 2012), 4–35 (2012)
9. Charnes, A., Cooper, W.W., Golany, B., Seiford, L., Stutz, J.: Foundations of data envelopment analysis for Pareto-Koopmans efficient empirical production functions. J. Econometr. **30**(1–2), 91–107 (1985)
10. Charnes, A., Cooper, W. W., Rhodes, E.: Measuring the efficiency of decision making units. Eur. j. of Oper. Res. **2**(6), 429–444 (1978)
11. Charnes, A, Cooper, W.W. .: Programming with linear fractional functionals. Naval Research Logistics Quarterly, **9**, 181–186 (1962)
12. Chen, Y., Cook, W.D., Li, N., Zhu, J.: Additive efficiency decomposition in two-stage DEA. Eur. J. Oper. Res. **196**(3), 1170–1176 (2009)
13. Chen, Y., Cook, W.D., Zhu, J.: Deriving the DEA frontier for two-stage processes. Eur. J. Oper. Res. **202**(1), 138–142 (2010)
14. Chen, Y., Zhu, J.: Measuring information technology's indirect impact on firm performance. Inf. Technol. Manage. **5**(1), 9–22 (2004)

15. Chugh, T.: Scalarizing functions in Bayesian multiobjective optimization. In: 2020 IEEE Congress on Evolutionary Computation (CEC), pp. 1–8. IEEE (2020, July)
16. Cook, W.D., Zhu, J., Bi, G., Yang, F.: Network DEA: additive efficiency decomposition. Euro. J. Oper. Res. **207**(2), 1122–1129 (2010)
17. Cooper, W.W., Park, K.S., Pastor, J.T.: RAM: a range adjusted measure of inefficiency for use with additive models, and relations to other models and measures in DEA. J. Prod. Anal. **11**(1), 5–42 (1999)
18. Cooper, W.W., Seiford, L.M., Tone, K.: Data Envelopment Analysis: A Comprehensive Text with Models, Applications, References and DEA-Solver Software. Springer, New York (2007)
19. Despotis, D.K., Koronakos, G., Sotiros, D.: Composition versus decomposition in two-stage network DEA: a reverse approach. J. Prod. Anal. **45**(1), 71–87 (2016)
20. Despotis, D.K., Koronakos, G., Sotiros, D.: The "weak-link" approach to network DEA for two-stage processes. Eur. J. Oper. Res. **254**(2), 481–492 (2016)
21. Despotis, D.K., Sotiros, D., Koronakos, G.: A network DEA approach for series multi-stage processes. Omega **61**, 35–48 (2016)
22. Doyle, J., Green, R.: Efficiency and cross-efficiency in DEA: derivations, meanings and uses. J. Oper. Res. Soc. **45**(5), 567–578 (1994)
23. Ebrahimnejad, A., Nasseri, S.H., Lotfi, F.H.: Bounded linear programs with trapezoidal fuzzy numbers. Int. J. Uncertain. Fuzziness Knowledge-Based Syst. **18**(03), 269–286 (2010)
24. Eguchi, S., Takayabu, H., Lin, C.: Sources of inefficient power generation by coal-fired thermal power plants in China: a metafrontier DEA decomposition approach. Renew. Sustain. Energy Rev. **138**, 110562 (2021)
25. Esmaeili, M.: An enhanced Russell measure in DEA with interval data. App. Math. Comput. **219**(4), 1589–1593 (2012)
26. Fare, R., Grosskopf, S.: Network DEA. Socioecon. Plann. Sci. **34**, 35–49 (2000)
27. Fare, R., Grosskopf, S., Lovell, C.A.K.: The Measurement of Efficiency of Production. Kluwer Academic Publisher (1985)
28. Farrell, M.J.: The measurement of productive efficiency. J. R. Stat. Soc. **120**, 253–281 (1957)
29. Henriques, C.O., Marcenaro-Gutierrez, O.D.: Efficiency of secondary schools in Portugal: a novel DEA hybrid approach. Socioecon. Plann. Sci. **74**, 100954 (2021)
30. Hosseinzadeh Lotfi, F., Jahanshahloo, G.R., Khodabakhshi, M., Rostamy-Malkhlifeh, M., Moghaddas, Z., Vaez-Ghasemi, M.: A review of ranking models in data envelopment analysis. J. Appl. Math. **2013** (2013)
31. Hosseinzadeh Lotfi, F., Ebrahimnejad, A., Vaez-Ghasemi, M., Moghaddas, Z.: Introduction to data envelopment analysis and fuzzy sets. In: Data Envelopment Analysis with R, pp. 1–17. Springer, Cham (2020)
32. Izadikhah, M., Saen, R.F.: Evaluating sustainability of supply chains by two-stage range directional measure in the presence of negative data. Transp. Res. Part D: Transp. Environ. **49**, 110–126 (2016)
33. Jahanshahloo, G.R., Junior, H.V., Lotfi, F.H., Akbarian, D.: A new DEA ranking system based on changing the reference set. Eur. J. Oper. Res. **181**(1), 331–337 (2007)
34. Jahanshahloo, G.R., Khodabakhshi, M., Lotfi, F.H., Goudarzi, M.M.: A cross-efficiency model based on super-efficiency for ranking units through the TOPSIS approach and its extension to the interval case. Math. Comput. Model. **53**(9–10), 1946–1955 (2011)
35. Jahanshahloo, G.R., Lotfi, F.H., Jafari, Y., Maddahi, R.: Selecting symmetric weights as a secondary goal in DEA cross-efficiency evaluation. Appl. Math. Model. **35**(1), 544–549 (2011)
36. Jahanshahloo, G.R., Lotfi, F.H., Khanmohammadi, M., Kazemimanesh, M., Rezaie, V.: Ranking of units by positive ideal DMU with common weights. Expert Syst. Appl. **37**(12), 7483–7488 (2010)
37. Jahanshahloo, G.R., Lotfi, F.H., Malkhalifeh, M.R., Namin, M.A.: A generalized model for data envelopment analysis with interval data. Appl. Math. Model. **33**(7), 3237–3244 (2009a)
38. Jahanshahloo, G.R., Lotfi, F.H., Rezaie, V., Khanmohammadi, M.: Ranking DMUs by ideal points with interval data in DEA. Appl. Math. Model. **35**(1), 218–229 (2011)

39. Jahanshahloo, G.R., Lotfi, F.H., Shoja, N., Tohidi, G., Razavyan, S.: Ranking using l1-norm in data envelopment analysis. Appl. Math. Comput. **153**(1), 215–224 (2004)
40. Jahanshahloo, G.R., Lotfi, F.H., Zohrehbandian, M.: Finding the efficiency score and RTS characteristic of DMUs by means of identifying the efficient frontier in DEA. Appl. Math. Comput. **170**(2), 985–993 (2005)
41. Jahanshahloo, G.R., Lotfi, F.H., Zohrehbandian, M.: Notes on sensitivity and stability of the classifications of returns to scale in data envelopment analysis. J. Prod. Anal. **23**(3), 309–313 (2005)
42. Jahanshahloo, G.R., Sanei, M., Rostamy-Malkhalifeh, M., Saleh, H.: A comment on "A fuzzy DEA/AR approach to the selection of flexible manufacturing systems." Comput. Ind. Eng. **56**(4), 1713–1714 (2009b)
43. Kao, C.: Efficiency measurement for parallel production systems. Eur. J. Oper. Res. **196**(3), 1107–1112 (2009)
44. Kao, C., Hwang, S.N.: Efficiency decomposition in two-stage data envelopment analysis: an application to non-life insurance companies in Taiwan. Eur. J. Oper. Res. **185**(1), 418–429 (2008)
45. Khaleghi, M., Jahanshahloo, G., Zohrehbandian, M., Lotfi, F.H.: Returns to scale and scale elasticity in two-stage DEA. Math. Comput. Appl. **17**(3), 193–202 (2012)
46. Kohl, S., Schoenfelder, J., Fügener, A., Brunner, J.O.: The use of data envelopment analysis (DEA) in healthcare with a focus on hospitals. Health Care Manage. Sci. **22**(2), 245–286 (2019)
47. Liu, S.T.: A fuzzy DEA/AR approach to the selection of flexible manufacturing systems. Comput. Ind. Eng. **54**(1), 66–76 (2008)
48. Lotfi, F.H., Eshlaghy, A.T., Saleh, H., Nikoomaram, H., Seyedhoseini, S.M.: A new two-stage data envelopment analysis (DEA) model for evaluating the branch performance of banks. Afr. J. Bus. Manage. **6**(24), 7230 (2012)
49. Lotfi, F.H., Jahanshahloo, G.R., Esmaeili, M.: An alternative approach in the estimation of returns to scale under weight restrictions. Appl. Math. Comput. **189**(1), 719–724 (2007)
50. Lotfi, F.H., Jahanshahloo, G.R., Vahidi, A.R., Dalirian, A.: Efficiency and effectiveness in multi-activity network DEA model with fuzzy data. Appl. Math. Sci. **3**(52), 2603–2618 (2009a)
51. Lotfi, F.H., Allahviranloo, T., Jondabeh, M.A., Alizadeh, L.: Solving a full fuzzy linear programming using lexicography method and fuzzy approximate solution. Appl. Math. Modelling **33**(7), 3151–3156 (2009b)
52. Maddahi, R., Jahanshahloo, G.R., Hosseinzadeh Lotfi, F., Ebrahimnejad, A.: Optimising proportional weights as a secondary goal in DEA cross-efficiency evaluation. Int. J. Oper. Res. **19**(2), 234–245 (2014)
53. Mahmoodirad, A., Allahviranloo, T., Niroomand, S.: A new effective solution method for fully intuitionistic fuzzy transportation problem. **23**(12), 4521–4530 (2019)
54. Pastor, J.T., Ruiz, J.L., Sirvent, I.: An enhanced DEA Russell graph efficiency measure. Eur. J. Oper. Res. **115**(3), 596–607 (1999)
55. Portela, M.C.A., Thanassoulis, E., Simpson, G.: Negative data in DEA: a directional distance approach applied to bank branches. J. Oper. Res. Soc. **55**(10), 1111–1121 (2004)
56. Ranjbar, H., Lotfi, F.H., Mozaffari, M., Gerami, J.: Finding defining hyperplanes with variable returns to scale technology. Int. J. Math. Anal. **3**(19), 943–954 (2009)
57. Rostamy Malkhlifeh, M., Ebrahimkhani, G.S., Saleh, H., Ebrahimkhani, G.N.: Congestion in DEA model with fuzzy data. Int. J. Appl. Oper. Res. **1**(2), 49–56 (2011)
58. Rostamy-Malkhalifeh, M., Sanei, M., Saleh, H.: A new method for solving fuzzy DEA models by trapezoidal approximation. J. Math. Extension **1**(4), 307–313 (2009)
59. Sahoo, B.K., Saleh, H., Shafiee, M., Tone, K., Zhu, J.: An alternative approach to dealing with the composition approach for series network production processes. Asia-Pacific J. Oper. Res. **38**(06), 2150004 (2021)
60. Saleh, H., Hosseinzade Lotfi, F., Rostamy, M.M., Shafiee, M.: Performance evaluation and specifying of Return to scale in network DEA international journal of industrial mathematics. J. Adv. Math. Modeling **10**(2), 309–340 (2021)

61. Saleh, H.I.L.D.A., Hosseinzadeh Lotfi, F., Rostmay-Malkhalifeh, M., Shafiee, M.: Provide a mathematical model for selecting suppliers in the supply chain based on profit efficiency calculations. J. New Res. Math. **7**(32), 177–186 (2021)
62. Saleh, H., Rostamy, M.M.: Performance evaluation in a bank branch with a two-stage DEA model. J. Syst. Manage. **1**(11), 17–33 (2013)
63. Saleh, H., Hosseinzadeh Lotfi, F., Toloie Eshlaghy, A., Shafiee, M.: A new two-stage DEA model for bank branch performance evaluation. In: 3rd National Conference on Data Envelopment Analysis, Islamic Azad University of Firoozkooh (2011)
64. Sanei, M., Rostami-Malkhalifeh, M., Saleh, H.: A new method for solving fuzzy DEA models. Int. J. Ind. Math. **1**(4), 307–313 (2009a)
65. Sanei, M., Noori, N., Saleh, H.: Sensitivity analysis with fuzzy Data in DEA. Appl. Math. Sci. **3**(25), 1235–1241 (2009b)
66. Sexton, T.R., Silkman, R.H., Hogan, A.J.: Data envelopment analysis: critique and extensions. New Directions Program Eval. **1986**(32), 73–105 (1986)
67. Shafiee, M., Hosseinzade Lotfi, F., Saleh, H.: Benchmark forecasting in data envelopment analysis for decision-making units. Int. J. Ind. Math. **13**(1), 29–42 (2021)
68. Shafiee, M., Saleh, H.: Evaluation of strategic performance with fuzzy data envelopment analysis. Int. J. Data Envelopment Anal. **7**(4), 1–20 (2019)
69. Shafiee, M., Saleh, H., Sanji, M.: Modifying the interconnecting activities through an adjusted dynamic DEA model: a slacks-based measure approach. J. Adv. Math. Modeling **10**(2), 309–340 (2020)
70. Shafiee, M., Saleh, H., Ziyari, R.: Projects efficiency evaluation by data envelopment analysis and balanced scorecard. J. Decis. Oper. Res. **6**(Special Issue), 1–19 (2022)
71. Stewart, T.J.: Goal directed benchmarking for organizational efficiency. Omega **38**(6), 534–539 (2010)
72. Tamaddon, L., Jahanshahloo, G.R., Lotfi, F.H., Mozaffari, M.R., Gholami, K.: Data envelopment analysis of missing data in crisp and interval cases. Int. J. Math. Anal. **3**(17–20), 955–969 (2009)
73. Tone, K.: A slacks-based measure of efficiency in data envelopment analysis. Eur. J. Oper. Res. **130**(3), 498–509 (2001)
74. Tone, K., Tsutsui, M.: Network DEA: a slacks-based measure approach. Eur. J. Oper. Res. **197**(1), 243–252 (2009)
75. Tone, K., Tsutsui, M.: An epsilon-based measure of efficiency in DEA—a third pole of technical efficiency. Eur. J. Oper. Res. **207**(3), 1554–1563 (2010)
76. Wang, Y.M., Greatbanks, R., Yang, J.B.: Interval efficiency assessment using data envelopment analysis. Fuzzy Sets Syst. **153**(3), 347–370 (2005)
77. Wu, J., Zhou, Z., Liang, L.: Measuring the performance of nations at Beijing summer Olympics using integer-valued DEA model. J. Sports Econ. **11**(5), 549–566 (2010a)
78. Wu, J., Liang, L., Song, H.: Measuring hotel performance using the integer DEA model. Tourism Econ. **16**(4), 867–882 (2010b)
79. Zhou, Z., Guo, X., Wu, H., Yu, J.: Evaluating air quality in China based on daily data: application of integer data envelopment analysis. J. Clean. Prod. **198**, 304–311 (2018)

Chapter 7
Supplier Selection

7.1 Introduction

Managers in all organizations want to optimally use the facilities and capacities in different departments. So the existence of a model to provide feedback to improve the performance of different departments of organizations and obtain a tool to meet this need. Managers, it seems very necessary and logical [61]. Nowadays, due to the continuous changes in economic conditions, the evaluation of the performance of industrial and economic units has become one of the most important factors to improve their position compared to other units [59]. Performance in any organization can be defined as a combination of efficiency and effectiveness [55]. In today's world, the complexity of information, the huge amount of data, and the influence of various factors make managers unable to know the performance of the units under their supervision without a proper scientific approach [53, 54]. Also, to evaluate the performance more accurately, unlike the traditional criteria, several criteria such as profit, liquidity, risk, and management strategies should be considered. Therefore, the evaluation of efficiency requires the use of different approaches to determine efficiency [56]. Therefore, the need to use scientific methods in dealing with the existing challenges in the way of evaluating the performance of organizations is felt in advance.

As explained in the previous chapters, due to the diversity of goods, tracking the flow of goods during the supply chains, and the short life of some products in the chains (for example, perishable food), these challenges seem more in the supply chain than in other fields. So creating a supply chain with the highest efficiency is an important issue [3].

Supplier selection is one of the most vital aspects of being responsive and competitive in ever-changing global supply chains. The success of a supply chain strongly depends on the selection of suppliers, so this is an essential part of production and logistics management for companies. Although the short-term goal of supply chain management is primarily to increase productivity and reduce total inventory and total cycle time, the long-term goal is to increase customer satisfaction, market share, and

© The Author(s), under exclusive license to Springer Nature Switzerland AG 2023
F. Hosseinzadeh Lotfi et al., *Supply Chain Performance Evaluation*,
Studies in Big Data 122, https://doi.org/10.1007/978-3-031-28247-8_7

profit for all organizations involved in the supply chain. Strong coordination between the organizations involved in the supply chain is required to achieve these goals. The reason for such coordination is that all members of the supply chain will be directly or indirectly affected by other members of the chain. For example, problems in sending defective raw materials to the upstream supplier lead to the downstream manufacturer's production of defective final products, reducing customer satisfaction [77]. In other words, choosing suitable suppliers is not only a cost-saving method but also an essential process for new product production. Also, more than 30% of the errors in the production process of a product are caused by the low quality of raw materials and delay in supply, which is due to the incorrect selection of suppliers due to the use of inappropriate selection strategies. Therefore, supplier selection is the basis of cooperation in the supply chain and is also a key factor for improving the supply chain's competitive power and thus the chain's success [16]. Supply chain design in general and supplier selection, in particular, require a significant effort in any organization. The organization must understand what is essential in choosing a specific supplier and clearly state its criteria [76]. For example, the purchase of high-tech products such as motor vehicles, railways, transportation equipment, machinery and equipment components, materials, and services included up to 80% of the total cost of the product. Therefore, choosing suitable suppliers is a critical procurement element and represents an excellent opportunity for companies to reduce costs [74]. Also, better participation among chain members is another potential application for evaluating the overall performance of suppliers [63]. Finally, evaluating suppliers provides solutions to improve the current supply chain management.

Since classic supplier selection techniques always require intuitive judgments, and also, and some methods need initial weights based on the opinion of chain managers supply chain; as a result, in some cases, these are choices biased [45].

Since DEA helps decision-makers make their choices without relying on intuitive judgments, in this section, we will explain and describe how to use DEA models to select superior suppliers.

7.2 Evaluation of Suppliers' Performance to Select the Best Supplier Using DEA Models

Suppliers are one of the components of the supply chain in organizations, and managing suppliers requires specialized skills because they are not part of the organization. Suppliers must be appropriately evaluated because they can have positive or negative effects on the overall performance of an organization. Supplier assessment is one of the most crucial issues that purchase managers face because several issues must be considered. The manner and extent of the importance of each of these issues will affect the choice of the final supplier. Therefore, choosing an efficient supplier in the supply chain is very important. According to the companies' strategies, different

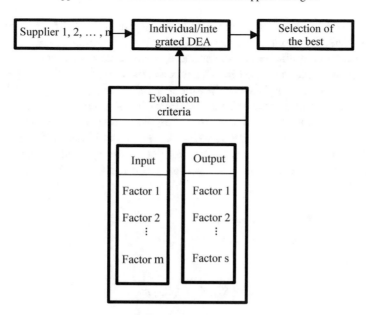

Fig. 7.1 Framework for supplier selection using DEA [16]

criteria can be used to evaluate and select the supplier. Various types of research indi-cate that raw materials cause problems related to organization product quality. As a result, careful selection of suppliers and how to evaluate them can minimize harmful effects and increase positive effects on the outputs of an organization. Therefore, assessing and choosing the right supplier is one of the essential components of the organizational structure. Dutta et al. [16] presented a general framework for using DEA models in supplier selection, as shown in Fig. 7.1.

In short, we describe the steps of selecting the best supplier as follows:

1. Identification of suppliers with whom it is possible to cooperate.
2. Determining effective indicators and criteria for decision-making concerning selecting the best supplier based on the expert opinion of the relevant field.

   2.1   Separation of selected indicators into two input and output groups.

3. Choosing the appropriate DEA model to evaluate suppliers.
4. Solving the model and analyzing the results to select the best suppliers.

In one of the first studies to evaluate the overall performance of suppliers, Weber et al. [70] proposed the use of the DEA models. In this study, six suppliers who supplied a product to a baby food manufacturer were evaluated. It showed that a significant cost reduction would occur if inefficient sellers could become efficient sellers based on the results obtained from an evaluation using DEA. It also reduces the delay in receiving the product, especially perishable products. Liu et al. [32] extended this research by using DEA to evaluate suppliers. Forker and [19] proposed a new

analytical method using DEA, which can help companies identify their most efficient suppliers, and suppliers move towards the target by imitating the best supplier.
Although the radial efficiency measurement, which changes the inputs or outputs
proportionally, is the most used in DEA models, the non-radial efficiency measurement allows the disproportionate adjustment of different inputs/outputs. Therefore,
non-radial models usually have higher discriminating power than the radial efficiency
measurement. Soltanifar and Sharafi [64] presented a method for selecting the best
supplier by focusing on non-radial models. Consider the RDM model presented in
Chap. 6. This model is a radial model, but due to the applications of non-radial
models, the RDM model was developed as a non-radial model, as shown in the
model (7.1).

$$\max\{\theta_i | i = 1, ..., m\}$$
$$\max\{\varphi_r | r = 1, ..., s\}$$
$$s.t$$
$$\sum_{j=1}^{n} \lambda_j x_{ij} \leq x_{io} - \theta_i R_{io}^-, \ i = 1, ..., m,$$
$$\sum_{j=1}^{n} \lambda_j y_{rj} \geq y_{ro} + \varphi_r R_{ro}^+, \ r = 1, ..., s,$$
$$\sum_{j=1}^{n} \lambda_j = 1,$$
$$\varphi_r, \theta_i, \lambda_j \geq 0, \ r = 1, ..., s, \ i = 1, ..., m, \ j = 1, ..., n. \qquad (7.1)$$

Then, using the weighted sum technique in Scalarizing Functions, which is
presented to solve multi-objective programming problems (MOLP), model (7.1)
is converted into model (7.2).

$$\max \frac{1}{m} \sum_{i=1}^{m} \theta_i + \frac{1}{s} \sum_{r=1}^{s} \varphi_r$$
$$s.t$$
$$\sum_{j=1}^{n} \lambda_j x_{ij} \leq x_{io} - \theta_i R_{io}^-, \ i = 1, ..., m,$$
$$\sum_{j=1}^{n} \lambda_j y_{rj} \geq y_{ro} + \varphi_r R_{ro}^+, \ r = 1, ..., s,$$
$$\sum_{j=1}^{n} \lambda_j = 1,$$
$$\varphi_r, \theta_i, \lambda_j \geq 0, \ r = 1, ..., s, \ i = 1, ..., m, \ j = 1, ..., n. \qquad (7.2)$$

Although using the model's weighting technique (7.1) has turned into a linear programming problem, the value of the objective function at the optimum point is not necessarily between 0 and 1. Therefore, this issue causes ambiguity in interpreting the results obtained from the model (7.2). To solve this problem, we first define an ideal and anti-ideal virtual DMU as follows:

$$x_{iI} = \min_{j}\{x_{ij}\}, \quad y_{rI} = \max_{j}\{y_{rj}\} \tag{7.3}$$

$$x_{iAI} = \max_{j}\{x_{ij}\}, \quad y_{rAI} = \min_{j}\{y_{rj}\} \tag{7.4}$$

Each DMU's input and output projection applies to (7.5) and (7.6).

$$x_{iI} \leq x_{io} - \theta_i R_{io}^- \leq x_{iAI}, \quad i = 1, \ldots, m \tag{7.5}$$

$$y_{rAI} \leq y_{ro} + \varphi_r R_{ro}^+ \leq y_{rI}, \quad r = 1, \ldots, s \tag{7.6}$$

As a result:

$$x_{iI} \leq x_{io} - \theta_i R_{io}^- \leq x_{iAI} \Rightarrow \frac{x_{io} - x_{iAI}}{R_{io}^-} \leq \theta_i \leq \frac{x_{io} - x_{iI}}{R_{io}^-} \tag{7.7}$$

$$x_{iAI} = \max_{j}\{x_{ij}\} \quad y_{rAI} = \min_{j}\{y_{rj}\} \tag{7.8}$$

Consequently:

$$x_{iI} \leq x_{io} - \theta_i R_{io}^- \leq x_{iAI} \Rightarrow \frac{x_{io} - x_{iAI}}{R_{io}^-} \leq \theta_i \leq \frac{x_{io} - x_{iI}}{R_{io}^-} \tag{7.9}$$

Therefore, the relation (7.10) obtains an upper bound for θ_i.

$$\bar{\theta}_i = \frac{x_{io} - x_{iI}}{R_{io}^-} \Rightarrow \theta_i \leq \bar{\theta}_i \Rightarrow \frac{\theta_i}{\bar{\theta}} \leq 1 \Rightarrow 0 \leq 1 - \frac{\theta_i}{\bar{\theta}} \leq 1 \tag{7.10}$$

Similarly to the mentioned process for θ_i, it can be shown that the upper bound of φ_r is obtained using the relation (7.11).

$$\bar{\varphi}_r = \frac{y_{rI} - y_{ro}}{R_{ro}^+} \Rightarrow \varphi_r \leq \bar{\varphi}_r \Rightarrow \frac{\varphi_r}{\bar{\varphi}_r} \leq 1 \Rightarrow 0 \leq 1 - \frac{\varphi_r}{\bar{\varphi}_r} \leq 1 \tag{7.11}$$

As a result, based on the process explained above, model (7.2) is written as the model (7.12):

$$\min \frac{1}{2}\left(\frac{1}{m}\sum_{i=1}^{m}\left(1-\frac{\theta_i}{\bar{\theta}_i}\right)+\frac{1}{s}\sum_{r=1}^{s}\left(1-\frac{\varphi_r}{\bar{\varphi}_r}\right)\right)$$

s.t

$$\sum_{j=1}^{n}\lambda_j x_{ij} \leq x_{io}-\theta_i R_{io}^{-}, i=1,\dots,m,$$

$$\sum_{j=1}^{n}\lambda_j y_{rj} \geq y_{ro}+\varphi_r R_{ro}^{+}, r=1,\dots,s,$$

$$\sum_{j=1}^{n}\lambda_j = 1,$$

$$\varphi_r, \theta_i, \lambda_j \geq 0, r=1,\dots,s, \ i=1,\dots,m, \ j=1,\dots,n. \qquad (7.12)$$

If $\bar{\theta}_i$ is equal to zero, the input of the under-evaluation DMU has the lowest value in the PPS, so reducing it will not be possible. Therefore, the corresponding term can be removed from the constraints and objective function. Also, if $\bar{\varphi}_r$ is equal to zero, it means that the output of the evaluated unit can produce the most significant amount in the set, so it will not be possible to increase it. Therefore, the corresponding sentence can be removed from the adverb and objective function.

If, in a particular case, the vector $R = (R_o^-, R_o^+)$ is considered as $R = (x_{io} - x_{il}, y_{rl} - y_{ro})$, , then model (7.12) can be rewritten as the model (7.13).

$$\min \frac{1}{2}(\frac{1}{m}\sum_{i=1}^{m}(1-\theta_i)+\frac{1}{s}\sum_{r=1}^{s}(1-\varphi_r))$$

s.t

$$\sum_{j=1}^{n}\lambda_j x_{ij} \leq x_{io}-\theta_i(x_{io}-x_{il}), \ \ i=1,\dots,m,$$

$$\sum_{j=1}^{n}\lambda_j y_{rj} \geq y_{ro}+\varphi_r(y_{rl}-y_{ro}), \ \ r=1,\dots,s,$$

$$\sum_{j=1}^{n}\lambda_j = 1,$$

$$\varphi_r, \theta_i, \lambda_j \geq 0, \ \ r=1,\dots,s, \ \ i=1,\dots,m, \ j=1,\dots,n. \qquad (7.13)$$

Also, the dual of the model (7.13) is obtained as the model (7.14).

$$\min \sum_{r=1}^{s}u_r y_{ro} - \sum_{i=1}^{m}v_{io}x_{io} + (\delta+1)$$

s.t

$$\sum_{r=1}^{s} u_r y_{rj} - \sum_{i=1}^{m} v_{ij} x_{ij} + \delta \leq 0, \quad j = 1, \ldots, n,$$

$$v_i(x_{io} - x_{il}) \geq \frac{1}{2m}, \quad i = 1, \ldots, m,$$

$$u_r(y_{rI} - y_{ro}) \geq \frac{1}{2s}, \quad r = 1, \ldots, s,$$

$$v_i, u_r \geq 0, \quad i = 1, \ldots, m, \quad r = 1, \ldots, s. \tag{7.14}$$

7.3   Supplier Selection in the Presence of Weight Restriction

Due to the complexity of the decision-making processes involved in supplier selection, existing methods in this field, including DEA, evaluate suppliers by weighting different criteria. The formulation of the DEA models is based on the fact that to maximize efficiency, the unit under evaluation can freely assign feasible weights to each input and output, provided that these weights are feasible for other DMUs. Saen [51] argues that one of the drawbacks of DEA models in supplier selection is complete freedom in assigning weights to inputs and outputs. In other words, no input or output is superior to any other input or output. Therefore, in some cases, by assigning inappropriate weights or even zero to some inputs and outputs of the suppliers under evaluation, the efficiency result is somewhat far from the mind. For example, suppose that in a problem of choosing the best supplier, the number of invoices sent without mistakes and the number of shipments that must arrive on time is considered as two outputs in a supplier. Suppose the suppliers are entirely free in choosing the weight. In this case, in a system, the number of invoices sent without mistakes may be considered essential and receive a higher weight, and in another evaluation system, the number of shipments that should be entered on time will be valued higher. On the other hand, the weighting ratio is also an important point and should be considered in some supplier selection systems. For example, in a decision-making committee in choosing a superior supplier, the issue may be raised that the number of shipments arriving on time should have a double value compared to the number of invoices sent without mistakes. One of the most common methods to avoid the mentioned problems is to use the weight restriction method in DEA models, which causes some managerial preferences to be applied in the model. Weight constraints may be applied directly to the weights of inputs and outputs or on the product of these weights with the corresponding input or output, called virtual input or virtual output. In the first view, based on the definition of Cooper et al. [12], the weight restriction in DEA models is expressed based on one of the following relations:

$$\delta_i \leq v_i \leq \beta_i, \quad \rho_r \leq u_r \leq \eta_r \tag{7.15}$$

$$\alpha_i \leq \frac{v_i}{v_{i+1}} \leq \psi_i, \quad \theta_r \leq \frac{u_r}{u_{r+1}} \leq \zeta_r \tag{7.16}$$

$$\varphi_i v_i \geq u_r \tag{7.17}$$

So that φ_i, δ_i, β_i, ρ_r, η_r, α_i, ψ_i, θ_r, ζ_r, are fixed values that system administrators add to models to reflect some limitations in decision-making. Wong and Beasley [71] suggested that virtual inputs and outputs can be normalized in the second viewpoint. In other words, it is assumed that:

$$a_r \leq \frac{u_r y_{rj}}{\sum_{r=1}^{s} u_r y_{rj}} \leq b_r \tag{7.18}$$

$$c_i \leq \frac{v_i x_{ij}}{\sum_{r=1}^{s} v_i x_{ij}} \leq d_i \tag{7.19}$$

To adopt weight restrictions based on the second point of view in DEA models, the following methods can be used:

1.   Restrictions are applied only to the virtual inputs and outputs of the DMU under evaluation.
2.   Restrictions are applied to all units' virtual inputs and outputs, in which case $2n(m+s)$ constraints are added to the model.
3.   Constraints (7.18) and (7.19) are only applied to weights of under-evaluation DMU, and, to prevent the freedom of weighting for other units, relations (7.20) and (7.21) are also added to the DEA model.

$$a_r \leq \frac{u_r \sum_{j=1}^{n} \frac{y_{rj}}{n}}{\sum_{r=1}^{s} u_r \sum_{j=1}^{n} \frac{y_{rj}}{n}} \leq b_r, r = 1, ..., s \tag{7.20}$$

$$c_i \leq \frac{v_i \sum_{j=1}^{n} \frac{x_{ij}}{n}}{\sum_{r=1}^{s} v_i \left(\sum_{j=1}^{n} \frac{x_{ij}}{n} \right)} \leq d_i, i = 1, ..., m. \tag{7.21}$$

Considering the weight restriction (7.19), [52] introduced the model (7.22) to select the best supplier by applying weight limitation.

$$\max \sum_{r=1}^{s} u_r y_{ro}$$

$$s.t$$

$$\sum_{i=1}^{m} v_i x_{io} = 1,$$

$$\sum_{r=1}^{s} u_r y_{rj} - \sum_{i=1}^{m} v_i x_{ij} \leq 0, \quad j = 1, ..., n,$$

$$c_i (\sum_{i=1}^{m} v_i x_{io}) - v_i x_{io} \leq 0, \quad i = 1, ..., m,$$

$$v_i x_{lo} - d_i (\sum_{i=1}^{m} v_i x_{io}) \leq 0, \quad i = 1, ..., m,$$

$$u_r, v_i \geq 0, \quad i = 1, ..., m, \quad r = 1, ..., s. \tag{7.22}$$

In short, the model (7.22) is a multiplier form of the DEA model in the input-oriented, which is used to select the best supplier by considering the weight restriction to apply managers' opinions in the evaluations. Similarly, weight restrictions can be applied in output-oriented models, too. Now suppose that α_i and β_i are dual variables, respectively, corresponding to the constraints of the weight restrictions in the model (7.22). So, the super-efficiency model with the weight restrictions was presented by [14] in the form of model (7.23).

$$\min \theta$$

$$s.t$$

$$\sum_{j=1, j \neq o}^{n} \lambda_j x_{ij} \leq \theta \, x_{io} + [(a_i - 1)\alpha_i + (1 - b_i)\beta_i] x_{io}$$

$$+ \sum_{j=1, j \neq o}^{n} (a_i \alpha_i + b_i \beta_i) x_{ij}, i = 1, ..., m,$$

$$\sum_{j=1, j \neq o}^{n} \lambda_j y_{rj} \leq \theta \, y_{ro}, r = 1, ..., s,$$

$$\sum_{j=1, j \neq o}^{n} \lambda_j (1 - d_i) w_{ij} \leq \theta \, w_{io}, i = 1, ..., m,$$

$$\sum_{j=1, j \neq o}^{n} \lambda_j d_i w_{ij} \geq w_{lo}, i = 1, ..., m,$$

$$\lambda_j \geq 0, j = 1, ..., n. \tag{7.23}$$

Although the use of weight restrictions is a suitable solution for applying the managers' opinions to the modeling process, this issue sometimes causes problems such as infeasibility. In many cases, this problem can be solved by some changes to the added restrictions. Still, these changes should be minimized to prevent dissatisfaction of decision makers. To solve this problem, [27] presented a method for selecting suppliers based on the way provided by [25]. First, suppose:

$$\delta_i \le v_i \le \beta_i, \qquad \rho_r \le u_r \le \eta_r \tag{7.24}$$

Thus:

$$A_t U \le b_t, t = 1, 2, ..., l_1 \quad and \quad C_h V \le d_h, h = 1, 2, ..., l_2 \tag{7.25}$$

So that A_t and C_h are $1 \times s$ and $1 \times m$ dimensional vectors, respectively. Also, $b_t \in R$ and $d_h \in R$. If the relations (7.25) are directly added to the CCR model, as seen in the model (7.26), the model may be infeasible.

$$\max \sum_{r=1}^{s} u_r y_{ro}$$

$$s.t$$

$$\sum_{i=1}^{m} v_i x_{io} = 1,$$

$$\sum_{r=1}^{s} u_r y_{rj} - \sum_{i=1}^{m} v_i x_{ij} \le 0, \quad j = 1, ..., n,$$

$$A_t U \le b_t, \quad t = 1, 2, ..., l_1,$$

$$C_h V \le d_h, \quad h = 1, 2, ..., l_2,$$

$$u_r, v_i \ge 0, \quad r = 1, ..., s, \quad i = 1, ..., m. \tag{7.26}$$

If b_t and d_h in the relations (7.25) are considered as a goal, then by defining the deviating variables η_t, η'_h, p'_h the relations (7.27) and (7.28) become as follows:

$$A_t U + \eta_t - p_t = b_t \tag{7.27}$$

$$C_h V + \eta'_h - p'_h = d_h \tag{7.28}$$

Since decision-makers always seek to reduce deviations. As a result, we aim to calculate the minimum possible values for p_t and p'_h by applying a penalty (M is considered a large number) to the objective function of the model (7.26). In other words, using a combined approach of Big-M and GP, first, the deviations related to each weight restriction are minimized, then the optimal solution of the model (7.26) is obtained.

$$\max \sum_{r=1}^{s} u_r y_{ro} - M \left(\sum_{t=1}^{l_1} p_t + \sum_{h=1}^{l_2} p'_h \right)$$

$$s.t$$

$$\sum_{i=1}^{m} v_i x_{io} = 1,$$

$$\sum_{r=1}^{s} u_r y_{rj} - \sum_{i=1}^{m} v_i x_{ij} \leq 0, \; j = 1, ..., n,$$

$$A_t U + \eta_t - p_t = b_t, \quad t = 1, 2, ..., l_1,$$

$$C_h V + \eta'_h - p'_h = d_h, \quad h = 1, 2, ..., l_2,$$

$$u_r, v_i \geq 0, \quad r = 1, ..., s, \quad i = 1, ..., m. \tag{7.29}$$

7.4 Supplier Selection with the Special Type of Data

In DEA models, it is assumed that the input and output indicators are non-negative real numbers. This assumption is not always possible in the real world, so in this sub-section, supplier selection in the presence of data Special is discussed.

7.4.1 Supplier Selection with Imprecise Data

Traditionally, supplier selection models focus on cardinal data and less on imprecise data, while many practical cases consider cardinal and imprecise data simultaneously. Although, in some situations, such factors can be represented quantitatively, providing an accurate quantitative measure that reflects the conditions for such elements is usually far from available. And this representation is only a tool to facilitate the modeling. For example, suppose a decision-making committee must choose the best supplier among five suppliers to sign a cooperation contract. The criteria for selection are:

- Total cost of shipments (TC)
- Number of shipments per month (NM)
- Supplier reputation (SR)
- Number of bills received from the supplier without errors (NB)
- Number of shipments to arrive on time (NOT).

A factor such as SR cannot be easily included in the modeling. According to the decision-making committee, the NOT index's importance is greater than that of NB, so this importance must be somehow included in the model. Also, there is no exact information about NB. This information is expressed based on each supplier's minimum and maximum incorrect invoices. Classic DEA models cannot evaluate and select suppliers with indicators as stated. As a result, [11, 28] classified imprecise data into the following four groups.

1. Bounded data
   If $\bar{x}_{ij}$, $\bar{y}_{rj}$, $\underline{x}_{ij}$, $\underline{y}_{rj}$ and are upper and lower bounds for inputs and outputs, respectively, then we have:

$$\bar{y}_{rj} \leq y_{rj} \leq \underline{y}_{rj}, \quad \bar{x}_{ij} \leq x_{ij} \leq \underline{x}_{ij} \quad for \ r \in Bo, \ i \in BI \tag{7.30}$$

So that BI and BO respectively represent the set of indices related to bounded inputs and outputs.

2. Ordinal data
   Without lose of generality, assume that:

$$x_{i1} \leq x_{i2} \leq \cdots \leq x_{ik} \leq \cdots \leq x_{in}, \quad i \in DI,$$
$$y_{r1} \leq y_{r2} \leq \cdots \leq y_{rk} \leq \cdots \leq y_{rm}, \quad r \in DO \tag{7.31}$$

So that DI and DO are, respectively, the set of indexes related to ordinal inputs and outputs (Weak ordinal data). Also, if the following ordinal relation exists between inputs and outputs, then it is said that there is a strong ordinal relation between the data.

$$x_{i1} < x_{i2} < \ldots < x_{ik} < \ldots < x_{in} \quad i \in SI,$$
$$y_{r1} < y_{r2} < \ldots < y_{rk} < \ldots < y_{rm} \quad r \in SO \tag{7.32}$$

And SI and SO include the indexes related to the inputs and outputs, with strong ordinal relations.

3. Ratio bounded data

$$L_{rj} \leq \frac{y_{rj}}{y_{ro}} \leq U_{rj}, \quad j \neq o, \quad r \in RO \tag{7.33}$$

$$G_{ij} \leq \frac{x_{ij}}{x_{io}} \leq H_{ij}, \quad j \neq o, \quad i \in RI \tag{7.34}$$

where L_{rj} and G_{ij} represent the lower bounds, and U_{rj} and H_{ij} represent the upper bounds. RO and RI represent the associated sets containing ratio bounded outputs and inputs, respectively.

Therefore, considering the importance of imprecise data in the evaluation and selection of the supplier and with the assumption that the supplier j consumes the inputs $(x_{1j}, \ldots, x_{mj})$ to produce the outputs $(y_{1j}, \ldots, y_{sj})$, [50] used the model (7.35) to evaluate suppliers in the presence of imprecise data.

$$\max \sum_{r=1}^{s} u_r y_{ro}$$

s.t

$$\sum_{r=1}^{s} u_r y_{rj} - \sum_{i=1}^{m} v_i x_{ij} \leq 0, \ j = 1, ..., n,$$

$$\sum_{i=1}^{m} v_i x_{io} = 1,$$

$$x_{ij} \in \Theta_i^-, \ i = 1, ..., m, j = 1, ..., n,$$

$$y_{rj} \in \Theta_r^+, \ r = 1, ..., s, \ j = 1, ..., n,$$

$$u_r, v_i \geq 0, \ r = 1, ..., s, \ i = 1, ..., m. \tag{7.35}$$

So that Θ_i^- and Θ_r^+ respectively represent one of the (7.30)–(7.34) relations. Also, using variable change $X_{ij} = w_i x_{ij}$ and $Y_{rj} = u_r y_{rj}$, which is presented by [75], the model (7.35) is converted to linear programming (7.36).

$$\max \sum_{r=1}^{s} Y_{ro}$$

$$s.t$$

$$\sum_{r=1}^{s} Y_{rj} - \sum_{i=1}^{m} X_{ij} \leq 0, \ j = 1, ..., n,$$

$$\sum_{i=1}^{m} X_{io} = 1,$$

$$X_{ij} \in \tilde{D}_i^-, \ \ i = 1, ..., m, \ j = 1, ..., n,$$

$$Y_{rj} \in \tilde{D}_r^+, \ \ r = 1, ..., s, \ j = 1, ..., n,$$

$$Y_{rj}, X_{ij} \geq 0, \ \ r = 1, ..., s, \ i = 1, ..., m, j = 1, ..., n. \tag{7.36}$$

So that, Θ_i^- and Θ_i^+ are transformed in $\tilde{D}_i^-$ and $\tilde{D}_r^+$ as follows:

1. Bounded data: $u_r \underline{y}_{rj} \leq Y_{rj} \leq u_r \bar{y}_{rj}$ and $v_i \underline{x}_{ij} \leq X_{ij} \leq v_i \bar{x}_{ij}$
2. Ordinal data: $X_{ij} \leq X_{ik}, \forall j \neq k$ and $Y_{rj} \leq Y_{rk}, \forall j \neq k$.
3. Ratio bounded data: $L_{rj} \leq \frac{Y_{rj}}{Y_{ro}} \leq U_{rj}$ and $G_{ij} \leq \frac{X_{ij}}{X_{io}} \leq H_{ij}$ so that: $j \neq o$.
4. Cardinal data: $X_{ij} = v_i \hat{x}_{ij}$ and $Y_{rj} = u_r \hat{y}_{rj}$, so that $\hat{x}_{ij}$ and $\hat{y}_{rj}$ are cardinal data.

7.4.2 Supplier Selection with Non-discretionary Data

As explained in the previous section, supplier selection is a multi-criteria optimization problem that depends on several indicators. However, sometimes in the issue of supplier selection, some criteria are beyond the management's control; these criteria are called non-discretionary data or exogenously fixed factors. For example, Banker and Murray [8] investigated the influence of non-discretionary factors on 60 fast food restaurants. This research studied three outputs and six inputs for each restaurant. So that the number of breakfast, lunch, and dinner meals was considered as the outputs of each system. Also, the inputs consist of:

- Expenditures for supplies and materials
- Expenditures related to labor
- Experience of each restaurant (According to the number of years of restaurant activity)
- Advertising expenditures (it is allocated to the store by headquarters)
- The store was located in an urban or rural area
- presence/absence of drive-in capability.

Except for the first two inputs, other inputs are non-discretionary. Also, in another study, [32] considered indicators such as the distance of suppliers to the destination and supply variety to select the best suppliers, while both discussed indicators are non-discretionary. Therefore, according to the importance of uncontrollable indicators in supplier selection, the definition of efficiency for units under evaluation with uncontrollable inputs was presented by Banker and Murray as follows:

$$E_{oj} = \frac{\sum_{r=1}^{s} u_r y_{rj} - \sum_{i \in IND} v_{iND} x_{ij}}{\sum_{i \in ID} v_{iD} x_{ij}} \tag{7.37}$$

Therefore, the model (7.38) was proposed to evaluate units with non-discretionary inputs as follows:

$$\min \sum_{r=1}^{s} u_r y_{ro} - \sum_{i \in IND} v_{iND} x_{io}$$

s.t

$$\sum_{i \in ID} v_{iD} x_{io} = 1,$$

$$\sum_{r=1}^{s} u_r y_{rj} - \left(\sum_{i \in ID} v_{iD} x_{ij} + \sum_{i \in IND} v_{iND} x_{ij} \right) \leq 0, j = 1, ..., n,$$

$$v_{iD} \geq 0, i \in ID,$$

$$v_{iND} \geq 0, i \in IND,$$

$$u_r \geq 0, r = 1, ..., s. \tag{7.38}$$

Next, Noorizadeh et al. [44] showed that since the expression $\sum_{r=1}^{s} u_r y_{rj} - \sum_{i \in IND} v_{iND} x_{ij}$ in the relation (7.37) is negative in some cases, the efficiency score of the model (7.38), contrary to the definition of efficiency of DEA models, will be negative. As a result, by adding a condition $\sum_{r=1}^{s} u_r y_{rj} - \sum_{i \in IND} v_{iND} x_{ij} \geq 0$ to model (7.38), model (7.39) evaluates suppliers with non-discretionary inputs.

$$\min \sum_{r=1}^{s} u_r y_{ro} - \sum_{i \in IND} v_{iND} x_{io}$$

$s.t$

$$\sum_{i \in ID} v_{iD} x_{io} = 1,$$

$$\sum_{r=1}^{s} u_r y_{rj} - \left(\sum_{i \in ID} v_{iD} x_{ij} + \sum_{i \in IND} v_{iND} x_{ij} \right) \leq 0, j = 1, ..., n,$$

$$\sum_{r=1}^{s} u_r y_{ro} - \sum_{i \in IND} v_{iND} x_{io} \geq 0,$$

$$v_{iD} \geq 0, \quad i \in ID,$$

$$v_{iND} \geq 0, \quad i \in IND,$$

$$u_r \geq 0, \quad r = 1, ..., s. \tag{7.39}$$

So that *IND* and *ID* are the set of indexes related to non-discretionary and discretionary inputs, respectively.

7.4.3   Supplier Selection with Undesirable Inputs and Outputs

The studies carried out in supplier selection with undesirable data are divided into two groups:

1. Input–Output exchange approach
2. Using the principle of weak disposability.

This section examines the modeling in these fields.

7.4.3.1   Supplier Selection with Undesirable Inputs and Outputs Based on Input–output Exchange

Let's assume that in an applied study that was conducted on the suppliers of automobile manufacturing, the following criteria were examined:

- Standardisation certificate
- Process and the product audit
- Price gap (The ratio of the target price to the price offered by the supplier)
- Order fulfillment (the ratio of items delivered by suppliers to total items ordered)
- On-time delivery
- Defective parts.

Although all the indicators mentioned above are system outputs, the system manager tends to increase the indicators 1–5 and decrease the sixth output. An indicator such as output 6 is called undesirable output.

Considering the undesirable outputs or inputs in selecting the best supplier is inevitable. Therefore, Mahdiloo et al. [39] developed the additive model for undesirable outputs to evaluate suppliers with undesirable outputs. In this case, undesirable outputs are considered like inputs. As a result, the model (7.40) was presented as follows.

$$\max \sum_{i=1}^{m} s_i^- \sum_{r \in O_U} s_r^{+U} + \sum_{r \in O_D} s_r^{+D}$$

$$s.t$$

$$\sum_{j=1}^{n} \lambda_j x_{ij} + s_i^- = x_{io}, i = 1, \ldots, m,$$

$$\sum_{j=1}^{n} \lambda_j y_{rj} - s_r^{+D} = y_{ro}, r \in O_D,$$

$$\sum_{j=1}^{n} \lambda_j y_{rj} + s_r^{+U} = y_{ro}, r \in O_U,$$

$$\sum_{j=1}^{n} \lambda_j = 1,$$

$$\lambda_j \geq 0, j = 1, \ldots, n,$$

$$s_i^- \geq 0, i = 1, \ldots, n,$$

$$s_r^{+U} \geq 0, r \in O_U,$$

$$s_r^{+D} \geq 0, r \in O_D. \tag{7.40}$$

So that O_D and O_U respectively refer to the set of indices related to desirable and undesirable outputs and $O = \{1, ..., S\} = O_U \cup O_D$. Also, y_{rj}^D and y_{rj}^U are symbols for introducing the desirable and undesirable outputs and s_r^{+D} and s_r^{+U} are the surplus variables corresponding to desirable and undesirable output. Consider again the practical case raised at the beginning of the discussion. In this problem, all the indicators are output for a system, so the problem is a practical example without input.

Theorem 7.1
In the absence of the constant return to scale principle, the additive model with a constant input is equivalent to an additive model without inputs.

Proof Without losing generality, suppose that a constant input is considered to equal 1 for all DMUs. Since $x_{ij} = 1, \sum_{j=1}^{n} \lambda_j x_{ij} + s_i^- = x_{io}, i = 1, ..., m$ and $\sum_{j=1}^{n} \lambda_j = 1$, so $s_i^- =$

0. Therefore constraint $\sum\limits_{j=1}^{n} \lambda_j x_{ij} + s_i^- = x_{io}$ in Model (7.40) is converted to $\sum\limits_{j=1}^{n} \lambda_j = 1$.

Hence in the absence of the constant return to scale principle, $\sum\limits_{j=1}^{n} \lambda_j x_{ij} + s_i^- = x_{io}$ in the additive model with a single constant input is a redundant constraint. So it can be deleted.

Therefore, Mahdiloo et al. [38] introduced the additive model without input, such as the model (7.41).

$$\max \sum_{r \in O_U} s_r^{+U} + \sum_{r \in O_D} s_r^{+D}$$

s.t

$$\sum_{j=1}^{n} \lambda_j y_{rj}^{D} + s_r^{+D} = y_{ro}^{D}, \ r \in O_D,$$

$$\sum_{j=1}^{n} \lambda_j y_{rj}^{U} + s_r^{+U} = y_{ro}^{U}, \ r \in O_U,$$

$$\sum_{j=1}^{n} \lambda_j = 1,$$

$$\lambda_j \geq 0, \ j = 1, ..., n,$$

$$s_i^- \geq 0, \ i = 1, ..., n,$$

$$s_r^{+U} \geq 0, \ r \in O_U,$$

$$s_r^{+D} \geq 0, \ r \in O_D. \tag{7.41}$$

As a result, the under-evaluation supplier by the additive model without inputs is efficient if only if $\sum_{r \in O_U} s_r^{+U*} + \sum_{r \in O_D} s_r^{+D*} = 0$ such that s_r^{+U*} and s_r^{+U*} are the optimal solution of the model (7.41). And finally, because some interpretations and analyses of the results are obtained only by multiplier models, the multiplier form of the additive model with undesirable outputs is expressed as the model (7.42).

$$\min \sum_{i=1}^{m} v_i x_{io} + \sum_{r \in O_U} u_r^U y_{ro}^U - \sum_{r \in O_D} u_r^D y_{ro}^D + w_o$$

s.t

$$\sum_{i=1}^{m} v_i x_{io} = 1,$$

$$\sum_{i=1}^{m} v_i x_{ij} + \sum_{r \in O_U} u_r^U y_{rj}^U - \sum_{r \in O_D} u_r^D y_{rj}^D + w_o \geq 0, j = 1, ..., n,$$

$$v_i \geq 1, \ i = 1, ..., m,$$

$$u_r^U \geq 1, \; r \in O_U,$$
$$u_r^D \geq 1, \; r \in O_D. \tag{7.42}$$

So that u_r^D and u_r^U are the weights corresponding to desirable and undesirable outputs. Although the additive model is a robust model for evaluating the suppliers with undesirable data and the power to distinguish efficient and inefficient units in this model is higher than in the classic DEA models, as stated in Chap. 6, the optimal value of the objective function in this model is not necessarily between 0 and 1. Therefore, for a better interpretation of the results, [77] suggested using the SBM model to evaluate suppliers with undesirable outputs.

$$\min \frac{1 - \frac{1}{m} \sum_{i=1}^m \frac{s_i^-}{x_{io}}}{1 + \frac{1}{s_1+s_2}\left(\sum_{r\in O_D}\frac{s_r^D}{y_{ro}^D} + \sum_{r\in O_U}\frac{s_r^U}{y_{ro}^U}\right)}$$

$$s.t$$

$$\sum_{j=1}^n \lambda_j y_{rj}^D - s_r^D = y_{ro}^D, \; r \in O_D,$$

$$\sum_{j=1}^n \lambda_j y_{rj}^U + s_r^U = y_{ro}^U, \; r \in O_U,$$

$$\sum_{j=1}^n \lambda_j x_{ij} + s_i^- = x_{io}, \; i = 1, ..., m,$$

$$\lambda_j \geq 0, \; j = 1, ..., n,$$
$$s_i^- \geq 0, \; i = 1, ..., n,$$
$$s_r^{+U} \geq 0, \; r \in O_U,$$
$$s_r^{+D} \geq 0, \; r \in O_D. \tag{7.43}$$

So that s_1 and s_2, respectively are the number of desirable and undesirable outputs of the under-evaluation suppliers. Similar to the classic SBM model, the linear form of the model (7.43) is the model (7.44).

$$\min t - \frac{1}{m} \sum_{i=1}^m \frac{S_i^-}{x_{io}}$$

$$s.t$$

$$1 = t + \frac{1}{s_1+s_2}\left(\sum_{r\in O_D}\frac{S_r^D}{y_{ro}^D} + \sum_{r\in O_U}\frac{S_r^U}{y_{ro}^U}\right)$$

$$\sum_{j=1}^{n} \Gamma_j y_{rj}^D - S_r^D = y_{ro}^D, \ r \in O_D,$$

$$\sum_{j=1}^{n} \Gamma_j y_{rj}^U + S_r^U = y_{ro}^U, \ r \in O_U,$$

$$\sum_{j=1}^{n} \Gamma_j x_{ij} + S_i^- = x_{io}, \ i = 1, ..., m,$$

$$t \geq 0,$$
$$S_i^- \geq 0, \ i = 1, ..., m$$
$$S_r^U \geq 0, \ r \in O_U, \ S_r^D \geq 0, \ r \in O_D,$$
$$\Lambda_j \geq 0, j = 1, ..., n. \tag{7.44}$$

And finally, the multiplier form of the SBM model with undesirable outputs is presented as the model (7.45).

$$\max \xi$$
$$s.t$$
$$\xi + v x_o + u^U y_o^U - u^D y_o^D = 1,$$
$$- v x_o - u^U Y^U + u^D Y^D \leq 0,$$
$$v \geq \frac{1}{m}\left[\frac{1}{x_o}\right],$$
$$u^U \geq \frac{\xi}{s}\left[\frac{1}{y_o^U}\right],$$
$$u^D \geq \frac{\xi}{s}\left[\frac{1}{y_o^D}\right]. \tag{7.45}$$

So that s in the model (7.45) equals the total number of outputs (desirable and undesired).

7.4.3.2 Supplier Selection with Undesirable Inputs and Outputs Based on the Weak Dispospobility Principle

With the increase in environmental awareness and cognition of the potential dangers of environmental destruction for humanity, attention to the undesirable production of factories, such as air pollutants and hazardous waste, known as undesirable factors, is increasing. For example, producing electricity from fossil fuels is always associated with producing undesirable outputs such as sulfur dioxide. As previously stated, in DEA, it is assumed that producing more outputs with fewer resources indicates a better system performance. In other words, technologies with more good (desirable) outputs and less bad (undesirable) outputs from fixed input sources are known to be more efficient compared to other technologies [13].

Also, as mentioned in the chapter (6), based on the disposability principle, if $(x, y) \in T$, $x' \geq x$, and $y' \leq y$ then $(x', y) \in T$ and also, $(x, y') \in T$.

While in many practical applications, this assumption is not necessarily valid. Therefore, based on the idea of [20], the principle of weak dispospobility was defined for undesirable outputs. In this case, the production possibility set is defined as follows:

$$T_e = \left\{ (x, y, u) \middle| \sum_{j=1}^{n} \lambda_j x_j \leq x, \ \sum_{j=1}^{n} \lambda_j y_j \geq y, \ \sum_{j=1}^{n} \lambda_j u_j = u, \ \lambda_j \geq 0, \ j = 1, ..., n \right\} \tag{7.46}$$

So that x is the system's input, and y and u are the desirable and undesirable outputs, respectively. And because this mode is usually used for environmental research, it is also called the environmental production possibility set. This idea can be similarly extended to undesirable inputs. This generalization is particularly useful when DMUs consume undesirable inputs such as carbon dioxide. According to the definition of T_e in relation (7.46), it is evident that if $u = 0$, the desirable outputs are equal to zero; in other words, the only way to eliminate all undesirable outputs is to end the production. Therefore, T_e better represents the real-world production process when both desirable and undesirable outputs exist simultaneously. In other words, the difference between T and T_e is that T_e is impossible to reduce only undesirable outputs, but the proportional reduction of desirable and undesirable outputs is possible. Another essential point to consider is that any economic production activity is a process consisting of the consumption of resources input (such as energy and water) and other non-resource inputs (such as capital and labor) to produce desirable outputs (goods produced in the production line factory) is accompanied by the release of pollutants (sewage, dust, solid waste). as a result:

$$T_e = \{(x, e, y, u) | (x, e) \text{ can produce } (y, u)\} \tag{7.47}$$

So that e and x are environmental and non-environmental factors, and y and u are desirable and undesirable outputs, respectively.

Therefore, energy efficiency or environmental efficiency alone is not acceptable in any production process because improving energy efficiency (or environmental efficiency) relies on improving all factors [10]. While some studies conducted in this field, such as [9, 72], only the reduction of environmental inputs that affect the reduction of emissions are considered, and therefore, [29] proposed model (7.48) to evaluate the efficiency of suppliers in the presence of environmental and non-environmental factors.

$$\min \left(w \frac{1}{m} \sum_{i=1}^{m} \theta_i + w' \frac{1}{L} \sum_{l=1}^{L} \beta_l + w'' \frac{1}{Q} \sum_{q=1}^{Q} \varphi_q \right)$$

$s.t$

$$\sum_{j=1}^{n} \lambda_j x_{ij} \leq \theta_i x_{io}, \; i = 1, ..., m,$$

$$\sum_{j=1}^{n} \lambda_j e_{lj} \leq \beta_l e_{lo}, \; l = 1, ..., L,$$

$$\sum_{j=1}^{n} \lambda_j y_{rj} \geq y_{ro}, r = 1, ..., s,$$

$$\sum_{j=1}^{n} \lambda_j u_{qj} = \varphi_q u_{qo}, q = 1, ..., Q,$$

$$\lambda_j \geq 0, j = 1, ..., n. \tag{7.48}$$

In another point of view, since desirable outputs (y^D) and undesirable (y^U) outputs are always produced by inputs x. The corresponding production technology is as follows.

$$T = \{(x, y^D, y^U) | (y^D, y^U) \text{ is produced by } x\} \tag{7.49}$$

So Shephard [62] believes that desirable and undesirable outputs follow the weak disposability axiom if and only if:

$$\{(x, y^D, y^U) | (x, \theta y^D, \theta y^U) \in T, \; 0 \leq \theta \leq 1\} \tag{7.50}$$

Based on this, Kuosmanen (2016) considered the principle of weak possibility for undesirable outputs in T_v and introduced T_v' as (7.51).

$$T_v' = \left\{ \begin{array}{l} (x, y^D, y^U) \left| \sum_{j=1}^{n} \lambda_j x_j \leq x, \; \sum_{j=1}^{n} \theta_j \lambda_j y_j^D \geq y^D, \; \sum_{j=1}^{n} \theta_j \lambda_j y_j^U = y^U \right. \\ 0 \leq \theta_j \leq 1, \sum_{j=1}^{n} \lambda_j = 1, \; \lambda_j \geq 0, \; j = 1, ..., n \end{array} \right\} \tag{7.51}$$

Considering that the production of undesirable output in many real cases is due to the production of undesirable output, the amount of undesirable output is a function of the production of desirable output. Therefore, the production possibility set (7.51) is written as (7.52) to consider this issue.

$$T'_v = \left\{ (x, y^D, y^U) \middle| \begin{array}{l} \displaystyle\sum_{j=1}^{n} \lambda_j x_j \leq x, \sum_{j=1}^{n} \theta_j \lambda_j y_j^D \geq y^D, \sum_{j=1}^{n} \theta_j \lambda_j y_j^U = y^U \\ \displaystyle\sum_{j=1}^{n} \theta_j \lambda_j y_j^U \leq f\left(\sum_{j=1}^{n} \lambda_j y_j^D\right), 0 \leq \theta_j \leq 1, \sum_{j=1}^{n} \lambda_j = 1, \lambda_j \geq 0, j = 1, ..., n \end{array} \right\} \quad (7.52)$$

So that, f is a function that shows the changes in the undesirable output based on the desired output. To avoid computational complexity, $\sum_{j=1}^{n} \theta_j \lambda_j y_j^U \leq f\left(\sum_{j=1}^{n} \lambda_j y_j^D\right)$ is expressed as $\sum_{j=1}^{n} \theta_j \lambda_j y_j^U \leq t \sum_{j=1}^{n} \lambda_j y_j^D$. Where t is a proportion of desirable outputs. So we have:

$$T'_v = \left\{ (x, y^D, y^U) \middle| \begin{array}{l} \displaystyle\sum_{j=1}^{n} \lambda_j x_j \leq x, \sum_{j=1}^{n} \theta_j \lambda_j y_j^D \geq y^D, \sum_{j=1}^{n} \theta_j \lambda_j y_j^U = y^U \\ \displaystyle\sum_{j=1}^{n} \theta_j \lambda_j y_j^U \leq t y_j^D, 0 \leq \theta_j \leq 1, \sum_{j=1}^{n} \lambda_j = 1, \lambda_j \geq 0, j = 1, ..., n \end{array} \right\} \quad (7.53)$$

To evaluate $DMU_j, j = 1, ..., n$ in the presence of undesirable outputs using the SBM model, Malmir [40] proposed the model (7.54).

$$\min \frac{1 - \frac{1}{m}\sum_{i=1}^{m} \frac{s_i^-}{x_{io}}}{1 + \frac{1}{\frac{1}{s_1}\left(\sum_{r \in O_D} \frac{s_r^D}{y_{ro}^D}\right)}}$$

$s.t$

$$\sum_{j=1}^{n} \lambda_j x_{ij} + s_i^- = x^{io}, \ i = 1, ..., m,$$

$$\sum_{j=1}^{n} \theta_j \lambda_j y_{rj}^D - s_r^D = y_{ro}^D, \ r \in O_D,$$

$$\sum_{j=1}^{n} \theta_j \lambda_j y_{rj}^U = y_{ro}^U, \ r \in O_U,$$

$$\sum_{j=1}^{n} \theta_j \lambda_j y_{rj}^U \leq t y_{rj}^D,$$

$$\sum_{j=1}^{n} \lambda_j = 1,$$

$$0 \leq \theta_j \leq 1, \lambda_j \geq 0, j = 1, ..., n,$$

$$s_i^- \geq 0, i = 1, ..., n, s_r^{+D} \geq 0, r \in O_D. \quad (7.54)$$

So that O_D and O_U correspond to the set of indexes related to desirable and undesirable outputs and s_1 indicates the number of desired outputs. The model (7.54) is a non-linear model. Therefore, the change of variables (7.55) and (7.56) are defined to perform the linearization process.

$$\theta_j \lambda_j = \rho_j \tag{7.55}$$

$$(1 - \theta_j)\lambda_j = \mu_j \Rightarrow \mu_j + \rho_j = \lambda_j \tag{7.56}$$

As a result, using the mentioned change of variables, the model (7.54) is transformed into the model (7.57).

$$\min \frac{1 - \frac{1}{m} \sum_{i=1}^{m} \frac{s_i^-}{x_{io}}}{1 + \frac{1}{s_1}\left(\sum_{r \in O_D} \frac{s_r^D}{y_{ro}^D}\right)}$$

$s.t$

$$\sum_{j=1}^{n} (\mu_j + \rho_j)x_{ij} + s_i^- = x_{io}, \ i = 1, \ldots, m,$$

$$\sum_{j=1}^{n} \rho_j y_{rj}^D - s_r^D = y_{ro}^D, \ r \in O_D,$$

$$\sum_{j=1}^{n} \rho_j y_{rj}^U = y_{ro}^U, \ r \in O_U,$$

$$\sum_{j=1}^{n} \rho_j y_{rj}^U \leq ty_{rj}^D,$$

$$\sum_{j=1}^{n} \mu_j + \rho_j = 1,$$

$$\mu_j, \ \rho_j \geq 0, j = 1, \ldots, n,$$

$$s_i^- \geq 0, i = 1, \ldots, n, s_r^{+D} \geq 0, r \in O_D. \tag{7.57}$$

7.4.4 Supplier Selection with Dual-Role Factors

When using models, it is assumed that it is clearly known which factor is the input and which is the output. For example, in conventional studies regarding the evaluation of suppliers' efficiency, indicators such as profit from purchases, the number of shipments sent by each supplier, outputs and resources such as the number of employees, and the cost of purchasing primary resources as conventional inputs DEA models are considered. However, in some situations, certain factors are allowed to play the role of input and output at the same time. For example, in one perspective, the research and development cost can be considered as an output because the increase in the research and development cost will increase the supplier's efficiency, but in another view, if the research and development cost is considered as an input, any reduction in the cost of research and development without reducing the output will increase efficiency.

In another example, Saen et al. [52] considered the following factors in choosing a third-party reverse logistics (3PL) provider.

- Total cost of shipments (TC)
- Revenue from the sale of recyclables (R)
- Service-quality experience (EXP)
- Service-quality credence (CRE).

It is easy to consider TC as input and R as output. But about EXP and CRE indicators, the matter becomes a little more complicated. Because from the point of view of the decision-makers to choose the best 3PL, it seems natural that the decision makers are looking to choose a 3PL that has a higher EXP, so it is reasonable to consider this indicator as an output, but from another point of view, considering that EXP is a tool to attract more customers. Therefore, the tendency is to attract more customers with less EXP, so it seems reasonable to assume this index as an input. In the same way, it can be argued that CRE can also play the role of input and output. Such indicators are called dual-role factors in DEA literature. Therefore, considering the importance of dual-role factors in evaluating suppliers and their impact on the performance of supply chains, [52] presented a model for selecting superior suppliers in the presence of dual-role factors. In other words, in this modeling, some factors are allowed to simultaneously play the role of input and output. Suppose that x_{ij} and x_{ij} are i-th input and r-the output. Also, w_{fj} is a dual-role factor. In this case, we have:

$$\max \frac{\sum_{r=1}^{s} u_r y_{ro} + \sum_{f=1}^{F} \gamma_f w_{fo} - \sum_{f=1}^{F} \beta_f w_{fo}}{\sum_{i=1}^{m} v_i x_{io}}$$

s.t

$$\frac{\sum_{r=1}^{s} u_r y_{rj} + \sum_{f=1}^{F} \gamma_f w_{fj} - \sum_{f=1}^{F} \beta_f w_{fj}}{\sum_{i=1}^{m} v_i x_{ij}} \leq 1, j = 1, ..., n,$$

$$u_r, v_i, \gamma_f, \beta_f \geq 0, r = 1, ..., s, i = 1, ..., m, f = 1, ..., F. \qquad (7.58)$$

Then, using Charnes and Cooper transformations, model (7.58) is converted to the linear programming problem (7.59).

$$\max \sum_{r=1}^{s} u_r y_{ro} + \sum_{f=1}^{F} \gamma_f w_{fo} - \sum_{f=1}^{F} \beta_f w_{fo}$$

s.t

$$\sum_{i=1}^{m} v_i x_{io} = 1,$$

$$\sum_{r=1}^{s} u_r y_{rj} + \sum_{f=1}^{F} \gamma_f w_{fj} - \sum_{f=1}^{F} \beta_f w_{fj} - \sum_{i=1}^{m} v_i x_{ij} \le 0, \quad j = 1, ..., n,$$

$$u_r, v_i, \gamma_f, \beta_f \ge 0, \quad r = 1, ..., s, \quad i = 1, ..., m, \quad f = 1, ..., F. \tag{7.59}$$

Now suppose that $\hat{\beta}_f$ and $\hat{\gamma}_f$ are the optimal solutions of the model (7.59). Therefore, one of the following three situations occurs:

1. $\hat{\gamma}_f - \hat{\beta}_f > 0$
   In this case, the desired dual-role factor plays the role of output in the supplier under evaluation, and its increase benefits the chain.
2. $\hat{\gamma}_f - \hat{\beta}_f < 0$
   Therefore, reducing the dual-role factor is helpful for system performance, and this index plays an input role in the desired supplier.
3. $\hat{\gamma}_f - \hat{\beta}_f = 0$
   In this case, reducing or increasing this index will have the same effect on efficiency. In other words, this index is in balance.

Also, Saen developed the (7.59) model with the presence of dual role data as well as non-discretionary inputs. If I_D and I_{ND} respectively are the set of non-discretionary and non-discretionary input indices, model (7.59) is developed to model (7.60) considering the non-discretionary inputs.

$$\max \sum_{r=1}^{s} u_r y_{ro} + \sum_{f=1}^{F} \gamma_f w_{fo} - \sum_{f=1}^{F} \beta_f w_{fo} - \sum_{i \in I_{ND}} v_{iND} x_{io}$$

s.t

$$\sum_{i \in I_D} v_{iD} x_{io} = 1,$$

$$\sum_{r=1}^{s} u_r y_{rj} + \sum_{f=1}^{F} \gamma_f w_{fj}$$

$$- \left(\sum_{f=1}^{F} \beta_f w_{fj} + \sum_{i \in I_D} v_{iD} x_{ij} + \sum_{i \in I_{ND}} v_{iND} x_{ij} \right) \le 0, j = 1, ..., n,$$

$$v_{iND} \ge 0, i \in I_{ND},$$

$$v_{iD} \ge 0, i \in I_D,$$

$$\beta_f, \gamma_f, u_r \ge 0, \quad f = 1, ..., F, r = 1, ..., s. \tag{7.60}$$

Therefore, it can be stated that in the method presented by [52], the dual-role factors are determined optimistically so that the supplier under evaluation gets a better efficiency score. While if this role changes, then the weight assigned to this factor may also vary. So, with these changes, the efficiency score, and in some cases, the classification of the desired supplier (efficient or inefficient), may change. In other words, this ambiguity in the role sometimes causes a change in the classification and completely different results. In this regard, [14] believe that to have more accurate results for evaluating the performance of suppliers with dual-role factors, it is better for the efficiency score of each supplier to be an aggregated performance score considering all possible situations dual-role factors. Therefore, they proposed a multi-step method to select the best supplier in the presence of dual-role factors.

Definition 7.1
$p = (d_1, d_2, ..., d_L)$ is called the vector of the partition of dual-role factors so that $d_i \in \{0, 1\}$. $d_i = 1$ means that the ith dual-role factor is considered output and if $d_i = 0$, then the index in question is defined as an input.

Suppose each supplier has L dual-role factors and two states have been considered for each of these indexes, so vector p has 2^L different forms. As a result, we define the set P as a set that includes all the states of p. Considering that CCR and BCC models suffer from the impossibility of complete ranking of units, Ding et al. used the super-efficiency model shown in the model (7.61) to evaluate and rank the DMUs under evaluation with the dual-role factors.

$$\min \ \theta$$

$$s.t$$

$$\sum_{j=1, j \neq o}^{n} \lambda_j x_{ij} \leq \theta \, x_{io}, \ i = 1, ..., m,$$

$$\sum_{j=1, j \neq o}^{n} \lambda_j y_{rj} \geq y_{ro}, \ r = 1, ..., s,$$

$$\sum_{j=1, j \neq o}^{n} \lambda_j (1 - d_l) w_{lj} \leq \theta \, w_{lo}, \ l = 1, ..., L,$$

$$\sum_{j=1, j \neq o}^{n} \lambda_j d_l w_{lj} \geq w_{lo}, \ l = 1, ..., L,$$

$$\lambda_j \geq 0, \ j = 1, ..., n. \tag{7.61}$$

In the model (7.61), the third and fourth constraints indicate dual-role factors, and the optimal solution will be determined based on the value of the binary variable that has already been defined.

As explained earlier, there are 2^L different states for the p, so to calculate the efficiency matrix, for each supplier under evaluation, the model (7.61) must be solved

2^L times for each DMUs. As a result, the efficiency matrix is a matrix $n \times 2^L$. Although this issue seems complex at first, in practical applications, the number of dual-role factors is limited, so L is a small number. As a result, the computational complexity of this algorithm is not high and can be ignored.

Definition 7.2
Suppose that the supplier k is under evaluation. The vector $f(p) =$ $(f_1(p), f_2(p), ..., f_n(p))$ is called Efficiency mapping such that $f_k(p)$ is the optimal solution of the model (54) for $p = (d_1, d_2, ..., d_L), p \in P$.

Definition 7.3
Efficiency matrix A is a matrix $n \times 2^L$ so that $a_{ij} = e_{ij} = f_i(p_j)$.
   Therefore, each row of the efficiency matrix represents the performance of a DMU under the different states for dual factors, and the columns of matrix A represent the performance of different units under the influence of choosing a specific form for dual factors. In other words, each matrix column A is the vector f(p) for $p \in P$.

Definition 7.4
Aggregate efficiency is equal to $E(\omega, f, p) = \sum_{j=1}^{n} \omega_j f_j(p)$ so that $\omega = (\omega_1, ..., \omega_n)^T$, $\omega_j \geq 0$, $\sum_{j=1}^{n} \omega_j = 1$ and $p \in P$. Such that ω_j represents the decision maker's preferences in the presence of DMUj. Considering that the decision-maker tends to choose ω and $p \in P$ in such a way that the efficiency of the system is maximized; therefore the model (7.62) was proposed to optimize the efficiency of the entire system:

$$\max_{\omega,p} E(\omega, f, p) = \sum_{j=1}^{n} \omega_j f_j(p)$$

s.t

$$\sum_{j=1}^{n} \omega_j = 1,$$

$$\omega_j \geq 0,$$

$$p \in P = \{p_1, p_2, ..., p_{2^L}\}. \tag{7.62}$$

7.4.5   Supplier Selection with Fuzzy Data

In supplier selection, decision-makers usually prefer to describe suppliers by linguistic terms instead of using the conventional numerical forms. For example, for an indicator such as supplier risk, experts often express their views in terms like high risk, low risk, and medium risk. So, using linguistic terms is a simple and concrete way for decision-makers to express their opinion. Although in some situations, such factors can be determined quantitatively, the quantitative criteria cannot correctly reflect such a factor. Generally, the results obtained are beyond reality, and such a quantity may often be considered only as a compulsion and for convenience in modeling. In addition, as stated in the previous sections, supplier selection is a multi-criteria decision-making problem. Suppose, five criteria have been considered for ranking suppliers: price, quality, on-time product delivery, and flexibility. Unlike the criteria stated in the previous sections, these criteria are qualitative. Since, in this selection, the supplier's performance in different indicators is described linguistically, the fuzzy set theory comes to mind to deal with this issue. So, in many supplier selection issues, it is necessary to consider the fuzzy data. Therefore, although using DEA models based on multiple inputs and outputs is one of the appropriate tools in ranking and choosing the best supplier, DEA models have an essential flaw when faced with such indicators. Therefore, some researchers suggested using fuzzy DEA (FDEA) models to deal with inaccuracy and ambiguity in some data and face qualitative data. Before starting the discussion about fuzzy DEA, we briefly review some basic concepts and definitions of fuzzy sets that are needed for future discussions.

7.4.5.1 Fuzzy Set Theory and Fuzzy Numbers

Although the set is one of the primary principles of mathematics and is undefinable, some definitions have been provided for a better understanding of a set. Therefore, a group of two-by-two distinct objects is called a set. A set is usually defined based on a property, and everything that has this property is a member of the set, and every member of the set has this property. So, if p is a particular property, the set A on X be displayed as (7.63).

$$A = \{x \in X \,|\, p(x)\} \tag{7.63}$$

So, for each member x of X, $p(x)$ can be true or false. If we represent the truth of the proposition with 1 and the falsehood with 0, then the fuzzy set $\tilde{A}$ can be defined as follows:

$$\tilde{A} = \left\{ (x, \mu_{\tilde{A}}(x)) \,|\, x \in X \right\} \tag{7.64}$$

So that μ is a function from X to the set $\{1,0\}$, in other words:

$$\mu(x) = \begin{cases} 1 & p(x) \;\; is \;\; true \\ 0 & p(x) \;\; is \;\; false \end{cases} \tag{7.65}$$

So, in defining a set, awareness of the value p is essential. Because members of a set are defined based on the value p. So p must be well-defined. For example, the members of the even numbers set are easily determined, but we have trouble picking the members of the large numbers set. Because being large is a vague feature, the identification of the members of such a set depends on the decision maker's opinion. Therefore, Professor Lotfi Zadeh introduced the theory of fuzzy sets in [73] as a generalization and expansion of the classic sets theory. Since fuzzy sets can describe many concepts and relations in the real world, this new branch of mathematics has made significant progress in recent years. Therefore, this subsection describes some preliminary but essential concepts on the fuzzy sets. Therefore, this subsection describes some preliminary but essential concepts on the fuzzy sets.

Definition 7.5
Suppose X is a nonempty set. Each fuzzy subset of X is defined by function $\mu_{\tilde{A}} : X \rightarrow$ [0, 1][0, 1], called the membership function. In other words, $\mu_{\tilde{A}}$ indicates the degree to which x belongs to the fuzzy set $\tilde{A}$. Thus, the fuzzy set $\tilde{A}$ is: $\tilde{A} = \{(x, \mu_{\tilde{A}})|x \in X\}$.

Definition 7.6
The support of the fuzzy set $\tilde{A}$ is: $Supp\tilde{A} = \{x|\mu_{\tilde{A}}(x) > 0\}$.

Definition 7.7
The set of the elements belonging to $\tilde{A}$ with membership degree α or greater than α is called $\alpha - cut$. So: $A_{\alpha} = \{x \in X | \mu_{\tilde{A}}(x) \geq \alpha\}$.

Definition 7.8
A fuzzy set $\tilde{A}$ is called a normal set if and only if: $\sup(\tilde{A}) = 1$.

Definition 7.9
A fuzzy set $\tilde{A}$ is called a convex set if and only if: $\mu_{\tilde{A}}(\lambda x_1 + (1 - \lambda)x_2) \geq \min\{\mu_{\tilde{A}}(x_1), \mu_{\tilde{A}}(x_2)\}, \lambda \geq 0$.

Definition 7.10
A convex and normal set $\tilde{A}$ is a fuzzy number when:

- There should be only one $x_0 \in \mathbb{R}$ such that: $\mu_{\tilde{A}}(x_0) = 1$.
- $\mu_{\tilde{A}}$ be a piecewise continuous function.

Based on the above discussion, fuzzy number $\tilde{A}$ with a membership function $\mu_{\tilde{A}}(x) = e^{-\frac{|x|}{2}}$ is illustrated in Fig. 7.2.

In Definition 7.10, if the condition "only one $x_0 \in \mathbb{R}$" is removed, the said definition becomes a fuzzy interval definition.

Attention to computational complexity is critical in solving all problems, including fuzzy issues. For this reason, in 1979, a new type of fuzzy number called LR number was introduced by Dubois and Prade. A fuzzy number is called an LR fuzzy number when the membership function is defined using the relation (7.66).

$$
\mu_{\tilde{A}}(x) = \begin{cases} L\left(\dfrac{m - x}{\alpha}\right) & x \leq m \\ R\left(\dfrac{x - m}{\beta}\right) & x > m \end{cases} \tag{7.66}
$$

Fig. 7.2 Membership
function of fuzzy number $\tilde{A}$

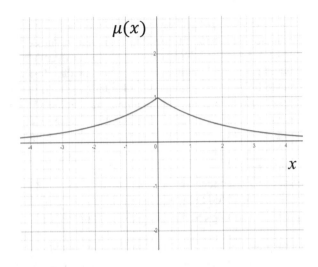

Fig. 7.2 Membership function of fuzzy number $\tilde{A}$

where L and R are non-increasing functions from $\mathbb{R}^+$ to [0,1] and $L(0) = R(0) = 1$. In this way, the LR fuzzy number is represented by the symbol $\tilde{A} = (m, \alpha, \beta)$ where m is the median value and α, β are called left and right spreads, respectively. If $\tilde{A} = (m, 0, 0)$, then $\tilde{A}$ is a real number. Among the fuzzy numbers, two types of these numbers are more important than other fuzzy numbers because they are a suitable way for describing many Linguistic terms, and calculations on these numbers are more straightforward. In the following, we introduce these numbers.

The fuzzy number $\tilde{A}$ is a triangular fuzzy number if:

$$\mu_{\tilde{A}}(x) = \begin{cases} \frac{x-m}{m-l} & l \leq x \leq m \\ \frac{u-x}{u-m} & m \leq x \leq u \\ 0 & otherwise \end{cases} \tag{7.67}$$

In other words, fuzzy triangular numbers are a particular type of LR number, so that $L(x) = R(x) = 1 - x$, $\alpha = m - l$ and $\beta = u - m$. Also, for ease of work, fuzzy triangular numbers are displayed with the symbol $\tilde{A} = (A^l, A^m, A^u)$. As shown in Fig. 7.3, A^l is considered the lowest or most pessimistic value, and A^u is the highest or most optimistic value for the triangular fuzzy number $\tilde{A}$.

If $\tilde{x} = (x^l, x^m, x^u)$ and $\tilde{y} = (y^l, y^m, y^u)$ are two non-negative fuzzy triangular numbers and λ is a non-negative real number, then algebraic operations on fuzzy triangular numbers are defined using relations (7.68)–(7.71).

$$\lambda\tilde{x} \simeq (\lambda x^l, \lambda x^m, \lambda x^u) \tag{7.68}$$

$$\tilde{x} + \tilde{y} \simeq (x^l + y^l, x^m + y^m, x^u + y^u) \tag{7.69}$$

Fig. 7.3 Fuzzy triangular number $\tilde{A}$

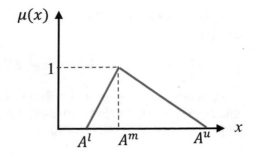

$$\tilde{x} - \tilde{y} \simeq (x^l - y^l, x^m - y^m, x^u - y^u) \tag{7.70}$$

$$\tilde{x} \times \tilde{y} \simeq (x^l y^l, x^m y^m, x^u y^u) \tag{7.71}$$

Another type of fuzzy number is the trapezoidal fuzzy number shown in Fig. 7.4.

$$\mu_{\tilde{A}}(x) = \begin{cases} 0 & x \le l \\ \frac{x - m_1}{m_1 - l} & l \le x \le m_1 \\ 1 & m_1 \le x \le m_2 \\ \frac{u - x}{u - m_2} & m_2 \le x \le u \\ 0 & x \ge u \end{cases} \tag{7.72}$$

For ease of work, similar to fuzzy triangular numbers, the trapezoidal fuzzy number is displayed with the symbol $\tilde{A} = (A^l, A^{m_1}, A^{m_2}, A^u)$.

It is necessary to explain that the trapezoidal fuzzy number is a fuzzy interval. Still, because of its many uses in describing natural phenomena, it has become famous as a fuzzy number. Similarly, with fuzzy triangular numbers, algebraic operators on trapezoidal numbers can also be defined.

Another essential issue to pay attention to when using fuzzy numbers is how to compare fuzzy numbers. In this field, there are many methods, and we briefly explain some of these methods. Suppose $\tilde{x} = (x^l, x^m, x^u)$ and $\tilde{y} = (y^l, y^m, y^u)$ are

Fig. 7.4 Fuzzy trapezoidal number $\tilde{A}$

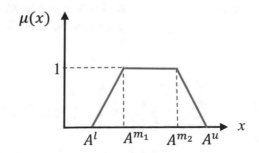

fuzzy triangular numbers. For comparison of $\tilde{x}$ and $\tilde{y}$, the relation is applied (7.73). (See [4]).

$$\tilde{x} \prec \tilde{y} \Leftrightarrow x^l < y^l, x^m < y^m, x^u < y^u \tag{7.73}$$

Since relation (7.73) is very restrictive and in many cases, it is not suitable for ranking some fuzzy numbers. So another type of ranking of fuzzy numbers was proposed in relation (7.74). (See [6]).

$$\tilde{x} \prec \tilde{y} \Leftrightarrow x^U < y^L \tag{7.74}$$

7.4.5.2 Fuzzy DEA Models for Supplier Selection

One of the first proposed methods for supplier selection in a fuzzy environment was introduced by [6]. In this method, firstly, the envelopment form of the input-oriented CCR model is considered with fuzzy input and output such as model (7.75) to select the best supplier.

$$\min \ \theta$$
$$s.t$$
$$\sum_{j=1}^{n} \lambda_j \tilde{x}_{ij} \leq \tilde{x}_{io}, \ i = 1, 2, \ldots, m,$$
$$\sum_{j=1}^{n} \lambda_j \tilde{y}_{rj} \geq \tilde{y}_{ro}, \ r = 1, 2, \ldots, s,$$
$$\lambda_j \geq 0. \tag{7.75}$$

So that $\tilde{x}_{ij}$ and $\tilde{y}_{rj}$ are the fuzzy input and output of the supplier o, respectively. As mentioned, there are different types of fuzzy numbers. Still, one of the widely used types of fuzzy numbers is fuzzy triangular numbers. Therefore, in this section, we assume that $\tilde{x}_{ij}$ and $\tilde{y}_{rj}$ are fuzzy triangular numbers such as: $\tilde{x}_{ij} = (x_{ij}^l, x_{ij}^m, x_{ij}^u)$ and $\tilde{y}_{rj} = (y_{rj}^l, y_{rj}^m, y_{rj}^u)$. By placing relationships $\tilde{x}_{ij}$ and $\tilde{y}_{rj}$ in the model (7.75), we have:

$$\min \ \theta$$
$$s.t$$
$$\sum_{j=1}^{n} \lambda_j (x_{ij}^l, x_{ij}^m, x_{ij}^u) \leq \theta (x_{io}^l, x_{io}^m, x_{io}^u), \ i = 1, 2, \ldots, m,$$
$$\sum_{j=1}^{n} \lambda_j (y_{rj}^l, y_{rj}^m, y_{rj}^u) \geq (y_{ro}^l, y_{ro}^m, y_{ro}^u), \ r = 1, 2, \ldots, s,$$

$$\lambda_j \geq 0, \, j = 1, ..., n. \tag{7.76}$$

Based on the algebraic operations on the fuzzy numbers expressed in relations (7.68)–(7.71), model (7.76) is converted into model (7.77).

$$\min \, \theta$$

$s.t$

$$\left(\sum_{j=1}^{n} \lambda_j x_{ij}^l, \, \sum_{j=1}^{n} \lambda_j x_{ij}^m, \, \sum_{j=1}^{n} \lambda_j x_{ij}^u \right) \leq \theta \left(x_{io}^l, x_{io}^m, x_{io}^u \right), i = 1, 2, ..., m,$$

$$\left(\sum_{j=1}^{n} \lambda_j y_{rj}^l, \, \sum_{j=1}^{n} \lambda_j y_{rj}^m, \, \sum_{j=1}^{n} \lambda_j y_{rj}^u \right) \geq \left(y_{ro}^l, y_{ro}^m, y_{ro}^u \right), r = 1, 2, ..., s,$$

$$\lambda_j \geq 0, j = 1, ..., n. \tag{7.77}$$

Also, by using the fuzzy inequality presented in relation (7.74), the inequalities of the model (7.77) are transformed as (7.78).

$$\min \, \theta$$

$s.t$

$$\sum_{j=1}^{n} \lambda_j x_{ij}^u \leq \theta x_{io}^l, \quad i = 1, 2, ..., m,$$

$$\sum_{j=1}^{n} \lambda_j y_{rj}^l \geq y_{ro}^u, \quad r = 1, 2, ..., s,$$

$$\lambda_j \geq 0, \quad j = 1, ..., n. \tag{7.78}$$

Similarly, the output-oriented of CCR model with fuzzy data is expressed as follows:

$$\max \, \varphi$$

$s.t$

$$\sum_{j=1}^{n} \lambda_j x_{ij}^u \leq x_{io}^l, \quad i = 1, ..., m,$$

$$\sum_{j=1}^{n} \lambda_j y_{rj}^u \geq \varphi \, y_{ro}^l, \quad r = 1, ..., s,$$

$$\lambda_j \geq 0, \, j = 1, ..., n. \tag{7.79}$$

As explained in Chap. 6, one of the applications of DEA models is to provide a benchmark for inefficient units. Still, in some cases, the benchmark provided by traditional DEA models is far from the goals of organization managers. In addition to inefficient units, this issue should be considered even for efficient DMUs. Thus, Stewart et al. [65] using the combination of GP and DEA, presented model (6.43) in Chap. 6 to present a projection as close as possible to the goals and ideals of the organization's managers [6], due to the importance of this issue in supplier selection, developed the model (6.43) to provide an ideal projection for suppliers in fuzzy environments. In this way, model (7.80) was introduced.

$$\min \Delta + \varepsilon \left[\sum_{i=1}^{m} w_{io}^I \delta_{io}^I + \sum_{r=1}^{s} w_{ro}^O \delta_{ro}^O \right]$$

$$s.t$$

$$\sum_{j=1}^{n} \lambda_j \tilde{x}_{ij} - \delta_{io}^I \leq \alpha \tilde{g}_{io} + (1-\alpha)\tilde{x}_{io}, \qquad i = 1, \ldots, m,$$

$$\sum_{j=1}^{n} \lambda_j \tilde{y}_{rj} + \delta_{ro}^O \geq \alpha \tilde{h}_{ro} + (1-\alpha)\tilde{y}_{ro}, \qquad r = 1, \ldots, s,$$

$$\Delta - w_{io}^I \delta_{io}^I \geq 0, \qquad\qquad i = 1, \ldots, m,$$

$$\Delta - w_{ro}^O \delta_{ro}^O \geq 0, \qquad\qquad r = 1, \ldots, s,$$

$$\lambda_j \geq 0, \qquad\qquad j = 1, \ldots, n. \qquad (7.80)$$

So that $\tilde{g}_{io}$ and $\tilde{h}_{ro}$ are the fuzzy ideals related to the ith input and the rth output; thus they are considered as fuzzy triangular numbers.

$$\min \Delta + \varepsilon \left[\sum_{i=1}^{m} w_{io}^I \delta_{io}^I + \sum_{r=1}^{s} w_{ro}^O \delta_{ro}^O \right]$$

$$s.t$$

$$\sum_{j=1}^{n} \lambda_j x_{ij}^u - \delta_{io}^I \leq \alpha g_{io}^l + (1-\alpha)x_{ro}^l, \qquad i = 1, \ldots, m,$$

$$\sum_{j=1}^{n} \lambda_j y_{rj}^l + \delta_{ro}^o \geq \alpha h_{ro}^u + (1-\alpha)y_{ro}^u, \qquad r = 1, \ldots, s,$$

$$\Delta - w_{io}^I \delta_{io}^I \geq 0, \qquad\qquad i = 1, \ldots, m,$$

$$\Delta - w_{ro}^O \delta_{ro}^O \geq 0, \qquad\qquad r = 1, \ldots, s,$$

$$\lambda_j \geq 0, \qquad\qquad j = 1, \ldots, n. \qquad (7.81)$$

Due to most of the economic interpretations being based on the optimal weights of the multiplier forms, [2, 4], for the selection of the best supplier in the fuzzy environment, focused on the multiplier forms of DEA models and introduced model (7.82).

$$\max \sum_{r=1}^{s} u_r \tilde{y}_{ro}$$

$$s.t$$

$$\sum_{i=1}^{m} v_i \tilde{x}_{io} = 1,$$

$$\sum_{r=1}^{s} u_r \tilde{y}_{rj} - \sum_{i=1}^{m} v_i \tilde{x}_{ij} \leq 0, \quad j = 1, ..., n,$$

$$u_r, v_i \geq 0. \tag{7.82}$$

The fuzzy numbers used in the model (7.82) are fuzzy triangular numbers. Therefore, similarly to what was said, model (7.82) becomes model (7.83):

$$\max \sum_{r=1}^{s} u_r (y_{ro}^l, y_{ro}^m, y_{ro}^u)$$

$$s.t$$

$$\sum_{i=1}^{m} v_i (x_{io}^l, x_{io}^m, x_{io}^u) = 1,$$

$$\sum_{r=1}^{s} u_r (y_{rj}^l, y_{rj}^m, y_{rj}^u) - \sum_{i=1}^{m} v_i (x_{ij}^l, x_{ij}^m, x_{ij}^u) \leq 0, \quad j = 1, ..., n,$$

$$u_r, v_i \geq 0, \quad r = 1, ..., s, \quad i = 1, ..., m. \tag{7.83}$$

Many researchers have studied fuzzy programming problems, especially fuzzy data DEA, and different techniques have been presented in this field, including Lotfi et al. [34, 35], Rostamy et al. [49], Sanei et al. [57, 58], Allahviranloo et al. [1], Ebrahimnejad et al. [17], Allahviranloo et al. (2013), Rostamy et al. [48]. Bagheri et al. [7], Nosrat et al. [46], Shafiee and Saleh [60], Hosseinzadeh et al. [21]. Although there are various methods related to solving fuzzy DEA models, one of the most popular techniques in FDEA is the approach based on α-cut [23, 26, 31]. The α-cut approach is a standard method for performing algebraic operations on fuzzy numbers. Using this method, a fuzzy programming problem at different levels of α becomes an interval programming problem.

Suppose $\tilde{x} = (x_{io}^l, x_{io}^m, x_{io}^u)$ is a fuzzy triangular number; as a result, in a level α, each member of the set α-cut $\tilde{x}$ belongs to the interval $[(1-\alpha)x_{io}^l + \alpha x_{io}^m, (1-\alpha)x_{io}^u + \alpha x_{io}^m]$. In this way, the model (7.83) becomes an IDEA problem at a level α pre-determined by the manager in the model (7.84).

$$\max \sum_{r=1}^{s} u_r((1-\alpha)y_{ro}^l + \alpha y_{ro}^m, (1-\alpha)y_{ro}^u + \alpha y_{ro}^m)$$

$s.t$

$$\sum_{i=1}^{m} v_i((1-\alpha)x_{io}^l + \alpha x_{io}^m, (1-\alpha)x_{io}^u + \alpha x_{io}^m) = 1,$$

$$\sum_{r=1}^{s} u_r((1-\alpha)y_{rj}^l + \alpha y_{rj}^m, (1-\alpha)y_{rj}^u + \alpha y_{rj}^m) -$$

$$\sum_{i=1}^{m} v_i((1-\alpha)x_{ij}^l + \alpha x_{ij}^m, (1-\alpha)x_{ij}^u + \alpha x_{ij}^m) \le 0, \ j = 1, ..., n,$$

$$u_r, v_i \ge 0, \ r = 1, ..., s, \ i = 1, ..., m. \tag{7.84}$$

There are several methods to solve the interval programming problem in the DEA. Among the strategies presented in this field are [22, 24, 23, 26, 66, 68].

In this section, the method provided by [5] is used to solve the model (7.84). In this method, the model (7.84) becomes two linear programming problems, so an upper and lower bound for efficiency are obtained. The maximum efficiency value occurs when the inputs have values equal to their minimum value, and the outputs have values equal to their maximum value.

$$\max \sum_{r=1}^{s} u_r((1-\alpha)y_{ro}^u + \alpha y_{ro}^m)$$

$s.t$

$$\sum_{i=1}^{m} v_i((1-\alpha)x_{io}^l + \alpha x_{io}^m) = 1,$$

$$\sum_{r=1}^{s} u_r((1-\alpha)y_{rj}^u + \alpha y_{rj}^m) - \sum_{i=1}^{m} v_i((1-\alpha)x_{ij}^l + \alpha x_{ij}^m) \le 0, \ j = 1, ..., n,$$

$$u_r, v_i \ge 0, r = 1, ..., s, i = 1, ..., m. \tag{7.85}$$

Also, as stated in the model (7.85), the lower bound of efficiency occurs so that inputs and outputs have values equal to their maximum and minimum value, respectively.

$$\max \sum_{r=1}^{s} u_r((1-\alpha)y_{ro}^l + \alpha y_{ro}^m)$$

$s.t$

$$\sum_{i=1}^{m} v_i((1-\alpha)x_{io}^u + \alpha x_{io}^m) = 1,$$

$$\sum_{r=1}^{s} u_r((1-\alpha)y_{rj}^l + \alpha y_{rj}^m)$$

$$-\sum_{i=1}^{m} v_i((1-\alpha)x_{ij}^u + \alpha x_{ij}^m) \leq 0, \quad j = 1, ..., n,$$

$$u_r, v_i \geq 0, \ r = 1, ..., s, \ i = 1, ..., m. \tag{7.86}$$

In the methods described so far, only inputs and outputs have been considered in a fuzzy number, and the coefficients are non-fuzzy. Tsai et al. [67] believe that if data are fuzzy numbers, then coefficients should also be fuzzy variables. Otherwise, the obtained results are invalid. Therefore, Tisai et al., Considering the multiplication of fuzzy numbers and a ranking method, defined the relative efficiency for suppliers with fuzzy inputs and outputs using relation (7.87).

$$\frac{\sum_{r=1}^{s} \tilde{u}_r \cdot \tilde{y}_{ro}}{\sum_{i=1}^{m} \tilde{v}_i \cdot \tilde{x}_{io}} \approx \frac{\sum_{r=1}^{s} (u_r^l y_{ro}^l + u_r^m y_{ro}^m + u_r^u y_{ro}^u)}{\sum_{i=1}^{m} (v_i^l x_{io}^l + v_i^m x_{io}^m + v_i^u x_{io}^u)} \tag{7.87}$$

So that $\tilde{u}_r$ and $\tilde{v}_i$ are the fuzzy coefficients corresponding to the r-th output and the i-th input, respectively. Also, in relation (7.87), the inputs and outputs and their coefficients were considered fuzzy triangular numbers. Therefore, the fractional form of the CCR model with fuzzy data is presented as the model (7.88) to select the best supplier in the fuzzy environment.

$$\max \frac{\sum_{r=1}^{s} (u_r^l y_{ro}^l + u_r^m y_{ro}^m + u_r^u y_{ro}^u)}{\sum_{i=1}^{m} (v_i^l x_{io}^l + v_i^m x_{io}^m + v_i^u x_{io}^u)}$$

s.t

$$\frac{\sum_{r=1}^{s} (u_r^l y_{rj}^l + u_r^m y_{rj}^m + u_r^u y_{rj}^u)}{\sum_{i=1}^{m} (v_i^l x_{ij}^l + v_i^m x_{ij}^m + v_i^u x_{ij}^u)} \leq 0, j = 1, ..., n,$$

$$0 \leq v_i^l, v_i^u \leq v_i^m, \ i = 1, ..., m,$$

$$0 \leq u_r^l, u_r^u \leq u_r^m, \ r = 1, ..., s. \tag{7.88}$$

We know that in fuzzy triangular numbers, left and right values are always smaller than or equal to the median. As a result, the constraints $0 \leq v_i^l, v_i^u \leq v_i^m$ and $0 \leq u_r^l, u_r^u \leq u_r^m$ are added to the model (7.88). Then, using Charnes and Cooper's transformations, the model (7.88) is transformed into the linear model (7.89).

$$\max \sum_{r=1}^{s} (u_r^l y_{ro}^l + u_r^m y_{ro}^m + u_r^u y_{ro}^u)$$

s.t

$$\sum_{i=1}^{m} (v_i^l x_{io}^l + v_i^m x_{io}^m + v_i^u x_{io}^u) = 1,$$

$$\sum_{r=1}^{s} (u_r^l y_{rj}^l + u_r^m y_{rj}^m + u_r^u y_{rj}^u)$$

$$- \sum_{i=1}^{m} (v_i^l x_{ij}^l + v_i^m x_{ij}^m + v_i^u x_{ij}^u) \le 0, \, j = 1, ..., n$$

$$0 \le v_i^l, v_i^u \le v_i^m, \, i = 1, ..., m,$$

$$0 \le u_r^l, u_r^u \le u_r^m, \, r = 1, ..., s. \tag{7.89}$$

And finally, we simplify the model (7.89) as the model (7.90).

$$\max \sum_{r=1}^{s} (u_r^l y_{ro}^l + u_r^m y_{ro}^m + u_r^u y_{ro}^u)$$

$$s.t$$

$$\sum_{i=1}^{m} (v_i^l x_{io}^l + v_i^m x_{io}^m + v_i^u x_{io}^u) = 1,$$

$$\sum_{r=1}^{s} (u_r^l y_{rj}^l + u_r^m y_{rj}^m + u_r^u y_{rj}^u)$$

$$- \sum_{i=1}^{m} (v_i^l x_{ij}^l + v_i^m x_{ij}^m + v_i^u x_{ij}^u) \le 0, j = 1, ..., n,$$

$$v_i^l - v_i^m \le 0, \, i = 1, ..., m,$$

$$v_i^u - v_i^m \le 0, \, i = 1, ..., m,$$

$$u_r^l - u_r^m \le 0, \, r = 1, ..., s,$$

$$u_r^u - u_r^m \le 0, \, r = 1, ..., s,$$

$$0 \le v_i^l, v_i^u, v_i^m, u_r^l, u_r^u, u_r^m, \, i = 1, ..., m, \, r = 1, ..., s. \tag{7.90}$$

7.5 Supplier Selection by Cross-Efficiency Models

In traditional DEA, each unit is free to choose its outputs and inputs weights, and even in some cases, all its priority and valuation may be focused on a specific output or input. In other words, since each DMU has its own set of weights, in some cases, the results obtained from DEA lead to an unrealistic weighting. One of the proposed solutions to solve this problem is to use cross-efficiency. So, Mahdilo et al. [38] proposed cross-efficiency models for Supplier selection when suppliers give volume discounts to encourage customers to purchase large quantities. Because in some circumstances, to encourage buyers to order more, suppliers may offer volume discounts.

As stated in Chap. 6, one of the problems of using the cross-efficiency method is the existence of alternative optimal solutions in calculating optimal weights and, as a result, the non-uniqueness of the value obtained from the relation (6.43). In this context, several articles have been presented to provide a secondary goal to prevent the existence of multiple weights. For further study, readers can refer to the sources of [15], Jahanshahloo et al. [22, 24], Hosseinzadeh [33, 37]. In this part, the aggressive method is applied to solve the problem of the non-uniqueness of weights, as presented by [15]. In this approach, the optimal weights of each unit are first calculated using the multiplier form of the CCR model, then, model (7.91) is solved.

$$\min \sum_{r=1}^{s} u_r \sum_{j \neq o} y_{rj}$$

$$s.t$$

$$\sum_{i=1}^{m} v_i \sum_{j \neq o} x_{ij} = 1,$$

$$\sum_{r=1}^{s} u_r y_{rj} - \sum_{i=1}^{m} v_i x_{ij} \leq 0, j = 1, \ldots, n, j \neq o$$

$$\sum_{r=1}^{s} u_r y_{ro} - E_{oo} \sum_{i=1}^{m} v_i x_{io} = 0$$

$$u_r, v_i \geq 0, \ i = 1, \ldots, m, \ r = 1, \ldots, s. \qquad (7.91)$$

It is necessary to explain that E_{oo} is the efficiency of the supplier o, which is obtained by the multiplier form of the CCR model. Suppose u_r^* and v_i^* are the optimal solution of the model (7.91), so the relative efficiency of supplier j based on the optimal weights of supplier o is represented by the symbol E_{oj} and be obtained from the relation (7.92):

$$E_{oj} = \frac{\sum_{r=1}^{s} u_r^* y_{rj}}{\sum_{i=1}^{m} v_i^* x_{ij}} \qquad (7.92)$$

As a result, the cross-efficiency of the supplier o is gotten using the relation (7.93):

$$\overline{E} = \frac{1}{n} \sum_{j=1}^{n} E_{oj} \qquad (7.93)$$

In this way, the following algorithm is used to select suppliers by considering volume discounts:

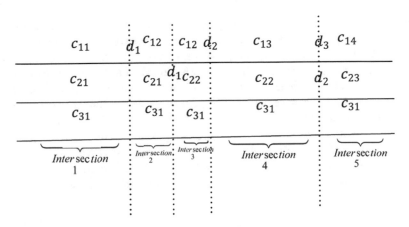

Fig. 7.5 Intersections of price breaks of three suppliers [38]

- Determine the quantity demanded by the buyer.
  At this stage, the buyer determines the number of materials using some method such as material requirement planning (MRP) and artificial neural networks.
- Determine the price vector for the desired quantity.
  For each supplier, the linear piecewise function of the material price is introduced so that per supplier's price in each interval is a constant value. (In Fig. 7.5, this issue is depicted for three suppliers).
- calculate the efficiency per supplier given the specified price at each intersection and other criteria such as quality and delivery performance.
- Run the model (7.91).
- Use the formula (7.93) to get the final score of supplier efficiency evaluation.
- Analysis of results.

According to the achieved results, determine the top suppliers at each intersection, then purchase the desired quantities at each intersection from the preferred supplier.

As already explained, the results obtained from DEA models (self-evaluation) and the results obtained through cross-efficiency calculation (peer-evaluation) are different. In other words, the efficiency score of some DMUs is better, and others get worse. So to identify the suppliers with the highest percentage of improvement, when the evaluation changes from the group evaluation to the self-evaluation, the false positive index (FPI) according to relation (7.94) is used.

$$FPI = \frac{E_{dd} - E_{jd}}{E_{jd}} \tag{7.94}$$

In another study, [44] developed the cross-efficiency technique to evaluate suppliers with non-discretionary inputs in the form of the model (7.95):

$$\min \sum_{r=1}^{s} u_r \sum_{j \neq o} y_{rj} - \sum_{i \in IND} v_{iND} \sum_{j \neq o} x_{ij}$$

$s.t$

$$\sum_{i \in ID} v_{iD} \sum_{j \neq o} x_{ij} = 1,$$

$$\sum_{r=1}^{s} u_r y_{rj} - \left(\sum_{i \in ID} v_{iD} x_{ij} + \sum_{i \in IND} v_{iND} x_{ij} \right) \leq 0, j = 1, ..., n, j \neq o,$$

$$\sum_{r=1}^{s} u_r y_{ro} - \sum_{i \in IND} v_{iND} x_{io} \geq 0,$$

$$\left(\sum_{r=1}^{s} u_r y_{ro} - \sum_{i \in IND} v_{iND} x_{io} \right) - E_{oo} \sum_{i \in ID} v_{iD} x_{io} = 0, j = 1, ..., n,$$

$$u_r, v_i \geq 0, i = 1, ..., m, r = 1, ..., s. \tag{7.95}$$

So that E_{oo} is the efficiency of the supplier o using the model (7.39). If u_r^*, v_{iND}^*, v_{iD}^* are the optimal solution of model (7.95), then:

$$E_{oj} = \frac{\sum_{r=1}^{s} u_r^* y_{rj} - \sum_{i \in IND} v_{iND}^* x_{ij}}{\sum_{i \in ID} v_{iD}^* x_{ij}} \tag{7.96}$$

Although using the mentioned methods, some problems in supplier selection have been solved, in these techniques, suppliers are considered independent DMUs, and competition between sellers is not considered. In other words, these methods only concentrate on the relationships between inputs and outputs. To overcome this problem, [69] proposed a combined DEA and game theory approach. Game theory, as one of the branches of mathematical science, was presented for the first time by John Von Neumann in 1928 and quickly spread in other sciences, especially economics. The game theory studies the behavior of decision-makers in different conditions and situations. Per decision-makers goal is to make a decision that leads to more profit at the end of the game. Of course, the profit per player, in addition to own decision, depends on the findings of other decision-makers. Each situation is called a "game," and each decision maker is named a "player". Also, the choices or decisions that each player can make are called the "strategy" of the player. According to different conditions, many classifications exist for all kinds of games. One of these general classifications is cooperative and non-cooperative games. In cooperative games, it is assumed that there is full cooperation between the players to achieve the optimal result for the whole system. In these games, the problem is changed from the several decision-makers and criteria to the one decision-maker. In other words, in cooperative games, decision-makers are integrated, and a combined objective function is created so that the problem is transformed from a multi-objective state to a single-objective one. In such issues, only one informed decision-maker is defined, and as a result, the final decision is made fairly and based on the criteria announced by all beneficiary parties. But in non-cooperative games, it is assumed that the primary concern of the players is to maximize their profit, and the players do not have any desire to exchange opinions or negotiate to reach a coalition or agree with each other. Therefore, the final results of non-cooperative games are affected by the decisions made by all players [41, 43, 47].

Suppose suppliers are considered game players, and their cross-efficiency is the payoff. Since each supplier intends to maximize their cross-efficiency, suppliers decide to have a non-cooperative game to maximize their profits.

Therefore, if the supplier selection process is argued with this approach, it is concluded that the existing evaluation methods have drawbacks. Therefore, studies have been done to solve this problem. Wang and Lee [69] used the Nash Bargaining Theory to obtain optimal weights in calculating the cross-efficiency of suppliers. In the Nash bargaining game, strategies are represented as ordered pairs (x, y) where x and y are chosen from the interval $[d, z]$, where z is the maximum inventory, and d is the opposition or threat point of the game (A point below which one of the players does not continue the game.). Let u and v be utility functions of players 1 and 2. In the Nash bargaining solution, players are trying to maximize $(u(x) - u(d)) * (v(y) - v(d))$, where $u(d)$ and $v(d)$ are profits that are obtained considering the situation where one of the players does not decide to continue. It is necessary to explain that in each solution of the bargaining game, each player will benefit according to their conditions and an amount considered for the opposite player (gain from cooperation).

Now suppose that $N = \{1, \ldots, n\}$ is the set of all players, and each payoff vector (achievement) is a member of the R, S is a subset of the payoff space, and b is the breakdown point (disagreement), and is a member of the payoff space. In this case, the bargaining problem can be represented using (N, S, b). In the general case, [42] showed that for the traditional bargaining problem for n players, there is a unique solution called the Nash solution, which can be obtained by solving the following maximization problem.

$$\max_{\substack{i=1 \\ \vec{u} \in S, \vec{u} \geq \vec{b}}} \prod_{i=1}^{n} (u_i - b_i) \tag{7.97}$$

So that u is the payment vector, u_i and b_i are respectively the ith component of the vector u and b. The following will explain the method of calculating cross-efficiency using the Nash bargaining approach.

As explained before, the efficiency obtained from DEA models is obtained from the maximum weighted sum of the output to the weighted sum of the input. Also, to obtain the maximum efficiency compared to other DMUs, each unit can freely assign the desired feasible weight to each index. Then it is concluded that the efficiency obtained by classic DEA models is an optimistic efficiency for each DMU under evaluation. Therefore, it is considered the maximum efficiency for each supplier in the production technology. Also, to obtain the minimum efficiency for each unit, model (7.98) is used.

$$E_{do}^* = \min \sum_{r=1}^{s} u_r y_{rd}$$

$s.t$

$$\sum_{i=1}^{m} v_i x_{id} = 1,$$

$$\sum_{r=1}^{s} u_r y_{rj} - \sum_{i=1}^{m} v_i x_{ij} \leq 0, j = 1, ..., n,$$

$$\sum_{r=1}^{s} u_r y_{ro} - E_{oo}^* \times \sum_{i=1}^{m} v_i x_{io} = 0,$$

$$u_r, v_i \geq 0, r = 1, ..., s, i = 1, ..., m. \tag{7.98}$$

Since that in the model (7.98), the minimum efficiency of the supplier d is obtained in the condition that the efficiency of the DMUo is in the best state (optimistic efficiency), so using the relation (7.99), the lower bound of efficiency of supplier j is obtained.

$$E_j^{\min} = \frac{1}{n-1} \sum_{\substack{d=1 \\ d \neq o}}^{n} E_{dj}^* \tag{7.99}$$

So by Nash bargaining, the model (7.100) is obtained.

$$\max \prod_{\substack{j=1 \\ j \neq l, l \in ES}}^{n} \left(\frac{\sum_{r=1}^{s} u_r y_{rj}}{\sum_{i=1}^{m} v_i x_{ij}} - E_j^{\min} \right) \left(E_j^{CCR} - \frac{\sum_{r=1}^{s} u_r y_{rj}}{\sum_{i=1}^{m} v_i x_{ij}} \right)$$

$s.t$

$$\frac{\sum_{r=1}^{s} u_r y_{rj}}{\sum_{i=1}^{m} v_i x_{ij}} \leq E_j^{CCR}, j = 1, ..., n,$$

$$\frac{\sum_{r=1}^{s} u_r y_{rj}}{\sum_{i=1}^{m} v_i x_{ij}} \geq E_j^{\min}, j = 1, ..., n, j \neq l,$$

$$u_r, v_i \geq 0, r = 1, ..., s, i = 1, ..., m. \tag{7.100}$$

So that ES is the index set of DMUs where $E_j^{\min} = E_j^{CCR}$. Therefore, there will be no bargaining for these DMUs. The relation (7.101) is used to calculate the cross efficiency of the supplier j:

$$E_j^* = \frac{\sum_{r=1}^{s} u_r^* y_{rj}}{\sum_{i=1}^{m} v_i^* x_{ij}}, j \neq l \tag{7.101}$$

(v^*, u^*) is the optimal solution of model (7.101) and also: $E_j^* = E_j^{\min} = E_j^{CCR}, j = l$.

In another research, based on the combined approach of game theory and cross-efficiency to select the best supplier, [36] proposed a new model, according to the idea of [30]. In this approach, considering the competition between suppliers, the set of proposed weights for calculating cross-efficiency is selected in such a way that it brings a unique Nash equilibrium. In simpler words, based on the Nash equilibrium, the leader chooses an optimal strategy for himself. As a result, others are forced to choose the best method for themselves according to the plan proposed by the leader. As a result, in the supplier selection process, supplier d chooses its optimal weight. Then other suppliers must choose their optimal weight strategy such that the efficiency of supplier d does not decrease. Therefore, the model (7.102) is used to calculate the efficiency of supplier d under the influence of the optimal weights of the supplier o (E_{sd}).

$$\max \sum_{r=1}^{s} u_{rl}^d y_{rl}$$

$s.t$

$$\sum_{i=1}^{m} v_{il}^d x_{il} = 1,$$

$$\sum_{r=1}^{s} u_{rl}^d y_{rj} - \sum_{i=1}^{m} v_{il}^d x_{ij} \leq 0, \quad j = 1, ..., n,$$

$$e_o \sum_{i=1}^{m} v_{il}^d x_{io} - \sum_{r=1}^{s} u_{rl}^d y_{ro} \leq 0,$$

$$u_{rl}^d, v_{il}^d \geq 0, \ i = 1, ..., m, \ r = 1, ..., s. \tag{7.102}$$

consequently:

$$\overline{CE}_j = \frac{\sum_{d=1}^{n} \sum_{r=1}^{s} u_{rj}^{d*}(e_d) y_{rj}}{n} \tag{7.103}$$

So that $u_{rj}^{d*}(e_d)$ is the optimal weight of the model (7.102).

As seen so far, the stated techniques for calculating the cross-efficiency and selecting the best suppliers are always trying to solve the problem of alternative optimal solutions and as a result, non-uniqueness of the cross-efficiency of the suppliers. And finally, to get the total efficiency, the arithmetic mean of the obtained results is used to calculate the final answer.

Soltanifar and Sharifi [64] believe that there is no convincing reason to use the arithmetic mean to aggregate the results and calculate the cross-efficiency of a supplier. Therefore, for more validity of the achieved results, they suggested using the fuzzy vikor method to introduce a secondary goal and aggregate the results to present the final answer.

In this method, due to the higher discrimination power of non-radial models compared to radial models, non-radial models have been used for initial evaluation. The following algorithm was proposed to calculate the cross-efficiency of the suppliers under the assessment.

1. Obtain the optimal solution of model (7.14) for supplier o, (σ_o^*).
2. Calculate s_k and R_k using relations (7.104) and (7.105).

$$s_k = \sum_{r=1}^{s} u_r \frac{y_{rI} - y_{rk}}{y_{rI} - y_{rAI}} + \sum_{i=1}^{m} v_i \frac{x_{ik} - x_{il}}{x_{iAI} - x_{il}} \qquad (7.104)$$

$$R_k = \max_{\substack{1 \le r \le s \\ 1 \le i \le m}} \left\{ u_r \left(\frac{y_{rI} - y_{rk}}{y_{rI} - y_{rAI}} \right), v_i \left(\frac{x_{ik} - x_{il}}{x_{iAI} - x_{il}} \right) \right\} \qquad (7.105)$$

3. Obtain the optimal solution of the model (7.106).

$$\min\{S_k, R_k\}$$

$$s.t$$

$$\sum_{r=1}^{s} u_r y_{rj} - \sum_{i=1}^{m} v_{ij} x_{ij} + \delta \le 0, \ j = 1, ..., n,$$

$$\sum_{r=1}^{s} u_r y_{ro} - \sum_{i=1}^{m} v_{io} x_{io} + (\delta + 1) = \sigma_o^*,$$

$$v_i(x_{lo} - x_{il}) \ge \frac{1}{2m}, \ i = 1, ..., m,$$

$$u_r(y_{rI} - y_{ro}) \ge \frac{1}{2s}, \ r = 1, ..., s,$$

$$v_i \ge 0, \ i = 1, ..., m,$$

$$u_r \ge 0, \ r = 1, ..., s. \qquad (7.106)$$

So that, S_k and R_k are obtained from the relations (7.104) and (7.105). As seen, model (7.106) is a bi-objective linear programming problem. Therefore, existing methods for solving the MOLP problem can be used. One of the most straightforward and practical methods for bi-objective issues is using the weighted sum of objective functions. Therefore, model (7.106) becomes model (7.107).

$$\min \alpha \, S_k + (1 - \alpha) R_k$$

$$s.t$$

$$\sum_{r=1}^{s} u_r y_{rj} - \sum_{i=1}^{m} v_{ij} x_{ij} + \delta \leq 0, j = 1, ..., n,$$

$$\sum_{r=1}^{s} u_r y_{ro} - \sum_{i=1}^{m} v_{io} x_{io} + (\delta + 1) = \sigma_o^*,$$

$$s_k = \sum_{r=1}^{s} u_r \frac{y_{rl} - y_{rk}}{y_{rl} - y_{rAI}} + \sum_{i=1}^{m} v_i \frac{x_{ik} - x_{il}}{x_{iAI} - x_{il}}$$

$$R_k \geq u_r \left(\frac{y_{rl} - y_{rk}}{y_{rl} - y_{rAI}} \right), \quad r = 1, ..., s,$$

$$R_k \geq v_i \left(\frac{x_{ik} - x_{il}}{x_{iAI} - x_{il}} \right), \quad i = 1, ..., m,$$

$$v_i(x_{io} - x_{il}) \geq \frac{1}{2m}, \quad i = 1, ..., m,$$

$$u_r(y_{rl} - y_{ro}) \geq \frac{1}{2s}, \quad r = 1, ..., s.$$

$$v_i \geq 0, \quad i = 1, ..., m,$$

$$u_r \geq 0, \quad r = 1, ..., s, \tag{7.107}$$

If $(v_1^*, ..., v_m^*, u_1^*, ..., u_s^*)$ is the optimal solution of the model, then: $\sum_{r=1}^{s} u_r^* y_{rk} - \sum_{i=1}^{m} v_i^* x_{ik} + (\delta^* + 1) = \sigma_{ko}, k = 1, ..., n$. It is necessary to explain that although $0 \leq \sigma_{oo} \leq 1$, $\sigma_{ko}, k \neq o$ is not necessarily between 0 and 1.

4. Obtain the cross-efficiency matrix.
5. Add the maximum value of each column to the members of that column. In this way, all the elements in each column become positive numbers.
6. Consider an interval for each column. So that the beginning of the interval is the minimum value, and the end of the interval is the maximum value of each column. This way, all the column members will be in the mentioned interval. Divide the introduced interval into eight equal pieces. The closer the desired numbers are to the end of the interval, it means they are in a better position. Then express the language terms equivalent to each part. For example, consider the interval [0 1] shown in Fig. 7.6. In this way, the cross-efficiency fuzzy matrix is obtained. Therefore, the fuzzy cross-efficiency matrix is a decision matrix in the fuzzy vikor method. In this case, DMUs play both the role of criterion and the role of alternative. In fact, in each row, the degree of eligibility of each supplier to be selected as the best supplier (the top alternative) is based on the existing criteria in each column (the evaluation results obtained from the optimal weights in the evaluation of other suppliers) is shown. It is also necessary to explain that there are different linguistic variables and scales for converting linguistic variables into fuzzy numbers. Many studies have been done in this field, including [18, 27], which can choose the appropriate method according to the conditions of the problem.

Fig. 7.6 Linguistic terms [64]

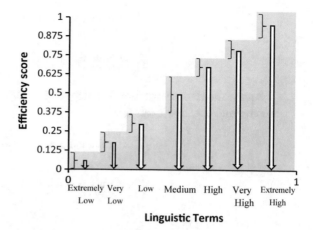

7. By using the fuzzy matrix obtained in step 5 and the fuzzy vikor method, the final ranking of the units is obtained.

The use of multi-criteria decision-making methods becomes very important when the goal of the problem is to choose the best option in the presence of multiple indicators. Some multi-criteria decision-making methods aim only to calculate the weights of indicators, while the second category of methods aims to rank the available options. The vikor method is among the second category methods; in other words, in this method, the goal is to rank options based on several criteria. This method is a consensus multi-criteria decision-making technique, and the ranking criterion in this method is based on their degree of nearness to the ideal point. This method can provide a maximum group utility for the majority and a minimum individual effect for the opposition. The fuzzy vikor method is similar to the classic vikor method, which is related to fuzzy numbers. Using the fuzzy environment makes the results more accurate due to overcoming the ambiguities and uncertainties of the problem. The steps of this method are expressed in the form of the following algorithm:

1. The first step in the fuzzy vikor method is to form the decision matrix $\tilde{X} = [\tilde{x}_{ij}]_{m \times n}$. Each row of the matrix corresponds to alternatives, and each column corresponds to criteria. The rows of the fuzzy matrix indicate the score of the desired criterion in each option.
2. Calculate the normalized matrix $\tilde{V} = [\tilde{v}_{ij}]_{m \times n}$ using the relation (7.108).

$$\tilde{v}_{ij} = \frac{\tilde{x}_{ij}}{\sum_{i=1}^{m} \tilde{x}_{ij}}, \quad i = 1, \ldots, m, \; j = 1, \ldots, n \qquad (7.108)$$

3. Calculate the positive and negative ideals. There are different ways to calculate positive and negative ideals. In one of the most common methods, if the desired criterion is a positive criterion (for example, profit), then the positive and negative ideals are:

$$\tilde{f}_j^+ = \max_{1 \le i \le m} \tilde{v}_{ij}, \, j = 1, 2, ..., n \qquad (7.109)$$

$$\tilde{f}_j^- = \min_{1 \le i \le m} \tilde{v}_{ij}, \, j = 1, 2, ..., n \qquad (7.110)$$

And if the desired criterion is a negative criterion (for example, cost), then positive and negative ideals are:

$$\tilde{f}_j^+ = \min_{1 \le i \le m} \tilde{v}_{ij}, \, j = 1, 2, ..., n \qquad (7.111)$$

$$\tilde{f}_j^- = \max_{1 \le i \le m} \tilde{v}_{ij}, \, j = 1, 2, ..., n \qquad (7.112)$$

4. Compute $\tilde{s}_i$ and $\tilde{R}_i$
5. Calculate the vikor index for each alternative.
   In this step, first, calculate the values of $\tilde{s}^+ = \max\{\tilde{s}_i | i = 1, ..., m\}$, $\tilde{s}^+ = \max\{\tilde{s}_i | i = 1, ..., m\}$, $\tilde{R}^- = \min\{\tilde{R}_i | i = 1, ..., m\}$ and $\tilde{R}^+ = \max\{\tilde{R}_i | i = 1, ..., m\}$. Then, the vikor index is obtained from the relation (7.115).

$$\tilde{Q}_i \approx v \left[\frac{\tilde{s}_i \ominus \tilde{s}^+}{\tilde{s}^- \ominus \tilde{s}^+} \right] + (1 - v) \left[\frac{\tilde{R}_i \ominus \tilde{R}^+}{\tilde{R}^- \ominus \tilde{R}^+} \right], \, i = 1, ..., m \qquad (7.115)$$

v is a number between 0 and 1, called maximum group utility. The larger v is, the decision maker is allowed to give more scores to group opinions, and conversely, the smaller v is, the more value is given to individual views.

6. Arrange alternatives based on $\tilde{s}, \tilde{R}, \tilde{Q}$ and, in ascending order. When sorting alternatives based on $\tilde{Q}$ values, suppose the alternatives are retagged as $(A^1, A^2, ..., A^n)$. Then A^1 is the best, provided that the following two conditions are met:

First: $\Omega(\tilde{Q}_{A^2}) + \Omega(\tilde{Q}_{A^1}) > \frac{1}{m-1}$ where $\Omega(.)$ is a defuzzification operator, and for a given triangular fuzzy number (a_1, a_2, a_3), $\Omega(.)$ is equal to: $\Omega(a_1, a_2, a_3) = \frac{a_1 + a_2 + a_3}{3}$. If this condition is not met, A^1 and A^2 will both be the best alternatives.

Second: Alternative A^1 is the first alternative in at least one of the sorts according to $\tilde{s}$ and $\tilde{R}$. If this condition is not met, A^1 and A^2 will both be the best alternatives.

7.6 Supplier Selection Using Cost and Profit Efficiency

Consider the supply chain network, which is shown in Fig. 7.7. The First stage includes suppliers, and the manufacturer is in the second stage. x^k, y^k are input

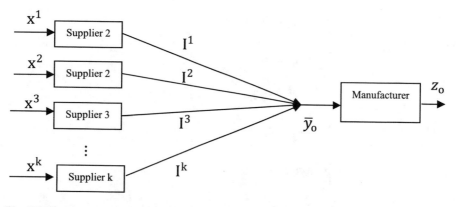

Fig. 7.7 Two-stage supply chain network (supplier-manufacturer)

and output vector corresponding to kth supplier and manufacturer o use input $\bar{y}_o$ to produce output z_o.

Suppose there are n manufacturers, and each manufacturer cooperates with k suppliers to supply material for producing the final production. Therefore, supplier selection plays a basic role in fulfilling the demand of the manufacturer. Suppose that the manufacturer o cooperates with $s_{i_1}, \ldots, s_{i_k}$ and uses y_o to produce z_o, and $p = (p_1, \ldots, p_k)$ is the price of the final output. Therefore, the sales revenue is equal to $p z_o$. In addition, the manufacturer should pay c_i^k to purchase one unit I_i^k from k-th suppliers so the cost of manufacturer o, for fulfilling its demand ($\bar{y}$), is equal to $\bar{c}_o = \sum_k c^k (\mu_k^o I^k)$ and revenue of manufacture is equal to $p z_o$. So that μ_k^o is the contribution of the k-th producer in completing the demand of the o-th producer.

The managers are interested in seeking other ways to increase the profit of their entity. For this purpose, the managers test their interactions with other suppliers. In such a situation, two questions arise for managers:

- Does the number of suppliers in the supply chain affect the performance of the supply chain and the efficiency of the producer's profit?
- Does the replacement of other suppliers in the chain affect the efficiency of the manufacturer's profit?

In this section, we respond to the above questions and present a new model. With the acceptance of the constant return to scale principle, supplier s_{i_k} uses $\mu_k^o x^k$ to produce $\mu_k^o I^k$. Now suppose that producer demand o is equal to $\bar{y}$ in this case $\sum_k \mu_k I^k = \bar{y}$. It is evident that if supplier k did not have any role in the production of $\bar{y}$, so $\mu_k^o = 0$. According to profit efficiency in classic DEA models and the mentioned discussion, model (7.116) is presented by Saleh et al. [53, 54] to calculate the profit efficiency of the supply chain as follows:

$$\max pz - \sum_k \mu_k c^k I^k$$

$$s.t$$

$$\sum_k \mu_k x^k \le \sum_k \mu_k^o x^k,$$

$$\sum_k \mu_k I^k = \bar{y},$$

$$\sum_k \mu_k c^k I^k \le \bar{c}_o,$$

$$\sum_j \lambda_j y_j \le \bar{y},$$

$$\sum_j \lambda_j z_j \ge z,$$

$$\mu_k \le 1, \ k = 1, ..., d,$$

$$\mu_k, \lambda_j, z \ge 0, \ j = 1, ..., n, \ k = 1, ..., d. \tag{7.116}$$

In model (7.116), the first, second, and third constraints seek another combination of suppliers to supply the manufacturer's demand. The second constraint also ensures that intermediate production equals the manufacturer's orders. The third inequality checks other combinations of suppliers to decrease the manufacturer's cost; the fourth and fifth inequalities are related to the revenue-efficiency of the manufacturer. So we rewrite the model (7.117) as follows:

$$\max pz - \sum_k \mu_k c^k I^k$$

$$s.t$$

$$\sum_k \mu_k x^k \le \sum_k \mu_k^o x^k,$$

$$\sum_k \mu_k c^k I^k \le \bar{c}_o,$$

$$\sum_j \lambda_j y_j \le \sum_k \mu_k I^k,$$

$$\sum_j \lambda_j z_j \ge z,$$

$$\mu_k \le 1, \ k = 1, ..., d,$$

$$\mu_k, \lambda_j, z \ge 0, \ j = 1, ..., n, \ k = 1, ..., d. \tag{7.117}$$

Therefore, model (7.117) can specify the number of suppliers and the type of suppliers in the supply chain synchronously and increase the profit efficiency of the manufacturer. In the actual case, according to the limitation of the source and some environmental conditions, suppliers confront some limitations, so they cannot produce any desired amount. We consider this limitation by adding constrain $\mu_k I^k \le \alpha_k$ to the model (7.117). The least of production is another limitation that suppliers encounter. Because, in some cases, based on production politics or economic politics, managers are not inclined to produce less than a definite value. Therefore, by considering this limitation, we can add constraint $\beta_k \leq \mu_k I^k$ to model (7.117) such that α_k and β_k are constant values and specified by the manager of the supplier.

Theorem 7.2

If (μ^*, λ^*, Z^*) is the optimal solution of model (7.117), constraint $\sum_j \lambda_j^* z_{rj} = z_r^*, \forall r$.

Proof Let (μ^*, λ^*, Z^*) be the optimal solution of model (7.117) and $\sum_j \lambda_j^* z_{lj} > z_l^*$. We define

$$\hat{z}_r = \begin{cases} \sum_j \lambda_j^* z_{lj} & r = l \\ z_r^* & o.w \end{cases} \tag{7.118}$$

It is evident that (μ^*, λ^*, Z^*) is the feasible solution of model (7.117) and $p\hat{z} > pz^*$. The value of the objective function for this feasible solution is equal to $p\hat{z} - \sum_k \mu_k^* c^k x^k$ and $p\hat{z} - \sum_k \mu_k^* c^k x^k > pz^* - \sum_k \mu_k^* c^k x^k$. Therefore, this is a contradiction, so $\sum_j \lambda_j^* z_{rj} = z_r^*$.

By using Theorem 7.2, model (7.117) is transformed into models (7.119) as follows:

$$\max p \left(\sum_j \lambda_j z_j \right) - \sum_k \mu_k c^k I^k$$

$s.t$

$$\sum_k \mu_k x^k \leq \sum_k \mu_k^o x^k,$$

$$\sum_k \mu_k c^k I^k \leq \bar{c}_o,$$

$$\sum_j \lambda_j y_j \leq \sum_k \mu_k I^k,$$

$$\mu_k \leq 1, \ k = 1, ..., d,$$

$$\mu_k, \lambda_j \geq 0, \ j = 1, ..., n, \ k = 1, ..., d. \tag{7.119}$$

Also, for a better understanding of the model (7.119), its extended form is expressed as follows:

$$\max \sum_r p_r \left(\sum_j \lambda_j z_{rj} \right) - \sum_k \mu_k \left(\sum_t c_t^k I_t^k \right)$$

$$s.t$$

$$\sum_k \mu_k x_i^k \leq \sum_k \mu_k^o x_i^k, \; i = 1, \ldots, m,$$

$$\sum_k \mu_k \left(\sum_t c_t^k I_t^k \right) \leq \bar{c}_o,$$

$$\sum_j \lambda_j y_{tj} \leq \sum_k \mu_k I_t^k, \; t = 1, \ldots, T,$$

$$\mu_k \leq 1, \; k = 1, \ldots, d,$$

$$\mu_k, \lambda_j \geq 0, \; j = 1, \ldots, n, \; k = 1, \ldots, d. \tag{7.120}$$

In the previous models, the manager seeks new relations with other suppliers to increase their profit efficiency. Suppose that the model (7.120) has alternative optimal solutions. For example, in the first optimal solution, $pz = 11$ and $\sum_k \mu_k c^k I^k = 3$ and in another optimal solution, $pz = 10$ and $\sum_k \mu_k c^k I^k = 2$. It is evident that in both cases, the profit is equal to 8 units. Still, in the first optimal solution, the manufacturer must spend three units, i.e., 1 unit more than the second case, so the manufacturer prefers the second strategy. In the real world, managers always look for more profit with less cost. To achieve this goal, we suggest the model (7.121).

$$\max \left[p \left(\sum_j \lambda_j z_j \right) - \sum_k \mu_k c^k I^k \right] - \varepsilon \sum_k \mu_k c^k I^k$$

$$s.t$$

$$\sum_k \mu_k x^k \leq \sum_k \mu_k^o x^k,$$

$$\sum_k \mu_k c^k I^k \leq \bar{c}_o,$$

$$\sum_j \lambda_j y_j \leq \sum_k \mu_k I^k,$$

$$\mu_k \leq 1, \; k = 1, \ldots, d,$$

$$\mu_k, \lambda_j \geq 0, \; j = 1, \ldots, n, \; k = 1, \ldots, d. \tag{7.121}$$

In the model (7.121), ε is a very small number, so, for achieving the maximum value of the objective function, at first, $p\left(\sum_j \lambda_j z_j\right) - \sum_k \mu_k c^k I^k$ (profit of manufacture) is increased and then $\sum_k \mu_k c^k I^k$ (cost of the manufacturer) is decreased. Also, in some cases, based on the condition of the project or the economic policy of an entity or market conditions, a manager tries to earn more revenue, and the project's cost is the second priority. Therefore, for this purpose, model (7.122) is presented:

$$\max p\left(\sum_j \lambda_j z_j\right) - \varepsilon \sum_k \mu_k c^k I^k$$

$$s.t$$

$$\sum_k \mu_k x^k \le \sum_k \mu_k^o x^k,$$

$$\sum_k \mu_k c^k I^k \le \bar{c}_o,$$

$$\sum_j \lambda_j y_j \le \sum_k \mu_k I^k,$$

$$\mu_k \le 1, \ k = 1, ..., d,$$

$$\mu_k, \lambda_j \ge 0, \ j = 1, ..., n, \ k = 1, ..., d. \tag{7.122}$$

Therefore, the model (7.122) first looks for the appropriate strategy to increase the producer's income; the next priority is reducing the producer's costs. Also, if there is no absolute priority between expenditure and revenue, the model (7.123) can be expressed as follows:

$$\max w_1\left(\sum_j \lambda_j z_j\right) - w_2 \sum_k \mu_k c^k I^k$$

$$s.t$$

$$\sum_k \mu_k x^k \le \sum_k \mu_k^o x^k,$$

$$\sum_k \mu_k c^k I^k \le \bar{c}_o,$$

$$\sum_j \lambda_j y_j \le \sum_k \mu_k I^k,$$

$$\mu_k \le 1, \ k = 1, ..., d,$$

$$\mu_k, \lambda_j \ge 0, \ j = 1, ..., n, \ k = 1, ..., d. \tag{7.123}$$

So that w_1 and w_2 are the weights determined by the manager and indicate the degree of importance of cost and income from the manager's point of view.

7.7 Conclusion

Supplier evaluation and selection is one of the most vital aspects of being responsive and competitive in today's ever-changing global supply chain. The success of a supply chain strongly depends on the selection of suppliers related to it, so this issue is an essential part of production and logistics management for many companies. Therefore, considering the importance of supplier selection to improve the competitive power of the supply chain and, as a result, the chain's success, therefore, in this chapter, supplier selection was investigated using DEA models. Because one of the crucial points in choosing a supplier is the performance and efficiency score of the suppliers under evaluation. Of course, based on the goal of supply chain managers, in addition to calculating technical efficiency, cost efficiency, profit, and income can also be used as criteria for choosing the best supplier. Therefore, this chapter explains the performance evaluation of suppliers using DEA models. Also another critical point in choosing DEA models for selecting a supplier is to pay attention to the type of input and output of each unit under evaluation. The data may be integer, fuzzy, negative, or interval. Also, attention to dual-role factors and undesirable data should be considered for a more accurate assessment. In short, we do the following steps to choose the best supplier:

- Determination of evaluation criteria
- Determining the inputs and outputs of the units under evaluation
- Determining the type of inputs and outputs of the units under evaluation, for example, integer, fuzzy, negative, or interval data.
- Specifying dual-role factors, if any.
- Specifying undesirable indicators, if any.

References

1. Allahviranloo, T., Hosseinzadeh Lotfi, F., AdabitabarFirozja, M.: Efficiency in fuzzy production possibility set. Iran. J. Fuzzy Syst. **9**(4), 17–30 (2012)
2. Amindoust, A., Saghafinia, A.: Supplier evaluation and selection using a FDEA model. In: Performance Measurement with Fuzzy Data Envelopment Analysis, pp. 255–269. Springer, Berlin, Heidelberg (2014)
3. Avkiran, N.K., Shafiee, M., Saleh, H., Ghaderi, M.: Benchmarking in the supply chain using data envelopment analysis. Theor. Econ. Lett. **8**(14), 2987 (2018)
4. Azadeh, A., Alem, S.M.: A flexible deterministic, stochastic and fuzzy data envelopment analysis approach for supply chain risk and vendor selection problem: simulation analysis. Expert Syst. Appl. **37**(12), 7438–7448 (2010)
5. Azadeh, A., Ghaderi, S.F., Javaheri, Z., Saberi, M.: A fuzzy mathematical programming approach to DEA models. Am. J. Appl. Sci. **5**(10), 1352–1357 (2008)
6. Azadi, M., Mirhedayatian, S.M., Saen, R.F.: A new fuzzy goal directed benchmarking for supplier selection. Int. J. Serv. Oper. Manage. **14**(3), 321–335 (2013)
7. Bagheri, M., Ebrahimnejad, A., Razavyan, S., Hosseinzadeh Lotfi, F., Malekmohammadi, N.: Solving the fully fuzzy multi-objective transportation problem based on the common set of weights in DEA. J. Intell. Fuzzy Syst. **39**(3), 3099–3124 (2020)
8. Banker, R.D., Morey, R.C.: Efficiency analysis for exogenously fixed inputs and outputs. Oper. Res. **34**(4), 513–521 (1986)
9. Bian, Y., Yang, F.: Resource and environment efficiency analysis of provinces in China: a DEA approach based on Shannon's entropy. Energy Policy **38**(4), 1909–1917 (2010)
10. Boyd, G.A., Pang, J.X.: Estimating the linkage between energy efficiency and productivity. Energy Policy **28**(5), 289–296 (2000)

11. Cooper, W.W., Park, K.S., Yu, G.: IDEA and AR-IDEA: models for dealing with imprecise data in DEA. Manage. Sci. **45**(4), 597–607 (1999)
12. Cooper, W.W., Seiford, L.M., Zhu, J.: Data envelopment analysis: history, models, and interpretations. In: Handbook on Data Envelopment Analysis, pp. 1–39. Springer, Boston, MA (2011)
13. Cooper, W.W., Seiford, L.M., Tone, K.: Data Envelopment Analysis: A Comprehensive Text with Models, Applications, References and DEA-Solver Software, vol. 2, p. 489. Springer, New York (2007)
14. Ding, J., Dong, W., Bi, G., Liang, L.: A decision model for supplier selection in the presence of dual-role factors. J. Oper. Res. Soc. **66**(5), 737–746 (2015)
15. Doyle, J., Green, R.: Efficiency and cross-efficiency in DEA: derivations, meanings and uses. J. Oper. Res. Soc. **45**(5), 567–578 (1994)
16. Dutta, P., Jaikumar, B., Arora, M.S.: Applications of data envelopment analysis in supplier selection between 2000 and 2020: a literature review. Ann. Oper. Res. 1–56 (2021)
17. Ebrahimnejad, A., Nasseri, S.H., Lotfi, F.H.: Bounded linear programs with trapezoidal fuzzy numbers. Int. J. Uncertain. Fuzziness Knowledge-Based Syst. **18**(03), 269–286 (2010)
18. Farokhnia, M., Beheshtinia, M.A.: A three-dimensional house: extending quality function deployment in two organizations. Manage. Decis. (2018)
19. Forker, L.B., Mendez, D.: An analytical method for benchmarking best peer suppliers. Int. J. Oper. Prod. Manage. (2001)
20. Färe, R., Grosskopf, S.: Modeling undesirable factors in efficiency evaluation: comment. Eur. J. Oper. Res. **157**(1), 242–245 (2004)
21. Hosseinzadeh Lotfi, F., Ebrahimnejad, A., Vaez-Ghasemi, M., Moghaddas, Z.: Introduction to data envelopment analysis and fuzzy sets. In: Data Envelopment Analysis with R, pp. 1–17. Springer, Cham (2020)
22. Jahanshahloo, G.R., Lotfi, F.H., Jafari, Y., Maddahi, R.: Selecting symmetric weights as a secondary goal in DEA cross-efficiency evaluation. Appl. Math. Model. **35**(1), 544–549 (2011)
23. Jahanshahloo, G.R., Lotfi, F.H., Malkhalifeh, M.R., Namin, M.A.: A generalized model for data envelopment analysis with interval data. Appl. Math. Model. **33**(7), 3237–3244 (2009)
24. Jahanshahloo, G.R., Lotfi, F.H., Rezaie, V., Khanmohammadi, M.: Ranking DMUs by ideal points with interval data in DEA. Appl. Math. Model. **35**(1), 218–229 (2011)
25. Jahanshahloo, G.R., Memariani, A., Hosseinzadeh, F., Shoja, N.: A feasible interval for weights in data envelopment analysis. Appl. Math. Comput. **160**(1), 155–168 (2005)
26. Jahanshahloo, G.R., Sanei, M., Rostamy-Malkhalifeh, M., Saleh, H.: A comment on "a fuzzy DEA/AR approach to the selection of flexible manufacturing systems." Comput. Ind. Eng. **56**(4), 1713–1714 (2009)
27. Jassbi, J., Saen, R.F., Lotfi, F.H., Hosseininia, S.S.: A new hybrid decision making system for supplier selection. RAIRO-Oper. Res. **50**(3), 645–664 (2016)
28. Kim, S.H., Park, C.G., Park, K.S.: An application of data envelopment analysis in telephone officesevaluation with partial data. Comput. Oper. Res. **26**(1), 59–72 (1999)
29. Krmac, E., Djordjević, B.: A new DEA model for evaluation of supply chains: a case of selection and evaluation of environmental efficiency of suppliers. Symmetry **11**(4), 565 (2019)
30. Liang, L., Wu, J., Cook, W.D., Zhu, J.: The DEA game cross-efficiency model and its Nash equilibrium. Oper. Res. **56**(5), 1278–1288 (2008)
31. Liu, S.T.: A fuzzy DEA/AR approach to the selection of flexible manufacturing systems. Comput. Ind. Eng. **54**(1), 66–76 (2008)
32. Liu, J., Ding, F.Y., Lall, V.: Using data envelopment analysis to compare suppliers for supplier selection and performance improvement. Supply Chain Manage. Int. J. **5**(3), 143–150 (2000)
33. Lotfi, F.H., Eshlaghy, A.T., Shafiee, M.: Providers ranking using data envelopment analysis model, cross efficiency and Shannon entropy. Appl. Math. Sci. **6**(4), 153–161 (2012)
34. Lotfi, F.H., Allahviranloo, T., Jondabeh, M.A., Alizadeh, L.: Solving a full fuzzy linear programming using lexicography method and fuzzy approximate solution. Appl. Math. Modell. **33**(7), 3151–3156 (2009b)

35. Lotfi, F.H., Jahanshahloo, G.R., Vahidi, A.R., Dalirian, A.: Efficiency and effectiveness in multi-activity network DEA model with fuzzy data. Appl. Math. Sci. **3**(52), 2603–2618 (2009a)
36. Ma, R., Yao, L., Jin, M., Ren, P.: The DEA game cross-efficiency model for supplier selection problem under competition. Appl. Math. Inf. Sci. **8**(2), 811 (2014)
37. Maddahi, R., Jahanshahloo, G.R., Hosseinzadeh Lotfi, F., Ebrahimnejad, A.: Optimising proportional weights as a secondary goal in DEA cross-efficiency evaluation. Int. J. Oper. Res. **19**(2), 234–245 (2014)
38. Mahdiloo, M., Farzipoor Saen, R., Tavana, M.: A novel data envelopment analysis model for solving supplier selection problems with undesirable outputs and lack of inputs. Int. J. Logist. Syst. Manage. **11**(3), 285–305 (2012)
39. Mahdiloo, M., Noorizadeh, A., Farzipoor Saen, R.: Suppliers ranking by cross-efficiency evaluation in the presence of volume discount offers. Int. J. Serv. Oper. Manage. **11**(3), 237–254 (2012)
40. Malmir, M.: Efficiency decomposition in network data envelopment analysis with undesirable, PhD thesis in applied mathematics, Karaj Branch, Islamic Azad University (2022)
41. Nash, J.: Non-cooperative games. Ann. Math. 286–295 (1951)
42. Nash, Jr., J.F.: The bargaining problem. Econom. J. Econometr. Soc. 155–162 (1950)
43. Neumann, J.V.: On the theory of game of strategy, in contributions to the theory of games. Annals Math. Stud. **IV**(40), 13–42 (1928)
44. Noorizadeh, A., Mahdiloo, M., Farzipoor Saen, R.: Using DEA cross-efficiency evaluation for suppliers ranking in the presence of non-discretionary inputs. Int. J. Shipping Transp. Logist. **5**(1), 95–111 (2013)
45. Noorizadeh, A., Mahdiloo, M., Saen, R.F.: Supplier selection in the presence of dual-role factors, non-discretionary inputs and weight restrictions. Int. J. Prod. Qual. Manage. **8**(2), 134–152 (2011)
46. Nosrat, A., Sanei, M., Payan, A., Hosseinzadeh Lotfi, F., Razavyan, S.: Using credibility theory to evaluate the fuzzy two-stage DEA: sensitivity and stability analysis. J. Intell. Fuzzy Syst. **37**(4), 5777–5796 (2019)
47. Peleg, B., Sudhölter, P.: Introduction to the Theory of Cooperative Games, vol. 34. Springer Science & Business Media (2007)
48. Rostamy, M.M., Ebrahimkhani, G.S., Saleh, H., Ebrahimkhani, G.N.: Congestion in DEA model with fuzzy data. Int. J. Appl. Oper. Res. **1**(2), 49–56 (2011)
49. Rostamy-Malkhalifeh, M., Sanei, M., Saleh, H.: A new method for solving fuzzy DEA models by trapeziodal approximation. J Math. Extension **1**(4), 307–313 (2009)
50. Saen, R.F.: Suppliers selection in the presence of both cardinal and ordinal data. Eur. J. Oper. Res. **183**(2), 741–747 (2007)
51. Saen, R.F.: Supplier selection by the new AR-IDEA model. Int. J. Adv. Manuf. Technol. **39**(11), 1061–1070 (2008)
52. Saen, R.F.: Restricting weights in supplier selection decisions in the presence of dual-role factors. Appl. Math. Model. **34**(10), 2820–2830 (2010)
53. Saleh, H., Hosseinzadeh Lotfi, F., Rostmay-Malkhalifeh, M., Shafiee, M.: Provide a mathematical model for selecting suppliers in the supply chain based on profit efficiency calculations. J. New Res. Math. **7**(32), 177–186 (2021)
54. Saleh, H., Shafiee, M., Hosseinzade Lotfi, F.: Performance evaluation and specifying of Return to scale in network DEA international journal of industrial mathematics. J. Adv. Math. Model. **10**(2), 309–340 (2021)
55. Saleh, H., Hosseinzadeh Lotfi, F., Toloie Eshlaghy, A., Shafiee, M.: A new two-stage DEA model for bank branch performance evaluation. In: 3rd National Conference on Data Envelopment Analysis. Islamic Azad University of Firoozkooh (2011)
56. Saleh, H., Rostamy, M.M.: Performance evaluation in a bank branch with a two-stage DEA model. J. Syst. Manage. **1**(11), 17–33 (2013)
57. Sanei, M., Noori, N., Saleh, H.: Sensitivity analysis with fuzzy data in DEA. Appl. Math. Sci. **3**(25), 1235–1241 (2009a)

58. Sanei, M., Rostami-Malkhalifeh, M., Saleh, H.: A new method for solving fuzzy DEA models. Int. J. Ind. Math. **1**(4), 307–313 (2009b)
59. Shafiee, M., Hosseinzade Lotfi, F., Saleh, H.: Benchmark forecasting in data envelopment analysis for decision making units. Int. J. Ind. Math. **13**(1), 29–42 (2021)
60. Shafiee, M., Saleh, H.: Evaluation of strategic performance with fuzzy data envelopment analysis. Int. J. Data Envelop. Analysis **7**(4), 1–20 (2019)
61. Shafiee, M., Saleh, H., Ziyari, R.: Projects efficiency evaluation by data envelopment analysis and balanced scorecard. J. Decis. Oper. Res. **6**(Special Issue), 1–19 (2022)
62. Shephard, R.: Theory of cost Production Function, Princeton University Press (1970)
63. Shin, H., Collier, D.A., Wilson, D.D.: Supply management orientation and supplier/buyer performance. J. Oper. Manage. **18**(3), 317–333 (2000)
64. Soltanifar, M., Sharafi, H.: A modified DEA cross efficiency method with negative data and its application in supplier selection. J. Combin. Optim. 1–32 (2021)
65. Stewart, T.J.: Goal directed benchmarking for organizational efficiency. Omega **38**(6), 534–539 (2010)
66. Tamaddon, L., Jahanshahloo, G.R., Lotfi, F.H., Mozaffari, M.R., Gholami, K.: Data envelopment analysis of missing data in crisp and interval cases. Int. J. Math. Analysis **3**(17–20), 955–969 (2009)
67. Tsai, C.M., Lee, H.S., Gan, G.Y.: A new fuzzy DEA model for solving the MCDM problems in supplier selection. J. Mar. Sci. Technol. **29**(1), 7 (2021)
68. Wang, Y.M., Greatbanks, R., Yang, J.B.: Interval efficiency assessment using data envelopment analysis. Fuzzy Sets Syst. **153**(3), 347–370 (2005)
69. Wang, M., Li, Y.: Supplier evaluation based on Nash bargaining game model. Expert Syst. Appl. **41**(9), 4181–4185 (2014)
70. Weber, C.A.: A data envelopment analysis approach to measuring vendor performance. Supply Chain Manage. Int. J. **1**(1), 28–39 (1996)
71. Wong, Y.H., Beasley, J.E.: Restricting weight flexibility in data envelopment analysis. J. Oper. Res. Soc. **41**(9), 829–835 (1990)
72. Wu, J., Zhu, Q., Yin, P., Song, M.: Measuring energy and environmental performance for regions in China by using DEA-based Malmquist indices. Oper. Res. Int. J. **17**(3), 715–735 (2017)
73. Zadeh, L.A.: Fuzzy sets. Inf. Control **8**(3), 338–353 (1965)
74. Zeydan, M., Çolpan, C., Çobanoğlu, C.: A combined methodology for supplier selection and performance evaluation. Expert Syst. Appl. **38**(3), 2741–2751 (2011)
75. Zhu, J.: Imprecise data envelopment analysis (IDEA): a review and improvement with an application. Eur. J. Oper. Res. **144**(3), 513–529 (2003)
76. Zolghadri, M., Eckert, C., Zouggar, S., Girard, P.: Power-based supplier selection in product development projects. Comput. Ind. **62**(5), 487–500 (2011)
77. Zoroufchi, K.H., Azadi, M., Saen, R.F.: Developing a new cross-efficiency model with undesirable outputs for supplier selection. Int. J. Ind. Syst. Eng. **12**(4), 470–484 (2012)

Chapter 8
Performance Evaluation of the Supply Chains Using DEA

8.1 Introduction

Today, companies are not units with a unique brand name that can operate independently. The complexity of goods and services in today's world is such that it rarely happens that an organization or institution can produce a product without the help of other organizations [29]. So as explained in the previous chapters, supply chain management has become one of the most critical topics in business literature in recent years. Supply chain management, like any management system, needs a performance measurement system to identify success factors, determine the degree of fulfillment of customer demands and help the organization understand the processes that these organizations were not aware of before [25]. Because in today's industrial world, in addition to dealing with organizational issues and internal resources, companies also need to manage and monitor related resources and events outside the organization to remain at the forefront of competition with other competitors. In other words, it is vital for supply chain management to know the capabilities of each company to compete effectively in the business environment, maintain its strengths and try to eliminate the negative points. One of the critical points in supply chain management is to measure performance and, in other words, to measure efficiency in competition with other competitors. The meaning of measurement is the measurement of all aspects of activity along the chain, such as the measurement of products, services, and planning processes in a company, how to use resources, how to produce and manufacture, and other activities in a chain [8]. In other words, evaluating supply chain performance is how to use quantitative and qualitative inputs and how to operate to produce quantitative and qualitative outputs. Therefore, measuring efficiency in supply chains is a fundamental issue for companies in creating regular competition on the playing field with other competitors. Because measuring the relative efficiency gives this information to the supply chain manager about where it stands among its competitors and how he/she can improve his position. Since the efficiency of the supply chain is the result of the coherence, performance, and interaction of all its parts, thus, its evaluation is a challenging activity.

© The Author(s), under exclusive license to Springer Nature Switzerland AG 2023
F. Hosseinzadeh Lotfi et al., *Supply Chain Performance Evaluation*, Studies in Big Data 122, https://doi.org/10.1007/978-3-031-28247-8_8

Techniques and approaches to evaluate the performance of supply chains be divided into periods. The first period, which lasted until the 1980s, relies on financial standards [19]. In other words, supply chains were traditionally managed and evaluated only by considering simple commercial criteria, such as the price/cost index as the performance evaluation index. And the producers were only looking for sources with lower costs and selling products with higher prices to achieve more profit. While the review of the literature on supply chain management and evaluation methods shows that this view has had problems. For example, buying raw materials at the lowest price does not necessarily mean minimizing the cost in the supply chain management cycle. Therefore, according to what has been said, it can be said that focusing solely on financial indicators provides poor results in performance evaluation and causes problems, such as the possibility of financial indicators conflicting with the organization's strategic goals [34].

Since the late 1980s, many changes have occurred concerning measuring the performance of modern supply chains, and researchers have accepted that performance is not just a financial matter. Traditional performance management ignores non-financial factors and seems inadequate in calculating all-round performance to meet the strategic development needs of companies. Financial indicators can only reflect the performance of companies in the past and do not accurately reflect the state of the company in the future [32]. Therefore, the evaluation results are suitable for use in the organization's current operations if the chain's structure is examined in multi-dimensional to provide the necessary conditions to keep up with the competition with other supply chains [5, 37].

For example, one of the essential issues in business is customer satisfaction, which is considered in the form of indicators such as customer response time, number of complaints, lead time, and so on. Another critical point in today's competitive environment is the flexibility of a company and its supply chain. Today's chains must be flexible regarding volume and combination of products [29]. Therefore, unlike the previous methods, new evaluation approaches were presented based on several indicators to evaluate the efficiency of modern supply chains. Hence, the latest techniques that entered the scope of supply chain efficiency evaluation suggested the using multi-factor performance measurement models, some of which are still in use. Among these methods are the balanced scorecard model and the performance excellence model. But these methods were presented for independent units and did not consider the complexity of the company's value chain. While financial indicators such as return on sales (ROS) and return on investment (ROI) are essential indicators for describing the overall efficiency of the supply chain [16, 23].

Therefore, evaluating the efficiency of a company's supply chain is a complex phenomenon, and instead of considering a single criterion, more cases should be considered. Knowing the position of a company in a specific industry compared to other competitors makes it possible to properly manage the company's expectations to increase its multiple outputs or reduce the various inputs to improve efficiency. In evaluating performance and calculating efficiency, one of the important goals is to identify sources of inefficiency, in other words, what should be the minimum use of initial inputs and the maximum production of final outputs for the system to work efficiently [28]. As a result, it is necessary to evaluate the performance of the supply chain in the form of a mathematical model using financial, knowledge, participation, and responsiveness indicators of the supply chain. Therefore, using DEA to evaluate supply chains became the interest of many researchers [25]. Although DEA, like other methods, has weaknesses, the DEA approach allows the analysis of supply chain performance in different dimensions by considering various indicators (input–output). This technique provides adequate information to evaluate the performance of supply chains by creating a frontier to compare the best position of each unit under evaluation with its current status and other units under assessment. In other words, DEA can measure the efficiency of supply chains with multiple inputs and outputs, like a multi-criteria decision-making tool (MCDM). Unlike some parametric evaluation methods, DEA models do not require any assumptions on the weights (value and importance of indicators), and managerial preferences are not considered in the modeling process. In other words, the advantage of using DEA compared to other MCDM methods is that it requires less information from decision-makers and analysts to value the indicators and rank the supply chains under evaluation. Also, DEA can simultaneously analyze performance in the face of quantitative and qualitative indicators and provide a complete classification for efficient and inefficient DMUs. Therefore, in this chapter, we explain and describe DEA models for evaluating the efficiency of supply chains.

8.2 Performance Evaluation of the Supply Chains Using DEA Models

To use DEA models to evaluate supply chain performance using DEA, we perform the following steps.

A. Determination of input and output

Although the indicators affecting the evaluation of supply chains in different industries are different, in general, to determine the inputs and outputs of supply chains, first by reviewing the literature and using the opinions of managers and experts in each part of the supply chain, the indicators affecting different parts of the supply chain are determined. Finally, they are divided into two groups based on experts' opinions, input and output.

B. Data gathering

Gathering data is one of the essential parts of any decision-making method. If this work is not done carefully and correctly, analyzing and conclusions from the final results will not have good credit. Usually, the data is already available at this stage. Because we are not looking for new information. We are only collecting information that has already existed, and in the documents of clinics, hospitals, civil registries, various centres of municipalities, and so on is available. Of course, in some cases, it will be challenging to obtain these data due to the passage of time or the confidentiality of information. Because of this, accuracy and patience are essential at this stage. Also, in some cases, we need to do calculations using formulas and the help of experts in the relevant field to access information. In the following, we'll go over examples of this case.

C. Choosing the DEA model

As explained in Chap. 6, we are faced with different types of models in DEA, and each model's use will have different results. In other words, incorrect use of DEA models causes incorrect classification (efficient-inefficient) and rankings for the under-evaluation supply chain. Therefore, based on the explanations in Chap. 6, the appropriate DEA model is selected according to the properties of the production possibilities set, the input and output type, and other models' properties, such as unit invariant.

D. Solving the DEA model to calculate the efficiency score

After selecting the appropriate model based on the input and output values for each of the inputs and outputs of the supply chains under evaluation, the desired model is solved once for each of the supply chains under assessment, and the efficiency scores of each chain supply are obtained. It is necessary to explain that software such as R, GAMS, LINDO, and LINGO use to solve the models.

E. Analysis of results

After solving the model and reporting the efficiency scores, supply chains are divided into efficient and inefficient groups based on the definitions of efficiency and inefficiency in different models. For example, in the input-oriented CCR model, if the optimal value of the objective function is equal to one, the unit under evaluation is efficient. The DMU is inefficient if the efficiency value is between zero and one. But in the output-oriented CCR model, the essential chain under evaluation is inefficient if the efficiency score is more than 1. Then, the supply chains are ranked based on the scores obtained and using ranking models (if needed).

In the next step, by obtaining the projection of inefficient units, a suitable benchmark for inefficient units can be presented so that the manager of inefficient units can strengthen the unit's position under management by determining the appropriate strategy. Some other benchmarking techniques in this chapter will be explained.

8.2.1 Performance Evaluation of the Healthcare Supply Chain

Healthcare is one of the essential industries in providing services, which plays a crucial role in human health and society's health in general. Providing medicine and quick access to medicine is one of the critical issues that significantly impact the healthcare sector, especially on the performance of the hospitals under evaluation. Therefore, the drug delivery system from the factory or the central depot to hospital

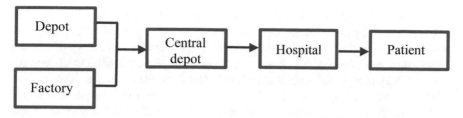

Fig. 8.1 Drug import supply chain

patients is considered a SC. Of course, it is one of the most critical and expensive supply chains due to its vital role in the recovery process of patients. Therefore, [2] focused on the drug supply chain.

Since the drugs needed by the patients are either entirely produced inside the country or imported from abroad, the drugs' suppliers are located outside the hospitals. Domestically produced drugs, after production, are sent to the central warehouse, and from there, they are sent to hospitals. Still, imported drugs are kept in warehouses in border cities and then sent to the central warehouse. In the next step, according to the demand of each hospital, medicines are sent from the main warehouse to the hospitals, and finally, the drugs are provided to the patients through the hospital.

Briefly, the drug supply chain is shown in Fig. 8.1. Using the opinions of managers and experts in each of the supply chain parts, the following indicators are considered vital and influential factors in the performance of the supply chain.

x_{1j} = Cost
x_{2j} = Time of delivering medicine to customers
y_{1j} = Physical strength
y_{2j} = Mental power
y_{3j} = Knowledge about situation assessment and situation analysis
y_{4j} = Business specifications and equipment
y_{5j} = Information and communication in the system
y_{6j} = Human resource management culture
y_{7j} = Teamwork

It is also necessary to explain that physical strength and mental power indicators are ergonomics indicators, and other indicators are classified as macro-ergonomics. After specifying the input and output indicators, the values of each indicator in each supply chain are collected in two ways.

- Historical data: By referring to documents, the values x_{1j} and x_{2j} are determined.
- Through the questionnaire: Preparing standard questionnaires and completing them by experts in drug production and supply and people in different hospital departments to determine the value of $y_{1j}, ..., y_{7j}, j = 1, ..., n$. First, each of the selected experts in this field will express their opinion regarding the desired index in each supply chain numerically between 0 and 10. Then, the experts' average opinion is considered the desired output value.

As explained and seen in Fig. 8.1, the drug supply chain has different parts, and even several parts with the same tasks may work in parallel. Therefore, each component of the chain is considered a DMU, and the average efficiency of all parts is the efficiency of the entire chain. After gathering data, it's time to choose the suitable DEA model. The output-oriented model is applied because the goal is to achieve maximum output for a specific input level. Then using the phase II model explained in detail in Chap. 6, the projection of each unit under evaluation is specified. To check the details and better analyze the results, it is possible to determine each index's impact on the chain's performance and determine which of the indicators has a more significant effect on the chain's performance.

As mentioned before, the output indicators in the drug supply chain are divided into macro-ergonomics and ergonomics. Therefore, to check the impact of macro-ergonomics and ergonomics on the performance of the entire supply chain, it is enough to calculate the efficiency of DMUs in the first step by only considering two inputs and five output indicators of macro-ergonomics, and finally, the average of these results state the impact of macro-ergonomics indicator on the pharmaceutical supply chain. In the next step, to show the effect of the ergonomics indicator, the performance of each DMU by considering two inputs and two ergonomics outputs is calculated, and the average obtained results in this step show the impact of the ergonomics indicators on the supply chain. Also, the DEA model can be implemented separately for each of the outputs to achieve more detailed results and determine which output indicators have a more significant impact on the chain's performance. For this purpose, it is enough to remove the desired index among the outputs and obtain the DMU performance using other outputs and inputs. In this way, we are looking for the effect of eliminating the desired output on the efficiency value.

8.2.2   The International Biomass Supply Chain: The Reconfiguration of the Chain Based on Different Processing Scenarios and Transportation Methods in the Presence of Undesirable Output

In most of the studies conducted in the field of the supply chain, supply chains are only considered with a specific origin, destination, and configuration. Although few researchers have investigated the different configurations of the supply chain between a particular origin and destination or a specific structure considering different origins, in some situations, it is necessary to examine other forms and consider different origins and destinations for the supply chain simultaneously. Among these issues, we can mention the biomass supply chain.

Greenhouse gases include carbon dioxide, methane, nitrogen oxide, and other gases that accumulate in the atmosphere and create a reflective layer for heat that keeps the earth at an acceptable temperature. These gases form an insulator that keeps the earth warm enough to sustain life. For thousands of years, the concentration of greenhouse gases and the earth's overall temperature remained relatively stable. Unfortunately, the sharp increase in the concentration of greenhouse gases due to industrial activities during the last two hundred years has thrown this system out of balance and caused global warming.

Greenhouse gases can be provided from two natural and artificial sources. Some important greenhouse gases such as carbon dioxide and methane, are naturally released by volcanoes, forest fires, and decomposed organic matter. But the main factors in producing greenhouse gases are factories, cars, and deforestation. Therefore, greenhouse gases do not allow sunlight to leave the earth, so the temperature of the earth's surface increases. This leads to global warming. Therefore, many countries of the world have committed to reducing greenhouse gases. For example, the European Union has committed to providing 20% of its energy consumption through renewable energy. Using biomass is one of the primary energy sources to support this process.

Biomass is a renewable organic material made from plants and animal waste. Biomass has been used for cooking food and keeping it warm since the beginning of wood burning. Wood is still the largest source of biomass energy today. Other sources include food products, herbaceous and woody plants, residues from agriculture or forestry, oil algae, and organic components of urban and industrial waste. Even vapours from landfills (which contain methane, the main component of natural gas) can be used as an energy source from biomass. But due to the limited nature of this resource, most of it is supplied through imports in some countries. For example, on the one hand, most Western European countries are facing a limited supply of domestic biomass and, on the other hand, a continuous increase in biomass demand. Meeting these needs requires significant international sources of biomass from outside Europe. In terms of current industrial practices, biomass provides mainly from the United States, Canada, and the East from forest residues, in the form of wood pellets or wood chips [36].

Most researchers have investigated international biomass supply chains with specific origins, destinations, and configurations. While countries of origin are using different technologies in terms of pre-processing of biomass, including investigation of pelletization, torrefaction, and torrefaction with pelletization. Also, considering other modes of transportation sometimes leads to checking different routes. Therefore, [26] investigate the biomass supply chain from Brazil to England. Brazil has an area of more than 8 million square kilometres.

More than 54.4% of its area is made up of forest areas. Therefore, in terms of forest areas, this country has the second-largest forest area in the world. And considering that most of its vegetation is related to pine and eucalyptus trees, this country is regarded as an excellent source for providing biomass resources for other countries. Among these countries is the UK. So [26], by identifying all the infrastructures, the main planting places and forests (origin of the chain), the main ports for export (intermediate destinations), and the UK as the final destination and using various methods of wood transportation such as truck and rail transportation, as well as checking different types of pre-processing technologies, different configurations were proposed for the biomass supply chain from Brazil to the UK. While in most previous studies, alternative potentials have not been considered. This way, four scenarios have been considered for different configurations in the biomass supply chain.

Scenario 1: Biomass is extracted from Brazilian forests and shipped to England in logs. Then it is transported to the final consumer's place, and at the final destination, it is turned torrefied and pelletized.

Scenario 2: Biomass is extracted from Brazilian forests, converted into white pellets at existing pelletizing facilities in Brazil and transported to the exporting port. The white pellets are then shipped to the UK and torrefied and pelletized at the end user location.

Scenario 3: Based on Scenario 2, Scenario 3 considers a combined torrefaction and pelletizing process at existing pellet mills in Brazil. The output is black pellets transported to the exporting ports to be shipped to the UK.

Scenario 4: In this scenario, wood is extracted, transported to existing pellet mills, pelletized, and shipped to Brazilian export ports. The port has a centralized centre where the biomass is melted and re-pelletized. Then it will be sent to the UK.

After specifying the scenario and subsequently specifying the biomass supply chain from Brazil to England, the critical criteria for evaluating the biomass supply chain are identified. These criteria include:

- Total costs (TC) as an input. So that, TC is transportation costs and process costs.
- Total energy consumption (TEC) as an input. Transportation energy consumption and process energy consumption.
- Total CO_2 emissions equivalent (TE) as an undesirable output. Transportation emissions and process emissions.

As seen in other supply chains, TC is a traditional measure to evaluate chains because it directly affects the business's profit. In addition to conventional indicators in evaluating supply chains, there is a need to introduce and use specific indicators for specific supply chains. This part of the evaluation process requires the use of experts in addition to using previous sources and studies. For example, in the biomass supply chain, TEC and TE are different criteria from what you have seen so far. Because a high fossil fuel energy input would limit the renewable nature of biomass as a fuel. Also, the amount of carbon dioxide emissions resulting from energy consumption is an important criterion to show the supply chain's contribution to preventing global warming. Therefore, TEC and TE have considered two influential indicators in the performance biomass supply chain. After introducing the indicators, in the next step, index values should be determined. As explained before, in some practical studies, the values of input and output indicators are directly available by referring to accounting files, documents, and databases.

But in some cases, it is necessary to use some analysis, calculations, and side formulas to obtain the exact values of the indicators. In such a situation, mathematical relationships and relevant sub-indices should be specified before any action using the opinion of experts in the relevant field. For example, in the biomass supply chain, the introduced input and output index values are related to the two parts of transportation and processing of materials. As a result:

$$TC = T_C + P_C \tag{8.1}$$

So that, transportation costs (T_C) is the total cost of transportation, including the cost of fuel and transportation by truck, rail, and sea transportation. Suppose the amount of cargo transported in step i to deliver one ton of black pellet biomass to the end user is represented by the symbol TMi. For calculations related to transportation, including cost, energy consumption, and emission of greenhouse gases, the capacity of trucks, trains, and ships was considered equal to the standard capacity of vehicles. Therefore, the total cost of transportation is obtained using the relation (8.2).

$$T_C = \sum_{\text{truck stages}} \left(T_{C_i}/TM_i \right) + \sum_{\text{train stages}} \left(T_{C_j}/TM_j \right)$$
$$+ \left(T_{C_{sheep}} + Port\ \text{fee} \right)/TM_{\text{ship stage}} \tag{8.2}$$

It should be explained that T_{C_i} and T_{C_j} represent the cost of transporting by truck and train. Also, the port fee includes the cost of fuel and other operating expenses that are usually carried out in ports. Next, it is time to extract information to calculate process costs (P_C). It includes pelletization costs $\left(P_{oc_{pelletisation}} \right)$, torrefaction and pelletization costs $\left(P_{oc_{torrefaction\ and\ pelletisation}} \right)$, , annual costs related to investment to create facilities and purchase new equipment, labour costs, and maintenance of equipment and facilities (I_a). P_C is calculated using the relation (8.3). So that, RC represents the reference capacity.

$$P_C = \frac{P_{oc_{pelletisation}}}{RC_{pelletisation}} + \frac{P_{oc_{torrefaction\ and\ pelletisation}}}{RC_{torrefaction\ and\ pelletisation}} + \frac{I_a}{RC_{pelletisation}} \tag{8.3}$$

If I, i, and N are the required amount of capital, discount rate (%), and the lifetime of the equipment in terms of years, then I_a is obtained from the Eq. (8.4).

$$I_a = I. \frac{i}{(i - (1 + i)^{-N})} \tag{8.4}$$

In the same way, the relation (8.5) is used to calculate the values of the TEC index.

$$TEC = T_{EC} + P_{EC} \tag{8.5}$$

T_{EC} means transportation energy consumption. Which is obtained similarly to the mentioned process for TC by using relation (8.6).

$$T_{EC} = \sum_{\text{truck stages}} \left(T_{EC_i}/TM_i \right) + \sum_{\text{train stages}} \left(T_{EC_j}/TM_j \right) + T_{EC_{sheep}}/TM_{\text{ship stage}} \tag{8.6}$$

P_{EC} includes the amount of electricity used for pelletizing, torrefaction per ton of cargo delivered, and considering the biomass consumption (for drying) per delivered t. So:

$$P_{EC} = P_{EC_{pelletisation}} + P_{EC_{torrefaction \ electicity}} + P_{EC_{torrefaction \ biomass}} \tag{8.7}$$

Finally, the relation (8.8) calculates TE.

$$TE = T_E + P_E \tag{8.8}$$

It is also used relations (8.9) and (8.10) to calculate T_E and P_E.

$$T_E = \sum_{truck \ stages} \left(T_{E_i}/TM_i \right) + \sum_{train \ stages} \left(T_{E_j}/TM_j \right) + T_{E_{sheep}}/TM_{ship \ stage} \tag{8.9}$$

and

$$P_E = P_{E_{pelletisation}} + P_{E_{torrefaction \ and \ pelletisation}} \tag{8.10}$$

$P_{E_{pelletisation}}$ is the emission rate of pollutants as a result of fuel consumption to produce energy in the pelletization process, according to the amount of energy consumption in each part and multiplying it by the electricity-specific emissions factor (kgCO$_2$eq/kWh) [4]. Also, the [10] method can be used to calculate the amount of CO$_2$ emissions in the torrefaction process $\left(P_{E_{torrefaction \ and \ pelletisation}} \right)$, assuming that the conditions are the same.

In the next step, evaluators must choose a suitable model to assess and rank different supply chains created based on different scenarios. Since when using the SBM model, outputs are maximized, and inputs are minimized simultaneously, Rentizelas et al. chose the SBM model to evaluate biomass supply chains. As explained in Chap. 6, more than one efficient DMU may be obtained when using DEA models. One of the most well-known ranking methods is the presentation method by Anderson and Patterson, which is obtained by removing the unit under evaluation from the PPS. Based on this idea, [35] introduced the model (8.11) to rank efficient supply chains obtained through the SBM model.

$$\rho^* = \min \frac{\frac{1}{m} \sum_{i=1}^{m} \frac{x_i}{x_{io}}}{\frac{1}{s} \sum_{r=1}^{s} \frac{y_r}{y_{ro}}}$$

$s.t$

$$\sum_{j=1, j \neq o}^{n} \lambda_j x_{ij} \leq x_i, \ i = 1, ..., m,$$

$$\sum_{j=1, j \neq o}^{n} \lambda_j y_{rj} \geq y_r, \ r = 1, \ldots, s,$$

$$x_o \leq x,$$

$$0 \leq y \leq y_o,$$

$$\lambda_j \geq 0, \ j = 1, \ldots, n. \tag{8.11}$$

Similarly, with the linearization technique stated for model (6.70), the linear form of model (8.11) is obtained as the model (8.12).

$$\min \frac{1}{m} \sum_{i=1}^{m} \frac{\bar{x}_i}{x_{io}}$$

$$s.t$$

$$\frac{1}{s} \sum_{r=1}^{s} \frac{\bar{y}_r}{y_{ro}} = 1,$$

$$\sum_{j=1, j \neq o}^{n} \gamma_j x_{ij} \leq \bar{x}_i, \ i = 1, \ldots, m,$$

$$\sum_{j=1, j \neq 0}^{n} \gamma_j y_{rj} \geq \bar{y}_r, \ r = 1, \ldots, s,$$

$$\bar{x} \geq t x_o, \bar{y} \leq t \bar{y}_o, \ j = 1, \ldots, n,$$

$$\gamma_j \geq 0, \ j = 1, \ldots, n,$$

$$t > 0. \tag{8.12}$$

Consider again the input and output indicators considered for the biomass supply chain. Contrary to expectations relevant to outputs, emissions should decrease. Therefore, as explained in Chap. 7, the emission is an undesirable output. Consequently, it cannot be directly entered into the model. When facing undesirable output, the following techniques can be used.

- Undesirable output should be considered as input.
- Suppose y_{ko} is an undesirable output. In this case, the inverse of the desired output $\left(\frac{1}{y_{ko}}\right)$ is used.
- Using transformation $y = \max_{y_{kj}} + \min_{y_{kj}} - DMU_{y_{ko}}$ for undesirable output.
- According to Tone's opinion, due to the nonlinearity of the $\frac{1}{y_{ko}}$ transformation, it is not appropriate to use this transformation in the SBM model. Also, transformation $(y = \max_{\text{emissions}} + \min_{\text{emissions}} - DMU_{\text{emissions}})$ in the SBM model is only possible if the principle of constant return to scale is removed from the set of the PPS.
- If the first point of view is used, all three indicators in this biomass supply chain (costs, energy, and emissions) appear as inputs. Therefore, a fixed output is considered for all supply chains.

8.2.3 Evaluating the Performance and Determining the Regression and Progress of Supply Chains Using the Malmquist Index Based on Success Indicators in the Financial Market in the Presence of Ratio Measures

As explained in the previous chapters, supply chain management with planning, directing, and controlling internal operations and coordinating jobs related to supply chain activity is the main driver of performance in various industries. References [7, 11] believe that a professional supply chain significantly contributes to creating shareholder value. While the review of the studies conducted in the field of DEA applications in various industries shows that the existing studies in the field of performance evaluation with DEA models are based on financial criteria and mainly focus on internal data based on accounting. And capital market data are less used [13]. In other words, they are rarely used in defining performance evaluation criteria to analyze financial success and operational SC performance drivers from the perspective of the stock market. Reference [42] is one of the few studies that considered the criteria for success in financial markets in industrial companies. Reference [3] also showed how assets and expenses affect airline companies' cash liabilities, market value, and market return. But in these studies, only companies are considered without considering their operational supply chain.

Reference [20] focused on the impact of success in financial markets on companies active in the US stock market in different industries. They examined the drivers of operational SC performance at both company and industry levels. We will explain these two views in the following.

1. Within-industry analysis: All companies (regardless of the relevant industry) compete not only for sales and market share in their industries but also for attracting investors in the financial markets. Therefore, in this viewpoint, the financial efficiency of firms in their respective industries and the relative importance of earn and turn for stock market performance are studied. It is necessary to explain that the performance of companies is measured in a certain period.

2. Cross-industry analysis: The aim is to compare supply chains in different industries according to the relative performance of the stock market and performance changes over time. For this purpose, in each year of the period under investigation, the supply chain indicators are obtained by aggregating the values of each input and output indicator in all participating companies in the supply chain in the relevant year.

It is necessary to explain that companies have been selected in a single jurisdiction due to the difference in some tax laws and different valuations in the financial market. Like what was mentioned at the beginning of the chapter, after specifying the units under evaluation (supply chains and participating companies in the chain), it is time to identify the indicators affecting the performance of the supply chain. Profitability and asset utilization represent the critical factors in the financial performance of manufacturing companies. Because these standards emphasize the usefulness and how to use assets as the two main drivers of the company's financial performance. Also, from the point of view of production technology, analysts are interested in an efficient combination of earn and turn. Similar criteria have been used by [17], which examine gross margin and inventory turnover and profitability on retailers' financial performance.

As a result, Hahn et al. evaluated the operational SC performance in 13 selected manufacturing industries, considering the factors related to profit and capital and the effects of the resulting value. For this purpose, stock market performance is considered as output to evaluate the financial performance of supply chains and related companies. High-level metric return on net working capital (RoNWC) in SCM is regarded as an input. As explained in the previous subsections, in some industries, according to the defined indicators, to obtain the values of each of the indicators, we need to use the opinions of experts in the relevant field and use side formulas. Therefore, we use the relation (8.13) to determine the input values.

$$RoNWC = \frac{Gross\ Profit}{NWC} \tag{8.13}$$

Gross profit equals sales revenue minus COGS (Cost of Goods Sold). Also, NWC means Net working capital. Based on the opinion of [15], RoNWC can be decomposed into two strategic levers in the supply chain called earns and turn, as shown in relation (8.14).

$$RoNWC = \frac{Gross\ Profit}{NWC} = \underbrace{\frac{Gross\ Profit}{Sales\ Revenue}}_{Earns} \times \underbrace{\frac{Sales\ Revenue}{NWC}}_{Turns} \tag{8.14}$$

Analysts are interested in an efficient combination of earn and turn to examine a firm's stock market performance. Based on the opinion of [21], Tobin's q is defined as the ratio of the market value of a company to the replacement value of its total assets, so the modified version of Tobin's q can depict financial market performance. Therefore, Tobin's q is considered an output index in the supply chain to check financial performance. To calculate Tobin's q, the relation (8.15) is used.

$$\tilde{q} = \frac{MVE + PS + DEBT - NCA}{NWC} \tag{8.15}$$

so that:

- NWC = Net Working Capital
- NCA = Non-Current Assets
- MVE = Market Value of Equity
- PS = Value of Preferred Stock
- DEBT = Debt Value.

According to the property of Isotonicity in DEA, an increase in any of the inputs should not lead to a decrease in any of the outputs. Therefore, the linear correlation between input and output indicators can be checked to check this issue among selected inputs and outputs. According to the definition of correlation, if the obtained values are negative, it means that isotonicity does not exist among the input and output indicators defined in the relevant industry.

After determining the effective criteria for the evaluation, it is time to choose the appropriate DEA model. Since the changes in the input and output are not proportional, therefore, using the BCC model is suggested. It is also evident that the company cannot directly affect the stock market's performance (output). Therefore, to provide a suitable model and optimize and restructure the supply chain, priority is given to input changes, so the BCC model is used in the input-oriented. Also, as explained in Chap. 6, the BCC model in input-oriented is invariant concerning output transmission, and this property solves the problem of facing negative data. Since the values of q may be negative in some cases, the input-oriented BCC model is suitable for this problem.

Also, the BCC model in input-oriented is used for cross-industry analysis, comparing supply chains with each other using aggregated data, similar to evaluating companies independently.

Another point that should be noted is that the indicators studied in the success of the financial market are presented as a ratio of two independent criteria. These indicators are called ratio measures. The use of ratio measure causes the property of convexity among these data to be violated [12, 24]. For example, consider Table 8.1:

School C is a convex combination of schools A and B. This school's actual pass rate (this indicator is a ratio measure) is 85%. In contrast, the convex combination of success rate in schools A and B is equal to 84%. Therefore, the property of convexity among this index has been violated. So far, several methods have been presented to solve the problem faced with ratio measures, which we explain as follows.

1. If the desired index is input, the numerator of the fraction in the input role and the denominator of the fraction in the output role are considered. Also, if the desired index is output, the numerator and denominator of the fraction are used as output and input, respectively, in the DEA model. Of course, if the number of ratio measures in DMUs under evaluation is high, there are more suitable techniques. Also, in some cases, because the information is classified, it is not possible to access the numerator and denominator values of the fractions as independent data, but the ratio of these two indices is available [12].

Table 8.1 DMUs with ratio measure

School	Student	Number of graduates	Success rate (%)
A	500	440	88
B	300	240	80
$C = \frac{1}{2} \times A + \frac{1}{2} \times B$	400	340	85

2. Another method is to replace the ratio measure with a volume index that partially reflects the status of the ratio measure. Of course, finding this index is not easy.

3. Suppose $y_{kj} = \frac{n_{kj}}{d_{kj}}, j = 1, ..., n$ are ratio measures. In this case, [12] defined the convex combination of ratio measures as $\sum_{j=1}^{n} \lambda_j y_{kj} = \dfrac{\sum_{j=1}^{n} \lambda_j n_{kj}}{\sum_{j=1}^{n} \lambda_j d_{kj}}$. Therefore, attention to this should be taken into account in DEA modeling.

4. Reference [9] suggest that a ratio measure could be multiplied by an appropriate measure of the size of the DMU so that the resulting measure could change in proportion to the size of the latter.

Readers can refer to references [29, 22, 33] for more information on ratio measures.

Hahn et al. considered the fourth point of view. So using the total assets of a DMU divided by the average of the total assets of all DMUs is a simple choice to solve the problem of the presence of a ratio measure when using DEA models to evaluate financial market success. As seen so far, using DEA models, efficiency values and the distance of each unit to the frontier are obtained in a period. In other words, the performance of supply chains is calculated without considering technological changes. Therefore, information about the regression or progress of the units under evaluation is not provided.

Therefore, an index called the Malmquist Productivity Index (MI) was presented in [14] by Fare and Grosskopf. With this index, the rate of progress or regression of each DMU is evaluated. The mathematical definition for calculating the Malmquist index is based on the distance function. The relation (8.16) is used to calculate MI.

$$MI_p = \underbrace{\frac{\theta^{t+1}(x_p^{t+1}, y_p^{t+1})}{\theta^t(x_p^t, y_p^t)}}_{Catch-up} \underbrace{\sqrt{\frac{\theta^t(x_p^{t+1}, y_p^{t+1}) \times \theta^t(x_p^t, y_p^t)}{\theta^{t+1}(x_p^t, y_p^t) \times \theta^{t+1}(x_p^{t+1}, y_p^{t+1})}}}_{Frontier-shif} \qquad (8.16)$$

Assume that the coordinates of DMU_p in the period t and t + 1 are represented by symbols (x_p^t, y_p^t) and (x_p^{t+1}, y_p^{t+1}). In this case, $\theta^l(x_p^k, y_p^k)$ represents the efficiency of DMU_p in the period k under production technology l. So that: $k, l \in \{t, t+1\}$. . Also, model (8.17) is used to calculate $\theta^l(x_p^k, y_p^k)$.

$$\theta^l(x_p^k, y_p^k) = \max \sum_{r=1}^{s} u_r y_{rp}^k + u_o$$

$s.t$

$$\sum_{i=1}^{m} v_i x_{rp}^k = 1,$$

$$\sum_{r=1}^{s} u_r y_{rp}^k - \sum_{i=1}^{m} v_i x_{rp}^k + u_o \leq 0,$$

$$\sum_{r=1}^{s} u_r y_{rj}^l - \sum_{i=1}^{m} v_i x_{rj}^l + u_o \leq 0, \ j = 1, ..., n, \ j \neq p$$

$$u_r, v_i \geq 0, \ i = 1, ..., m, \ r = 1, ..., s. \tag{8.17}$$

Therefore, by placing the values of k and l in the model (8.17), the values of $\theta^t(x_p^t, y_p^t), \theta^{t+1}(x_p^{t+1}, y_p^{t+1}), \theta^{t+1}(x_p^t, y_p^t), \theta^t(x_p^{t+1}, y_p^{t+1})$ are obtained.

If $MI_p > 1$, then DMU_p at period t relative to period t + 1 has progressed.
If $MI_p < 1$, then DMU_p at period t relative to period t + 1 has regressed.
If $MI_p = 1$, then DMU_p at period t relative to period t + 1 has not changed.

Therefore, concerning the evaluation of success in the financial market, the changes in the catchup correspond to the relative changes in the company's ability to convert SC performance into market value, while the frontier shift shows the relative changes in evaluating the operational SC performance from the perspective of the financial market.

8.2.4 Oil Import Supply Chain: Performance Evaluation of Supply Chain Based on Supply Chain Risk

Since the twentieth century, oil has played a crucial role in the field of energy, but due to the imbalance of supply and demand, many crises have occurred in the field of export and import of this vital product in different parts of the world, because most of the significant oil consumers Crude, they are not able to supply the oil they need on their own, and they provide their needs through imports. Crude oil importing countries supply most of their imports through the Middle East and North African countries. Since these areas are known as high-risk areas due to some wars and numerous political challenges, when importing oil from different oil routes, there are potential risks and threats at each stage of oil import that affect the oil performance supply chain. One of the essential importers in the field of crude oil is China. Since 1993, this dependence on imports has increased, so in 2011, 59% of China's oil was supplied through imports. Therefore, assessing the supply chain risk of Chinese oil imports in different import routes is an essential issue in this industry. Therefore, [39], to assess the risk of China's oil chain, first identified different sources of risk, as shown in Fig. 8.2.

Also, each of the import routes is considered a DMU. For example, the supply routes for China's oil imports are the Middle East, North Africa, West Africa, Asia Pacific, Central & South America, and the former Soviet Union region. It is necessary to explain that the impact of the risks of different parts on the security of the

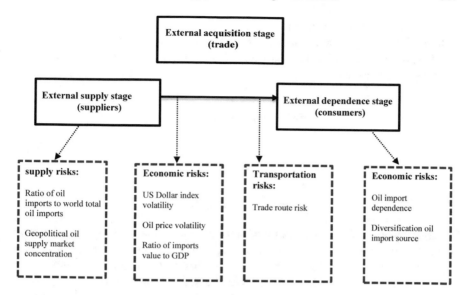

Fig. 8.2 Indicators for each stage of oil import [39]

entire supply chain is different. Therefore, the risk assessment of different parts independently does not provide accurate information regarding the overall risk of the oil
chain. As a result, it is necessary to integrate the risk indicators of different stages for
the overall chain assessment. Considering that the risks and dangers expressed for
different parts of the oil supply chain are various in different economic conditions
and international environments, so valuing and determining the related weights with
varying risks in different paths is a crucial point to be considered. Therefore, to get
away from any neglect of some risks or pay too much attention to some other risks
that personal comments may cause, the DEA technique is used for weighing and
measuring the security of the oil supply chain.

Consider the risk indicators again at different stages of the oil chain. The value
assigned to the indicators is such that their increase causes to increase in the efficiency
of DMUs or obtains a better rank. As a result, a DEA model without input or with a
fixed input, for example, one as a virtual input, is applied to evaluate the risk of oil
import supply chains. In the following, the risk of supply chains is assessed using a
two-phase method.

Phase I:

After receiving the information and specifying the risk of different parts in different
years, the data using the relation (8.18) is normalized to prevent the indices from
becoming zero.

$$z_{po} = 0.1 + 0.9 \frac{x_{ij} - x_{ij}^-}{x_{ij}^+ - x_{ij}^-} \qquad (8.18)$$

Although there are several methods for data normalization, [41] showed that the choice of the normalization method does not affect the final results. After normalizing the data, using the model (8.19), the result of the security assessment of the oil import supply chain is obtained.

$$VES = \max \sum_{p=1}^{P} w_p z_{po}$$

$$\sum_{p=1}^{P} w_p z_{pj} \leq 1, \, j = 1, ..., n,$$

$$w_p \geq 0, \, p = 1, ..., P. \tag{8.19}$$

Thus, by repeatedly solving this model for each year, a set of weights for different indicators in different years is calculated. As was stated in Chap. 6, DEA models seek to obtain the efficiency of the units in the best state. As a result, more weight may be assigned to indicators with a higher value, and some indicators with lower values are ignored. Therefore, to overcome this problem, weight restriction methods are used, as explained in Chap. 7. So, the constraint $L_p \leq \frac{w_p z_{po}}{\sum_{j=1}^{n} w_p z_{pj}} \leq U_p$ is added to the model (8.19), and then the model (8.19) is rewritten as the model (8.20).

$$VES_o = \max \sum_{p=1}^{P} w_p z_{po}$$

$$\sum_{p=1}^{P} w_p z_{pj} \leq 1, \, j = 1, ..., n,$$

$$L_p \leq \frac{w_p z_{po}}{\sum_{j=1}^{n} w_p z_{pj}} \leq U_p, \, p = 1, ...P,$$

$$w_p \geq 0, \, p = 1, ...P. \tag{8.20}$$

Therefore, VES_o shows the overall risk of the oil import supply chain, and a higher VES indicates a higher risk.

Phase II:

As seen in phase I, the model (8.20) is only looking for an optimistic assessment of the security of supply chains in different years. In other words, the valuation of risk sources and, as a result determining the overall risk of the chain in different years are considered independent of each other. Of course, this problem is not far-fetched in DEA models. To solve this, [27] evaluated the performance of the units under evaluation using a common set of weights. But when using CSW models, all information obtained from models with variable weights is discarded. Therefore, to achieve an agreed set of weights for long-term planning and determining the overall risk of the chain model (8.21) was proposed by [39].

$$\min \sum_{j=1}^{n} d_i$$

$s.t$

$$\sum_{p=1}^{P} w_p z_{pj} + d_j = VES_j, \ j = 1, ..., n,$$

$$w_p \geq \varepsilon, \ p = 1, ...P, \ d_j \geq 0, \ j = 1, ..., n. \tag{8.21}$$

8.2.5 Evaluating the Impact of Different Information-Sharing Scenarios for Reconfiguring the Supply Chain Using Cross-Efficiency

As explained in the previous chapters, supply chain management considers the processes between each chain component as a complete integrated system. In other words, to increase internal and extra-organisational integration, establishing communication between the inside and outside of each organization and companies involved in the supply chain is one of the most critical responsibilities of supply chain management. Therefore, implementing information flow techniques and using information systems allow organizations to strive to improve their supply chain and integrate their processes. In other words, companies can manage product flow and information related to issues such as production capacity, customer demand, and inventory at lower costs through information technology. As a result, information sharing can significantly improve supply chain performance. In short, information is essential for making critical decisions at all levels of the supply chain. Without information, a manager does not have an understanding of what customers want, the amount of inventory in the warehouse, forecasting demand, and how to produce and ship products. The information allows managers to monitor all supply chain operations and make decisions to improve supply chain performance. Therefore, the need for information systems in today's supply chains is quite apparent. Although the information and communication systems available to organizations today collect and store a large amount of data, the accumulation and storage of data are only helpful in some cases if the information is shared in a timely, accurate, and managed manner. In this case, this information can become a powerful tool for the supply chain management. Also, if supply chain components know how to benefit from information sharing, they will be more willing to share information.

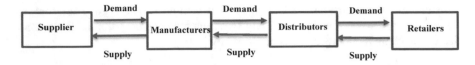

Fig. 8.3 Information-sharing in supply chain

In short, the performance of a supply chain can be affected by many factors, among which the way of sharing information is one of the essential things that supply chain managers must consider. Because by sharing information between business partners and coordinating with them, it is possible to reduce costs further and improve customer service. However, only some studies have focused on how different combinations of information sharing affect supply chain performance.

Consider the supply chain shown in Fig. 8.3. Consider the supply chain shown in Fig. 8.3. This supply chain starts with the shipment of raw materials by the suppliers and ends with the distribution of the product by the retailer. Reference [38] investigated the chain's performance under different information scenarios regarding the importance of information sharing in chain performance. For this purpose, the indicators affecting the performance of the supply chain were first introduced as follows.

- Shortage costs
- Holding costs
- Order costs
- Total costs
- Fulfillment rate
- Customer service level
- Order cycle time.

Then they defined nine information-sharing scenarios as follows:

- No information will be shared between the entities (N)
- Partial information sharing, which consists of six combinations:

   o Capacity information sharing (C)
   o Demand information sharing (D)
   o Inventory information sharing (I)
   o Demand and capacity information sharing (D&C)
   o Demand and inventory information sharing (D&I)
   o Capacity and inventory information sharing (C&I)

- Full information sharing with capacity, demand, and inventory (F)
- The strategic alliance of the supply chain (vendor-managed inventory, VMI, is adopted herein).

To compare different scenarios, a simulation technique is used to determine the values of the indicators in each of the selected scenarios by using parameters suitable for each system and considered scenarios. After calculating the values of the supply chain indicators under the influence of different scenarios, we are now faced with a multi-criteria evaluation problem that is formulated by considering a set of options and a set of criteria. Therefore, at this stage, the DEA technique is used to evaluate the efficiency of different scenarios to identify the efficient scenario and determine how to share information based on that.

When using DEA models to evaluate different information-sharing scenarios, each DMU can assign the most favourable weights to input and output indicators to obtain a better ranking. Therefore, a unit may reach a relative efficiency score of 1 by considering a high weight for a few desirable inputs and outputs and completely ignoring other inputs and outputs in which it is weak. Thus, the results do not necessarily indicate the best way to combine information-sharing strategies. Therefore, as explained in Chap. 6, cross-efficiency models can be used to solve this problem. Also, to solve the problem of the existence of an alternative optimal solution when calculating the cross-efficiency using the relation (6.91), the aggressive approach explained in Chap. 7 or the benevolent approach like the model (8.22) can be used.

$$\max \sum_{r=1}^{s} u_r \sum_{\substack{j=1 \\ j \neq o}}^{n} y_{rj}$$

$$s.t$$

$$\sum_{i=1}^{m} v_i \sum_{\substack{j=1 \\ j \neq o}}^{n} x_{ij} = 1,$$

$$\sum_{r=1}^{s} u_r y_{rj} - \sum_{i=1}^{m} v_i x_{ij} \leq 0, \ j = 1, \ldots, n, \ j \neq o$$

$$\sum_{r=1}^{s} u_r y_{ro} - E_{oo} \sum_{i=1}^{m} v_i x_{io} = 0,$$

$$u_r, v_i \geq 0, \ r = 1, \ldots, s, \ i = 1, \ldots, m. \tag{8.22}$$

It is necessary to explain that E_{oo} is the efficiency of DMU_o which is obtained by solving the multiplier form of the CCR model. Suppose that u_r^* and v_i^* are the optimal weights of the CCR model. Therefore, the relative efficiency of DMU_o, which is obtained based on the optimal weights of DMU_j and is represented by the symbol E_{oj}, is obtained from the relation (8.23):

$$E_{oj} = \frac{\sum_{r=1}^{s} u_r^* y_{rj}}{\sum_{i=1}^{m} v_i^* x_{ij}} \tag{8.23}$$

As a result, the cross efficiency of DMU_o based on the benevolent approach is obtained using the relation (8.24):

$$\overline{E} = \frac{1}{n} \sum_{o=1}^{n} E_{oj} \qquad (8.24)$$

Of course, according to some researchers' ideas, considering that we aim to find the best way to share information in a supply chain, the aggressive method is better than the benevolent method.

8.3 Aggregation of Indicators in the Evaluation of Supply Chain Performance

As was explained at the beginning of this chapter, the first step in using DEA models is to select appropriate indicators as input or output in the units under evaluation. These indicators include financial and non-financial factors concerning the organization's activities. So many indicators may be extracted to evaluate the performance of supply chains, while some of these indicators can be classified into a more general category. For example, consider the supply chain of pharmaceutical products introduced by ref. [6]. The effective indicators of this supply chain, based on Cherfi et al.'s opinion, are total supply chain management cost (TSCMC), budgetary gap (BG), cost per unit (CPU), expenses to net revenue (ENR), inventory turnover (IT), rate of loss due to obsolescence (RLO), order fulfilment cycle time (OFCT), social benefit (SB) -number of the patient served-, Health satisfaction index (HSI) -patient satisfaction-, perfect order fulfilment (POF), short term availability of health products (STA), consumption exactitude (CE), compliance of health products with standards (CS), the total density of health premises per 100,000 population (TD), total capacity, the number of facilities, distance travelled.

Some of these indicators can be placed in a more general category. For example:

- The Cost based metrics: TSCMC, BG, CPU, ENR, IT, and RLO.
- The responsiveness metric: OFCT.
- The effectiveness metrics are: SB, HIS, POF, STA, CE, CS, TD,
- The design-based metrics: The number of facilities, Total storage capacity, and the transportation metric (the distance travelled).

Although these indicators are somewhat similar, they have different amounts, and the value and affection of these indicators on the chain's performance are not the same. On the one hand, reducing the number of indicators leads to a reduction in the volume of calculations; on the other hand, ignoring some of these indicators leads to missing the impact of these indicators on the performance of the supply chain. Therefore, we need to use techniques to aggregate indicators by considering the value and importance of each of these indicators. Several techniques in this field have been presented by different researchers, including the method proposed by ref. [40]. In this method, a weighted combination is presented based on each indicator's best and worst weights to integrate the indicators as an aggregate index to evaluate the performance of unit j. To do this work, the indicators are first categorized. Then based on each category's indicators, the best weight is calculated based on the indicators of group k using the model (8.25).

$$gI_o^k = \max \sum_{p=1}^{P} w_p^g I_{po}^k$$

$$\sum_{p=1}^{P} w_p^g I_{po}^k \leq 1, \, j = 1, ..., n,$$

$$w_p^g \geq 0, \, p = 1, ..., P. \tag{8.25}$$

Next, using model (8.26), the worst weighting is determined for the indicators of group k.

$$bI_o^k = \min \sum_{p=1}^{P} w_p^b I_{po}^k$$

$$\sum_{p=1}^{P} w_p^b I_{pj}^k \leq 1, \, j = 1, ..., n,$$

$$w_p^b \geq 0, \, p = 1, ..., P. \tag{8.26}$$

And finally, using the relation (8.27), an aggregated value is calculated for the indicators of supply chain o that belong to group k.

$$CI_o^k(\alpha) = \alpha \frac{gI_o^k - gl^-}{gl^* - gl^-} + (1 - \alpha) \frac{bI_o^k - bl^-}{bl^* - bl^-} \tag{8.27}$$

Such that:

$$gl^* = \max\{gI_j^k | j = 1, ..., n\} \tag{8.28}$$

$$bl^* = \max\{bI_j^k | j = 1, ..., n\} \tag{8.29}$$

also:

$$gl^- = \min\{gI_j^k | j = 1, ..., n\} \tag{8.30}$$

$$bl^- = \min\{bI_j^k | j = 1, ..., n\} \tag{8.31}$$

It is necessary to explain that: $0 \leq \alpha \leq 1$. If $\alpha = 1$, then $I_i(\alpha)$ represents the normalized value of the best weighting and $\alpha = 0$ represents the worst normalized weighting. Also, for balanced decision-making, $\alpha = 0.5$ is considered. Therefore, the following algorithm is used to evaluate the supply chains under evaluation.

1. Determining all the indicators needed to evaluate the supply chains under evaluation.
2. Classification of indicators.
3. Solving the model (8.25) and (8.26) and using the relation (8.27) to calculate the composite index $(CI_j^k(\alpha), j = 1, ..., n, k = 1, ..., n)$.
4. Separation of inputs and outputs.
5. Determining relative efficiency using DEA models. In this step, we use the composite index obtained in step 3 for the inputs and outputs required in DEA models.

8.4 Benchmarking for Supply Chain

Although data envelopment analysis is a powerful technique to evaluate the performance of supply chains, the results obtained from DEA models are based on the system's past. In other words, the performance evaluation scores are obtained based on past data. Therefore, the efficiency improvement solutions for inefficient units provided through the projection of efficient units may not be valid in some situations because the efficient unit may perform better in the future compared to its current position. As a result, improving efficiency for the inefficient unit may not be effective in the future. Therefore, to overcome this weakness of DEA models, in this subsection, we describe the modeling technique presented by refs. [1, 31]. In this technique, by gathering information related to the past of a supply chain and using a simulation technique such as system dynamics simulation, artificial neural network, and so on, the amount of inputs and outputs of each supply chain is predicted in the future. Then, to evaluate the performance of the unit under evaluation, using DEA models, the chain under assessment is considered in the current position and other units in the simulated place in the future. On the other hand, DMU_o is regarded at period k and $DMU_j, j = 1, ..., n, j \neq o$ is considered at period k + 1. That's mean:

$$DMU_1^{k+1} : \left(x_{11}^{k+1}, ..., x_{1m}^{k+1}, z_{11}^{k+1}, ..., z_{1d}^{k+1} y_{11}^{k+1}, ..., y_{1s}^{k+1}\right)$$

$$\vdots$$

$$DMU_o^K : \left(x_{o1}^k, ..., x_{om}^k, z_{11}^k, ..., z_{1d}^k y_{o1}^k, ..., y_{os}^k\right)$$

$$\vdots$$

$$DMU_n^{K+1} : \left(x_{n1}^{K+1}, ..., x_{nm}^{k+1}, z_{11}^{k+1}, ..., z_{1d}^{k+1} y_{n1}^{k+1}, ..., y_{ns}^{k+1}\right)$$

And in this way, efficiency improvement solutions become efficiency improvement strategies, and conditions are provided competitively for all units. Therefore, we explain the benchmarking process for inefficient supply chains using DEA as follows.

1. Identifying all the inputs and outputs affecting the efficiency of the supply chain.

At this stage, by referring to the research literature and using experts' opinions, essential inputs and outputs are identified for the performance evaluation of the supply chains.

2. Identifying cause and effect relationships between indicators.

At this step, the way inputs and outputs affect each other should be shown as a mathematical relation. For example, a regression equation can establish the relationship between indicators. Another solution is to extract the relationships using the DEMATEL technique and then express the relationships quantitatively through regression or other mathematical models. It is necessary to explain that according to the simulation technique used, if there is no need to know the relationships between the indicators, this step can be ignored.

3. Data collection and analysis by software.

Now, the input and output information of the supply chain is collected in the past few periods.

4. Forecasting the inputs and outputs of the units under evaluation in the future based on the appropriate simulation technique.

The collected data and cause-and-effect relationships (if needed) are entered into the simulation software. Different software can be selected in this field depending on the simulation technique used. For example, Vensim software can be used when using the DS technique.

5. Choosing the appropriate DEA model and its implementation.

In this section, the appropriate DEA model is selected according to the structure of the supply chain and the type of indicators used in the chain, which have been explained in detail earlier.

6. Solving the DEA model and analysis of the obtained results.

Finally, a suitable benchmark is introduced for each inefficient unit based on the obtained results and the classification of efficient and inefficient chains.

8.5 Conclusion

This chapter explains that evaluating supply chain performance is not just a financial matter. Unlike the traditional methods, the new techniques consider all supply chain activities, such as the quality of products and services, planning processes in a company, the way of using resources, and the method of production. In other words, new techniques for performance evaluation of the supply chain consider how to use quantitative and qualitative inputs and produce quantitative and qualitative outputs. Therefore, the researchers suggested evaluating supply chain performance using mathematical models by considering financial and non-financial indicators. Therefore, using DEA models was proposed to evaluate supply chains.

As seen in this chapter, although there are slight differences in evaluating supply chains in different industries using DEA models, the steps of assessing supply chains using DEA models are almost the same. When using DEA models, the following steps are performed.

- Determining the input and output indicators by reviewing the literature, using experts' opinions, and aggregating the indicators (If necessary).
- Determining the input and output values using the companies' documents and also using some formulas if needed.
- Choosing the suitable DEA model and solving it.
- Analysis of results.
- Providing a benchmark.

Therefore, although some supply chains in specific industries have been described in this chapter, these steps can be used to evaluate the performance of supply chains in other industries [30].

References

1. Avkiran, N.K., Shafiee, M., Saleh, H., Ghaderi, M.: Benchmarking in the supply chain using data envelopment analysis. Theor. Econ. Lett. **8**(14), 2987 (2018)
2. Azadeh, A., Haghighi, S.M., Gaeini, Z., Shabanpour, N.: Optimization of healthcare supply chain in context of macro-ergonomics factors by a unique mathematical programming approach. Appl. Ergon. **55**, 46–55 (2016)
3. Bowlin, W.F.: Financial analysis of civil reserve air fleet participants using data envelopment analysis. Eur. J. Oper. Res. **154**(3), 691–709 (2004)
4. Brander, M., Sood, A., Wylie, C., Haughton, A., Lovell, J.: Technical paper|electricity-specific emission factors for grid electricity. Ecometrica, Emission Factors.com (2011)
5. Charnes, A., Cooper, W.W.: Measuring the efficiency of decision-making units. Eur. J. Oper. Res. **2**(6), 429–444 (1978)
6. Chorfi, Z., Benabbou, L., Berrado, A.: A two stage DEA approach for evaluating the performance of public pharmaceutical products supply chains. In: 2016 3rd International Conference on Logistics Operations Management (GOL), pp. 1–6. IEEE (May 2016)
7. Christopher, M., Ryals, L.: Supply chain strategy: its impact on shareholder value. Int. J. Logist. Manage. **10**(1), 1–10 (1999)
8. Christopher, M.: Logistics and Supply Chain Management: Strategies for Reducing Cost and Improving Service Financial Times: Pitman Publishing. London (1999)
9. Dyson, R.G., Allen, R., Camanho, A.S., Podinovski, V.V., Sarrico, C.S., Shale, E.A.: Pitfalls and protocols in DEA. Eur. J. Oper. Res. **132**(2), 245–259 (2001)
10. Ehrig, R., Behrendt, F.: Co-firing of imported wood pellets–An option to efficiently save CO_2 emissions in Europe? Energy Policy **59**, 283–300 (2013)

11. Ellinger, A., Shin, H., Northington, W.M., Adams, F.G., Hofman, D., O'Marah, K.: The influence of supply chain management competency on customer satisfaction and shareholder value. Supply Chain Manage.: Int. J. (2012)
12. Emrouznejad, A., Amin, G.R.: DEA models for ratio data: Convexity consideration. Appl. Math. Model. **33**(1), 486–498 (2009)
13. Emrouznejad, A., Yang, G.L.: A survey and analysis of the first 40 years of scholarly literature in DEA: 1978–2016. Socioecon. Plann. Sci. **61**, 4–8 (2018)
14. Fare, R., Grosskopf, S.: Malmquist productivity indexes and fisher ideal indexes. Econ. J. **102**(410), 158–160 (1992)
15. Feroz, E. H., Kim, S., & Raab, R. L.: Financial statement analysis: A data envelopment analysis approach. J. Oper. Res. Soc. **54**, 48–58 (2003)
16. Folan, P., Browne, J.: Development of an extended enterprise performance measurement system. Prod. Plan. Control **16**(6), 531–544 (2005)
17. Gaur, V., Fisher, M.L., Raman, A.: An econometric analysis of inventory turnover performance in retail services. Manage. Sci. **51**(2), 181–194 (2005)
18. Gerami, J., Kiani Mavi, R., Farzipoor Saen, R., Kiani Mavi, N.: A novel network DEA-R model for evaluating hospital services supply chain performance. Ann. Oper. Res. 1–26 (2020)
19. Ghalayini, A.M., Noble, J.S.: The changing basis of performance measurement. Int. J. Oper. Prod. Manag. (1996)
20. Hahn, G.J., Brandenburg, M., Becker, J.: Valuing supply chain performance within and across manufacturing industries: a DEA-based approach. Int. J. Prod. Econ. **240**, 108203 (2021)
21. Lang, L.H., Stulz, R.M.: Tobin's q, corporate diversification, and firm performance. J. Polit. Econ. **102**(6), 1248–1280 (1994)
22. Mozaffari, M.R., Kamyab, P., Jablonsky, J., Gerami, J.: Cost and revenue efficiency in DEA-R models. Comput. Ind. Eng. **78**, 188–194 (2014)
23. Neely, A., Gregory, M., Platts, K.: Performance measurement system design: a literature review and research agenda. Int. J. Oper. Prod. Manag. **25**(12), 1228–1263 (2005)
24. Olesen, O.B., Petersen, N.C., Podinovski, V.V.: Efficiency analysis with ratio measures. Eur. J. Oper. Res. **245**(2), 446–462 (2015)
25. Olfat, L., Bamdad Soufi, J., Amiri, M., Ebrahimpour Azbari, M.: A model for supply chain performance evaluation using by network data envelopment analysis model (Case of: supply chain of pharmaceutical companies in Tehran stock exchange. J. Ind. Manag. Stud. **26**, 9–26 (2012)
26. Rentizelas, A., Melo, I.C, Junior, P.N.A., Campoli, J.S., do Nascimento Rebelatto, D.A.: Multi-criteria efficiency assessment of international biomass supply chain pathways using data envelopment analysis. J. Cleaner Prod. **237**, 117690 (2019)
27. Roll, Y., Cook, W.D., Golany, B.: Controlling factor weights in data envelopment analysis. IIE Trans. **23**(1), 2–9 (1991)
28. Sahoo, B.K., Saleh, H., Shafiee, M., Tone, K., Zhu, J.: An alternative approach to dealing with the composition approach for series network production processes. Asia Pac. J. Oper. Res. **38**(06), 2150004 (2021)
29. Saleh, H., Hosseinzadeh Lotfi, F., Rostmay-Malkhalifeh, M., Shafiee, M.: Provide a mathematical model for selecting suppliers in the supply chain based on profit efficiency calculations. J. New Res. Math. **7**(32), 177–186 (2021)
30. Saleh, H., Shafiee, M., Hosseinzade Lotfi, F.: Performance evaluation and specifying of Return to scale in network DEA. Int. J. Ind. Math. J. Adv. Math. Model. **10**(2), 309–340 (2021)
31. Shafiee, M., Hosseinzade Lotfi, F., Saleh, H.: Benchmark forecasting in data envelopment analysis for decision making units. Int. J. Ind. Math. **13**(1), 29–42 (2021)
32. Shafiee, M., Saleh, H.: Evaluation of strategic performance with fuzzy data envelopment analysis. Int. J. Data Envelopment Anal. **7**(4), 1–20 (2019)
33. Shafiee, M., Saleh, H., Sanji, M.: Modifying the interconnecting activities through an adjusted dynamic DEA model: a slacks-based measure approach. J. Adv. Math. Model. **10**(2), 309–340 (2020)

34. Shafiee, M., Saleh, H., Ziyari, R.: Projects efficiency evaluation by data envelopment analysis and balanced scorecard. J. Decisions Oper. Res. **6**(Special Issue), 1–19 (2022)
35. Tone, K.: A slacks-based measure of super-efficiency in data envelopment analysis. Eur. J. Oper. Res. **143**, 32–41 (2002)
36. Uslu, A., Faaij, A.P., Bergman, P.C.: Pre-treatment technologies, and their effect on international bioenergy supply chain logistics. Techno-economic evaluation of torrefaction, fast pyrolysis and pelletisation. Energy **33**(8), 1206–1223 (2008)
37. Wagner, W.P., Chung, Q.B., Baratz, T.: Implementing corporate intranets: lessons learned from two high-tech firms. Ind. Manag. Data Syst. **102**(3), 140–145 (2002)
38. Yu, M.M., Ting, S.C., Chen, M.C.: Evaluating the cross-efficiency of information sharing in supply chains. Expert Syst. Appl. **37**(4), 2891–2897 (2010)
39. Zhang, H.Y., Ji, Q., Fan, Y.: An evaluation framework for oil import security based on the supply chain with a case study focused on China. Energy Econ. **38**, 87–95 (2013)
40. Zhou, P., Ang, B.W., Poh, K.L.: A mathematical programming approach to constructing composite indicators. Ecol. Econ. **62**, 291–297 (2007)
41. Zhou, P., Ang, B.W., Zhou, D.Q.: Weighting and aggregation in composite indicator construction: a multiplicative optimization approach. Soc. Indic. Res. **96**(1), 169–181 (2010)
42. Zhu, J.: Multi-factor performance measure model with an application to Fortune 500 companies. Eur. J. Oper. Res. **123**(1), 105–124 (2000)

Chapter 9
Supply Chain Evaluation by Network DEA

9.1 Introduction

As explained in the previous chapters, supply chain management has become one of the topics discussed in business literature today. Supply chain management includes a company's products, services, and processes. It is a fundamental cornerstone for companies to create sustainable competition and remain at the forefront of the playing field. Therefore, supply chain operations management is critical to compete effectively in the business environment. In the process of supply chain management, measuring its performance is a significant activity that should be done because it helps the management to understand where the company stands among its competitors and how it can improve further. Supply chains have traditionally been managed as a series of simple, segmented business functions led by manufacturers who manage and control the pace of product development, production, and distribution.

Measuring the efficiency of traditional supply chains can be done easily and using a straightforward method. However, in current supply chains, evaluating the efficiency of the supply chain needs to examine the multi-dimensional structure that takes into account other indicators in addition to financial indicators [56].

As explained in Chap. 7, DEA is a powerful optimization technique that can be used to evaluate the performance of any supply chain by considering multiple indicators. But this technique also has certain limitations. In traditional DEA models, units under evaluation in general and supply chains in specific are considered black boxes, and only their initial inputs and final outputs are included in the evaluation, and attention is not paid to their internal structure. While the supply chain is a type of decision-making unit that not only has input and output indicators but also uses intermediary indicators that flow from the previous stage to the next stage, and each stage may have its independent inputs and outputs. Therefore, due to the networked or multi-stage nature of the supply chain, the traditional DEA models cannot wholly and correctly evaluate the supply chain, provide reliable results and accurately identify the sources of inefficiency in a supply chain. Therefore, identifying sources of inefficiency in units under evaluation using DEA may not be done correctly.

© The Author(s), under exclusive license to Springer Nature Switzerland AG 2023
F. Hosseinzadeh Lotfi et al., *Supply Chain Performance Evaluation*,
Studies in Big Data 122, https://doi.org/10.1007/978-3-031-28247-8_9

Therefore, to solve the defects of classical DEA models, Fare and Grosskopf [13] introduced Network Data Envelopment Analysis (NDEA) models, which evaluate the system's performance based on the activity of all processes and components. Due to the importance of multi-stage structures in practical cases, this DEA branch quickly attracted researchers' attention, and extensive studies have been conducted in the field. So other researchers, including (Lewis and Sexton [26], Kao and Hwang [21], Liang et al. [28], Despotis et al. [9], Sahoo et al. [43]), developed network models. Therefore, this chapter will discuss the performance evaluation of supply chains using NDEA models.

9.2 Performance Evaluation of Supply Chains Using Radial Models in NDEA

Chapters 6 and 8 explain that a general category of DEA models for performance evaluation of under-assessment supply chains is the envelopment models. Although the envelopment form of DEA models has many strengths and applications, due to ignoring internal interactions, the envelopment form of NDEA models for assessing supply chain performance is introduced in this section.

9.2.1 Definition of New PPS for Performance Evaluation of Supply Chain

Suppose there is n two-member supply chain (supplier-manufacturer) as shown in Fig. 9.1. In supply chain j ($j = 1, ..., n$), the supplier consumes primary input (x) to produce intermediate output (z). Then, the manufacturer uses the output produced by the supplier to create the system's final output (y).

Although in recent years, many researchers evaluated the performance of supply chains using DEA and NDEA (Khalili-Damghani et al. [24], Torabi et al. [55], Shoja et al. [49], Shafiee et al. [47]). In many of these studies, a clear definition of production possibility set (PPS) was not provided to evaluate supply chains. As explained in Chap. 6, Lotfi et al. [31] presented a new PPS to assess the performance of networks under evaluation. Still, due to some unique features in a multi-member supply chain structure, this PPS is unsuitable for evaluating supply chains.

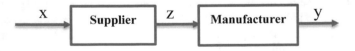

Fig. 9.1 Two-member supply chain

In research conducted by Yang et al. [58], they believe that the performance of all supply chain members and the relationship between members is effective on the chain's performance. Also, they describe viewing that the PPS of the supply chain is characterized not only by the performance of the existing supply chains but also by the displacement of members of other supply chains and the creation of a new combination. In other words, in some cases, a supply chain may be recognized as efficient by DEA models (for example, the CCR model). In contrast, suppliers of this chain may be inefficient compared to the suppliers of other chains. In such a case, this supply chain can improve its performance by replacing an efficient supplier instead of the current supplier. Also, this issue may occur for the manufacturers, too. This means that an efficient supply chain using the CCR model can be dominated by virtual supply chains built simply by inserting new members into the chain from another supply chain without using more powerful technology. Therefore, Yang et al. [58] defined a new PPS for the performance evaluation of supply chains. Also, in addition to accepting the existing principles in DEA, they regarded the following new assumptions:

- Both members of the chain have the same returns to scale.
- A supply chain is detachable. In other words, without considering additional cost, a subsystem can be replaced with a homogeneous subsystem from other supply chains.

Therefore, the supply chain PPS is provided not only by the performance of the existing supply chains but also by the cooperation of the different members of the other supply chains in various situations. Therefore, the definition of a virtual supply chain is theoretically permissible. Thus, the production possibilities set for two-member supply chains under conditions of constant return to scale were presented by Yang et al. [58] as (9.1).

$$
T_{SC}^{CRS} = \left\{ (x, y) \left| \begin{array}{l} \sum_{j=1}^{n} \left(\theta_{Sj}^{*} x_j\right)\lambda_j \leq x, \\[2mm] \sum_{j=1}^{n} z_j \lambda_j \geq z, \\[2mm] \sum_{j=1}^{n} z_j \lambda_j \leq z, \\[2mm] \sum_{j=1}^{n} \left(\dfrac{y_j}{\theta_{Mj}^{*}}\right)\lambda_j \geq y, \quad \lambda_j \geq 0, \quad j = 1, \ldots, n \end{array} \right. \right\} \tag{9.1}
$$

So that, θ_{Sj}^{*} and θ_{Mj}^{*} are the efficiency of supplier and manufacturer in a supply chain, respectively. In this case, they are obtained using the independent input-oriented CCR models. Therefore, to get the efficiency of a supply chain, model (9.2) was proposed by Yang et al. [58] as follows:

$$\min \ \theta$$

s.t.

$$\sum_{j=1}^{n} \lambda_j x_j^* \leq \theta x_o,$$

$$\sum_{j=1}^{n} \lambda_j y_j^* \geq y_o,$$ (9.2)

$$\left(x_j^*, y_j^*\right) \in T_{SC}^{CRS}, \quad j = 1, \ldots, n$$

$$\lambda_j \geq 0, \quad j = 1, \ldots, n.$$

Next, using the relation (9.1), model (9.2) becomes model (9.3).

$$\min \ \theta$$

s.t.

$$\sum_{j=1}^{n} x_j^* \lambda_j \leq \theta x_o, \quad \sum_{j=1}^{n} y_j^* \lambda_j \geq y_o,$$

$$\sum_{j=1}^{n} \left(\theta_{Sj}^* x_j\right) \lambda_j' \leq x_j^*, \quad \sum_{j=1}^{n} z_j \lambda_j' \geq z_j^*,$$ (9.3)

$$\sum_{j=1}^{n} z_j \lambda_j' \leq z_j^*, \quad \sum_{j=1}^{n} \left(\frac{y_j}{\theta_{Mj}^*}\right) \lambda_j' \geq y_j^*,$$

$$\lambda_j, \lambda_j' \geq 0, \quad j = 1, \ldots, n.$$

Although in the method presented by Yang et al. [58], unlike other studies conducted in this field, a new PPS for two-member supply chains has been given, and also by using this method can easily be introduced a benchmark for inefficient supply chains, the model (9.3) is a non-linear programming problem. Since the existing algorithms for solving non-linear models are usually time-consuming and, in some cases, the solution obtained from non-linear algorithms is the local optimal solution and not the global optimal solution. Therefore, this makes the use of non-linear models problematic in some situations. Also, the model (9.3) has been only expressed for the constant scale return scale. Thus, to solve the mentioned problems, Saleh et al. [44] introduced a new PPS, such as (9.4), to evaluate two-member supply chains based on the idea proposed by Yang et al. [58].

$$T_{SC}^{CRS} = \left\{ (x, y) \left| \begin{array}{l} \sum_{j=1}^{n} \left(\theta_{Sj}^* x_j\right) \lambda_j \leq x, \\ \sum_{j=1}^{n} \left(\varphi_{Mj}^* y_j\right) \lambda_j \geq y, \quad \lambda_j \geq 0, \quad j = 1, \ldots, n \end{array} \right. \right\}$$ (9.4)

So that θ_{Sj}^* and φ_{Mj}^* are calculated using the input-oriented CCR model for the supplier and the output-oriented CCR model for the manufacturer, respectively. It is evident that: $\theta_{Sj}^* \leq 1$ and $\varphi_{Mj}^* \geq 1$. Therefore, to calculate the efficiency of the supply chain o according to the PPS T_{SC}^{CRS}, model (9.5) is introduced to evaluate the performance of the two-member supply chain:

$$\theta_N^* = \min \ \theta_N$$
$$s.t. \tag{9.5}$$
$$(\theta_N x_o, y_o) \in T_{SC}^{CRS}$$

Using the definition of T_{SC}^{CRS}, model (9.5) is converted into the model (9.6).

$$\min \theta$$
$$s.t.$$
$$\sum_{j=1}^{N} \lambda_j \left(\theta_{Sj}^* x_j \right) \leq \theta x_o,$$
$$\sum_{j=1}^{N} \lambda_j \left(\varphi_{Mj}^* y_j \right) \geq y_o, \tag{9.6}$$
$$\lambda_j \geq 0, \quad j = 1, ..., n.$$

Now suppose T_{BB}^{CRS} is a PPS for supply chains without taking into account intermediate links (black box), so T_{BB}^{CRS} is expressed as follows:

$$T_{BB}^{CRS} = \left\{ (x, y) \middle| \sum_{j=1}^{n} x_j \lambda_j \leq x, \sum_{j=1}^{n} y_j \lambda_j \geq y, \lambda_j \geq 0, j = 1, \dots, n \right\} \tag{9.7}$$

Therefore, to evaluate the supply chain o without considering intermediate links, the model (9.8) is applied.

$$\theta_{BB}^* = \min \ \theta_{BB}$$
$$s.t. \tag{9.8}$$
$$(\theta_{BB} x_o, y_o) \in T_{BB}^{CRS}$$

Theorem 9.1 $T_{BB}^{CRS} \subseteq T_{SC}^{CRS}$.

Proof Suppose that $(x_o, y_o) \in T_{BB}^{CRS}$. Then, using the definition of T_{BB}^{CRS}, it holds that $\sum_j x_j \lambda_j \leq \theta x_o$ and $\sum_j y_{qj} \lambda_j \geq y_{qo}$. Since $\theta_{Sj}^* \leq 1$ and $\varphi_{Mj}^* \geq 1$, one can easily deduce that $\sum_j \left(\theta_{Sj}^* x_j \right) \lambda_j \leq \sum_j x_j \lambda_j \leq \theta x_o$ and $\sum_j \left(\varphi_{Mj}^* y_j \right) \lambda_j \geq \sum_j y_j \lambda_j \geq y_o$ and hence, $(x_o, y_o) \in T_{SC}^{CRS}$. This concludes the proof.

Corollary 9.1 The efficiency of the supply chain o in the network technology is no more than that in the black-box technology, i.e., $\theta_N^* \leq \theta_{BB}^*$.

Proof The proof is like that of Theorem 9.1.

Now consider model (9.9):

$$\min \theta$$

$$\sum_{j=1}^{n} x_j^* \lambda_j \leq \theta x_o,$$

$$\sum_{j=1}^{n} y_j^* \lambda_j \geq y_o,$$

$$\sum_{j=1}^{n} \left(\theta_{\text{Sj}}^* x_j \right) \lambda_j' \leq x_j^*, \ j = 1, ..., n, \quad (9.9)$$

$$\sum_{j=1}^{n} \left(\varphi_{\text{Mj}}^* y_j \right) \lambda_j' \geq y_j^*, \ j = 1, ..., n,$$

$$\lambda_j, \lambda_j' \geq 0, \ j = 1, ..., n.$$

Lemma 9.1 The optimal objective function values of models (9.3) and (9.9) are equal.

Proof Let $\left(\hat{\theta}, \hat{\lambda}, \hat{\lambda}', \hat{x}^*, \hat{y}^* \right)$ and $\left(\tilde{\theta}, \tilde{\lambda}, \tilde{\lambda}', \tilde{x}^*, \tilde{y}^*, \tilde{z}^* \right)$ be the optimal solution vector of models (9.9) and (9.3), respectively. It is evident that the vector $\left(\tilde{\theta}, \tilde{\lambda}, \tilde{\lambda}', \tilde{x}^*, \tilde{y}^* \right)$ is also a feasible solution for the model (9.9), and hence $\tilde{\theta} \geq \hat{\theta}$. Similarly, one can prove that $\left(\hat{\theta}, \hat{\lambda}, \hat{\lambda}', \hat{x}^*, \hat{y}^*, \hat{z}^* \left(= \sum_j z_j \hat{\lambda}_j' \right) \right)$ is a feasible solution vector of the model (9.3), as a result $\hat{\theta} \geq \tilde{\theta}$. Consequently, $\hat{\theta} = \tilde{\theta}$, which completes the proof.

Now consider the inequality $\sum_{j=1}^{n} \left(\theta_{\text{Sj}}^* x_j \right) \lambda_j' \leq x_j^*, \ j = 1, ..., n$ in the model (9.3). By multiplying the sides of this inequality by $\lambda_{j'}$ and then, by adding these n new inequalities, we have the obtained inequality on the index j':

$$\sum_{j'=1}^{n} \sum_{j=1}^{n} \theta_{\text{Sj}}^* x_j \lambda_j' \lambda_{j'} \leq \sum_{j'=1}^{n} x_j^* \lambda_{j'} \quad (9.10)$$

We know that: $\sum_{j'=1}^{n} x_j^* \lambda_{j'} \leq \theta x_o$. So: $\sum_{j'=1}^{N} \sum_{j=1}^{N} \theta_{\text{Sj}}^* x_j \lambda_j' \lambda_{j'} \leq \theta x_o$. Similarly, it can be proved that $\sum_{j'=1}^{n} \sum_{j=1}^{n} y_j \big/ \theta_{\text{Mj}}^* y_j \lambda_j' \lambda_{j'} \geq y_o$. It is evident that:

$$\sum_{j'=1}^{n} \sum_{j=1}^{n} \left(\theta_{\text{Sj}}^* x_j \lambda_j' \right) \lambda_{j'} = \sum_{j=1}^{n} \theta_{\text{Sj}}^* x_j \lambda_j' \left(\sum_{j'=1}^{n} \lambda_{j'} \right) \quad (9.11)$$

And also:

$$\sum_{j'=1}^{n} \sum_{j=1}^{n} \frac{y_j}{\theta_{\text{Mj}}^*} \lambda_j' \lambda_{j'} = \sum_{j=1}^{n} \frac{y_j}{\theta_{\text{Mj}}^*} y_j \lambda_j' \left(\sum_{j=1}^{n} \lambda_j \right) \quad (9.12)$$

Therefore, by defining $\mu_j = \lambda'_j\left(\sum_{j'=1}^{N} \lambda_{j'}\right)$ and placing relations (9.11) and (9.12) in the model (9.3), model (9.13) is obtained as follows:

$$\min \theta$$
$$s.t.$$
$$\sum_{j=1}^{n} \mu_j\left(\theta_{Sj}^* x_j\right) \leq \theta x_o,$$
$$\sum_{j=1}^{n} \mu_j\left(\frac{y_j}{\theta_{Mj}^*}\right) \geq y_o,$$
$$\mu_j \geq 0, \quad j = 1, ..., n.$$

(9.13)

As you know, in the CCR model $\varphi_{Mj}^* = 1/\theta_{Sj}^*$, the model (9.13) becomes as follows.

$$\min \theta$$
$$s.t.$$
$$\sum_{j=1}^{N} \mu_j\left(\theta_{Sj}^* x_j\right) \leq \theta x_o,$$
$$\sum_{j=1}^{N} \mu_j\left(\varphi_{Mj}^* y_j\right) \geq y_o,$$
$$\mu_j \geq 0, \quad j = 1, ..., n$$

(9.14)

Based on the process mentioned above, we state Theorem 9.2.

Theorem 9.2 The optimal value of the objective function of models (9.9) and (9.14) are equal.

Proof Suppose $\left(\tilde{\theta}, \tilde{\lambda}, \tilde{\lambda}', \tilde{x}^*, \tilde{y}^*\right)$ and (θ^*, μ^*) are the optimal solution of models (9.9) and (9.14), respectively. Therefore, according to the mentioned process, $\left(\tilde{\theta}, \tilde{\lambda}' \sum_{j=1}^{n} \tilde{\lambda}_j\right)$ is a feasible solution for the model (9.14). As a result: $\theta^* \leq \tilde{\theta}$. On the other hand, it can be easily shown that $\left(\theta, \lambda_j, \lambda'_j, x_{pj}^*, y_{pj}^*\right) = \left(\theta^*, 0, \mu_j^*, x_{pj}, 0\right)$ is a feasible solution for the model (9.9). Thus $\tilde{\theta} \leq \theta^*$. Consequently $\tilde{\theta} = \theta^*$.

Using Theorem 9.2 and Lemma 9.1, conclude that the optimal value of the objective function of models (9.3) and (9.14) are equal. So model (9.14) can be used instead of model (9.3).

Model (9.3) is non-linear, while model (9.14) is linear. Since approximate algorithms or heuristic methods are usually used to obtain the optimal solution in non-linear models, the optimal solution obtained by these algorithms usually has a calculation error. On the other hand, existing algorithms for solving linear models usually get local optimal solutions. Also, the number of constraints and variables is one of the crucial factors in the modeling process. Because the more the number of constraints in a problem increases, the computational complexity of the algorithms to solve the problem increases. Therefore, if the number of DMUs under evaluation or the number of inputs and outputs of the units under assessment are large, the time and, as a result, the cost of performing calculations will increase significantly. Obviously, the number of constraints of the model (9.14) is less than model (9.3). Another advantage of using model (9.14) is that, unlike model (9.3), it can be easily extended for VRS mode. The PPS $T_{\text{SC}}^{\text{VRS}}$ is introduced based on the idea presented in the previous subsection and by removing the principle of constant return to scale from $T_{\text{SC}}^{\text{CRS}}$.

$$T_{\text{SC}}^{\text{VRS}} = \left\{ (x, y) \middle| \begin{array}{l} \sum_{j=1}^{n} \left(\theta_{\text{Sj}}^{*} x_j \right) \lambda_j \leq x, \\[2mm] \sum_{j=1}^{n} \left(\varphi_{\text{Mj}}^{*} y_j \right) \lambda_j \geq y, \\[2mm] \sum_{j=1}^{n} \lambda_j = 1, \quad \lambda_j \geq 0, \quad j = 1, ..., n \end{array} \right\} \tag{9.15}$$

So that, θ_{Sj}^{*} and φ_{Mj}^{*} are obtained using the input-oriented and out-oriented BCC model for supplier and manufacturer, respectively. The geometric interpretation is shown in Fig. 9.2.

The PPS on the left shows the PPS of the supplier, the PPS on the right shows the PPS of the manufacturer. Finally, the PPS at the bottom of Fig. 9.2 illustrates the PPS of the supply chain without considering the constant return to scale principle.

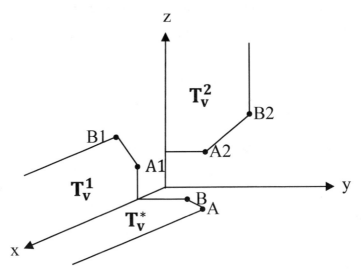

Fig. 9.2 Geometric interpretation of $T_{\text{SC}}^{\text{VRS}}$

Using the definition T_{SC}^{VRS}, the model (9.16) is expressed as follows.

$$\min \theta$$

$$s.t.$$

$$\sum_{j=1}^{N} \lambda_j \left(\theta_{Sj}^* x_{ij} \right) \leq \theta x_{io}, \quad i = 1, ..., m$$

$$\sum_{j=1}^{N} \lambda_j \left(\varphi_{Mj}^* y_{Rj} \right) \geq y_{ro}, \quad r = 1, ..., s \tag{9.16}$$

$$\sum_{j=1}^{N} \lambda_j = 1,$$

$$\lambda_j \geq 0, \quad j = 1, ..., n.$$

Dual of the model (9.16) is expressed as (9.17):

$$\phi_N^* = \max \sum_{r=1}^{s} u_r y_{ro} + \omega_o$$

$$s.t.$$

$$\sum_{i=1}^{m} v_i x_{io} = 1, \tag{9.17}$$

$$\sum_{r=1}^{s} u_r \left(\varphi_j^* y_{rj} \right) - \sum_{i=1}^{m} v_i \left(\theta_j^* x_{ij} \right) + \omega_o \leq 0, \quad j = 1, ..., n$$

$$u_r, v_i \geq 0, \quad i = 1, ..., m, \quad r = 1, ..., s.$$

Similar to the models presented in the evaluation of single-stage DMUs (black box), the second phase of the model (9.17) is expressed as follows to calculate the combined inefficiency of the units under the assessment.

$$\max \sum_{i=1}^{m} s_i^- + \sum_{r=1}^{s} s_r^+$$

$$s.t.$$

$$\sum_{j=1}^{N} \left(\theta_{Sj}^* x_{ij} \right) \lambda_j + s_i^- = \phi_N^* x_{io}, \quad i = 1, ..., m,$$

$$\sum_{j=1}^{N} \left(\varphi_{Mj}^* y_{rj} \right) \lambda_j - s_r^+ = y_{ro}, \quad r = 1, ..., s, \tag{9.18}$$

$$\sum_{j=1}^{N} \lambda_j = 1,$$

$$\lambda_j \geq 0, \quad j = 1, ..., n,$$

$$s_i^-, s_r^+ \geq 0, \quad i = 1, ..., m, \quad r = 1, ..., s.$$

Definition 9.1 The projection of the network o is defined as follows:

$$\hat{x}_{io} = \phi_N^* x_{io} - s_i^{-*} \tag{9.19}$$

$$\hat{y}_{ro} = y_{ro} + s_r^{+*} \tag{9.20}$$

So that ϕ_N^* is the optimal solution of model (9.17) and s_i^{-*} and s_r^{+*} are the optimal solution of the model (9.18).

Theorem 9.3 The projection of the network o $\left(\hat{x}_o, \hat{y}_o\right)$ is efficient.

Proof To evaluate $\left(\hat{x}_o, \hat{y}_o\right)$, the model (9.21) is used.

$$\min \hat{\theta}$$

$$s.t.$$

$$\sum_{j=1}^{N} \left(\theta_{Sj}^* x_{ij}\right) \hat{\lambda}_j + \hat{s}_i^- = \hat{\theta} \hat{x}_{io}, \quad i = 1, \dots, m,$$

$$\sum_{j=1}^{N} \left(\varphi_{Mj}^* y_{rj}\right) \hat{\lambda}_j - \hat{s}_r^+ = \hat{y}_{ro}, \quad r = 1, \dots, s, \tag{9.21}$$

$$\sum_{j=1}^{N} \hat{\lambda}_j = 1,$$

$$\lambda_j \geq 0, \quad j = 1, \dots, n,$$

$$\hat{s}_i^-, \hat{s}_r^+ \geq 0, \quad i = 1, \dots, m, \quad r = 1, \dots, s.$$

Using relations (9.19) and (9.20), we have:

$\sum_{j=1}^{N} \left(\theta_{Sj}^* x_{ij}\right) \hat{\lambda}_j = \left(\hat{\theta} \phi_N^*\right) x_{io} - \left(\hat{\theta} s_i^{-*} + \hat{s}_i^-\right), i = 1, \dots, m$

and $\sum_{j=1}^{N} \left(\varphi_{Mj}^* y_{rj}\right) \hat{\lambda}_j = y_{ro} + \left(s_r^{+*} + \hat{s}_r^+\right), r = 1, \dots, s.$ Also, we define:

$\tilde{\theta} = \hat{\theta} \phi_N^*, \tilde{s}_r^+ = s_r^{+*} + \hat{s}_r^+$ and $\tilde{s}_i^- = \hat{\theta} s_i^{-*} + \hat{s}_i^-.$ It is evident that $\left(\hat{\lambda}, \tilde{\theta}, \tilde{s}^-, \tilde{s}^+\right)$ is a feasible solution for the model (9.16). If $\hat{\theta} < 1$, then $\tilde{\theta} = \hat{\theta} \theta^* < \theta^*$ and this is a contradiction. So $\hat{\theta} = 1$. As a result: $\sum_{i=1}^{m} \tilde{s}_i^- + \sum_{r=1}^{s} \tilde{s}_r^+ = \left(\sum_{i=1}^{m} \hat{s}_i^- + \sum_{r=1}^{s} \hat{s}_r^+\right) + \left(\sum_{i=1}^{m} s_i^{-*} + \sum_{r=1}^{s} s_r^{+*}\right).$ If $\sum_{i=1}^{m} \hat{s}_i^- + \sum_{r=1}^{s} \hat{s}_r^+ = 0$, then the proof is complete. Otherwise, $\sum_{i=1}^{m} \tilde{s}_i^- + \sum_{r=1}^{s} \tilde{s}_r^+ > \sum_{i=1}^{m} s_i^{-*} + \sum_{r=1}^{s} s_r^{+*}.$ This is a contradiction.

9.2.2 Performance Evaluation of Centralized and Decentralized Supply Chains

Supply chains have different types in terms of size, internal interactions, style of leadership, and decisions made by chain managers, which multiplies the complexity

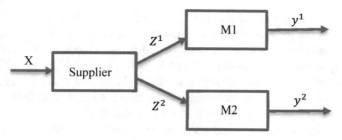

Fig. 9.3 Supplier–manufacturer supply chain

of chain management. Therefore, Chen and Yan [9] first raised two critical questions to evaluate supply chains as follows:

- What is the definition of supply chain efficiency?
- How should the internal interactions of the supply chain be considered?

To answer the first question, Chen et al. used Pareto-Koopman's definition of efficiency considering intermediate products. So that, a supply chain should be regarded as an efficient chain if and only if none of the initial inputs or final outputs improves without worsening at least one of the other inputs or outputs of the chain. Also, in response to the second question, the point of view of the Organization mechanism in the chain is used. So to evaluate the chain's performance, in addition to considering the multi-member structures of the chain and the type of indicators used, the kind of leadership and management should also be considered.

To better understand this issue, consider the supply chain shown in Fig. 9.3. If all chain members are guided by a decision-maker and the manager has permission to access the information of all departments, this supply chain is named the centralized chain. In other words, in a centralized supply chain, a manager supervises the performance of the entire chain.

For example, consider the supply chain of Iran's wood and paper industry (Chouka). This factory has forests in its possession to produce the required wood. The resulting wood is first cut in different sizes and thicknesses in the sawmills under the supervision of the main factory. The timber is sent to the main factory to produce the final products, and finally, the products made are sent to sales agents.

But if each member of the chain has its strategy and the strategy of each part is considered separate from the other part, then this chain is decentralized. Among the types of decentralized supply, we can mention the bread production supply chain. In this chain, wheat is first purchased from the farmer, and then the purchased grain is converted into flour. The obtained flour is distributed among the bakeries, and finally, the bread is provided to the customer. In this supply chain, the farmer, the flour factory, the baker, and the final customer are independent.

Of course, according to the development of global supply chains, it is rarely seen that a supply chain is only affected by centralized or decentralized management. In other words, some parts of the supply chain are under the leadership of a decision-maker, and other parts operate independently. To better understand this issue, consider the supply chain of sending parts to automobile factories. To better understand this issue, consider the parts supply chain to automobile factories. To supply the parts, automobile factories produce some of them in their workshops under their supervision and buy other parts from other domestic or foreign parts factories. Therefore, some of the members of this supply chain are under the control of single management, while others operate independently and only try to make more profit for themselves. In such a situation, the supply chain is under mixed control.

Therefore, the modeling process should be considered according to the chain management type for each approach. In the following, we will describe each type of model used for every kind of viewpoint.

Without losing the generality of the matter, consider a two-member supply chain, including the supplier (S) and the manufacturer (M), as shown in Fig. 9.3. So that x is the system's initial input, and z^1 and z^2 are the intermediate products of the chain. And finally, y^1 and y^2 are taken as the manufacturers' products and m and n as the final outputs of the system.

As you know, global supply chains' main goal is to meet consumer demand. Therefore, to achieve this goal, decision-makers in different parts of the chain prioritize reducing the inputs used for a fixed level of demand. As a result, input-oriented DEA models are used in the modeling proposed in this sub-section. But the critical point is that to achieve this goal, how should intermediate products be considered in classical DEA models?

In the centralized supply chain approach, since the entire chain is under the decision of single management, the activities of the producer and supplier are aimed at increasing the system's efficiency. Therefore, the only thing that should be emphasized is that for intermediate products, it should be noted that the products sent by the supplier should not be less than the amount of the manufacturer's demand. As a result, PPS for centralized chains is expressed as (9.22).

$$
T_{\text{central}} = \left\{ \left(x, y^1, y^2 \right) \left|
\begin{array}{l}
\sum\limits_{j=1}^{n} x_j \lambda_j^1 \leq x, \\[2mm]
\sum\limits_{j=1}^{n} \lambda_j^1 z_j^1 \geq \sum\limits_{j=1}^{n} \lambda_j^2 z_j^1, \\[2mm]
\sum\limits_{j=1}^{n} \lambda_j^1 z_j^2 \geq \sum\limits_{j=1}^{n} \lambda_j^3 z_j^2, \\[2mm]
\sum\limits_{j=1}^{n} \lambda_j^2 y_j^1 \geq y^1, \\[2mm]
\sum\limits_{j=1}^{n} \lambda_j^3 y_j^2 \geq y^2, \quad \lambda_j^1, \lambda_j^2, \lambda_j^3 \geq 0, \quad j = 1, ..., n
\end{array}
\right. \right\} \tag{9.22}
$$

Suppose that the centralized supply chain o is under evaluation, so it is used to evaluate the supply chain's performance and calculate the model's technical efficiency (9.23).

$$\min \ \theta_{\text{central}}$$
$$s.t.$$
$$\left(\theta_{\text{central}} x_o, z^1, z^2, y^1, y^2 \right) \in T_{\text{central}}$$

(9.23)

Therefore, using T_{central}, model (9.23) converts to (9.24).

$$\min \ \theta_{\text{central}}$$
$$s.t.$$
$$\sum_{j=1}^{n} x_j \lambda_j^1 \leq \theta_{\text{central}} x_o,$$
$$\sum_{j=1}^{n} \lambda_j^1 z_j^1 \geq \sum_{j=1}^{n} \lambda_j^2 z_j^1, \ \sum_{j=1}^{n} \lambda_j^1 z_j^2 \geq \sum_{j=1}^{n} \lambda_j^3 z_j^2,$$
$$\sum_{j=1}^{n} \lambda_j^2 y_j^1 \geq y_o^1, \ \sum_{j=1}^{n} \lambda_j^3 y_j^2 \geq y_o^2,$$
$$\lambda_j^1, \lambda_j^2, \lambda_j^3 \geq 0, \ j = 1, ..., n.$$

(9.24)

So it is evident that a centralized supply chain o is weakly efficient if and only if $\theta_{\text{central}}^* = 1$.

Unlike centralized supply chains, in decentralized supply chains, due to the absence of integrated leadership, each component of the chain independently tries to achieve its interests. Therefore, the current input of suppliers (i.e. z_o^1 and z_o^2) should not be greater than the current level available to them. Otherwise, these suppliers will not agree to the continuation of this activity. This way, the PPS of the decentralized supply chain is introduced as (9.25).

$$T_{\text{Decentral}} = \left\{ \left(x, y^1, y^2 \right) \left| \begin{array}{l} \sum_{j=1}^{n} x_j \lambda_j^1 \leq x, \\[2mm] \sum_{j=1}^{n} \lambda_j^1 z_j^1 \geq \sum_{j=1}^{n} \lambda_j^2 z_j^1, \\[2mm] \sum_{j=1}^{n} \lambda_j^1 z_j^2 \geq \sum_{j=1}^{n} \lambda_j^3 z_j^2, \\[2mm] \sum_{j=1}^{n} \lambda_j^2 z_j^1 \leq z^1, \\[2mm] \sum_{j=1}^{n} \lambda_j^3 z_j^2 \leq z^2 \\[2mm] \sum_{j=1}^{n} \lambda_j^2 y_j^1 \geq y^1, \\[2mm] \sum_{j=1}^{n} \lambda_j^3 y_j^2 \geq y^2, \ \lambda_j^1, \lambda_j^1, \lambda_j^1 \geq 0, \ j = 1, ..., n \end{array} \right. \right\}$$

(9.25)

In this way, with a process similar to the method mentioned earlier for centralized supply chains, the model (9.26) was proposed to calculate the efficiency score of each decentralized supply chain and separate efficient and inefficient chains. Constraints fourth and sixth show the selfish and self-interested aspects of the chain members.

$$\min \ \theta_{\text{decentral}}$$

$$s.t.$$

$$\sum_{j=1}^{n} x_j \lambda_j^1 \leq \theta_{\text{decentral}} x_o,$$

$$\sum_{j=1}^{n} \lambda_j^1 z_j^1 \geq \sum_{j=1}^{n} \lambda_j^2 z_j^1, \ \sum_{j=1}^{n} \lambda_j^1 z_j^2 \geq \sum_{j=1}^{n} \lambda_j^3 z_j^2, \tag{9.26}$$

$$\sum_{j=1}^{n} \lambda_j^2 z_j^1 \leq z_o^1, \ \sum_{j=1}^{n} \lambda_j^2 y_j^1 \geq y_o^1,$$

$$\sum_{j=1}^{n} \lambda_j^3 z_j^2 \leq z_o^2, \ \sum_{j=1}^{n} \lambda_j^3 y_j^2 \geq y_o^2,$$

$$\lambda_j^1, \lambda_j^2, \lambda_j^3 \geq 0, \ j = 1, ..., n.$$

As decentralized supply chain o is called weak efficient if $\theta_{\text{decentral}}^* = 1$.

Now suppose supplier S and manufacturer M^2 are under the leadership of a decision maker. Still, the manufacturer M^1 acts independently, so the chain under evaluation is a mixed supply chain. As a result, the production possibility set under the mixed strategy is introduced as (9.27).

$$T_{\text{mix}}$$

$$= \left\{ (x, y^1, y^2) \left| \begin{array}{l} \sum_{j=1}^{n} x_j \lambda_j^1 \leq x, \\[2mm] \sum_{j=1}^{n} \lambda_j^1 z_j^1 \geq \sum_{j=1}^{n} \lambda_j^2 z_j^1, \\[2mm] \sum_{j=1}^{n} \lambda_j^1 z_j^2 \geq \sum_{j=1}^{n} \lambda_j^3 z_j^2, \\[2mm] \sum_{j=1}^{n} \lambda_j^2 z_j^1 \leq z^1, \\[2mm] \sum_{j=1}^{n} \lambda_j^2 y_j^1 \geq y^1, \ \sum_{j=1}^{n} \lambda_j^3 y_j^2 \geq y^2, \\[2mm] \lambda_j^1, \lambda_j^1, \lambda_j^1 \geq 0, \ j = 1, ..., n \end{array} \right. \right\} \tag{9.27}$$

Therefore, using the definition T_{mix}, model (9.28) was introduced to evaluate the performance of mixed supply chains as follows.

$$\min \ \theta_{\text{mix}}$$

$$s.t.$$

$$\sum_{j=1}^{n} x_j \lambda_j^1 \leq \theta_{\text{mix}} x_o,$$

$$\sum_{j=1}^{n} \lambda_j^1 z_j^1 \geq \sum_{j=1}^{n} \lambda_j^2 z_j^1, \ \sum_{j=1}^{n} \lambda_j^1 z_j^2 \geq \sum_{j=1}^{n} \lambda_j^3 z_j^2, \tag{9.28}$$

$$\sum_{j=1}^{n} \lambda_j^2 z_j^1 \leq z_o^1, \ \sum_{j=1}^{n} \lambda_j^2 y_j^1 \geq y_o^1,$$

$$\sum_{j=1}^{n} \lambda_j^3 y_j^2 \geq y_o^2,$$

$$\lambda_j^1, \lambda_j^2, \lambda_j^3 \geq 0, \ j = 1, ..., n.$$

Thus mixed supply chain is called weak efficient if $\theta_{\text{mix}}^* = 1$.

In most industries, the manufactured product has different parts that must be supplied through other suppliers. In some cases, lack of proper management, and low quality or defects in parts, cause an imbalance between supply and demand. So raw materials in the buyer's warehouses or the seller are stashed. As a result, this imbalance causes waste of internal resources and creates waste. Therefore, there is a need to reconfigure the chain so that the producer's demand is still met by reducing the consumption resources in the supply chain. Also, from the point of view of performance evaluation, it is clear that the appropriate reduction of intermediate products does not harm the efficiency level of the chain. Therefore, to calculate the amount of waste of intermediate resources in the decentralized chain, model (9.29) is introduced.

$$\min \ \theta_{\text{decentral}} - \varepsilon \begin{pmatrix} e_{1+}^T s_{o1}^+ \\ +e_{11-}^T s_{o1}^{1-} \\ +e_{12-}^T s_{o1}^{2-} \\ +e_{2+}^T s_{o2}^+ \\ +e_{2-}^T s_{o2}^- \\ +e_{3+}^T s_{o3}^+ \\ +e_{3-}^T s_{o3}^- \end{pmatrix}$$

$s.t.$

$$\sum_{j=1}^{n} x_j \lambda_j^1 + s_{o1}^+ = \theta_{\text{decentral}} x_o,$$

$$\sum_{j=1}^{n} \lambda_j^1 z_j^1 - s_{o1}^{1-} = \sum_{j=1}^{n} \lambda_j^2 z_j^1,$$

$$\sum_{j=1}^{n} \lambda_j^1 z_j^2 - s_{o1}^{2-} = \sum_{j=1}^{n} \lambda_j^3 z_j^2,$$ (9.29)

$$\sum_{j=1}^{n} \lambda_j^2 z_j^1 + s_{o2}^+ = z_o^1,$$

$$\sum_{j=1}^{n} \lambda_j^2 y_j^1 - s_{o2}^- = y_o^1,$$

$$\sum_{j=1}^{n} \lambda_j^3 z_j^2 + s_{o3}^+ = z_o^2,$$

$$\sum_{j=1}^{n} \lambda_j^3 y_j^2 - s_{o3}^- = y_o^2,$$

$$\left(s_{o1}^+, s_{o1}^{1-}, s_{o1}^{2-}, s_{o2}^+, s_{o2}^-, s_{o3}^+, s_{o3}^-\right) \geq 0,$$

$$\lambda_j^1, \lambda_j^2, \lambda_j^3 \geq 0, \quad j = 1, ..., n.$$

So that $s_{o1}^+, s_{o1}^{1-}, s_{o1}^{2-}, s_{o2}^+, s_{o2}^-, s_{o3}^+, s_{o3}^-$ are the slack and surplus variables for the first to seventh constraints, and e is a vector whose all components are equal to 1. The number of elements of e is similar to the number of elements of its coefficient vector.

Thus it is evident that a supply chain is called input efficient using model (9.29) if $\theta_{\text{decentral}}^* = 1$ and $s_{o1}^{+*} = 1$.

In the following, we explain the waste of resources in the chain from two perspectives. In the first case, the loss of intermediate resources occurs in the current supply chain. Still, the loss of intermediate resources in the effective supply chain is explained in the second case. In the following, we describe the waste of resources in the chain from two perspectives. In the first case, the loss of intermediate resources occurs in the current supply chain. Also, the loss of intermediate resources in the efficient supply chain is explained in the second case.

A: Waste of Intermediate Resources in the Current Supply Chain

Consider Fig. 9.4. By consuming x, the supplier s produces two intermediate products (z_o^1 and z_o^2). In the next, to manufacture the final products (y_o^1 and y_o^2), z_o^1 and z_o^2 are entirely consumed by manufacturers 1 and 2.

While the two producers, $M1$ and $M2$, with the minimum amount of $\sum_{j=1}^{n} \lambda_j^* z_o^1$ and $\sum_{j=1}^{n} \lambda_j^* z_o^2$, can also produce final products in the amount of y_o^1 and y_o^2. Therefore, the amount of waste from $M1$ products is equal to $z_o^1 - \sum_{j=1}^{n} \lambda_j^{2*} z_o^1 = s_{1o}^{+*} \geq 0$. If $s_{1o}^{+*} = 0$, then there is no waste of resources to produce this product. But if $s_{1o}^{+*} > 0$, then due to the lack of balance in the amount of demand and the amount of production, s_{1o}^{+*} indicates the amount of product wastage. In the same way, s_{2o}^{+*} can also be interpreted.

B: Waste of Intermediate Resources in Efficient Supply Chain

To check the amount of product wastage inefficient supply chain, consider Fig. 9.5. Similar to the previous case, s_{1o}^{+*} and s_{2o}^{+*} indicate the amount of product wastage.

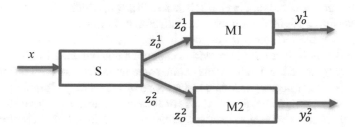

Fig. 9.4 Current supply chain

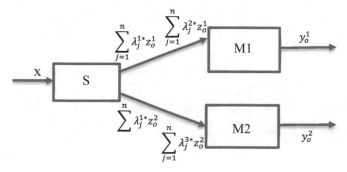

Fig. 9.5 Input efficient supply chain

Suppose that $\sum_{j=1}^{n} \lambda_j^{1*} z_o^1$ is the desired output of suppliers after becoming an efficient chain and $\sum_{j=1}^{n} \lambda_j^{2*} z_o^1$ is the amount of demand required by $M1$ to produce output y_o^1. Therefore, if $\sum_{j=1}^{n} \lambda_j^{1*} z_o^1 - \sum_{j=1}^{n} \lambda_j^{2*} z_o^1 = s_{1o}^{+*} > 0$, there is product waste in the efficient supply chain.

9.3   Performance Evaluation of Supply Chains Using Non-radial Models in NDEA

The NDEA models presented in the previous sub-section for evaluating supply chains align with the idea used for radial DEA models. They are based on the assumption that all indicators (input–output) change proportionally. In other words, all inputs (outputs) decrease (increase) at the same rate. While in some evaluations, it is not reasonable to expect the same and proportionate changes for all inputs (outputs). As an example, suppose that in a two-stage supply chain, in the first stage, using labor and capital, a commodity is produced and sent to the second stage. In the second stage, the producer manufacture and sells the final product using his independent workforce. Therefore, the proportional reduction of all inputs means that the percentage of the changes in labor and capital should be considered the same. This does not seem logical. Now consider the supply chain shown in Fig. 9.6, which relates to the supply chain of consumer medical equipment manufacturers in Iran.

By checking the indicators introduced in Fig. 9.6, you will quickly understand that unlike the indicators used in previous supply chains, where all indicators have values greater than or equal to 0, in this chain, some indicators are only positive, and some are positive or negative. Consider the "Rate of increasing of number of green products" as an example. Positive values for these indicators mean that the supply chain has spent more money to produce green products than before, and negative values indicate that the supply chain has paid less to produce green products than before. In the same way, other indicators can also be interpreted.

In DEA models, all indicators are considered positive. But this assumption is not necessarily valid in many practical applications. Therefore, based on the idea proposed by [7] and the development of the model (6.101), Izidikhah and Sean [54]

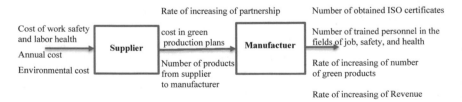

Fig. 9.6 The supply chain of consumable medical equipment

introduced the non-radial model (9.30) to calculate the efficiency score of the two-stage supply chains, as shown in Fig. 9.6. This model can be generalized to evaluate two-member supply chains with negative data.

$$\min \; w_1\left(\frac{1}{m}\sum_{i=1}^{m}\frac{\theta_i}{\bar{\theta}_o}+\frac{1}{D}\sum_{d=1}^{D}\frac{\varphi_d}{\bar{\varphi}_o}\right)+w_2\left(\frac{1}{D}\sum_{d=1}^{D}\frac{\alpha_d}{\bar{\alpha}_o}+\frac{1}{s}\sum_{r=1}^{s}\frac{\beta_r}{\bar{\beta}_o}\right)$$

s.t.

$$\sum_{j=1}^{n}\lambda_j x_{ij}\le x_{io}-\theta_i|x_{io}|, \; i=1,...,m,$$

$$\sum_{j=1}^{n}\lambda_j z_{dj}\ge \tilde{z}_{do}+\varphi_d|\tilde{z}_{do}|, \; d=1,...,D,$$

$$\sum_{j=1}^{n}\mu_j z_{dj}\le \tilde{z}_{do}-\alpha_d|\tilde{z}_{do}|, \; d=1,...,D,$$

$$\sum_{j=1}^{n}\mu_j y_{rj}\ge y_{ro}+\beta_r|y_{ro}|, \; r=1,...,s,$$

$$\sum_{j=1}^{n}\lambda_j=1, \; \sum_{j=1}^{n}\mu_j=1,$$

$$\lambda_j,\mu_j\ge 0, \; j=1,...,n$$

(9.30)

So that $\bar{\theta}_o,\bar{\alpha}_o,\bar{\varphi}_o,\bar{\beta}_o$ are obtained using relations (9.31–9.34).

$$\bar{\theta}_o=\max\left\{\frac{x_{io}-x_{iI}}{|x_{io}|}|x_{io}\ne 0, \; i=1,...,m\right\} \quad (9.31)$$

$$\bar{\alpha}_o=\max\left\{\frac{z_{do}-z_{dI'}}{|z_{do}|}|z_{do}\ne 0, \; d=1,...,D\right\} \quad (9.32)$$

$$\bar{\varphi}_o=\max\left\{\frac{z_{dI}-z_{do}}{|z_{do}|}|z_{do}\ne 0, \; d=1,...,D\right\} \quad (9.33)$$

$$\bar{\beta}_o=\max\left\{\frac{y_{rI'}-y_{ro}}{|y_{ro}|}|y_{ro}\ne 0, \; r=1,...,s\right\} \quad (9.34)$$

Also: $x_{iI}=\min\{x_{ij}|j=1,...,n\}$, $z_{dI'}=\min\{z_{dj}|j=1,...,n\}$, $z_{dI}=\max\{z_{dj}|j=1,...,n\}$, $y_{rI'}=\max\{y_{rj}|j=1,...,n\}$.

As explained in the previous chapters, the weights used in DEA models indicate the importance and priority of each indicator over the others. Similarly, in the model (9.30), w_1 and w_2 represent the preferences of stages 1 and 2 over each other, which the manager specifies, and of course, using methods such as pairwise comparisons and eigenvalue theory presented by Saaty [34] or using techniques such as point allocation, paired comparisons, trade-off analysis, suitable values can be calculated (Ebrahimnejad et al. [47]; Kleindorfer et al. [8]). Of course, it should be noted that $w_1 + w_2 = 1$. Also, if stages 1 and 2 have the same importance, in this case, w_1 and w_2 will have the same values.

Suppose that $\theta_i^*, \alpha_d^*, \varphi_d^*, \beta_r^*$ are the optimal solutions obtained from model (9.30). Thus, the overall efficiency of the chain is obtained using Eq. (9.35).

$$\Theta^* = 1 - \left[\begin{array}{c} \dfrac{w_1}{2}\left(\dfrac{1}{m}\sum_{i=1}^{m}\dfrac{\theta_i^*}{\bar{\theta}_o} + \dfrac{1}{D}\sum_{d=1}^{D}\dfrac{\varphi_d^*}{\bar{\varphi}_o}\right) \\[4mm] + \dfrac{w_2}{2}\left(\dfrac{1}{D}\sum_{d=1}^{D}\dfrac{\alpha_d^*}{\bar{\alpha}_o} + \dfrac{1}{s}\sum_{r=1}^{s}\dfrac{\beta_r^*}{\bar{\beta}_o}\right) \end{array} \right] \tag{9.35}$$

Also, relations (9.36) and (9.37) are used to obtain the efficiency of stages 1 and 2, respectively.

$$\Theta_1^* = 1 - \frac{1}{2}\left(\frac{1}{m}\sum_{i=1}^{m}\frac{\theta_i^*}{\bar{\theta}_o} + \frac{1}{D}\sum_{d=1}^{D}\frac{\varphi_d^*}{\bar{\varphi}_o}\right) \tag{9.36}$$

$$\Theta_2^* = 1 - \frac{1}{2}\left(\frac{1}{D}\sum_{d=1}^{D}\frac{\alpha_d^*}{\bar{\alpha}_o} + \frac{1}{s}\sum_{r=1}^{s}\frac{\beta_r^*}{\bar{\beta}_o}\right) \tag{9.37}$$

And finally, it can be easily shown that: $0 \le \Theta_1^*, \Theta_2^* \le 1$.

9.4 Performance Evaluation of the Supply Chains Using Hybrid Models in NDEA

In hybrid (radial and non-radial) NDEA models, the difference between changeable and semi-fixed factors has not been considered. For example, inputs such as labor can change quickly, but semi-fixed indicators, such as assets, cannot be adjusted rapidly or proportionally to other changes. While in radial and non-radial models, the difference between these indicators has been ignored. Although few researchers have tried to include this difference in DEA models in the black box case (Tone and Tsutsui [54]) and apply a combined approach, including radial and non-radial factors, this difference has not been considered in NDEA models. Therefore, hybrid models are introduced in this sub-section to measure the efficiency of supply chains.

9.4.1 Performance Evaluation E of Two-member Series Supply Chain Based on Short-term and Long-term Decision-making

Consider a two-member supply chain, as shown in Fig. 9.7. Assume that x_o^R and x_o^{NR} are the radial and non-radial inputs, respectively. Also, the radial and non-radial outputs of stage 1 are introduced with the symbols z_o^R and z_o^{NR}.

In this case, Nikfarjam et al. [36] introduced the model (9.38) to evaluate the suppliers of under-evaluation supply chains.

$$
v^{\text{Supplier}} = \min \frac{1 - \frac{1}{P_2+1}\left[(1 - \theta^s) + \sum_{p_2=1}^{P_2} \frac{s_{p_2}^{NR-}}{x_{p_2o}^{NR}}\right]}{1 + \frac{1}{K_2+1}\left[(\varphi^s - 1) + \sum_{k_2=1}^{K_2} \frac{s_{k_2}^{NR+}}{z_{k_2o}^{NR}}\right]}
$$

$s.t.$

$$
\sum_{j=1}^{n} \lambda_j^s x_{p_1 j}^R \leq \theta^s x_{p_1 o}^R, \quad p_1 = 1, ..., P_1,
$$

$$
\sum_{j=1}^{n} \lambda_j^s x_{p_2 j}^{NR} + s_{p_2}^{NR-} = x_{p_2 o}^R, \quad p_2 = 1, ..., P_2,
$$

$$
\sum_{j=1}^{n} \lambda_j^s z_{k_1 j}^R \geq \varphi^s z_{k_1 o}^R, \quad k_1 = 1, ..., K_1,
$$

$$
\sum_{j=1}^{n} \lambda_j^s z_{k_2 j}^{NR} - s_{k_2}^{NR+} = z_{k_2 o}^{NR}, \quad k_2 = 1, ..., K_2,
$$

$$
\lambda_j^s \geq 0, \quad j = 1, ..., n,
$$

$$
\theta^s \leq 1, \quad \varphi^s \geq 1,
$$

$$
s_{p_2}^{NR-} \geq 0, \quad s_{k_2}^{NR+} \geq 0, \quad p_2 = 1, ..., P_2, \quad k_2 = 1, ..., K_2.
$$

(9.38)

So that P_1 and P_2 are equal to the number of radial and non-radial inputs of the supplier, respectively. So that $P_1 + P_2 = P$.

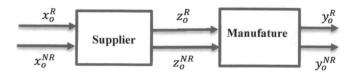

Fig. 9.7 A two-member supply chain with radial and non-radial indicators

Also, K_1, K_2, and K indicate the number of radials and non-radial outputs and the total number of supplier outputs. Also, $K = K_1 + K_2$. Suppose the manufacturer consumes input $z = \left(z_o^R, z_o^{NR} \right)$ to produce Q_1 radial $\left(y_o^R \right)$ and Q_2 non-radial outputs y_o^{NR}. So that: $Q = Q_1 + Q_2$. So, to evaluate the producer in the supply chain, model (9.39) is introduced as follows.

$$
\nu^{\text{Manufactor}} = \min \frac{1 - \frac{1}{K_2+1} \left[(1 - \theta^m) + \sum_{k_2=1}^{K_2} \frac{s_{k_2}^{NR-}}{z_{k_2 o}^{NR}} \right]}{1 + \frac{1}{Q_2+1} \left[(\varphi^m - 1) + \sum_{q_2=1}^{Q_2} \frac{s_{k_2}^{NR+}}{y_{q_2 o}^{NR}} \right]}
$$

$s.t.$

$$
\sum_{j=1}^{n} \lambda_j^m z_{k_1 j}^R \leq \theta^m z_{k_1 o}^R, \quad k_1 = 1, ..., K_1,
$$

$$
\sum_{j=1}^{n} \lambda_j^m z_{k_2 j}^{NR} + s_{k_2}^{NR-} = z_{k_2 o}^{NR}, \quad k_2 = 1, ..., K_2, \tag{9.39}
$$

$$
\sum_{j=1}^{n} \lambda_j^m y_{k_1 j}^R \geq \varphi^m y_{k_1 o}^R, \quad q_1 = 1, ..., Q_1,
$$

$$
\sum_{j=1}^{n} \lambda_j^m y_{k_2 j}^{NR} - s_{k_2}^{NR+} = y_{k_2 o}^{NR}, \quad q_2 = 1, ..., Q_2,
$$

$$
\lambda_j^m \geq 0, \quad j = 1, ..., n,
$$

$$
\theta^m \leq 1, \quad \varphi^m \geq 1,
$$

$$
s_{k_2}^{NR-} \geq 0, \quad s_{q_2}^{NR+} \geq 0, \quad k_2 = 1, ..., K_2, \quad q_2 = 1, ... Q_2.
$$

In the same way, model (9.40) will be used to evaluate the overall performance of the supply chain.

$$
v^{\text{Supply chain}} = \min \frac{1 - \frac{1}{P_2+K_2+2}\left[(1-\theta) + \sum\limits_{p_2=1}^{P_2} \frac{s_{p_2}^{\text{NR}-}}{x_{p_2o}^{\text{NR}}} + (1-\theta^{m^*}) + \sum\limits_{k_2=1}^{K_2} \frac{s_{k_2}^{*\text{NR}-}}{z_{k_2o}^{\text{NR}}}\right]}{1 + \frac{1}{Q_2+K_2+2}\left[(\varphi^{s^*}-1) + \sum\limits_{k_2=1}^{K_2} \frac{s_{k_2}^{\text{NR}+^*}}{z_{k_2o}^{\text{NR}}} + (\varphi-1) + \sum\limits_{q_2=1}^{Q_2} \frac{s_{k_2}^{\text{NR}+}}{y_{q_2o}^{\text{NR}}}\right]}
$$

s.t.

$$
\sum_{j=1}^{n} \lambda_j^s x_{p_1 j}^R \le \theta x_{p_1 o}^R, \quad p_1 = 1, \dots, P_1,
$$

$$
\sum_{j=1}^{n} \lambda_j^s x_{p_2 j}^{\text{NR}} + s_{p_2}^{\text{NR}-} = x_{p_2 o}^R, \quad p_2 = 1, \dots, P_2,
$$

$$
\sum_{j=1}^{n} \lambda_j^s z_{k_1 j}^R \ge \varphi^{*s} z_{k_1 o}^R, \quad k_1 = 1, \dots, K_1, \qquad (9.40)
$$

$$
\sum_{j=1}^{n} \lambda_j^s z_{k_2 j}^{\text{NR}} - s_{k_2}^{\text{NR}+^*} = z_{k_2 o}^{\text{NR}}, \quad k_2 = 1, \dots, K_2,
$$

$$
\sum_{j=1}^{n} \lambda_j^m z_{k_1 j}^R \le \theta^{m^*} z_{k_1 o}^R, \quad k_1 = 1, \dots, K_1,
$$

$$
\sum_{j=1}^{n} \lambda_j^m z_{k_2 j}^{\text{NR}} + s_{k_2}^{\text{NR}-^*} = z_{k_2 o}^{\text{NR}}, \quad k_2 = 1, \dots, K_2,
$$

$$
\sum_{j=1}^{n} \lambda_j^m y_{q_1 j}^R \ge \varphi y_{q_1 o}^R, \quad q_1 = 1, \dots, Q_1,
$$

$$
\sum_{j=1}^{n} \lambda_j^m y_{q_2 j}^{\text{NR}} - s_{q_2}^{\text{NR}+} = y_{q_2 o}^{\text{NR}}, \quad q_2 = 1, \dots, Q_2,
$$

$$
\lambda_j^m, \lambda_j^s \ge 0, \quad j = 1, \dots, n,
$$

$$
\theta \le 1, \quad \varphi \ge 1,
$$

$$
\left(s_{k_2}^{\text{NR}-}, s_{q_2}^{\text{NR}+}, s_{p_2}^{\text{NR}-}, s_{k_2}^{\text{NR}+}\right) \ge 0,
$$

$$
k_2 = 1, \dots, K_2, \quad p_2 = 1, \dots, P_2, \quad q_2 = 1, \dots, Q_2.
$$

So that $\left(\theta^{s^*}, \varphi^{s^*}, s_{p_2}^{\mathrm{NR}-*}, s_{k_2}^{\mathrm{NR}+*}\right)$ and $\left(\theta^{m^*}, \varphi^{m^*}, s_{k_2}^{\mathrm{NR}-*}, s_{k_2}^{\mathrm{NR}+*}\right)$ are respectively the optimal solutions for evaluating the supplier and manufacturer performance using models (9.38) and (9.39). In the model (9.40), the chain's overall performance is calculated based on the optimal solutions of stages 1 and 2. In other words, by keeping the efficiency of stages 1 and 2 (supplier and manufacturer) constant, the efficiency of the entire chain is calculated. So, the supply chain under evaluation is efficient using model (9.40) if and only if $\theta^* = \varphi^* = \theta^{*^m} = \varphi^{m^*} = \theta^{*^s} = \varphi^{s^*} = 1$ and all auxiliary variables are equal to zero.

In the real world, managers must consider many conditions, including short-term and long-term policies of the organization, to make sensible decisions. Also, one of the most critical issues in analyzing short-term performance evaluation or long-term policies is understanding the system's behavior. For example, extending the building lease for another year or extending the one-year employment contracts of employees are examples of short-term decisions. But system management must adopt long-term planning to change and use new technologies. Therefore, some inputs are changeable in short-term decision-making, and some are semi-fixed. In this subsection, radial inputs are considered changeable inputs, and non-radial inputs as semi-fixed inputs. So it is evident that only changes in radial inputs are considered in short-term decision-making. In this way, the model (9.41) has been expressed to evaluate the supplier's performance in the short term.

$$v_{\text{Short - term}}^{\text{Supplier}} = \min \frac{1 - \frac{1}{1}(1 - \theta^s)}{1 + \frac{1}{K_2+1}\left[(\varphi^s - 1) + \frac{1}{K}\sum_{k_2=1}^{K_2} \frac{s_{k_2}^{\mathrm{NR}+}}{z_{k_2 o}^{\mathrm{NR}}}\right]}$$

$s.t.$

$$\sum_{j=1}^{n} \lambda_j^s x_{p_1 j}^R \le \theta^s x_{p_1 o}^R, \ p_1 = 1, ..., P_1,$$

$$\sum_{j=1}^{n} \lambda_j^s x_{p_2 j}^{\mathrm{NR}} \le x_{p_2 o}^R, \ p_2 = 1, ..., P_2,$$

$$\sum_{j=1}^{n} \lambda_j^s z_{k_1 j}^R \ge \varphi^s z_{k_1 o}^R, \ k_1 = 1, ..., K_1,$$ (9.41)

$$\sum_{j=1}^{n} \lambda_j^s z_{k_2 j}^{\mathrm{NR}} - s_{k_2}^{\mathrm{NR}+} = z_{k_2 o}^{\mathrm{NR}}, \ k_2 = 1, ..., K_2,$$

$$\lambda_j^s \ge 0, \ j = 1, ..., n,$$

$$\theta^s \le 1, \ \varphi^s \ge 1,$$

$$s_{k_2}^{\mathrm{NR}+} \ge 0, \ k_2 = 1, ..., K_2.$$

In the same way, the model (9.42) is used to evaluate the performance of the supplier in the long term. So to evaluate the supplier in the long term, changes in all inputs are considered equally in the modeling process.

$$v_{\text{Long - term}}^{\text{Supplier}} = \min \frac{\theta^s}{1 + \frac{1}{K_2+1}\left[(\varphi^s - 1) + \sum\limits_{k_2=1}^{K_2} \frac{s_{k_2}^{\text{NR+}}}{z_{k_2 o}^{\text{NR}}}\right]}$$

s.t.

$$\sum_{j=1}^{n} \lambda_j^s x_{pj}^R \leq \theta^s x_{po}^R, \quad p = 1, ..., P,$$

$$\sum_{j=1}^{n} \lambda_j^s z_{k_1 j}^R \geq \varphi^s z_{k_1 o}^R, \quad k_1 = 1, ..., K_1, \tag{9.42}$$

$$\sum_{j=1}^{n} \lambda_j^s z_{k_2 j}^{\text{NR}} - s_{k_2}^{\text{NR+}} = z_{k_2 o}^{\text{NR}}, \quad k_2 = 1, ..., K_2,$$

$$\lambda_j^s \geq 0, \quad j = 1, ..., n,$$

$$\theta^s \leq 1, \quad \varphi^s \geq 1,$$

$$s_{k_2}^{\text{NR+}} \geq 0, \quad k_2 = 1, ..., K_2.$$

Similar to the short-term supplier evaluation process, model (9.43) is presented for the short-term evaluation of the manufacturer.

$$v_{\text{Short - term}}^{\text{Manufacture}} = \min \frac{1 - \frac{1}{1}(1 - \theta^m)}{1 + \frac{1}{Q_2+1}\left[(\varphi^m - 1) + \sum\limits_{q_2=1}^{Q_2} \frac{s_{k_2}^{\text{NR+}}}{y_{q_2 o}^{\text{NR}}}\right]}$$

s.t.

$$\sum_{j=1}^{n} \lambda_j^m z_{k_1 j}^R \leq \theta^m z_{k_1 o}^R, \quad k_1 = 1, \ldots, K_1,$$

$$\sum_{j=1}^{n} \lambda_j^m z_{k_2 j}^{\text{NR}} \leq z_{k_2 o}^{\text{NR}}, \quad k_2 = 1, \ldots, K_2,$$

$$\sum_{j=1}^{n} \lambda_j^m y_{q_1 j}^R \geq \varphi^m y_{q_1 o}^R, \quad q_1 = 1, \ldots, Q_1, \tag{9.43}$$

$$\sum_{j=1}^{n} \lambda_j^m y_{q_2 j}^{\text{NR}} - s_{q_2}^{\text{NR+}} = y_{q_2 o}^{\text{NR}}, \quad q_2 = 1, \ldots, Q_2,$$

$$\lambda_j^m \geq 0, \quad j = 1, \ldots, n,$$

$$\theta^m \leq 1, \quad \varphi^m \geq 1,$$

$$s_{q_2}^{\text{NR+}} \geq 0, \quad q_2 = 1, \ldots, Q_2.$$

In the following, model (9.44) is presented for evaluating the manufacturer in the long term.

$$v_{\text{Long - Run}}^{\text{Manufacture}} = \min \frac{\theta^m}{1 + \frac{1}{Q_2+1}\left[(\varphi^m - 1) + \sum_{q_2=1}^{Q_2} \frac{s_{k_2}^{\text{NR+}}}{y_{q_2 o}^{\text{NR}}}\right]}$$

$s.t.$

$$\sum_{j=1}^{n} \lambda_j^m z_{kj}^R \leq \theta^m z_{ko}^R, \ k = 1, \ldots, K,$$

$$\sum_{j=1}^{n} \lambda_j^m y_{q_1 j}^R \geq \varphi^m y_{q_1 o}^R, \ q_1 = 1, \ldots, Q_1, \qquad (9.44)$$

$$\sum_{j=1}^{n} \lambda_j^m y_{q_2 j}^{\text{NR}} - s_{q_2}^{\text{NR+}} = y_{q_2 o}^{\text{NR}}, \ q_2 = 1, \ldots, Q_2,$$

$$\lambda_j^m \geq 0, \ j = 1, \ldots, n,$$

$$\theta^m \leq 1, \ \varphi^m \geq 1,$$

$$s_{q_2}^{\text{NR+}} \geq 0, \ q_2 = 1, \ldots, Q_2.$$

Suppose that θ^{*m} and $\left(\varphi^{s*}, s_{k_2}^{\text{NR+}*}\right)$ are the optimal solutions obtained from the short-term evaluation of the supplier and manufacturer with models (9.43) and (9.41), respectively. The model (9.45) is used to evaluate the supply chain's overall performance in the short term.

$$v_{\text{Short - term}}^{\text{Supply chain}} = \min \frac{1 - \frac{1}{2}\left[(1 - \theta) + \left(1 - \theta^{m^*}\right)\right]}{1 + \frac{1}{K_2+Q_2+2}\left[\begin{array}{c} (\varphi^{s^*} - 1) \\ + \sum_{k_2=1}^{K_2} \frac{s_{k_2}^{\text{NR}+^*}}{z_{k_2o}^{\text{NR}}} + (\varphi - 1) + \sum_{q_2=1}^{Q_2} \frac{s_{k_2}^{\text{NR}+}}{y_{q_2o}^{\text{NR}}} \end{array} \right]}$$

s.t.

$$\sum_{j=1}^{n} \lambda_j^s x_{p_1 j}^R \leq \theta x_{p_1 o}^R, \quad p_1 = 1, \, ..., \, P_1,$$

$$\sum_{j=1}^{n} \lambda_j^s x_{p_2 j}^{\text{NR}} \leq x_{p_2 o}^R, \quad p_2 = 1, \, ..., \, P_2,$$

$$\sum_{j=1}^{n} \lambda_j^s z_{k_1 j}^R \geq \varphi^{s^*} z_{k_1 o}^R, \quad k_1 = 1, \, ..., \, K_1,$$

$$\sum_{j=1}^{n} \lambda_j^s z_{k_2 j}^{\text{NR}} - s_{k_2}^{\text{NR}+^*} = z_{k_2 o}^{\text{NR}}, \quad k_2 = 1, \, ..., \, K_2,$$ (9.45)

$$\sum_{j=1}^{n} \lambda_j^m z_{k_1 j}^R \leq \theta^{m^*} z_{k_1 o}^R, \quad k_1 = 1, \, ..., \, K_1,$$

$$\sum_{j=1}^{n} \lambda_j^m z_{k_2 j}^{\text{NR}} \leq z_{k_2 o}^{\text{NR}}, \quad k_2 = 1, \, ..., \, K_2,$$

$$\sum_{j=1}^{n} \lambda_j^m y_{q_1 j}^R \geq \varphi y_{q_1 o}^R, \quad q_1 = 1, \, ..., \, Q_1,$$

$$\sum_{j=1}^{n} \lambda_j^m y_{q_2 j}^{\text{NR}} - s_{q_2}^{\text{NR}+} = y_{q_2 o}^{\text{NR}}, \quad q_2 = 1, \, ..., \, Q_2,$$

$$\lambda_j^m, \lambda_j^s \geq 0, \quad j = 1, \, ..., \, n,$$

$$\theta \leq 1, \, \varphi \geq 1,$$

$$\left(s_{k_2}^{\text{NR}-}, s_{q_2}^{\text{NR}+}, s_{p_2}^{\text{NR}-}, s_{k_2}^{\text{NR}+}\right) \geq 0,$$

$$k_2 = 1, \, ..., \, K_2, \, p_2 = 1, \, ..., \, P_2, \, q_2 = 1, \, ..., \, Q_2.$$

And finally, to evaluate the overall efficiency of the supply chain in the long term, the model (9.45) is introduced. So that, $\left(\varphi^{s^*}, s_{k_2}^{\text{NR}+^*}\right)$ and θ^{*^m} are the optimal solutions of models (9.42) and (9.44).

$$v_{\text{Long - term}}^{\text{Supply chain}} = \min \frac{\frac{1}{2}\left(\theta + \theta^{m^*}\right)}{1 + \frac{1}{K_2+Q_2+2}\left[\begin{array}{c}\left(\varphi^{s^*} - 1\right) \\ + \displaystyle\sum_{k_2=1}^{K_2} \frac{s_{k_2}^{\text{NR}+^*}}{z_{k_2 o}^{\text{NR}}} + (\varphi - 1) + \displaystyle\sum_{q_2=1}^{Q_2} \frac{s_{k_2}^{\text{NR}+}}{y_{q_2 o}^{\text{NR}}}\end{array}\right]}$$

$s.t.$

$$\sum_{j=1}^{n} \lambda_j^s x_{\text{p}j} \leq \theta x_{\text{po}}, \; p = 1, \ldots, P,$$

$$\sum_{j=1}^{n} \lambda_j^s z_{k_1 j}^R \geq \varphi^{s^*} z_{k_1 o}^R, \; k_1 = 1, \ldots, K_1,$$

$$\sum_{j=1}^{n} \lambda_j^s z_{k_2 j}^{\text{NR}} - s_{k_2}^{\text{NR}+^*} = z_{k_2 o}^{\text{NR}}, \; k_2 = 1, \ldots, K_2,$$

$$\sum_{j=1}^{n} \lambda_j^m z_{k_1 j}^R \leq \theta^{m^*} z_{k_1 o}^R, \; k_1 = 1, \ldots, K_1, \qquad (9.46)$$

$$\sum_{j=1}^{n} \lambda_j^m z_{k_2 j}^{\text{NR}} \leq z_{k_2 o}^{\text{NR}}, \; k_2 = 1, \ldots, K_2,$$

$$\sum_{j=1}^{n} \lambda_j^m y_{q_1 j}^R \geq \varphi y_{q_1 o}^R, \; q_1 = 1, \ldots, Q_1,$$

$$\sum_{j=1}^{n} \lambda_j^m y_{q_2 j}^{\text{NR}} - s_{q_2}^{\text{NR}+} = y_{q_2 o}^{\text{NR}}, \; q_2 = 1, \ldots, Q_2,$$

$$\lambda_j^m, \lambda_j^s \geq 0, \; j = 1, \ldots, n,$$

$$\theta \leq 1, \; \varphi \geq 1,$$

$$\left(s_{k_2}^{\text{NR}-}, s_{q_2}^{\text{NR}+}, s_{p_2}^{\text{NR}-}, s_{k_2}^{\text{NR}+}\right) \geq 0,$$

$$k_2 = 1, \ldots, K_2,$$

$$p_2 = 1, \ldots, P_2,$$

$$q_2 = 1, \ldots, Q_2.$$

9.4.2   Performance Evaluation of Mixed Multi-member Supply Chains

Before starting the discussion in this subsection, we will describe a practical problem. Tourism is considered one of the most profitable industries in today's world. Therefore, many studies have been carried out on the performance of the tourism supply chains. For example, the tourist supply chain in China is shown by Huang [15] in Fig. 9.8.

This process is shown as a mixed (series–parallel) four-member supply chain (sourcing, supply, delivery, and efficiency). Tourist education is considered the first stage of this chain, which are teachers of tourism education and the schools of tourism education as input for this stage, and skilled employees for other stages are the outputs of this stage.

In the second stage, there is the tourist Hotel, where employees and fixed assets are defined as inputs, and the revenues of hotels and service capacity (rooms provided for rent) are the output of this stage. The delivery stage is related to travel agencies' provision of tourism services. The inputs of this stage include employees (sent to this stage by the education department), service and hotel capacity (created by the tourist hotel department), and fixed assets. Also, this stage's outputs include the number of passengers attracted by the travel agency and the revenue of the travel agency. The final stage is tourism destinations, which shows the efficiency of the tourism supply chain manager. Therefore, the number of travelers attracted by travel agencies and the number of tourist spots at each point is defined as the input. Revenue from tourism is the final output of this supply chain.

In the education stage, the number of teachers is a radial input, and the number of schools is a non-radial and semi-fixed input. Because unlike the number of schools, the number of teachers can be proportionally changed. Accordingly, the amount

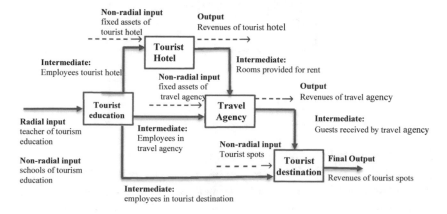

Fig. 9.8 Graphical illustration of factors within the tourism service supply chain (Huang [15])

of assets of hotels is a non-radial input for suppliers, but the number of trained employees is a radial input for travel agencies. Therefore, a hybrid (radial and non-radial) model should be used to evaluate this supply chain. Although the model presented by Nikfarjam et al. [36] is a hybrid model for measuring the efficiency of radial and non-radial supply chains, this model is only expressed for two-member series chains. Therefore, it is not suitable for evaluating the performance of supply chains, as shown in Fig. 9.8. Thus, Huang [15] focused on solving this problem and presented a model for tourism supply chain evaluation that can be easily applied to other practical applications. According to the explanations above, inputs are divided into radial and non-radial types in the education stage, represented by the symbols $x_{b_1}^{E(R)} \in R^+$, $b_1 = 1, ..., B_1$ and $x_{b_2}^{E(NR)} \in R^+$, $b_2 = 1, .., B_2$, respectively. The outputs of this stage, which are the inputs for the stages of hotels, tourist destinations, and travel agencies, are shown with the symbols $z_{c_1}^{E(R)(H)} \in R^+$, $c_1 = 1, ..., C_1$ and $z_{c_3}^{E(R)(D)} \in R^+$, $c_3 = 1, ..., C_3$, and $z_{c_2}^{E(R)(A)} \in R^+$, $c_2 = 1, ..., C_2$ respectively. For ease of work, the education stage's input and output (intermediate) are shown with symbols $x^{E(\bullet)}$, $z^{E(\bullet)}$. Therefore, the PPS for the education stage is introduced in (9.47) without considering the constant return to scale principle.

$$
T^E = \begin{cases}
\left(x^{E(\bullet)}, z^{E(\bullet)}\right): \\
\sum_{j=1}^{n} \lambda_j^E x_j^{E(R)} \leq x^{E(R)}, \\
\sum_{n=1}^{n} \lambda_j^E x_j^{E(NR)} \leq x^{E(NR)}, \\
\sum_{j=1}^{n} \lambda_j^E z_j^{E(R)(H)} \geq z^{E(R)(H)}, \\
\sum_{j=1}^{n} \lambda_j^E z_j^{E(R)(A)} \geq z^{E(R)(A)}, \\
\sum_{j=1}^{n} \lambda_j^E z_j^{E(R)(D)} \geq z^{E(R)(D)}, \\
\sum_{j=1}^{n} \lambda_j^E = 1, \lambda_j^E \geq 0, j = 1, \ldots, n.
\end{cases} \tag{9.47}
$$

In the same way, in the travel hotels stage, $x_g^{H(NR)} \in R_+^G$, $g = 1, \ldots, G$, $y_{i_2}^{H(R)} \in R^+$, $i_2 = 1, \ldots, I_2$, $z_{c_1}^{E(R)(H)} \in R^+$, $c_1 = 1, \ldots, C_1$ and $z_{i_1}^{H(R)} \in R^+$, $i_1 = 1, \ldots, I_1$ respectively represent non-radial independent inputs, radial independent outputs, radial intermediate values that are entered into this stage as output from the previous stage (education), and the radial intermediate values that enter the third stage (travel agencies) from this stage. As seen, independent radial input is not required at this stage. As a result, the production possibility set at this stage is defined as (9.48).

$$
T^H = \left\{
\begin{aligned}
& \left(x^{H(\bullet)}, z^{E(\bullet)}, z^{H(\bullet)}, y^{H(\bullet)} \right) : \\
& \sum_{j=1}^{N} \lambda_j^H x_j^{H(NR)} \le x^{H(NR)}, \\
& \sum_{j=1}^{N} \lambda_j^H z_j^{E(R)(H)} \le z^{E(R)(H)}, \\
& \sum_{j=1}^{N} \lambda_j^H z_j^{H(R)} \ge z^{H(R)}, \\
& \sum_{j=1}^{N} \lambda_j^H y_j^{H(R)} \ge y^{H(R)}, \\
& \sum_{j=1}^{N} \lambda_j^H = 1, \lambda_j^H \ge 0, \ j = 1, \ldots, n.
\end{aligned}
\right\}
\tag{9.48}
$$

In the travel agencies stage, the indicators are classified into three categories. The first category is the non-radial inputs $x_k^{A(NR)} \in R^+$, $k = 1, \ldots, K$, the second category is the radial intermediate values $\left(z_{l_1}^{A(R)} \in R^+, l_1 = 1, \ldots, L_1 \right)$ that enter into this stage as an output from the education stage, and the third category is the independent outputs of this stage, which are displayed with the symbol $y_{l_2}^{A(R)} \in R^+$, $l_2 = 1, \ldots, L_2$. As a result, the production possibility set at this stage is defined as

$$T^A = \left\{ \begin{array}{l} \left(x^{A(\bullet)}, z^{E(\bullet)}, z^{H(\bullet)}, z^{A(\bullet)}, y^{A(\bullet)} \right): \\[2mm] \sum_{j=1}^{N} \lambda_j^A z_j^{H(R)} \leq z^{H(R)}, \\[4mm] \sum_{j=1}^{N} \lambda_j^A z_j^{E(R)(A)} \leq z^{E(R)(A)}, \\[4mm] \sum_{j=1}^{N} \lambda_j^A x_j^{A(NR)} \leq x^{A(NR)}, \\[4mm] \sum_{j=1}^{N} \lambda_j^A z_j^{A(R)} \geq z^{A(R)}, \\[4mm] \sum_{j=1}^{N} \lambda_j^A y_j^{A(R)} \geq y^{A(R)}, \\[4mm] \sum_{j=1}^{N} \lambda_j^A = 1, \lambda_j^A \geq 0, \ j = 1, \ldots, n \end{array} \right\} \tag{9.49}$$

In stage 4, we introduce the PPS for tourism destinations $\left(T^D \right)$ as follows:

$$T^D = \left\{ \begin{array}{l} \left(x^{D(\bullet)}, z^{E(\bullet)}, z^{A(\bullet)}, y^{D(\bullet)} \right): \\[2mm] \sum_{j=1}^{N} \lambda_j^D z_j^{A(R)} \leq z^{A(R)}, \\[4mm] \sum_{j=1}^{N} \lambda_j^D z_j^{E(R)(D)} \leq z^{E(R)(D)}, \\[4mm] \sum_{j=1}^{N} \lambda_j^D x_j^{D(NR)} \leq x^{D(NR)}, \\[4mm] \sum_{j=1}^{N} \lambda_j^D y_j^{D(R)} \geq y^{D(R)}, \\[4mm] \sum_{j=1}^{N} \lambda_j^D = 1, \lambda_j^D \geq 0, \ j = 1, \ldots, n. \end{array} \right\} \tag{9.50}$$

So that $x^{D(\bullet)}$ and $y^{D(\bullet)}$ is the symbol of non-radial inputs and is the independent output in tourism destinations. Therefore, to assess the overall efficiency of the tourism supply chain, the model (9.55) is introduced.

$$\min \frac{1}{4} \left(\frac{1 - \frac{1}{B_2+1}\left[\left(1-\varphi^I\right) + \sum_{b_2=1}^{B_2} \frac{s_{b_2 o}^{E-}}{x_{b_2 o}^{E(NR)}} \right]}{\varphi^o} \right.$$

$$+ \frac{1 - \frac{1}{G+1}\left[\left(1-\eta^I\right) + \sum_{g=1}^{G} \frac{s_g^{H-}}{x_{go}^{E(NR)}} \right]}{\eta^o} +$$

$$\frac{1 - \frac{1}{K+1}\left[\left(1-\delta^I\right) + \sum_{k=1}^{K} \frac{s_k^{A-}}{x_{ko}^{A(NR)}} \right]}{\delta^o}$$

$$\left. + \frac{1 - \frac{1}{K+1}\left[\left(1-\rho^I\right) + \sum_{k=1}^{K} \frac{s_p^{D-}}{x_{po}^{D(NR)}} \right]}{\rho^o} \right)$$

$s.t.$

$$\sum_{j=1}^{N} \lambda_j^E x_j^{E(R)} \leq \varphi^I x_o^{E(R)},$$

$$\sum_{n=1}^{N} \lambda_j^E x_j^{E(NR)} = x_o^{E(NR)} - s_{b_2}^{E-}, \tag{9.51}$$

$$\sum_{j=1}^{N} \lambda_j^E z_j^{E(R)(H)} \geq \varphi^o z_o^{E(R)(H)},$$

$$\sum_{j=1}^{N} \lambda_j^E z_j^{E(R)(A)} \geq \varphi^o z_o^{E(R)(A)},$$

$$\sum_{j=1}^{N} \lambda_j^E z_j^{E(R)(D)} \geq \varphi^o z_o^{E(R)(D)},$$

$$\sum_{j=1}^{N} \lambda_j^E = 1, \lambda_j^E \geq 0,$$

$$\sum_{j=1}^{N} \lambda_j^H x_j^{H(NR)} = x_o^{H(NR)} - s_g^{H-},$$

$$\sum_{j=1}^{N} \lambda_j^H z_j^{E(R)(H)} \leq \eta^I z_o^{E(R)(H)},$$

$$\sum_{j=1}^{N} \lambda_j^H z_j^{H(R)} \geq \eta^o z_o^{E(R)},$$

$$\sum_{j=1}^{N} \lambda_j^H y_j^{H(R)} \geq \eta^o y_o^{H(R)},$$

$$\sum_{j=1}^{N} \lambda_j^H = 1, \ \lambda_j^H \geq 0, \ \sum_{j=1}^{N} \lambda_j^A z_j^{H(R)} \leq \delta^I z_o^{H(R)},$$

$$\sum_{j=1}^{N} \lambda_j^A z_j^{E(R)(A)} \leq \delta^I z_o^{E(R)(A)},$$

$$\sum_{j=1}^{N} \lambda_j^A x_j^{A(NR)} = x_o^{A(NR)} - s_k^{A-},$$

$$\sum_{j=1}^{N} \lambda_j^A z_j^{A(R)} \geq \delta^o z_o^{A(R)},$$

$$\sum_{j=1}^{N} \lambda_j^A y_j^{A(R)} \geq \delta^o y_o^{A(R)},$$

$$\sum_{j=1}^{N} \lambda_j^A = 1, \lambda_j^A \geq 0$$

$$\sum_{j=1}^{N} \lambda_j^D z_j^{A(R)} \leq \rho^I z_o^{A(R)},$$

$$\sum_{j=1}^{N} \lambda_j^D z_j^{E(R)(D)} \leq \rho^I z_o^{E(R)(D)},$$

$$\sum_{j=1}^{N} \lambda_j^D x_j^{D(NR)} = x_o^{D(NR)} - s_p^{D-},$$

$$\sum_{j=1}^{N} \lambda_j^D y_j^{D(R)} \geq \rho^o y_o^{D(R)},$$

$$\sum_{j=1}^{N} \lambda_j^D = 1, \ \lambda_j^D \geq 0.$$

$$\left(s_{b_2}^{E-}, s_g^{H-}, s_k^{A-}, s_p^{D-} \right) \geq 0,$$

$$1 \leq \varphi^o, \eta^o, \delta^o, \rho^o,$$

$$0 \leq \varphi^I, \eta^I, \delta^I, \rho^I \leq 1.$$

If φ^{I*}, $s_{b_2o}^{E-*}$, and φ^{o*} are the optimal values obtained from model (9.51), and the total number of inputs, radial and non-radial inputs at this stage is equal to B, B_1, and B_2, in this case, we use Eq. (9.52) to calculate the efficiency of the tourist education stage.

$$e_o^{E*} = \frac{1 - \frac{1}{B_2+1}\left((1 - \varphi^{I*}) + \sum_{b_2=1}^{B_2} \frac{s_{b_2o}^{E-*}}{x_{b_2o}^{E(NR)}}\right)}{\varphi^{o*}} \qquad (9.52)$$

Now suppose that G and K are the numbers of independent inputs of the hotel and travel agencies stage, and C_1 and C_2 are equal to the number of inputs sent from the training stage to the hotel stage and from the training stage to the travel agencies stage. And also represents the number of intermediate indicators between the hotel sector and travel agencies. Therefore, relations (9.53) and (9.54) are used to calculate tourist hotel and travel agencies' performance.

$$e_o^{H*} = \frac{1 - \frac{1}{G+1}\left[(1 - \eta^{I*}) + \sum_{g=1}^{G} \frac{s_g^{H-*}}{x_{go}^{E(NR)}}\right]}{\eta^{o*}} \qquad (9.53)$$

Such that η^{I*}, s_g^{H-*}, η^{o*}, δ^{I*}, s_k^{A-*}, δ^{o*} are the optimal solution of the model (9.51).

$$e_o^{A*} = \frac{1 - \frac{1}{K+1}\left[(1 - \delta^{I*}) + \sum_{k=1}^{K} \frac{s_k^{A-*}}{x_{ko}^{A(NR)}}\right]}{\delta^{o*}} \qquad (9.54)$$

And finally, if ρ^{I*}, s_p^{D-*}, and ρ^{o*} are the optimal solution obtained from model (9.51) and P is the number of independent inputs in this stage, L_1 and C_3 are the numbers of inputs sent from travel agencies, and the education stage to tourism destinations, then the efficiency of tourism destinations is obtained using the relation (9.55).

$$e_o^{D*} = \frac{1 - \frac{1}{P+1}\left[(1 - \rho^{I*}) + \sum_{p=1}^{P} \frac{s_p^{D-*}}{x_{po}^{D(NR)}}\right]}{\rho^{o*}} \qquad (9.55)$$

9.4.3 Performance Evaluation of Supply Chain Using Network Epsilon-based DEA Model

Although in the hybrid models introduced in the previous sub-section, the simultaneous effect of changes on the radial and non-radial indicators is considered, the importance and influence of these indicators on the efficiency score are considered the same, and all the indicators are considered at a level of significance. As stated in Chap. 6, the EBM model considering the radial and non-radial indicators and the importance of each of these indicators were presented by Tone and Tsutsui [54]. But, in this model, the internal relationships of the system are ignored. Tavana et al. [52] developed a new model for evaluating supply chains by developing the EBM model to solve this problem. They called the new model NEBM. Among the advantages of the NEBM model compared to the previous models presented for the evaluation of the supply chain are from:

- Considering internal relationships in the supply chain.
- Considering radial and non-radial changes in indexes.
- Considering the relative importance of each index in calculating the efficiency of the supply chain.

Consider the multi-member supply chain in general, as shown in Fig. 9.9. Therefore, the NEBM model is introduced as the model (9.56).

$$\gamma^* = \min \sum_{h=1}^{k} W_h \left(\theta^h - \varepsilon_x^h \sum_{i=1}^{m_h} \frac{w_i^{h-} s_i^{h-}}{x_{io}^h} \right)$$

$s.t.$

$$\sum_{i=1}^{m_h} x_{ij}^h \lambda_j^h + s_i^{h-} = \theta^h x_{io}^h, \ i = 1, ..., m_h, \ h = 1, ..., k,$$

$$\sum_{i=1}^{m_h} y_{rj}^h \lambda_j^h \geq y_{ro}^h, \ r = 1, ..., s_h, \ h = 1, ..., k, \qquad\qquad (9.56)$$

$$\sum_{j=1}^{n} z_{f_{(h,h')}j}^{(h,h')} \lambda_j^h = \sum_{j=1}^{n} z_{f_{(h,h')}j}^{(h,h')} \lambda_j^{h'}, \ f_{(h,h')} = 1, ..., F_{(h,h')}, \ \forall(h, h'),$$

$$\theta^h \leq 1, \ h = 1, ..., k,$$

$$\lambda_j^h \geq 0, \ j = 1, ..., n, \ h = 1, ..., k,$$

$$s_i^{h-} \geq 0, \ i = 1, ..., m_h, \ h = 1, ..., k.$$

w_i^{h-} and s_i^{h-} respectively indicate the weight (importance) and the auxiliary variable corresponding to the i-th input sent from stage h. It should be noted that ε_x^h is determined based on the degree of dispersion of parameters related to stage h. And finally, W_h is the weight corresponds to stage h, which is determined by the

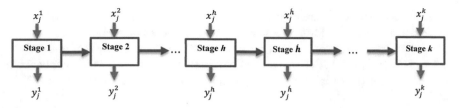

Fig. 9.9 General structure of a series multi-stage supply chain

manager. Constraints 1, 2, and 3, respectively, correspond to the inputs of the stage h and the intermediate products between the stages h and h'. The important point in using model (9.56) is how to determine ε_x^h. For this purpose, the following algorithm is suggested:

1. Forming the diversity matrix: The diversity matrix is displayed with a symbol $D^h = [D_{i,k}^h]_{m_h \times m_h}$. Suppose that $x_i^h = (x_{i1}^h, x_{i2}^h, ..., x_{in}^h)$ and $x_k^h = (x_{k1}^h, x_{k2}^h, ..., x_{kn}^h)$. In this case, $D_{i,k}^h$ is calculated using relation (9.57):

$$D_{i,k}^h = D(x_i^h, x_k^h) = \frac{\sum_{j=1}^{n} \left| c_j^h - c^{-h} \right|}{n \left(c_{max}^h - c_{min}^h \right)} \tag{9.57}$$

such that: $c_j^h = \ln X_i^h / X_k^h$ and $c^{-h} = \sum_{j=1}^{n} c_j^h / n$. Also, $c_{max}^h = \max_j \{ c_j^h \}$ and $c_{min}^h = \min_j \{ c_j^h \}$

2. Form the affinity matrix represented by the symbol $S^h = [S_{i,k}^h]_{m_h \times m_h}$, $i, k = 1, ..., m_h$. So that $S_{i,k}^h = 1 - 2 \times D_{i,k}^h$, $i, k = 1, ..., m_h$ indicates the degree of affinity between x_i^h and x_k^h.
3. Calculate the values of ε_x^h and w_i^{h-} using relations (9.58) and (9.59).

$$\varepsilon_x^h = \begin{cases} \frac{m_h - \rho_x^h}{m_h - 1}, & m_h > 1 \\ 0, & m_h = 1 \end{cases} \tag{9.58}$$

and

$$w_i^{h-} = \frac{w_{ix}^{h-}}{\sum_{i=1}^{m_h} w_{ix}^h} \tag{9.59}$$

ρ_x^h is the largest eigenvalue of the matrix S^h and $W_x^h = (w_{1x}^h, w_{2x}^h, ..., w_{m_h x}^h)$ is the eigenvector corresponding to ρ_x^h.

After calculating ε_x^h based on the mentioned process, the NEBM model can be used to evaluate the performance of supply chains under evaluation.

It is evident that supply chain o using model (9.56) is efficient if and only if $\gamma^* = 1$. If $\left(\theta^{h^*}, s^{h-*}\right)$ is obtained from model (9.56), then stage h of supply chain o using model (9.56) is efficient if and only if $\gamma^{h^*} = 1$. So that: $\gamma^{h^*} = \theta^{h^*} - \varepsilon_x^h \sum_{i=1}^{m_h} w_i^{h-} s_i^{h-*} / x_{io}^h$.

As explained earlier, ε_x^h indicates the degree of dispersion of the inputs, and the larger this value shows, the more dispersion among the inputs and vice versa. Also, if the value ε_x^h gets closer to zero for all stages, model (9.56) becomes a radial network model in the form of model (9.60).

$$\theta^* = \min \sum_{h=1}^{k} W_h \theta^h$$

$s.t.$

$$\sum_{i=1}^{m_h} x_{ij}^h \lambda_j^h \le \theta^h x_{io}^h, \ i = 1, ..., m_h, \ h = 1, ..., k,$$

$$\sum_{i=1}^{m_h} y_{rj}^h \lambda_j^h \ge y_{ro}^h, \ r = 1, ..., s_h, \ h = 1, ..., k, \tag{9.60}$$

$$\sum_{j=1}^{n} z_{f_{(h,h')}j}^{(h,h')} \lambda_j^h = \sum_{j=1}^{n} z_{f_{(h,h')}j}^{(h,h')} \lambda_j^{h'}, \ f_{(h,h')} = 1, ..., F_{(h,h')}, \ \forall (h, h'),$$

$$\theta^h \le 1, \ h = 1, ..., k,$$

$$\lambda_j^h \ge 0, \ j = 1, ..., n, \ h = 1, ..., k.$$

Also, if the value of ε_x^h gets closer to 1 for all stages, then the model (9.56) becomes the NSBM model, such as model (9.61).

$$\phi^* = \min \sum_{h=1}^{k} W_h \left(1 - \frac{1}{m_h} \sum_{i=1}^{m_h} \frac{w_i^{h-} s_i^{h-}}{x_{io}^h} \right)$$

$s.t.$

$$\sum_{i=1}^{m_h} x_{ij}^h \lambda_j^h + s_i^{h-} = \theta^h x_{io}^h, \ i = 1, ..., m_h, \ h = 1, ..., k,$$

$$\sum_{i=1}^{m_h} y_{rj}^h \lambda_j^h \ge y_{ro}^h, \ r = 1, ..., s_h, \ h = 1, ..., k, \tag{9.61}$$

$$\sum_{j=1}^{n} z_{f_{(h,h')}j}^{(h,h')} \lambda_j^h = \sum_{j=1}^{n} z_{f_{(h,h')}j}^{(h,h')} \lambda_j^{h'}, \ f_{(h,h')} = 1, ..., F_{(h,h')}, \ \forall (h, h'),$$

$$\lambda_j^h \ge 0, \ j = 1, ..., n, \ h = 1, ..., k,$$

$$s_i^{h-} \ge 0, \ i = 1, ..., m_h, \ h = 1, ..., k.$$

It can be easily shown that: $\phi^* \le \gamma^* \le \theta^*$.

9.5 Performance Evaluation of Supply Chains Using Multiplier Models

Envelopment models are widely used in DEA and NDEA literature, and many studies have been conducted on them. Since the essence of envelopment models is based on the definition of the PPS, as a result, because of the problems in the description of the PPS for networks, decision-makers are skeptical about the results obtained from the envelopment models. While multiplier models are obtained based on the definition of efficiency. Thus, using the definition provided for network efficiency, the modeling process can be easily presented to evaluate the performance of network units. Also, it is easier to apply some restrictions, such as weight restriction, through multiplier models. Therefore, in this sub-section, we focus on the multiplier NDEA models.

9.5.1 Performance Evaluation of Supply Chains Using Frontier-shift

Although the models presented in the field of supply chain performance evaluation are often expressed in a general way and can be used for various practical problems, specific and appropriate modeling for a phenomenon should be presented in some cases. One of these topics is the energy supply chain and the "carbon credit" issue. In the late 1990s, due to the increase in greenhouse gases and their destructive effects, the United Nations asked the member countries to reduce the amount of carbon dioxide they produce in the form of treaties called the "Kyoto Treaty". The "Kyoto Treaty", which has been signed and recognized by 161 countries, divides the countries of the world into two categories: "developed" and "developing". All developed countries recognized this treaty must reduce the amount of carbon dioxide produced between 2008 and 2012 to less than 5% of the total carbon dioxide produced in 1990. But developing countries are not required to reduce the amount of carbon dioxide they produce. Still, if these countries implement projects that reduce the amount of carbon dioxide, they will receive carbon credit, which can be sold to developed countries. Therefore, developed countries can reduce their carbon dioxide emissions by applying new technologies or getting permission to emit more carbon dioxide by paying money and helping to reduce carbon dioxide in developing countries. The Kyoto Treaty also encourages the "carbon trading" policy at the factory. In other words, developing countries that have reduced their carbon dioxide production by using new technologies will be able to sell the received carbon credits to factories whose output still does not meet the carbon standards. The basis of the carbon credit is determined based on the value of hydrocarbon energy carriers such as oil and gas. As the global price of oil increases, the carbon credit also increases.

Sellers and buyers of carbon credits can trade carbon credits on a trading platform such as the stock exchange. Accordingly, paying attention to this issue and managing

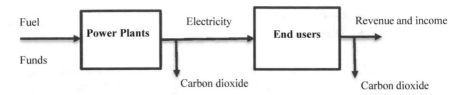

Fig. 9.10   A two stage energy supply chain

carbon credit is the main topic of many articles, including supply chain management. Therefore, modeling and addressing this issue in many supply chains must consider certain constraints. Zhai et al. [60] believe that due to the importance of energy, looking at the energy supply chain (ESC) should be different from traditional views, so Zhai et al. [60] used frontier-shift methodology to evaluate the ecological efficiency of the electricity industry in Anhui, China and presented a new NDEA model to assess the performance of two-member supply chains. Four principles have been considered to perform this modeling.

1. Considering the carbon credit ceiling in each city, it is assumed that no production activity can be carried out without granting appropriate emission rights.
2. If emitters improve their energy efficiency by reducing carbon consumption, these carbon credits will be stored, and they can sell them in the market for more profit. Therefore, factories try to reduce their carbon emissions.
3. The expenses incurred by emitters due to carbon reduction should not be more than the profit they can get from selling carbon credit. Otherwise, the credit transaction will not take place. Since carbon emission permits are limited, organizations that need additional emissions must purchase them from those that need fewer credits.
4. Since ESC often produces limited products, only one intermediate product, electricity, exists in the examined supply chain, as shown in Fig. 9.10.

As shown in Fig. 9.10, in stage 1, the power plant produces electricity using fossil fuels and making necessary investments. Of course, due to the use of fossil fuels to produce electricity, Carbon monoxide will also be created. Of course, only electricity is transferred to the next stage, i.e., the final consumer. At this stage, the final consumer, by consuming electricity in production centers, on the one hand, creates income and profit for the energy supply chain; on the other hand, it causes emissions, including carbon. Although there is a desire to increase the production of electricity, it is necessary to avoid the production of waste products in the environment as much as possible. Therefore, in this supply chain, carbon dioxide is an undesirable output. As a result, in the modeling process in ESC, in addition to considering the four assumptions mentioned above, the impact of undesirable outputs should also be considered.

Consider the supply chain in Fig. 9.11. z^D and z^U represent desirable and undesirable intermediate products so that z^D enters stage 2 as input but z^U leaves the system

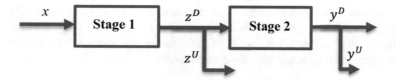

Fig. 9.11 A two-stage system structure with undesirable outputs

as an independent output. Also, y^D and y^U indicate desirable and undesirable final products in the supply chain.

Based on assumption 2, the emitters tend to improve their efficiency with the minimum amount of carbon emission. Suppose the maximum amount of carbon reduced in the l-th intermediate index for DMUo in stage 1 is equal to A_l. In that case, this additional credit can be divided among other units. Now suppose the amount of extra credit transferred to DMUj in stage 1 is represented by the symbol α_{1j}, in this case: $\sum_{j=1}^{n} \alpha_{1j} = A_l$.

As a result, the model (9.62) is introduced to evaluate the performance of stage 1 in the energy supply chain by considering the carbon credit as follows.

$$\theta_{\text{new}-I} = \max \frac{\sum_{k=1}^{K} w_k^D z_{ko}^D}{\sum_{i=1}^{m} v_i x_{io} + \sum_{l=1}^{L} w_l^U \left(z_{lo}^U - A_l \right)}$$

$s.t.$

$$\frac{\sum_{k=1}^{K} w_k^D z_{ko}^D}{\sum_{i=1}^{m} v_i x_{io} + \sum_{l=1}^{L} w_l^U \left(z_{lo}^U - A_l \right)} \le 1,$$

$$\frac{\sum_{l=1}^{L} w_l^D z_{1j}^D}{\sum_{i=1}^{m} v_i x_{ij} + \sum_{l=1}^{L} w_l^U \left(z_{1j}^U + \alpha_{1j} \right)} \le 1, \, j = 1, ..., n, \, j \ne o, \tag{9.62}$$

$$\sum_{j=1}^{n} \alpha_{1j} = A_l, \, l = 1, ..., L.$$

$$v_i, w_l^U, w_k^D \ge 0, \, i = 1, ..., m, \, l = 1, ..., L, \, k = 1, ..., K.$$

$$\alpha_{1j} \ge 0, \, z_{1j}^U \ge A_l, \, l = 1, ..., L, \, j = 1, ..., n.$$

It is necessary to explain that in this section, the point of view described in subsection (7.4.3.1) was used to consider the impact of undesirable output. Note that $\alpha_{1j} = 0$ means no carbon exchange between DMU$_o$ and DMU$_j$ in this indicator.

Suppose that the maximum amount that can be reduced for DMU$_o$ in stage 2 in the r-th intermediate index is equal to B_r. In this case, this additional credit amount

can be divided among other units. Also, the extra credit that can be transferred to
DMU$_j$ in stage 2 is displayed with the symbol β_{rj}, in this case: $\sum_{j=1}^{n} \beta_{rj} = B_r$. As
a result, the model (9.63) is proposed to evaluate the performance of stage 2 in the
energy supply chain, considering the carbon credit.

$$\theta_{\text{new - II}} = \max \frac{\sum_{q=1}^{Q} u_q^D y_{qo}^D}{\sum_{k=1}^{K} w_k'^D z_{ko}^D + \sum_{r=1}^{s} u_r^U \left(y_{rj}^U - B_r \right)}$$

s.t.

$$\frac{\sum_{q=1}^{Q} u_q^D y_{qo}^D}{\sum_{k=1}^{K} w_k'^D z_{ko}^D + \sum_{r=1}^{s} u_r^U \left(y_{rj}^U - B_r \right)} \leq 1,$$

$$\frac{\sum_{q=1}^{Q} u_q^D y_{qj}^D}{\sum_{k=1}^{K} w_k^D z_{ko}^D + \sum_{r=1}^{s} u_r^U \left(y_{rj}^U + \beta_{rj} \right)} \leq 1, \ j = 1, ..., n, \ j \neq o,$$ (9.63)

$$\sum_{j=1}^{n} \beta_{rj} = B_r, r = 1, ..., s,$$

$$u_r^U, u_q^D, w_k'^D \geq 0, r = 1, ..., s, q = 1, ..., Q, k = 1, ..., K.$$

$$\beta_{rj} \geq 0, B_r \geq y_{ro}^U, r = 1, ..., s, j = 1, ..., n.$$

Now, to evaluate the performance of the two-stage energy supply chain, model
(9.64) is obtained based on the constraints of models (9.62) and (9.63). So that the
weights corresponding to the intermediate products as output from stage 1 and as
input in stage 2 are considered the same. Also, ω^1 and ω^2 are the weights corresponding to stages 1 and 2, which reflect the importance of stages 1 and 2 and are
determined by the manager.

$$\theta_{\text{new}} = \max \omega^1 \theta_{\text{new - I}} + \omega^2 \theta_{\text{new - II}}$$

s.t.

Constraints of (9.62), (9.64)
Constraints of (9.63),
$$w^D = w'^D.$$

Therefore, based on the said process, emitters seek to improve their efficiency
by reducing carbon emissions at each stage, but this reduction is impossible without
creating new costs. In other words, to reduce the emission of greenhouse gases, it
is necessary to have an expert workforce and use advanced technologies. Therefore,
chain management should be able to find an approach through which it can reduce
its carbon footprint at the lowest cost. Thus, to achieve this goal, model (9.65) is

presented for stage 1.

$$\min \sum_{l=1}^{L} w_l^U A_l$$

s.t.

$$\frac{\sum_{k=1}^{K} w_k^D z_{ko}^D}{\sum_{i=1}^{m} v_i x_{io} + \sum_{l=1}^{L} w_l^U \left(z_{lo}^U - A_l\right)} \le 1,$$

$$\frac{\sum_{l=1}^{L} w_l^D z_{lj}^D}{\sum_{i=1}^{m} v_i x_{ij} + \sum_{l=1}^{L} w_l^U \left(z_{lj}^U + \alpha_{lj}\right)} \le 1, \ j = 1, ..., n, \ j \neq o, \tag{9.65}$$

$$\sum_{j=1}^{n} \alpha_{lj} = A_l,$$

$$v_i, w_l^U, w_k^D \ge 0, \ i = 1, ..., m, \ l = 1, ..., L, \ k = 1, ..., K.$$

$$\alpha_{lj} \ge 0, z_{lj}^U \ge A_l, \ l = 1, ..., L, \ j = 1, ..., n.$$

Also, model (9.66) is used for the second stage of the energy supply chain.

$$\max \sum_{r=1}^{s} u_r^U B_r$$

s.t.

$$\frac{\sum_{q=1}^{Q} u_q^D y_{qo}^D}{\sum_{k=1}^{K} w_k'^D z_{ko}^D + \sum_{r=1}^{s} u_r^U \left(y_{rj}^U - B_r\right)} \le 1,$$

$$\frac{\sum_{q=1}^{Q} u_q^D y_{qj}^D}{\sum_{k=1}^{K} w_k^D z_{ko}^D + \sum_{r=1}^{s} u_r^U \left(y_{rj}^U + \beta_{rj}\right)} \le 1, \ j = 1, ..., n, \ j \neq o, \tag{9.66}$$

$$\sum_{j=1}^{n} \beta_{rj} = B_r, r = 1, ..., s,$$

$$w_k'^D, u_r^U, u_q^D \ge 0, \ k = 1, ..., K, \ r = 1, ..., s, \ q = 1, ..., Q.$$

$$\beta_{rj} \ge 0, B_r \ge y_{ro}^U, \ r = 1, ..., s, \ j = 1, ..., n.$$

Fig. 9.12 Illustration of
CCR frontier and the shifted
frontier of Stage I or II for
DMU_o

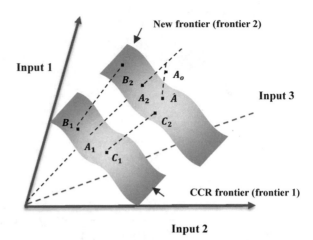

Assume that: $A = \left[\alpha_{1j}\right]_{n^U \times N}$ and $B = \left[\beta_{rj}\right]_{s^U \times N}$ so that $\beta_{ro} = \alpha_{1o} = 0$. If the value of $\hat{\alpha}_{1j}$ and $\hat{\beta}_{rj}$ is the optimal amount of carbon required for city j in stages 1 and 2, city j becomes efficient after reducing this amount. So the indicator values in

stage 1 will change from $\begin{pmatrix} x \\ z^D \\ z^U \end{pmatrix}_{(m+K+L)\times N}$ to $\begin{pmatrix} x \\ z^D - \hat{A} \\ z^U \end{pmatrix}_{(m+K+L)\times N}$. And also, in

stage 2, it varies from $\begin{pmatrix} z^D \\ y^D \\ y^U \end{pmatrix}_{(K+Q+S)\times N}$ to $\begin{pmatrix} z^D \\ y^D \\ y^U - \hat{B} \end{pmatrix}_{(K+Q+S)\times N}$.

In this way, with the new data, the frontier of the PPS will shift, as shown in Fig. 9.12.

As a result, to calculate the performance of supply chains under evaluation in stages 1 and 2, based on the new frontier, models (9.67) and (9.68) are used.

$$\theta_{\text{shift - I}} = \max \frac{\displaystyle\sum_{k=1}^{K} w_k^D z_{ko}^D}{\displaystyle\sum_{i=1}^{m} v_i x_{io} + \sum_{l=1}^{L} w_l^U z_{lo}^U}$$

$s.t.$

$$\frac{\displaystyle\sum_{k=1}^{K} w_k^D z_{ko}^D}{\displaystyle\sum_{i=1}^{m} v_i x_{io} + \sum_{l=1}^{L} w_l^U z_{lo}^U} \le 1, \tag{9.67}$$

$$\frac{\displaystyle\sum_{l=1}^{L} w_l^D z_{1j}^D}{\displaystyle\sum_{i=1}^{m} v_i x_{ij} + \sum_{l=1}^{L} w_l^U \left(z_{1j}^U + \hat{\alpha}_{1j} \right)} \le 1, \quad j = 1, ..., n, \quad j \ne o,$$

$$v_i, w_l^U, w_k^D \ge 0, \quad i = 1, ..., m, \quad l = 1, ..., L, \quad k = 1, ..., K.$$

Therefore, $\theta_{\text{shift - I}}^*$ and $\theta_{\text{shift - II}}^*$ indicate the efficiency of stages 1 and 2 in ESC based on the new frontier.

$$\theta_{\text{shift - II}} = \max \frac{\displaystyle\sum_{q=1}^{Q} u_q^D y_{qo}^D}{\displaystyle\sum_{k=1}^{K} w_k'^D z_{ko}^D + \sum_{r=1}^{s} u_r^U \left(y_{rj}^U \right)}$$

$s.t.$

$$\frac{\displaystyle\sum_{q=1}^{Q} u_q^D y_{qo}^D}{\displaystyle\sum_{k=1}^{K} w_k'^D z_{ko}^D + \sum_{r=1}^{s} u_r^U \left(y_{rj}^U \right)} \le 1, \tag{9.68}$$

$$\frac{\displaystyle\sum_{q=1}^{Q} u_q^D y_{qj}^D}{\displaystyle\sum_{k=1}^{K} w_k^D z_{ko}^D + \sum_{r=1}^{s} u_r^U \left(y_{rj}^U + \hat{\beta}_{rj} \right)} \le 1, \quad j = 1, ..., n, \quad j \ne o,$$

$$u_r^U, u_q^D, w_k'^D \ge 0, \quad r = 1, ..., s, \quad q = 1, ..., Q, \quad k = 1, ..., K.$$

And finally, model (9.69) is used to evaluate the chain's performance based on the new boundary.

$$\theta_{shift} = \max \omega^1 \theta_{shift\text{-}I} + \omega^2 \theta_{shift\text{-}II}$$

$$s.t.$$

Constraints of (9.67)),   (9.69)
Constraints of (9.68),

$$w^D = w'^D.$$

9.5.2   Performance Evaluation of Three-member Supply Chains with a Focus on the Education Supply Chain

Schools and universities play an essential role in producing knowledge in such a way that advanced societies are based on educated and creative people, so implementing a special mechanism inside schools to establish a relationship between schools and universities and training students to enter universities is essential. Considering the importance of educational systems in developing countries, many studies have evaluated educational systems in different countries. So due to the particular importance of education, we focus exclusively on the education supply chain in this sub-section.

Although the overall efficiency of the education system is the product of individual efficiency, this efficiency is directly dependent on the efficiency of primary education, secondary education, and higher education. So this highlights the interrelationship of these three parts of the process in the global education system. Thus to evaluate the efficiency of the educational system, evaluation of all academic stages is necessary simultaneously.

Therefore, Ramzi [38] believes that indicators related to the internal linking of education should also be considered to evaluate education systems. Thus, to assess the educational system of Tunisia, the supply chain shown in Fig. 9.13 was evaluated.

The primary input in the entire educational system is the number of schools and students enrolled in the first year of primary education. The output produced from primary education is the number of graduate students in primary school who are entered as input in the secondary education system. The number of graduates in this stage is considered the output. And finally, university graduates are considered the final output of the education system.

So far, supply chain management includes various issues such as purchasing, production, etc. However, there is little evidence about the application of supply

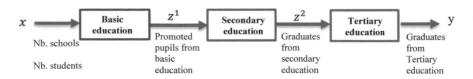

Fig. 9.13 The education supply chain as a three stage-process [38]

chains in the service sector, including education, while educational centers are suppliers for other sectors in societies. Therefore, educational systems are also a supply chain. Chains in which human, physical and financial resources are considered input, and each member of this chain is a supplier of another member. For example, students in secondary education are supplied by graduates of primary education, and finally, the workforce needed by society is provided by the graduates of the higher education department. Therefore, considering the interrelationship between primary, secondary, and higher education levels, education is regarded as a three-members supply chain. Improving performance in all three members helps improve the overall performance of the education supply chain.

Therefore, to evaluate the education supply chain, the development of the Kao model is used to assess the 3-member supply chain according to Fig. 9.13. Because the Kao model provides the possibility of analyzing the educational system's inefficiency and identifying the reasons for the inefficiency of the entire educational system as a solution for future improvements. As a result, model (9.70) is used to evaluate the performance of three-member supply chains, especially the education supply chain.

$$
\begin{aligned}
&\max \ u y_o \\
&s.t. \\
&v x_o = 1, \\
&u y_j - v x_j \le 0, \ j = 1, ..., n, \\
&w^1 z_j^1 - v x_j \le 0, \ j = 1, ..., n, \\
&w^2 z_j^2 - w^1 z_j^1 \le 0, \ j = 1, ..., n, \\
&u y_j - w^2 z_j^2 \le 0, \ j = 1, ..., n, \\
&\left(u, w^1, w^2, v \right) \ge 0.
\end{aligned}
\tag{9.70}
$$

Also, relations (9.71), (9.72), and (9.73) are used for the efficiency of stages 1, 2, and 3, respectively.

$$
e_o^1 = \frac{w^{1*} z_o^1}{v^* x_o}
\tag{9.71}
$$

$$
e_o^2 = \frac{w^{2*} z_o^2}{w^{1*} z_o^1}
\tag{9.72}
$$

$$
e_o^3 = \frac{u^* y_o}{w^{2*} z_o^2}
\tag{9.73}
$$

So that $\left(v^*, w^{1*}, w^{2*}, u^* \right)$ is the optimal solution of the model (9.70). It is necessary to explain that if the model (9.70) has alternative optimal solutions, then it is possible to calculate an interval of efficiency changes for each stage by using different subproblems on the optimal solution of the model (9.70).

9.5.3   Performance Evaluation of Mixed Supply Chains

To start the discussion in this section, consider the supply chain of goods transportation in shipping lines as shown in Fig. 9.14. This supply chain includes three parts supplying (providing basic needs for starting and maintaining the ship), Producing (production of services), and distribution (distribution of services). And each stage contains several sub-stages, as shown in Fig. 9.14.

The supplier department includes five sub-departments: Ship Finance, Ship Manning, Technical Provision, Technical Repairs, Technical Supply, and Maintenance. After receiving appropriate inputs from the suppliers of the shipping industry, a ship is fully prepared for the transportation of goods and passengers, that is, the production of services. Therefore, the production stage in the shipping supply chain includes two sub-stage: passenger service and container service containing required goods. Finally, the products of the previous stage are sold or distributed by distribution representatives and sales staff in the distribution stage.

As a result, as you can see in Fig. 9.14, the shipping supply chain is mixed (parallel and series). Therefore, with the development of Kao's [20] and Kao and Hwung's [21] model, model (9.74) for the evaluation of mixed supply chains according to Fig. 9.14 was presented by Omrani and Keshavarz [37]. Although the model (9.74) is shown only for evaluating mixed supply chains according to Fig. 9.14, it can be easily developed for other mixed supply chains.

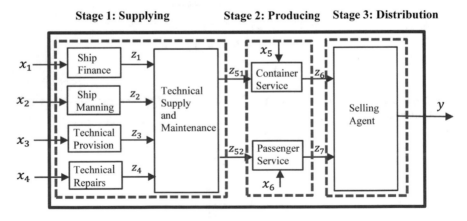

Fig. 9.14 Supply chain of shipping line in Iran

$$\max \ u y_o$$

$$s.t.$$

$$v^1 x_o^1 + v^2 x_o^2 + v^3 x_o^3 + v^4 x_o^4 + v^5 x_o^5 + v^6 x_o^6 = 1,$$

$$u y_j - \left(v^1 x_j^1 + v^2 x_j^2 + v^3 x_j^3 + v^4 x_j^4 + v^5 x_j^5 + v^6 x_j^6 \right) \leq 0, \ j = 1, ..., n,$$

$$w^1 z_j^1 - v^1 x_j^1 \leq 0, \ j = 1, ..., n,$$

$$w^2 z_j^2 - v^2 x_j^2 \leq 0, \ j = 1, ..., n,$$

$$w^3 z_j^3 - v^3 x_j^3 \leq 0, \ j = 1, ..., n,$$

$$w^4 z_j^4 - v^4 x_j^4 \leq 0, \ j = 1, ..., n, \tag{9.74}$$

$$w^{51} z_j^{51} + w^{52} z_j^{52} - \left(v^1 x_j^1 + v^2 x_j^2 + v^3 x_j^3 + v^4 x_j^4 \right) \leq 0, \ j = 1, ..., n,$$

$$w^6 z_j^6 - \left(w^{51} z_j^{51} + v^5 x_j^5 \right) \leq 0, \ j = 1, ..., n,$$

$$w^7 z_j^7 - \left(w^{52} z_j^{52} + v^6 x_j^6 \right) \leq 0, \ j = 1, ..., n,$$

$$u y_j - \left(w^6 z_j^6 + w^7 z_j^7 \right) \leq 0, \ j = 1, ..., n,$$

$$\left(v^1, ..., v^6 \right) \geq 0, \ u \geq 0,$$

$$\left(w^1, ..., w^7, w^{51}, w^{52} \right) \geq 0.$$

So that the second constraint is related to the overall efficiency of the supply chain, constraints 3 to 6 are related to the efficiency of each of the sub-stages in the supplier stage, and constraint 7 is related to the supplier's efficiency. Also, constraints 8 and 9 correspond to the efficiency of stages 1 and 2 in the production stage. Finally, constraint 10 is related to the distribution department.

9.5.4 Performance Evaluation of Supply Chains in the Presence of Returnable Relations and Identifying the Unit of Relative Efficient

Suppose in a supply chain of merchandise, outputs, in addition to final merchandise, contain iron filings and wastewater. The made products are delivered to the customer and, in other words, leave the chain. Still, the iron filings and wastewater are returned to the system for purification, recycling, and reuse. Therefore, in such a situation, we face returnable relations in the supply chain, as shown in Fig. 9.15. While this issue has been ignored in the previous models.

Also, as seen in the previous sections, one of the strengths of DEA models is the identification of relatively efficient units, while this is not necessarily the case when using NDEA models. Therefore, Shafiee et al. [48] presented a new model for evaluating supply chains to solve the mentioned problems.

z_{rj}^{pq} r-th $\left(r = 1, ..., s_{pq} \right)$ element of a s_{pq}-D vector of DMU$_j$ that leaves stage p and enters stage q.

Fig. 9.15 General structure of the supply chain with returnable relation

u_r^{pq} The weight is related to the r-th element of the output vector that exits stage p and enters stage q.

v_r^{qp} The weight is associated with the r-th element of the input vector of stage q and the output of stage p.

So the efficiency of stage p of DMU$_j$ is defined as relation (9.75).

$$\theta_{pj} = \frac{\sum_q \sum_r u_r^{pq} z_{rj}^{pq}}{\sum_q \sum_r v_r^{qp} z_{rj}^{pq}} \tag{9.75}$$

Also, the overall efficiency of network j is defined as a convex combination of the efficiency of the stages as follows.

$$\theta_j = \sum_{j=1}^{p} w_p \theta_{pj}, \; \sum_{j=1}^{p} w_p = 1 \tag{9.76}$$

As you know, the choice of weights w_p is a critical issue in evaluating the performance of a network. Therefore, we define the appropriate weights as the ratio of the inputs used in stage p to the total inputs used in the network j, similar to the idea presented by Cook et al. [8]. Therefore:

$$w_p = \frac{\sum_q \sum_r u_r^{pq} z_{rj}^{pq}}{\sum_p \sum_q \sum_r v_r^{qp} z_{rj}^{pq}}, \; p = 1, ..., k \tag{9.77}$$

As a result, the efficiency of the network j is:

$$\theta_j = \frac{\sum_p \sum_q \sum_r u_r^{pq} z_{rj}^{pq}}{\sum_p \sum_q \sum_r v_r^{qp} z_{rj}^{pq}}, \; j = 1, ..., n \tag{9.78}$$

Therefore, model (9.79) is introduced to evaluate the supply chain performance.

$$\max \ \theta_o$$

$$s.t.$$

$$\theta_j \leq 1, \ j = 1, ..., n, \tag{9.79}$$

$$\theta_{pj} \leq 1, \ j = 1, ..., n, \ p = 1, ..., P,$$

$$u_r^{pq} \geq 0, v_i^{qp} \geq 0, r = 1, ..., s_{pq}, \ p = 1, ..., P, \ q = 1, ..., Q.$$

Therefore, by replacing relations (9.77) and (9.78) in model (9.79), model (9.80) is obtained.

$$\max \ \frac{\sum_p \sum_q \sum_r u_r^{pq} z_{ro}^{pq}}{\sum_p \sum_q \sum_r v_r^{qp} z_{ro}^{pq}}$$

$$s.t.$$

$$\frac{\sum_p \sum_q \sum_r u_r^{pq} z_{rj}^{pq}}{\sum_p \sum_q \sum_r v_r^{qp} z_{rj}^{pq}} \leq 1, \ j = 1, ..., n, \tag{9.80}$$

$$\frac{\sum_q \sum_r u_r^{pq} z_{rj}^{pq}}{\sum_q \sum_r v_r^{qp} z_{rj}^{pq}} \leq 1, \ j = 1, ..., n, \ p = 1, ..., P,$$

$$u_r^{pq} \geq 0, v_i^{qp} \geq 0, r = 1, ..., s_{pq}, \ p = 1, ..., P, \ q = 1, ..., Q.$$

Therefore, using Charnes and Cooper transformations, the model (9.80) linearizes as (9.81).

$$\max \sum_p \sum_q \sum_r u_r^{pq} z_{ro}^{pq}$$

$$s.t.$$

$$\sum_p \sum_q \sum_r v_r^{qp} z_{ro}^{pq} = 1,$$

$$\sum_p \sum_q \sum_r u_r^{pq} z_{rj}^{pq} - \sum_p \sum_q \sum_r v_r^{qp} z_{rj}^{pq} \leq 0, j = 1, ..., n, \tag{9.81}$$

$$\sum_q \sum_r u_r^{pq} z_{rj}^{pq} - \sum_q \sum_r v_r^{qp} z_{rj}^{pq} \leq 0, j = 1, ..., n, \ p = 1, ..., P,$$

$$u_r^{pq} \geq 0, v_i^{qp} \geq 0, r = 1, ..., s_{pq}, \ p = 1, ..., P, \ q = 1, ..., Q.$$

Lemma 9.2 If DMU$_o$ is the unit under evaluation and (u^*, v^*) is an optimal solution, then there exists an index like l so that: $\sum_p \sum_q \sum_r u_r^{pq*} z_{rl}^{pq} - \sum_p \sum_q \sum_r v_r^{qp*} z_{rl}^{pq} = 0.$

Proof We write model (9.82) as follows:

$$\max \sum_{p} \sum_{q} \sum_{r} u_r^{pq} z_{ro}^{pq}$$

$$s.t.$$

$$\sum_{p} \sum_{q} \sum_{r} v_r^{qp} z_{ro}^{pq} = 1, \tag{9.82}$$

$$\sum_{p} \sum_{q} \sum_{r} u_r^{pq} z_{rj}^{pq} - \sum_{p} \sum_{q} \sum_{r} v_r^{qp} z_{rj}^{pq} \leq 0, j = 1, ..., n,$$

$$u_r^{pq} \geq 0, v_i^{qp} \geq 0, r = 1, ..., s_{pq}, \ p = 1, ..., P, \ q = 1, ..., Q.$$

It is evident that the feasible region of the model (9.81) is a subset of the feasible region of the model (9.82), but the value of the objective function in both of them is equal. Therefore, the optimal solution of model (9.81) is also the optimal solution of model (9.82). Let $(\overline{u}, \overline{v})$ be the optimal solution for model (9.82). If an optimal solution exists, then an optimal extreme also exists. Let (u^*, v^*) be the optimal extreme solution for model (9.82), so it lies on $k = 2 \sum_q \sum_p s^{pq}$ linearly independent hyperplanes. We prove that all variables are not binding in the optimal solution.

On the one hand, every v_i^{qp*} cannot be equal to zero because, in this case: $\sum_p \sum_q \sum_r v_r^{qp*} z_{ro}^{pq} = 0 \neq 1$. On the other hand: every u_r^{pq*} cannot be equal to zero because the value of the objective function is equal to 0. So at most, the $(k-2)$ of variables in optimality are binding.

Moreover $\sum_p \sum_q \sum_r v_r^{qp} z_{ro}^{pq} = 1$ is binding at every feasible solution, so this constraint is binding at the optimal solution. Consequently, a least one of the constraints $\sum_p \sum_q \sum_r u_r^{pq} z_{rj}^{pq} - \sum_p \sum_q \sum_r v_r^{qp} z_{rj}^{pq} \leq 0$, in optimality should be binding. So $\sum_p \sum_q \sum_r u_r^{pq*} z_{rl}^{pq} - \sum_p \sum_q \sum_r v_r^{qp*} z_{rl}^{pq} = 0$ and since (u^*, v^*) is the optimal solution for model (9.82), it is an optimal solution for (9.81). Therefore, the proof is complete.

Theorem 9.4 If $DMU_1, \ldots, DMU_n$ be DMU_s which are under evaluation, then there exists at least one relative efficient DMU.

Proof Let (u^*, v^*) be the optimal solution for model (9.81) when DMU_o is under evaluation. By using Lemma 9.2, we know that $\sum_p \sum_q \sum_r u_r^{pq*} z_{rl}^{pq} - \sum_p \sum_q \sum_r v_r^{qp*} z_{rl}^{pq} = 0$. The efficiency of DMU_l is gained by solving the following model:

$$\max \sum_{p} \sum_{q} \sum_{r} u_r^{pq} z_{rl}^{pq}$$

$$s.t.$$

$$\sum_{p} \sum_{q} \sum_{r} v_r^{qp} z_{rl}^{pq} = 1, \tag{9.83}$$

$$\sum_{p} \sum_{q} \sum_{r} u_r^{pq} z_{rj}^{pq} - \sum_{p} \sum_{q} \sum_{r} v_r^{qp} z_{rj}^{pq} \leq 0, j = 1, ..., n,$$

$$u_r^{pq} \geq 0, v_i^{qp} \geq 0, \ r = 1, ..., s_{pq}, \ p = 1, ..., P, \ q = 1, ..., Q.$$

If $\sum_p \sum_q \sum_r v_r^{qp^*} z_{rl}^{pq} = 1$, the proof is evident. Otherwise $\sum_p \sum_q \sum_r v_r^{qp^*} z_{rl}^{pq} = \alpha > 0$. So we have: $\sum_p \sum_q \sum_r \left(\frac{v_r^{qp^*}}{\alpha}\right) z_{rl}^{pq} = 1$ and $\sum_p \sum_q \sum_r \left(\frac{u_r^{pq^*}}{\alpha}\right) z_{rj}^{pq} - \sum_p \sum_q \sum_r \left(\frac{v_r^{qp^*}}{\alpha}\right) z_{rj}^{pq} \leq 0.$

These relations imply that $(u^*/\alpha, v^*/\alpha)$ is a feasible solution for model (9.82) and since the optimal solution for model (9.82) is the optimal solution in the model (9.81); the objective value is equal to 1 for model (9.82) and (9.81). So DMU$_l$ is efficient.

9.6 Performance Evaluation of Supply Chains Using Game Theory

Game theory, as one of the branches of mathematical science, was presented by Neumann [35] for the first time and quickly spread to other sciences, especially economics. The game theory studies the behavior of decision-makers in different conditions and situations. And it is assumed that every decision-maker behaves entirely logically and wise. Also, the goal of each decision-maker is to make a decision that leads to more profit at the end of the game. Of course, the benefit that each player receives at each stage of each game depends not only on his own decision but also on the decisions of other decision-makers. Each situation is referred to as a "game", and each decision maker is referred to as a "player". Also, the choices or decisions that each player can make are called the "strategy" of that player. According to different conditions, there are many classifications for all kinds of games in this science. One of these general classifications is cooperative and non-cooperative games. In cooperative games, it is assumed that there is full cooperation between the players to achieve the optimal result for the whole system. In these games, the problem is changed from the state of several decision-makers and several criteria to the state of one decision-maker. In other words, in cooperative games, decision-makers are integrated, and a combined objective function is created so that the problem is transformed from a multi-objective state to a single-objective one. In such issues, only one informed decision-maker is defined, and as a result, the final decision is made fairly and based on the criteria announced by all interested parties. But in non-cooperative games, the default is that the main concern of the players is to maximize their profit, and the players do not have any desire to exchange opinions or negotiate to reach a coalition or agree with each other. Therefore, the decisions made by all players affect the final results of non-cooperative games, but in non-cooperative games, the default is that the primary concern of the players is to maximize their profit, and the players do not have any desire to exchange opinions or negotiate to reach a coalition or agree with each other. Due to the increasing expansion of game theory in various sciences, the use of game theory to provide an analytical framework for analyzing and configuring the supply chain is also of interest to researchers. Also, several studies have been conducted using game theory in DEA and evaluating the performance of the units under evaluation, including (Lozano [32], Rezaee et al. [39], Lozano et al. [33], Yaya et al. [59]). Therefore, in this part, we will focus on assessing the supply chain using NDEA and the approach of cooperative and non-cooperative games.

9.6.1 Performance Evaluation Using NDEA Models with the Approach of Non-cooperative Games

Consider a two-stage supply chain, buyer and seller, as shown in Fig. (9.16). So that x^1 is the seller's input vector, and z is the seller's output vector (products manufactured by the seller), which the buyer uses as input. The vector y is the buyer's output vector, produced by consuming the vectors z, and x^2 (independent buyer input). As an example, in the research conducted by Liang et al. [29], it is assumed that the seller consumes inputs such as labor and operating costs (transportation costs, etc.) to send the products (z^1) to the buyer. In the next step, the buyer, with the participation of his independent labor (z^2), sells the products received from the seller (z^1) to obtain the final profit y.

In most practical matters, each buyer and seller usually prioritize their profit, so we face a non-cooperative game in such a situation. Therefore, two paths are possible for this non-cooperative game. In the first case, the seller is considered the leader of the chain and the buyer the follower, but in the second case, the buyer is regarded as the leader and policy-determiner of the chain, and the seller is required to follow it.

(A) **Seller (leader)—Buyer (follower)**

According to Li et al. [27], the interaction between the seller and the buyer is a two-stage non-cooperative game where the seller is the leader, and the buyer is the follower. Because in many real-world problems, the buyer has no control over the seller, and the seller determines the optimal strategy (optimal weights for intermediate criteria). For example, consider a non-cooperative relationship between the leader and the follower of the Toyota car company. Suppose Toyota decides to introduce a new product. Therefore, it allocates an amount for advertising to its local agencies. Then, based on the amount received and their local strategies, each agency decides how to adjust the cost of advertising. Therefore, in such a situation, we need to use NDEA models with

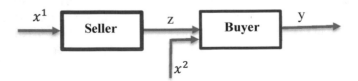

Fig. 9.16 A series seller-buyer supply chain

a non-cooperative approach to model the relationship between the buyer and the seller. As a result, according to the adoption of a non-cooperative strategy by the seller and the buyer, the weight of the intermediate products used as the output of the first stage and the input of the second stage should be considered differently from each other.

Liang et al. [29], to evaluate the performance of the supply chain shown in Fig. (9.16) with a non-cooperative approach, first calculated the performance of the seller (leader) using the CCR model. Suppose $e^{*seller}$ is the optimal solution of the CCR model, so to evaluate the buyer (follower) model (9.84) was proposed by Liang et al. [29]

$$e^{Buyer \text{ - } Seller} = \max \frac{\sum\limits_{r=1}^{s} u_r y_{ro}}{\sum\limits_{i=1}^{m} v_i^2 x_{io}^2 + \sum\limits_{d=1}^{D} w_d z_{do}}$$

$s.t.$

$$\frac{\sum\limits_{r=1}^{s} u_r y_{rj}}{\sum\limits_{i=1}^{m} v_i^2 x_{ij}^2 + \sum\limits_{d=1}^{D} w_d z_{dj}} \leq 1, \ j = 1, ..., n,$$

$$\frac{\sum\limits_{d=1}^{D} w_d z_{do}}{\sum\limits_{i=1}^{m} v_i^1 x_{io}^1} = e^{*seller}, \tag{9.84}$$

$$\frac{\sum\limits_{d=1}^{D} w_d z_{dj}}{\sum\limits_{i=1}^{m} v_i^1 x_{ij}^1} \leq 1, \ j = 1, ..., n,$$

$$v_i^1, w_d, v_k^2, u_r \geq 0,$$

$$i = 1, ..., m, \ d = 1, ..., D, \ k = 1, ..., K, \ r = 1, ..., s.$$

In the following, using Charnes-Cooper transformations and the definition of $Q = \sum_{i=1}^{m} v_i^1 x_{io}^1$, the model (9.85) is obtained.

$$e^{\text{Buyer - Seller}} = \max \frac{\sum\limits_{r=1}^{s} u_r y_{ro}}{\sum\limits_{i=1}^{m} v_i^2 x_{io}^2 + Q \times \sum\limits_{d=1}^{D} w_d z_{do}}$$

$s.t.$

$$\frac{\sum\limits_{r=1}^{s} u_r y_{rj}}{\sum\limits_{i=1}^{m} v_i^2 x_{ij}^2 + Q \times \sum\limits_{d=1}^{D} w_d z_{dj}} \leq 1, \; j = 1, ..., n,$$

$$\sum\limits_{d=1}^{D} w_d z_{do} = e^{*\text{seller}},$$

$$\sum\limits_{i=1}^{m} v_i^1 x_{io}^1 = 1,$$

$$\sum\limits_{d=1}^{D} w_d z_{dj} - \sum\limits_{i=1}^{m} v_i^1 x_{ij}^1 \leq 0, \; j = 1, ..., n,$$

$$v_i^1, w_d, v_k^2, u_r, Q \geq 0,$$

$$i = 1, ..., m, \; d = 1, ..., D, \; k = 1, ..., K, \; r = 1, ..., s.$$

(9.85)

Note that in the model (9.85), the efficiency of the buyer is determined based on the optimal weights of the seller. In other words, the buyer's efficiency is calculated so that the efficiency of the seller (leader) remains equal to $e^{*\text{seller}}$. In the following, using Charnes-Cooper transformations again, the model (9.85) is transformed into (9.86).

$$e^{\text{Buyer - Seller}} = \max \sum\limits_{r=1}^{s} \mu_r y_{ro}$$

$s.t.$

$$\sum\limits_{i=1}^{m} \gamma_i^2 x_{io}^2 + q \times \sum\limits_{d=1}^{D} w_d z_{do} = 1,$$

$$\sum\limits_{r=1}^{s} u_r y_{rj} - \left(\sum\limits_{i=1}^{m} \gamma_i^2 x_{ij}^2 + q \times \sum\limits_{d=1}^{D} w_d z_{dj} \right) \leq 0, \; j = 1, ..., n,$$

$$\sum\limits_{d=1}^{D} w_d z_{do} = e^{*\text{seller}},$$

$$\sum\limits_{i=1}^{m} \gamma_i^1 x_{io}^1 = 1,$$

$$\sum\limits_{d=1}^{D} w_d z_{dj} - \sum\limits_{i=1}^{m} \gamma_i^1 x_{ij}^1 \leq 0, \; j = 1, ..., n,$$

$$\gamma_i^1, w_d, \gamma_k^2, \mu_r, q \geq 0,$$

$$i = 1, ..., m, \; d = 1, ..., D, \; k = 1, ..., K, \; r = 1, ..., s.$$

(9.86)

As seen, due to the existence of q, model (9.86) cannot be solved easily. Therefore, this problem can be solved by determining a range for q and considering an initial value for q. It is evident that $\sum_{i=1}^{m} \gamma_i^2 x_{io}^2 \geq 0$ and $\sum_{d=1}^{D} w_d z_{do} \geq 0$. As a result: $0 \leq q \sum_{d=1}^{D} w_d z_{do} \leq 1 \Rightarrow 0 \leq q \leq \frac{1}{w_d z_{do}} = \frac{1}{e^{*seller}}$.

Therefore, to solve the model (9.86), we consider the value of q as a parameter in the interval $\left[0, \frac{1}{e^{*seller}}\right]$. So, we define: $q_t = \frac{1}{e^{*seller}} - \varepsilon \times t$, so that ε is a very small positive number. As a result, in each iteration t for a specific value of q_t, we calculate the value of $e^{\text{Buyer - Seller}}$ and display it with the symbol $e^{\text{Buyer - Seller}}(q_t)$ and then $e^{*\text{Buyer - Seller}} = \max\{e^{\text{Buyer - Seller}}(q_t)\}$. Finally, the overall efficiency of the supply chain is equal to:

$$e^{\text{Total}} = \frac{e^{*\text{Buyer - Seller}} + e^{*\text{Seller}}}{2} \tag{9.87}$$

(B) Buyer (Leader)-Seller (Follower)

Although in most practical issues, the buyer is the leader and the seller is the follower, nowadays, in some cases, this role is sometimes changed due to the buyers' bargaining power growth. One of the most successful examples in this field is Walmart. In a report published on October 3, 2003, Walmart has much influence over its suppliers and dominates them. Therefore, contrary to the previous part, the buyer is present as the leader and the seller as the follower in such a situation.

Let $e^{*\text{Buyer}}$ be the buyer's efficiency obtained by the CCR model. Therefore, the model (9.88) is proposed to evaluate the seller's efficiency such that the seller's efficiency is obtained under the influence of optimal weights for assessing the buyer.

$$e^{\text{Seller - Buyer}} = \max \frac{Q \times \sum_{d=1}^{D} w_d z_{do}}{\sum_{i=1}^{m} v_i^1 x_{io}^1}$$

$s.t.$

$$\frac{Q \times \sum_{d=1}^{D} w_d z_{dj}}{\sum_{i=1}^{m} v_i^1 x_{ij}^1} \leq 1, \ j = 1, \ldots, n,$$

$$\sum_{r=1}^{s} u_r y_{ro} = e^{*\text{Buyer}},$$

$$\sum_{d=1}^{D} w_d z_{do} + \sum_{k=1}^{K} v_k^2 x_{ko}^2 = 1,$$

$$\sum_{r=1}^{s} u_r y_{rj} - \left(\sum_{d=1}^{D} w_d z_{dj} + \sum_{k=1}^{K} v_k^2 x_{kj}^2 \right) \leq 0, \ j = 1, \ldots, n,$$

$$v_i^1, \ w_d, \ v_k^2, \ u_r, \ Q \geq 0,$$

$$i = 1, \ldots, m, \ d = 1, \ldots, D, \ k = 1, \ldots, K, \ r = 1, \ldots, s.$$

(9.88)

Next, with a process similar to what was mentioned for model (9.84), model (9.88) is rewritten as the model (9.89).

$$e^{\text{Seller - Buyer}} = \max q \times \sum_{d=1}^{D} w_d z_{do}$$

$s.t.$

$$\sum_{i=1}^{m} \gamma_i^1 x_{io}^1 = 1,$$

$$q \times \sum_{d=1}^{D} w_d z_{dj} - \sum_{i=1}^{m} \gamma_i^1 x_{ij}^1 \leq 0, \ j = 1, \ldots, n,$$

$$\sum_{r=1}^{s} u_r y_{ro} = e^{*\text{Buyer}},$$

$$\sum_{d=1}^{D} w_d z_{do} + \sum_{k=1}^{K} v_k^2 x_{ko}^2 = 1,$$

$$\sum_{r=1}^{s} u_r y_{rj} - \left(\sum_{d=1}^{D} w_d z_{dj} + \sum_{k=1}^{K} v_k^2 x_{kj}^2 \right) \leq 0, \ j = 1, \ldots, n,$$

$$\gamma_i^1, \ w_d, \ v_k^2, \ u_r, \ q \geq 0,$$

$$i = 1, \ldots, m, \ d = 1, \ldots, D, \ k = 1, \ldots, K, \ r = 1, \ldots, s.$$

(9.89)

Model (9.89) is a non-linear programming problem. Therefore, this problem can be solved by determining a range for q and considering an initial value for

q. For this purpose, first, obtain the optimal value of the model (9.90).

$$\max \frac{Q \times \sum_{d=1}^{D} w_d^* z_{do}}{\sum_{i=1}^{m} v_i^1 x_{io}^1}$$

s.t.

$$\frac{Q \times \sum_{d=1}^{D} w_d^* z_{dj}}{\sum_{i=1}^{m} v_i^1 x_{ij}^1} \leq 1, \ j = 1, \ldots, n, \qquad (9.90)$$

$$v_i^1, \ w_d, \ Q \geq 0, \ i = 1, \ldots, m, \ d = 1, \ldots, D.$$

So that w^* is the optimal answer of the CCR model for evaluating the buyer. Also, (9.90) is rewritten as the model (9.91).

$$E = \max q \times \sum_{d=1}^{D} w_d^* z_{do}$$

s.t.

$$\sum_{i=1}^{m} \gamma_i^1 x_{io}^1 = 1, \qquad (9.91)$$

$$q \times \sum_{d=1}^{D} w_d^* z_{dj} - \sum_{i=1}^{m} \gamma_i^1 x_{ij}^1 \leq 0, \ j = 1, \ldots, n,$$

$$\gamma_i^1, \ w_d, \ q \geq 0,$$

$$i = 1, \ldots, m, \ d = 1, \ldots, D.$$

We know that $q \times \sum_{d=1}^{D} w_d z_{do} \geq E^*$. Also $\sum_{d=1}^{D} w_d z_{do} + \sum_{k=1}^{K} v_k^2 x_{ko}^2 = 1$, $\sum_{d=1}^{D} w_d z_{do} \geq 0$ and $\sum_{k=1}^{K} v_k^2 x_{ko}^2 \geq 0$. As a result $\sum_{d=1}^{D} w_d z_{do} \leq 1$, so $q \geq E^*$.

Therefore, the starting point of movement for parameter q equals $q = E^*$. Now consider the change of variables $\lambda_d = q w_d$ and $g = \frac{1}{q}$. So based on the stated change of variables, the model (9.89) is converted into a model (9.92) based on the g parameter.

$$e^{\text{Seller - Buyer}} = \max \sum_{d=1}^{D} \lambda_d z_{do}$$

$$s.t.$$

$$\sum_{i=1}^{m} \gamma_i^1 x_{io}^1 = 1,$$

$$\sum_{d=1}^{D} \lambda_d z_{dj} - \sum_{i=1}^{m} \gamma_i^1 x_{ij}^1 \leq 0, \ j = 1, \ldots, n,$$

$$\sum_{r=1}^{s} u_r y_{ro} = e^{*\text{Buyer}}, \qquad\qquad (9.92)$$

$$g \sum_{d=1}^{D} \lambda_d z_{do} + \sum_{k=1}^{K} v_k^2 x_{ko}^2 = 1,$$

$$\sum_{r=1}^{s} u_r y_{rj} - \left(g \sum_{d=1}^{D} \lambda_d z_{dj} + \sum_{k=1}^{K} v_k^2 x_{kj}^2 \right) \leq 0, \ j = 1, \ldots, n,$$

$$\gamma_i^1, \ \lambda_d, \ v_k^2, \ u_r \geq 0,$$

$$i = 1, \ldots, m, \ d = 1, \ldots, D, \ k = 1, \ldots, K, \ r = 1, \ldots, s,$$

$$g \geq 0.$$

According to the relation $q \geq E^*$, it is easily concluded that $0 \leq g \leq \frac{1}{q}$. In this way, by setting the parameter g, the optimal value of $e^{*\text{Seller - Buyer}}$ and $e^{\text{Total}} = e^{*\text{Buyer}} + e^{*\text{Seller - Buyer}} / 2$ is obtained with a process similar to part (A).

9.6.2 Performance Evaluation Using NDEA Models with the Approach of Cooperative Games

In some issues, it is assumed that the seller and the buyer have the same power to influence the supply chain system and the cooperation of the two is beneficial to the supply chain. In such a case, contrary to the process mentioned in the previous subsection, we are facing a cooperative game, which we will explain in detail in the following.

Again, consider the buyer–seller supply chain as shown in Fig. 9.16. Suppose that the seller and the buyer have the same power to influence the supply chain system, and the cooperation benefits the supply chain. Therefore, the modeling process seeks to maximize the efficiency of the seller and the buyer to achieve a powerful supply chain. In such a situation, the cooperation between the members (buyer and seller) benefits the chain. So the weight of intermediate products when it is the output of the first stage is considered equal to its weight when it is the input of the second stage.

Therefore, the efficiency of the buyer and seller is equal to:

$$e^{\text{Seller}} = \frac{\sum\limits_{d=1}^{D} w_d z_{do}^1}{\sum\limits_{i=1}^{m} v_i^1 x_{io}^1} \tag{9.93}$$

and

$$e^{\text{Buyer}} = \frac{\sum\limits_{r=1}^{s} u_r y_{ro}}{\sum\limits_{d=1}^{D} w_d z_{do}^1 + \sum\limits_{k=1}^{K} v_k^2 x_{ko}^2} \tag{9.94}$$

In the same way, total efficiency is defined as the average efficiency of buyer and seller:

$$e^{\text{Total}} = \frac{1}{2} \left[\frac{\sum\limits_{d=1}^{D} w_d z_{do}^1}{\sum\limits_{i=1}^{m} v_i^1 x_{io}^1} + \frac{\sum\limits_{r=1}^{s} u_r y_{ro}}{\sum\limits_{d=1}^{D} w_d z_{do}^1 + \sum\limits_{k=1}^{K} v_k^2 x_{ko}^2} \right] \tag{9.95}$$

As a result, the model (9.96) was introduced by [29] to evaluate the performance of the supply chain shown in Fig. 9.16:

$$\max \frac{1}{2} \left[\frac{\sum\limits_{d=1}^{D} w_d z_{do}^1}{\sum\limits_{i=1}^{m} v_i^1 x_{io}^1} + \frac{\sum\limits_{r=1}^{s} u_r y_{ro}}{\sum\limits_{d=1}^{D} w_d z_{do}^1 + \sum\limits_{k=1}^{K} v_k^2 x_{ko}^2} \right]$$

$$s.t.$$

$$\frac{\sum\limits_{d=1}^{D} w_d z_{dj}^1}{\sum\limits_{i=1}^{m} v_i^1 x_{ij}^1} \leq 1, \ j = 1, \ldots, n, \tag{9.96}$$

$$\frac{\sum\limits_{r=1}^{s} u_r y_{rj}}{\sum\limits_{d=1}^{D} w_d z_{dj}^1 + \sum\limits_{k=1}^{K} v_k^2 x_{kj}^2} \leq 1, j = 1, \ldots, n,$$

$$v_i^1, w_d, v_k^2, u_r \geq 0,$$

$$i = 1, \ldots, m, \ d = 1, \ldots, D, \ k = 1, \ldots, K, \ r = 1, \ldots, s.$$

Using Charnes and Cooper transformations, define: $t_1 = 1 \Big/ \sum\limits_{i=1}^{m} v_i^1 x_{io}^1$ and $t_2 =$

$1 \Big/ \sum\limits_{d=1}^{D} w_d z_{do}^1 + \sum\limits_{k=1}^{K} v_k^2 x_{ko}^2$. As a result, by changing the variable $\gamma^1 = t_1 v^1$, $\gamma^2 = t_2 v^2$, $\omega^1 = t_1 w^1$, and $\mu = t_2 u$, it can be shown that $\omega^2 = \lambda \omega^1$, $\lambda \geq 0$. Therefore, using the mentioned changes, the model (9.95) becomes model (9.96).

$$\max \frac{1}{2}\left[\sum_{d=1}^{D} \omega_d^1 z_{do} + \sum_{r=1}^{s} \mu_r y_{ro} \right]$$

$s.t.$

$$\sum_{i=1}^{m} \gamma_i^1 x_{io}^1 = 1,$$

$$\sum_{d=1}^{D} \lambda \omega_d^1 z_{do} + \sum_{k=1}^{K} \gamma_k^2 x_{ko}^2 = 1, \qquad\qquad (9.97)$$

$$\sum_{d=1}^{D} \omega_d^1 z_{dj} - \sum_{i=1}^{m} \gamma_i^1 x_{ij}^1 \leq 0, \ j = 1, \ldots, n,$$

$$\sum_{r=1}^{s} \mu_r y_{rj} - \left(\sum_{d=1}^{D} \lambda \omega_d^1 z_{dj} + \sum_{k=1}^{K} \gamma_k^2 x_{kj}^2 \right) \leq 0, \ j = 1, \ldots, n,$$

$$\gamma_i^1, \ \omega_d^1, \ \gamma_k^2, \ \mu_r, \ \lambda \geq 0,$$

$$i = 1, \ldots, m, \ d = 1, \ldots, D, \ k = 1, \ldots, K, \ r = 1, \ldots, s.$$

Note that $\lambda = \dfrac{1 - \sum\limits_{k=1}^{K} \gamma_k^2 x_{ko}^2}{\sum\limits_{d=1}^{D} \omega_d^1 z_{do}} \leq \dfrac{1}{\sum\limits_{d=1}^{D} \omega_d^1 z_{do}}$, on the other hand, suppose that the

optimal solution of model (9.89) is equal to $e^{*\text{Seller - Buyer}}$, in this case: $\sum\limits_{d=1}^{D} \omega_d^1 z_{do} \geq$

$e^{*\text{Seller - Buyer}}$. As a result: $0 \leq \lambda \leq \frac{1}{e^{*\text{Seller - Buyer}}}$, therefore, it can be solved by using parametric programming of the model (9.97).

9.7 Performance Evaluation of Supply Chains in the Presence of Fuzzy Data Using FNDEA

In the supply chains discussed in the previous sections, it is assumed that the values of all independent inputs, final outputs, and intermediate products in the supply chain are expressed as definite values. This assumption in many practical problems in the real world is very illogical and cumbersome. For example, consider the supply chain

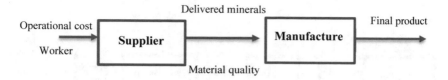

Fig. 9.17 The supply chain with fuzzy index

shown in Fig. 9.17. Unlike other indicators, material quality is a qualitative index. As explained in the previous chapters, in these cases, the fuzzy theory is used to describe the supply chain.

Therefore, the models presented in this section are not suitable for evaluating the performance of supply chains with fuzzy indicators. Although there are several methods for performance evaluation of DMUs with fuzzy data using FDEA, including Rostamy et al. [40], Liu [30], Jahanshahloo et al. [19], Sanei et al. [45], Rostamy et al. [41], Ghobadi et al. [14], Allahviranloo et al. [3], Bagheri et al. [5], Shafiee and Saleh [46], these methods are inappropriate for assessing multi-members supply chains with internal interactions. Thus, in the following, we describe models that focus on supply chain evaluation using NDEA models with fuzzy data.

9.7.1 Performance Evaluation of Supply Chains in the Presence of Fuzzy Data Using Envelopment Form of NDEA Models

One industry that has proliferated in recent years is the semiconductor industry. Due to the increasing use of semiconductor products in many industries, such as microprocessors, microcontrollers, etc., semiconductor products have received attention and importance. On the other hand, the supply chain, production, and distribution related to semiconductor products have many complexities due to multiple layers of communication and internal activities. Consider Fig. 9.18 as an example.

Independent inputs, final outputs, and productions in each stage are stated in Table 9.1. Also, intermediate products in the mentioned supply chain include:

- $z_j^{(h,h+1)}$: Product flow: The forward logistics transferred from stage h to stage $(h + 1)$ such that $h = 1, 2, 3$.
- $z_j^{(h+1,h)}$: Demand forecast: The reverse logistics transferred from stage$)h + 1($ to stage $(h + 1)$ and $(h = 1, 2, 3)$.

Although real numbers can easily express some of the indicators introduced in this supply chain, inputs such as On-time delivery, Number of stoppages, price compared with the other supplier, and the number of interruptions in production cannot be expressed with real numbers. Therefore, fuzzy numbers should be used to explain

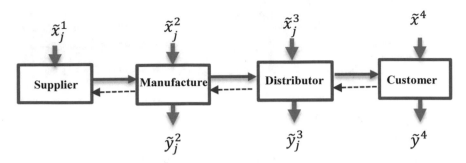

Fig. 9.18 The semiconductor industry supply chain

Table 9.1 Inputs and outputs of the semiconductor industry supply chain

Stage	Input		Output	
Supplier	x_{1j}^1	On-time delivery: The standard deviation of the delivery times (days)	–	
	x_{2j}^1	Location: The geographical distance to the manufacturers (kilometers)		
	$\tilde{x}_{3j}^1$	Price: The price compared with the other suppliers		
Manufacturer	$\tilde{x}_{1j}^2$	Number of stoppages: The number of interruptions in production	$\tilde{y}_{1j}^2$	Flexibility: The flexibility to change the production plan
	x_{2j}^2	Number of laborers	$\tilde{y}_{2j}^2$	Flexibility: The flexibility to change the production plan
	x_{3j}^2	Setup time: The setup time of the production facility (hours)		
Distributor	x_{1j}^3	Cost per dollar revenue: The distribution cost per dollar of revenue	$\tilde{y}_{1j}^3$	Sales average: The distributer's sales amount
	x_{2j}^3	On-time delivery: The standard deviation of the delivery time (days)	$\tilde{y}_{2j}^3$	Service level: The level of service provided to customers
Customer	x_{1j}^4	Order cancellations: The percentage of customers canceling their orders	$\tilde{y}_{1j}^4$	Performance history: The percentage of fulfilled orders

these indicators, as explained in the Chap. 7. Similarly, the outputs of flexibility, equipment technology level, sales average, service level, and performance history are also qualitative values that need to be explained by fuzzy theory.

On the other hand, the intermediate products in the mentioned supply chain have returnable relations, so we need to introduce a new model to evaluate the supply chain

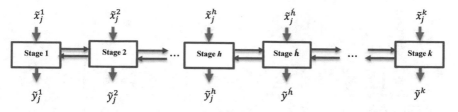

Fig. 9.19 General structure of the supply chain with returnable relations

with fuzzy data and returnable relations. In general, consider the supply chain shown in Fig. 9.19. Suppose that $\tilde{x}_j^h$ and $\tilde{y}_j^h$ represent the fuzzy input and output of stage h of the supply chain j. Intermediate productions are also indicated with the symbol $\tilde{z}_j^{(h,h')}$. Thus $\tilde{z}_j^{(h,h')}$ leaves stage h and enters the stage h'. It is necessary to explain that the intermediate products are considered crisp numbers in this subsection.

To evaluate the performance of supply chains in the presence of fuzzy data using envelopment models, Momeni et al. [34] considered the NEBM model with fuzzy data as the model (9.98).

$$\tilde{\rho}_o = \min \sum_{h=1}^{k} w_h \left(1 - \frac{1}{m_h} \sum_{i=1}^{m_h} \frac{\tilde{s}_i^{h-}}{\tilde{x}_{io}^h} \right)$$

$s.t.$

$$\sum_{j=1}^{n} \tilde{x}_{ij}^h \lambda_j^h + \tilde{s}_i^{h-} = \tilde{x}_{io}^h, \ i = 1, ..., m_h, \ h = 1, ..., k,$$

$$\sum_{i=1}^{m_h} \tilde{y}_{rj}^h \lambda_j^h \geq \tilde{y}_{ro}^h, \ r = 1, ..., s_h, \ h = 1, ..., k,$$

$$\sum_{j=1}^{n} z_{f_{(h,h')}j}^{(h,h')} \lambda_j^h = \sum_{j=1}^{n} z_{f_{(h,h')}j}^{(h,h')} \lambda_j^{h'}, \ f_{(h,h')} = 1, ..., F_{(h,h')}, \ \forall (h, h'),$$

$$\sum_{j=1}^{n} z_{f_{(h,h')}j}^{(h',h)} \lambda_j^h = \sum_{j=1}^{n} z_{f_{(h,h')}j}^{(h',h)} \lambda_j^{h'}, \ f_{(h,h')} = 1, ..., F_{(h,h')}, \ \forall (h, h'),$$

$$\lambda_j^h \geq 0, \ j = 1, ..., n, \ h = 1, ..., k,$$

$$\tilde{s}_i^{h-} \geq 0, \ i = 1, ..., m_h, \ h = 1, ..., k.$$

(9.98)

Also, to mathematically define constraints 1 and 2 correctly, the auxiliary variables $\left(\tilde{s}_i^{h-} \right)$ must be considered in fuzzy variables.

Now suppose that $\tilde{x}_{ij}^h, \tilde{y}_{rj}^h, \tilde{s}_i^{h-}$ fuzzy triangular numbers, so as described in Chap. 7 in detail: $\tilde{x}_{ij}^h = \left(x_{ij}^{h^l}, x_{ij}^{h^m}, x_{ij}^{h^u} \right)$, $\tilde{y}_{rj}^h = \left(y_{rj}^{h^l}, y_{rj}^{h^m}, y_{rj}^{h^u} \right)$, $\tilde{s}_i^{h-} = \left(s_i^{h-l}, s_i^{h-m}, s_i^{h-u} \right)$. As a result, based on the algebraic relations (7.68–7.71), the constraints of the model (9.98) are rewritten as relations (9.99).

$$\sum_{j=1}^{n} \left(x_{ij}^{h^l}, x_{ij}^{h^m}, x_{ij}^{h^u} \right) \lambda_j^h + \left(s_i^{h-^l}, s_i^{h-^m}, s_i^{h-^u} \right) = \left(x_{io}^{h^l}, x_{io}^{h^m}, x_{io}^{h^u} \right)$$

$$\Rightarrow \left(\sum_{j=1}^{n} x_{ij}^{h^l} \lambda_j^h + s_i^{h-^l}, \sum_{j=1}^{n} x_{ij}^{h^m} \lambda_j^h + s_i^{h-^m}, \sum_{j=1}^{n} x_{ij}^{h^u} \lambda_j^h + s_i^{h-^u} \right) \tag{9.99}$$

$$= \left(x_{io}^{h^l}, x_{io}^{h^m}, x_{io}^{h^u} \right)$$

As already explained, several relations have been introduced to compare fuzzy numbers. In this part, the relation (7.73) is used. Therefore, based on the relation (7.73), we have:

$$\sum_{j=1}^{n} x_{ij}^{h^l} \lambda_j^h + s_i^{h-^l} = x_{io}^{h^l} \tag{9.100}$$

$$\sum_{j=1}^{n} x_{ij}^{h^m} \lambda_j^h + s_i^{h-^m} = x_{io}^{h^m} \tag{9.101}$$

$$\sum_{j=1}^{n} x_{ij}^{h^u} \lambda_j^h + s_i^{h-^u} = x_{io}^{h^u} \tag{9.102}$$

Similarly, we simplify the second constraint of the model (9.98) as the relation (9.103).

$$\sum_{j=1}^{n} \left(y_{rj}^{h^l}, y_{rj}^{h^m}, y_{rj}^{h^u} \right) \lambda_j^h \geq \left(y_{ro}^{h^l}, y_{ro}^{h^m}, y_{ro}^{h^u} \right)$$

$$\Rightarrow \left(\sum_{j=1}^{n} y_{rj}^{h^l} \lambda_j^h, \sum_{j=1}^{n} y_{rj}^{h^m} \lambda_j^h, \sum_{j=1}^{n} y_{rj}^{h^u} \lambda_j^h \right) \geq \left(y_{ro}^{h^l}, y_{ro}^{h^m}, y_{ro}^{h^u} \right) \tag{9.103}$$

$$\Rightarrow \sum_{j=1}^{n} y_{rj}^{h^m} \lambda_j^h \geq y_{ro}^{h^m}$$

Also, the objective function of the model (9.98) is expressed as the relation (9.104).

$$\sum_{h=1}^{k} w_h \left(1 - \frac{1}{m_h} \sum_{i=1}^{m_h} \frac{\tilde{s}_i^{h-}}{\tilde{x}_{io}^h} \right)$$

$$= \sum_{h=1}^{k} w_h \left(1 - \frac{1}{m_h} \sum_{i=1}^{m_h} \frac{\left(s_i^{h-l}, s_i^{h-m}, s_i^{h-u} \right)}{\left(x_{io}^{h l}, x_{io}^{h m}, x_{io}^{h u} \right)} \right)$$

$$= \begin{bmatrix} \sum_{h=1}^{k} w_h \left(1 - \frac{1}{m_h} \sum_{i=1}^{m_h} \frac{s_i^{h-u}}{x_{io}^{h l}} \right), \\ \sum_{h=1}^{k} w_h \left(1 - \frac{1}{m_h} \sum_{i=1}^{m_h} \frac{s_i^{h-m}}{x_{io}^{h m}} \right), \\ \sum_{h=1}^{k} w_h \left(1 - \frac{1}{m_h} \sum_{i=1}^{m_h} \frac{s_i^{h-l}}{x_{io}^{h u}} \right) \end{bmatrix} \qquad (9.104)$$

Thus the objective function of the model (9.104) is expressed as a fuzzy number $\tilde{\rho}_o = \left(\rho_o^l, \rho_o^m, \rho_o^u \right)$. So that:

$$\rho_o^l = \sum_{h=1}^{k} w_h \left(1 - \frac{1}{m_h} \sum_{i=1}^{m_h} \frac{s_i^{h-u}}{x_{io}^{h l}} \right) \qquad (9.105)$$

$$\rho_o^m = \sum_{h=1}^{k} w_h \left(1 - \frac{1}{m_h} \sum_{i=1}^{m_h} \frac{s_i^{h-m}}{x_{io}^{h m}} \right) \qquad (9.106)$$

$$\rho_o^u = \sum_{h=1}^{k} w_h \left(1 - \frac{1}{m_h} \sum_{i=1}^{m_h} \frac{s_i^{h-l}}{x_{io}^{h u}} \right) \qquad (9.107)$$

Therefore, to calculate the optimal value of ρ_o^l, i.e., ρ_o^{l*}, we will have:

$$\rho_o^l = \min \sum_{h=1}^{k} w_h \left(1 - \frac{1}{m_h} \sum_{i=1}^{m_h} \frac{s_i^{h-^u}}{x_{io}^{h^l}} \right)$$

$s.t.$

$$\sum_{j=1}^{n} x_{ij}^{h^l} \lambda_j^h + s_i^{h-^l} = x_{io}^{h^l}, \ i = 1, ..., m_h, \ h = 1, ..., k,$$

$$\sum_{j=1}^{n} x_{ij}^{h^m} \lambda_j^h + s_i^{h-^m}, \ i = 1, ..., m_h, \ h = 1, ..., k,$$

$$\sum_{j=1}^{n} x_{ij}^{h^u} \lambda_j^h + s_i^{h-^u} = x_{io}^{h^u}, \ i = 1, ..., m_h, \ h = 1, ..., k, \qquad (9.108)$$

$$\sum_{j=1}^{n} y_{rj}^{h^m} \lambda_j^h \geq y_{ro}^{h^m}, \ r = 1, ..., s_h, \ h = 1, ..., k,$$

$$\sum_{j=1}^{n} z_{f_{(h,h')}j}^{(h,h')} \lambda_j^h = \sum_{j=1}^{n} z_{f_{(h,h')}j}^{(h,h')} \lambda_j^{h'}, \ f_{(h,h')} = 1, ..., F_{(h,h')}, \ \forall (h, h'),$$

$$\lambda_j^h \geq 0, \ j = 1, ..., n, \ h = 1, ..., k,$$

$$s_i^{h-} \geq 0, \ i = 1, ..., m_h, \ h = 1, ..., k.$$

Similarly, to calculate ρ_o^{m*} and $\rho_o^{u^*}$ it is enough to replace the objective function of the model (9.108) with relations (9.106) and (9.107).

It is evident that supply chain o is efficient if and only if: $\rho_o^{l^*} = \rho_o^{m^*} = \rho_o^{u^*} = 1$. Also, the efficiency of stage h of supply chain o is equal to $\tilde{\rho}_o^{h^*} = \left(\rho_o^{hl^*}, \rho_o^{hm^*}, \rho_o^{hu^*} \right)$ so that:

$$\rho_o^{hl*} = 1 - \frac{1}{m_h} \sum_{i=1}^{m_h} \frac{s_i^{h-u*}}{x_{io}^{hl}} \qquad (9.109)$$

$$\rho_o^{hm*} = 1 - \frac{1}{m_h} \sum_{i=1}^{m_h} \frac{s_i^{h-m*}}{x_{io}^{hm}} \qquad (9.110)$$

$$\rho_o^{hu*} = 1 - \frac{1}{m_h} \sum_{i=1}^{m_h} \frac{s_i^{h-l*}}{x_{io}^{hu}} \qquad (9.111)$$

Therfore stage h of supply chain o is efficient if and only if:
$\rho_o^{hl*} = \rho_o^{hm*} = \rho_o^{hu*} = 1$.

As stated before, although the main goal in using DEA and NDEA models is to calculate the efficiency and evaluate the performance of the units under evaluation, comparing the efficiency scores and ranking of supply chains is also essential.

Although many ranking techniques have been presented so far, including (Abbasbandy and Asady [1], Ezzati et al. [12], Akyar et al. [2]), Momeni et al. [34] used the Trust matrix to classify performance results. Assume that $\tilde{\rho}_i^* = \left(\rho_i^{l^*}, \rho_i^{m^*}, \rho_i^{u^*}\right)$ and $\tilde{\rho}_j^* = \left(\rho_j^{l^*}, \rho_j^{m^*}, \rho_j^{u^*}\right)$ are the efficiency score of the supply chain i and j, respectively. So: $\tilde{\rho}_i^* \geq \tilde{\rho}_j^*$ if $\bar{t}_i \geq \bar{t}_j$. In other words, the supply chain with a larger $\bar{t}$ has a better ranking. To calculate $\bar{t}_i$, proceed as follows:

1. Calculate $T\left(\tilde{\rho}_i^* \geq \tilde{\rho}_j^*\right)$ so that:

$$T\left(\tilde{\rho}_i^* \geq \tilde{\rho}_j^*\right) = \sup\left\{\min\left(\mu_{\tilde{\rho}_i^*}(x), \mu_{\tilde{\rho}_j^*}(y)\right), x \geq y\right\} \quad (9.112)$$

2. Calculate the values of t_{ij} using (9.113).

$$t_{ij} = T\left(\tilde{\rho}_i^* \geq \tilde{\rho}_j^*\right) = \begin{cases} 1, & \rho_i^{m^*} \geq \rho_j^{m^*} \\ 0, & \rho_i^{u^*} \geq \rho_j^{l^*} \\ \dfrac{\rho_j^{l^*} - \rho_i^{u^*}}{\left(\rho_i^{m^*} - \rho_i^{u^*}\right) - \left(\rho_j^{m^*} - \rho_j^{l^*}\right)}, & \text{otherwise} \end{cases} \quad (9.113)$$

3. Compute the Trust matrix as shown in (9.114).

$$T_{n \times n} = [t_{ij}]_{n \times n} = \begin{array}{c} \\ \tilde{\rho}_1^* \\ \vdots \\ \tilde{\rho}_j^* \\ \vdots \\ \tilde{\rho}_n^* \end{array} \begin{array}{c} \tilde{\rho}_1^* \ \cdots \ \tilde{\rho}_j^* \ \cdots \ \tilde{\rho}_n^* \\ \begin{bmatrix} 1 & \cdots & t_{1j} & \cdots & t_{1n} \\ \vdots & & \vdots & & \vdots \\ t_{j1} & & 1 & \cdots & t_{jn} \\ \vdots & & \vdots & & \vdots \\ t_{n1} & \cdots & t_{nj} & \cdots & \end{bmatrix} \end{array} \quad (9.114)$$

4. Compute $\bar{t}_i$ so that: $\bar{t}_j = \sum_{j=1}^n t_{ij} \Big/ n$.

9.7.2 Performance Evaluation of Supply Chains in the Presence of Fuzzy Data Using Multiplier Form of NDEA Models

Consider a two-stage supply chain, as shown in Fig. 9.20. $\tilde{x}_j$, $\tilde{y}_j$, and $\tilde{z}_j$ are this supply chain's initial input, intermediate products, and final output, which is considered a fuzzy number.

Khalili-Damghani et al. [23] developed Kao et al. model to evaluate supply chains' performance in the presence of fuzzy data.

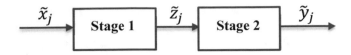

Fig. 9.20 A two-stage supply chain with fuzzy data

$$\max \sum_{r=1}^{s} u_r \tilde{y}_{ro}$$

$$s.t.$$

$$\sum_{r=1}^{s} u_r \tilde{y}_{rj} - \sum_{d=1}^{D} w_d \tilde{z}_{dj} \leq 0, \ j = 1, ..., n,$$

$$\sum_{d=1}^{D} w_d \tilde{z}_{dj} - \sum_{i=1}^{m} v_i \tilde{x}_{ij} \leq 0, \ j = 1, ..., n, \qquad (9.115)$$

$$\sum_{i=1}^{m} v_i \tilde{x}_{io} = 1,$$

$$u_r, w_d, v_i \geq 0, \ r = 1, ..., n, \ d = 1, ..., D, \ i = 1, ..., m.$$

So that $\tilde{x}_{ij}$, $\tilde{y}_{rj}$, and $\tilde{z}_{dj}$ are trapezoidal fuzzy numbers. So $\tilde{x}_{ij} = \left(x_{ij}^l, x_{ij}^{m_1}, x_{ij}^{m_2}, x_{ij}^u \right)$, $\tilde{y}_{rj} = \left(y_{rj}^l, y_{rj}^{m_1}, y_{rj}^{m_2}, y_{rj}^u \right)$, $\tilde{z}_{dj} = \left(z_{dj}^l, z_{dj}^{m_1}, z_{dj}^{m_2}, z_{dj}^u \right)$. Also, in a particular case, $\tilde{x}_{ij}$, $\tilde{y}_{rj}$, and $\tilde{z}_{dj}$ can also be considered fuzzy triangular numbers. According to the definition of $\alpha - cut$ in Chap. 7, for an arbitrary value of α, the member of the set $\alpha - cut$ for the trapezoidal number $\tilde{x}_{ij}$ are displayed in Fig. 9.21.

It can be easily shown that any arbitrary $\alpha_i \in [0, 1]$, values $\left(x_{ij}^l \right)_{\alpha_i}$ and $\left(x_{ij}^u \right)_{\alpha_i}$ can be obtained using relations (9.116) and (9.117).

Fig. 9.21 Geometric interpretation $\alpha - cut$ for a trapezoidal fuzzy numbers

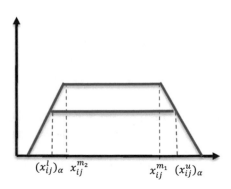

$$\left(x_{ij}^l\right)_{\alpha_i} = x_{ij}^l + \alpha_i\left(x_{ij}^{m_1} - x_{ij}^l\right), \ i = 1, ..., m, \ j = 1, ..., n. \qquad (9.116)$$

$$\left(x_{ij}^u\right)_{\alpha_i} = x_{ij}^u - \alpha_i\left(x_{ij}^u - x_{ij}^{m_2}\right), \ i = 1, ..., m, \ j = 1, ..., n. \qquad (9.117)$$

In the same way, we obtain the values of $\left(y_{rj}^l\right)_{\alpha_r}$, $\left(y_{rj}^u\right)_{\alpha_r}$, $\left(z_{dj}^l\right)_{\alpha_d}$, and $\left(z_{dj}^u\right)_{\alpha_d}$ using (9.118–9.121) relations.

$$\left(y_{rj}^l\right)_{\beta_r} = y_{rj}^l + \beta_r\left(y_{rj}^{m_1} - y_{rj}^l\right), \ r = 1, ..., s, \ j = 1, ..., n. \qquad \beta_r \in [0, 1], \quad (9.118)$$

$$\left(y_{rj}^u\right)_{\beta_r} = y_{rj}^u - \beta_r\left(y_{rj}^u - y_{rj}^{m_2}\right), \ r = 1, ..., s, \ j = 1, ..., n. \qquad (9.119)$$

$$\left(z_{dj}^l\right)_{\gamma_d} = z_{dj}^l + \gamma_d\left(z_{dj}^{m_1} - z_{dj}^l\right), \ d = 1, ..., D, \ j = 1, ..., n. \ \gamma_d \in [0, 1], \quad (9.120)$$

$$\left(z_{dj}^u\right)_{\gamma_d} = z_{dj}^u - \gamma_d\left(z_{dj}^u - z_{dj}^{m_2}\right), \ d = 1, ..., D, \ j = 1, ..., n. \qquad (9.121)$$

This way, the input, output, and intermediate values are displayed at each level with intervals, $\left[\left(x_{ij}^l\right)_{\alpha_i}, \left(x_{ij}^l\right)_{\alpha_i}\right]$, $\left[\left(y_{rj}^l\right)_{\beta_r}, \left(y_{rj}^u\right)_{\beta_r}\right]$ and $\left[\left(z_{dj}^l\right)_{\gamma_d}, \left(z_{dj}^u\right)_{\gamma_d}\right]$, respectively. Therefore, to calculate the efficiency values at each level, it is enough to calculate the upper and lower efficiency bounds at each level using the interval DEA (Wang et al. [25], Jahanshahloo et al. [15], Jahanshahloo et al. [52], Tamaddon et al. [6]).

Therefore, models (9.122) and (9.123) are used to obtain upper bound (e_o^u) and lower bound (e_o^l) respectively.

$$e_o^u = \max \sum_{r=1}^{s} u_r \left(y_{ro}^u - \alpha_r \left(y_{ro}^u - y_{ro}^{m_2} \right) \right)$$

$s.t.$

$$\sum_{r=1}^{s} u_r \left(y_{ro}^u - \alpha_r \left(y_{ro}^u - y_{ro}^{m_2} \right) \right) - \sum_{d=1}^{D} w_d \left(z_{do}^l + \alpha_d \left(z_{do}^{m_1} - z_{do}^l \right) \right) \leq 0,$$

$$\sum_{r=1}^{s} u_r \left(y_{rj}^l + \alpha_r \left(y_{rj}^{m_1} - y_{rj}^l \right) \right) - \sum_{d=1}^{D} w_d \left(z_{dj}^u - \alpha_d \left(z_{dj}^u - z_{dj}^{m_2} \right) \right) \leq 0,$$

$$j = 1, ..., n, \ j \neq o,$$

$$\sum_{d=1}^{D} w_d \left(z_{dj}^l + \alpha_d \left(z_{dj}^{m_1} - z_{dj}^l \right) \right) - \sum_{i=1}^{m} v_i \left(x_{ij}^u - \alpha_i \left(x_{ij}^u - x_{ij}^{m_2} \right) \right) \leq 0,$$

$$j = 1, ..., n, \ j \neq o,$$

$$\sum_{d=1}^{D} w_d \left(z_{do}^u - \alpha_d \left(z_{do}^u - z_{do}^{m_2} \right) \right) - \sum_{i=1}^{m} v_i \left(x_{io}^l + \alpha_i \left(x_{io}^{m_1} - x_{io}^l \right) \right) \leq 0,$$

$$\sum_{i=1}^{m} v_i \left(x_{io}^l + \alpha_i \left(x_{io}^{m_1} - x_{io}^l \right) \right) = 1,$$

$$u_r, w_d, v_i \geq 0, \quad r = 1, ..., n, \quad d = 1, ..., D, \quad i = 1, ..., m.$$

(9.122)

In the model (9.122), using the method provided by Despotis and Smirlis [10] to calculate the upper bounds of efficiency (optimistic efficiency), it is enough to consider the output of the supply chain under evaluation in each stage at the upper bounds and its input at the lowest value. In other supply chains, the lowest value for inputs and the highest value for the outputs be regarded. In the same way, for the calculation $\left(e_o^l \right)$ in the model (9.123), We do the opposite of the said process for $\left(e_o^u \right)$.

$$e_o^l = \max \sum_{r=1}^{s} u_r \left(y_{ro}^l + \beta_r \left(y_{ro}^{m_1} - y_{ro}^l \right) \right)$$

s.t.

$$\sum_{r=1}^{s} u_r \left(y_{ro}^l + \beta_r \left(y_{ro}^{m_1} - y_{ro}^l \right) \right) - \sum_{d=1}^{D} w_d \left(z_{do}^u - \gamma_d \left(z_{do}^u - z_{do}^{m_2} \right) \right) \leq 0,$$

$$\sum_{r=1}^{s} u_r \left(y_{rj}^u - \beta_r \left(y_{rj}^u - y_{rj}^{m_2} \right) \right) - \sum_{d=1}^{D} w_d \left(z_{dj}^l + \gamma_d \left(z_{dj}^{m_1} - z_{dj}^l \right) \right) \leq 0,$$

$$j = 1, ..., n, \ j \neq o,$$

$$\sum_{d=1}^{D} w_d \left(z_{do}^l + \gamma_d \left(z_{do}^{m_1} - z_{do}^l \right) \right) - \sum_{i=1}^{m} v_i \left(x_{io}^u - \alpha_i \left(x_{io}^u - x_{io}^{m_2} \right) \right) \leq 0, \qquad (9.123)$$

$$\sum_{d=1}^{D} w_d \left(z_{dj}^u - \gamma_d \left(z_{dj}^u - z_{dj}^{m_2} \right) \right) - \sum_{i=1}^{m} v_i \left(x_{ij}^l + \alpha_i \left(x_{ij}^{m_1} - x_{ij}^l \right) \right) \leq 0,$$

$$j = 1, ..., n, \ j \neq o,$$

$$\sum_{i=1}^{m} v_i \left(x_{io}^u - \alpha_i \left(x_{io}^u - x_{io}^{m_2} \right) \right) = 1,$$

$$u_r, w_d, v_i \geq 0,$$

$$r = 1, ..., n, \quad d = 1, ..., D, \quad i = 1, ..., m.$$

without loose of generality, for ease of calculations, suppose that $\alpha = \alpha_i$, $\beta = \beta_r$, $\gamma = \gamma_d$. So that α, β, and γ are constant values.

It is obvious that (9.122) and (9.123) are non-linear, so using changing the variable $\eta_r = u_r \beta$, $\theta_d = w_d \gamma$, $\lambda_i = v_i \alpha$ the model (9.122) becomes a linear programming problem (9.124).

$$e_o^{\mathrm{u}} = \max \sum_{r=1}^{s} \left(\mathrm{u}_r y_{\mathrm{ro}}^{\mathrm{u}} - \eta_r \left(y_{\mathrm{ro}}^{\mathrm{u}} - y_{\mathrm{ro}}^{m_2} \right) \right)$$

$s.t.$

$$\sum_{r=1}^{s} \left(\mathrm{u}_r y_{\mathrm{ro}}^{\mathrm{u}} - \eta_r \left(y_{\mathrm{ro}}^{\mathrm{u}} - y_{\mathrm{ro}}^{m_2} \right) \right) - \sum_{d=1}^{D} \left(\mathrm{w}_d z_{\mathrm{do}}^{l} + \theta_d \left(z_{\mathrm{do}}^{m_1} - z_{\mathrm{do}}^{l} \right) \right) \leq 0,$$

$$\sum_{r=1}^{s} \left(\mathrm{u}_r y_{\mathrm{rj}}^{l} + \eta_r \left(y_{\mathrm{rj}}^{m_1} - y_{\mathrm{rj}}^{l} \right) \right) - \sum_{d=1}^{D} \left(\mathrm{w}_d z_{\mathrm{dj}}^{\mathrm{u}} - \theta_d \left(z_{\mathrm{dj}}^{\mathrm{u}} - z_{\mathrm{dj}}^{m_2} \right) \right) \leq 0,$$

$$j = 1, \dots, n, \ j \neq o,$$

$$\sum_{d=1}^{D} \left(\mathrm{w}_d z_{\mathrm{dj}}^{l} + \theta_d \left(z_{\mathrm{dj}}^{m_1} - z_{\mathrm{dj}}^{l} \right) \right) - \sum_{i=1}^{m} \left(\mathrm{v}_i x_{\mathrm{ij}}^{\mathrm{u}} - \lambda_i \left(x_{\mathrm{ij}}^{\mathrm{u}} - x_{\mathrm{ij}}^{m_2} \right) \right) \leq 0, \qquad (9.124)$$

$$j = 1, \dots, n, \ j \neq o,$$

$$\sum_{d=1}^{D} \left(\mathrm{w}_d z_{\mathrm{do}}^{\mathrm{u}} - \theta_d \left(z_{\mathrm{do}}^{\mathrm{u}} - z_{\mathrm{do}}^{m_2} \right) \right) - \sum_{i=1}^{m} \left(\mathrm{v}_i x_{\mathrm{io}}^{l} + \lambda_i \left(x_{\mathrm{io}}^{m_1} - x_{\mathrm{io}}^{l} \right) \right) \leq 0,$$

$$\sum_{i=1}^{m} \left(\mathrm{v}_i x_{\mathrm{io}}^{l} + \lambda_i \left(x_{\mathrm{io}}^{m_1} - x_{\mathrm{io}}^{l} \right) \right) = 1,$$

$$\mathrm{u}_r \geq \eta_r \geq 0, \quad r = 1, \dots, n,$$

$$\mathrm{w}_d \geq \theta_d \geq 0, \quad d = 1, \dots, D,$$

$$\mathrm{v}_i \geq \lambda_i \geq 0, \quad i = 1, \dots, m.$$

As described above, model (9.123) is also converted to model (9.125).

$$e_o^l = \max \sum_{r=1}^{s} \left(u_r y_{ro}^l + \eta_r \left(y_{ro}^{m_1} - y_{ro}^l \right) \right)$$

$s.t.$

$$\sum_{r=1}^{s} \left(u_r y_{ro}^l + \eta_r \left(y_{ro}^{m_1} - y_{ro}^l \right) \right) - \sum_{d=1}^{D} \left(w_d z_{do}^u - \theta_d \left(z_{do}^u - z_{do}^{m_2} \right) \right) \le 0,$$

$$\sum_{r=1}^{s} \left(u_r y_{rj}^u - \eta_r \left(y_{rj}^u - y_{rj}^{m_2} \right) \right) - \sum_{d=1}^{D} \left(w_d z_{dj}^l + \theta_d \left(z_{dj}^{m_1} - z_{dj}^l \right) \right) \le 0,$$

$$j = 1, ..., n, \ j \ne o,$$

$$\sum_{d=1}^{D} \left(w_d z_{do}^l + \theta_d \left(z_{do}^{m_1} - z_{do}^l \right) \right) - \sum_{i=1}^{m} \left(v_i x_{io}^u - \lambda_i \left(x_{io}^u - x_{io}^{m_2} \right) \right) \le 0, \qquad (9.125)$$

$$\sum_{d=1}^{D} \left(w_d z_{dj}^u - \theta_d \left(z_{dj}^u - z_{dj}^{m_2} \right) \right) - \sum_{i=1}^{m} \left(v_i x_{ij}^l + \lambda_i \left(x_{ij}^{m_1} - x_{ij}^l \right) \right) \le 0,$$

$$j = 1, ..., n, \ j \ne o,$$

$$\sum_{i=1}^{m} \left(v_i x_{io}^u - \alpha_i \left(x_{io}^u - x_{io}^{m_2} \right) \right) = 1,$$

$$u_r \ge \eta_r \ge 0, \quad r = 1, ..., n,$$
$$w_d \ge \theta_d \ge 0, \quad d = 1, ..., D,$$
$$v_i \ge \lambda_i \ge 0, \quad i = 1, ..., m.$$

As explained earlier, when using Kao and Hwang model (model (6.104)) to calculate the efficiency related to stages 1 and 2, we face some problems due to the existence of alternative optimal solutions and multiple weights. So by developing the presented method by Kao and Hwang to calculate the efficiency of stage 1 in the presence of fuzzy data, we calculate the upper and lower bounds of the efficiency of stage 1 using models (9.126) and (9.127). e_o^{u*} and e_o^{l*} are the optimal solution of the model (9.124) and (9.125). In this way, by keeping the upper bounds of total efficiency constant and searching among multiple optimal weights, calculate the upper bounds of efficiency for stage 1 as $\left[e_o^{1+} \right]^u$.

$$\left[e_o^{1+}\right]^u = \max \sum_{d=1}^{D}\left(w_d z_{dj}^u - \theta_d\left(z_{dj}^u - z_{dj}^{m_2}\right)\right)$$

$s.t.$

$$\sum_{r=1}^{s} u_r\left(y_{ro}^u - \eta_r\left(y_{ro}^u - y_{ro}^{m_2}\right)\right) = e_o^{u*}$$

$$\sum_{d=1}^{D}\left(w_d z_{dj}^l + \theta_d\left(z_{dj}^{m_1} - z_{dj}^l\right)\right) - \sum_{i=1}^{m}\left(v_i x_{ij}^u - \lambda_i\left(\left(x_{ij}^u - x_{ij}^{m_2}\right)\right)\right) \leq 0,$$

$j = 1, ..., n, \ j \neq o,$

$$\sum_{d=1}^{D}\left(w_d z_{do}^u - \theta_d\left(z_{do}^u - z_{do}^{m_2}\right)\right) - \sum_{i=1}^{m}\left(v_i x_{io}^l + \lambda_i\left(\left(x_{io}^{m_1} - x_{io}^l\right)\right)\right) \leq 0, \tag{9.126}$$

$$\sum_{r=1}^{s}\left(u_r y_{ro}^u - \eta_r\left(y_{ro}^u - y_{ro}^{m_2}\right)\right) - \sum_{d=1}^{D}\left(w_d z_{do}^l + \theta_d\left(z_{do}^{m_1} - z_{do}^l\right)\right) \leq 0,$$

$$\sum_{r=1}^{s}\left(u_r y_{rj}^l + \eta_r\left(y_{rj}^{m_1} - y_{rj}^l\right)\right) - \sum_{d=1}^{D}\left(w_d z_{dj}^u - \theta_d\left(z_{dj}^u - z_{dj}^{m_2}\right)\right) \leq 0,$$

$j = 1, ..., n, \ j \neq o,$

$$\sum_{i=1}^{m}\left(v_i x_{io}^l + \lambda_i\left(x_{io}^{m_1} - x_{io}^l\right)\right) = 1,$$

$u_r \geq \eta_r \geq 0, \quad r = 1, ..., n,$

$w_d \geq \theta_d \geq 0, \quad d = 1, ..., D,$

$v_i \geq \lambda_i \geq 0, \quad i = 1, ..., m.$

In the following, by keeping the lower bound of the total efficiency constant and using model (9.127), the lower bound of efficiency of stage 1 is calculated. In this way, an efficiency interval for the efficiency score of stage 1 is obtained as $\left[\left[e_o^{1+}\right]^l, \left[e_o^{1+}\right]^u\right]$.

$$\left[e_o^{1+}\right]^l = \max \sum_{d=1}^{D}\left(w_d z_{dj}^1 + \theta_d\left(z_{dj}^2 - z_{dj}^1\right)\right)$$

s.t.

$$\sum_{r=1}^{s}\left(u_r y_{ro}^l + \eta_r\left(y_{ro}^{m_1} - y_{ro}^l\right)\right) = e_o^{l*},$$

$$\sum_{d=1}^{D}\left(w_d z_{dj}^u - \theta_d\left(z_{dj}^u - z_{dj}^{m_2}\right)\right) - \sum_{i=1}^{m}\left(v_i x_{ij}^l + \lambda_i\left(x_{ij}^{m_1} - x_{ij}^l\right)\right) \le 0,$$

$$j = 1, ..., n, \ j \ne o,$$

$$\sum_{d=1}^{D}\left(w_d z_{do}^l + \theta_d\left(\left(z_{do}^{m_1} - z_{do}^l\right)\right)\right) - \sum_{i=1}^{m}\left(v_i x_{io}^u - \lambda_i\left(x_{io}^u - x_{io}^{m_2}\right)\right) \le 0,$$

$$\sum_{r=1}^{s}\left(u_r y_{rj}^u - \eta_r\left(y_{rj}^u - y_{rj}^{m_2}\right)\right) - \sum_{d=1}^{D}\left(w_d z_{dj}^l + \theta_d\left(\left(z_{dj}^{m_1} - z_{dj}^l\right)\right)\right) \le 0,$$ (9.127)

$$j = 1, ..., n, \ j \ne o,$$

$$\sum_{r=1}^{s}\left(u_r y_{ro}^l + \eta_r\left(y_{ro}^{m_1} - y_{ro}^l\right)\right) - \sum_{d=1}^{D}\left(w_d z_{do}^u - \theta_d\left(z_{do}^u - z_{do}^{m_2}\right)\right) \le 0,$$

$$\sum_{i=1}^{m}\left(v_i x_{io}^u - \lambda_i\left(x_{io}^u - x_{io}^{m_2}\right)\right) = 1,$$

$$u_r \ge \eta_r \ge 0, \quad r = 1, ..., n,$$
$$w_d \ge \theta_d \ge 0, \quad d = 1, ..., D,$$
$$v_i \ge \lambda_i \ge 0, \quad i = 1, ..., m.$$

Similarly, we obtain the efficiency interval $\left[\left[e_o^{1+}\right]^l, \left[e_o^{1+}\right]^u\right]$ for the efficiency score of stage 2 of the supply chain using the models (9.128) and (9.129).

$$\left[e_o^{2+}\right]^{\mathrm{u}} = \max \sum_{r=1}^{s} \left(\mathrm{u}_r y_{\mathrm{ro}}^{\mathrm{u}} - \eta_r \left(y_{\mathrm{ro}}^{\mathrm{u}} - y_{\mathrm{ro}}^{m_2}\right)\right)$$

$s.t.$

$$\sum_{r=1}^{s} \left(\mathrm{u}_r y_{\mathrm{ro}}^{\mathrm{u}} - \eta_r \left(y_{\mathrm{ro}}^{\mathrm{u}} - y_{\mathrm{ro}}^{m_2}\right)\right) - e_o^{\mathrm{u}*} \times \sum_{i=1}^{m} \left(\mathrm{v}_i x_{\mathrm{io}}^{l} + \lambda_i \left(x_{\mathrm{io}}^{m_1} - x_{\mathrm{io}}^{l}\right)\right)$$

$$\sum_{d=1}^{D} \left(\mathrm{w}_d z_{\mathrm{dj}}^{l} + \theta_d \left(z_{\mathrm{dj}}^{m_1} - z_{\mathrm{dj}}^{l}\right)\right) - \sum_{i=1}^{m} \left(\mathrm{v}_i x_{\mathrm{ij}}^{\mathrm{u}} - \lambda_i \left(x_{\mathrm{ij}}^{\mathrm{u}} - x_{\mathrm{ij}}^{m_2}\right)\right) \leq 0,$$

$$j = 1, \ldots, n, \ j \neq o,$$

$$\sum_{d=1}^{D} \left(\mathrm{w}_d z_{\mathrm{do}}^{\mathrm{u}} - \theta_d \left(z_{\mathrm{do}}^{\mathrm{u}} - z_{\mathrm{do}}^{m_2}\right)\right) - \sum_{i=1}^{m} \left(\mathrm{v}_i x_{\mathrm{io}}^{l} + \lambda_i \left(x_{\mathrm{io}}^{m_1} - x_{\mathrm{io}}^{l}\right)\right) \leq 0,$$

$$\sum_{r=1}^{s} \left(\mathrm{u}_r y_{\mathrm{ro}}^{\mathrm{u}} - \eta_r \left(y_{\mathrm{ro}}^{\mathrm{u}} - y_{\mathrm{ro}}^{m_2}\right)\right) - \sum_{d=1}^{D} \left(\mathrm{w}_d z_{\mathrm{do}}^{l} + \theta_d \left(z_{\mathrm{do}}^{m_1} - z_{\mathrm{do}}^{l}\right)\right) \leq 0,$$

$$\sum_{r=1}^{s} \left(\mathrm{u}_r y_{\mathrm{rj}}^{l} + \eta_r \left(y_{\mathrm{rj}}^{m_1} - y_{\mathrm{rj}}^{l}\right)\right) - \sum_{d=1}^{D} \left(\mathrm{w}_d z_{\mathrm{dj}}^{\mathrm{u}} - \theta_d \left(z_{\mathrm{dj}}^{\mathrm{u}} - z_{\mathrm{dj}}^{m_2}\right)\right) \leq 0,$$

$$j = 1, \ldots, n, \ j \neq o,$$

$$\sum_{d=1}^{D} \left(\mathrm{w}_d z_{\mathrm{do}}^{l} + \theta_d \left(z_{\mathrm{do}}^{m_1} - z_{\mathrm{do}}^{l}\right)\right) = 1,$$

$$\mathrm{u}_r \geq \eta_r \geq 0, \quad r = 1, \ldots, n,$$

$$\mathrm{w}_d \geq \theta_d \geq 0, \quad d = 1, \ldots, D,$$

$$\mathrm{v}_i \geq \lambda_i \geq 0, \quad i = 1, \ldots, m.$$

(9.128)

So that $e_o^{\mathrm{u}*}$ and e_o^{l*} are the optimal values obtained from (9.124) and (9.125) models.

$$\left[e_o^{2+}\right]^l = \max \sum_{r=1}^{s} \left(u_r y_{ro}^l + \eta_r\left(y_{ro}^{m_1} - y_{ro}^l\right)\right)$$

s.t.

$$\sum_{r=1}^{s} \left(u_r y_{ro}^l + \eta_r\left(y_{ro}^{m_1} - y_{ro}^l\right)\right) - e_o^{l*} \times \sum_{d=1}^{D} \left(w_d z_{do}^u - \theta_d\left(z_{do}^u - z_{do}^{m_2}\right)\right) \le 0,$$

$$\sum_{d=1}^{D} \left(w_d z_{do}^l + \theta_d\left(z_{do}^{m_1} - z_{do}^l\right)\right) - \sum_{i=1}^{m} \left(v_i x_{io}^u - \lambda_i\left(x_{io}^u - x_{io}^{m_2}\right)\right) \le 0,$$

$$\sum_{d=1}^{D} \left(w_d z_{dj}^u - \theta_d\left(z_{dj}^u - z_{dj}^{m_2}\right)\right) - \sum_{i=1}^{m} \left(v_i x_{ij}^l + \lambda_i\left(x_{ij}^{m_1} - x_{ij}^l\right)\right) \le 0,$$

$$j = 1, \ldots, n, \ j \ne o,$$

$$\sum_{r=1}^{s} \left(u_r y_{ro}^l + \eta_r\left(y_{ro}^{m_1} - y_{ro}^l\right)\right) - \sum_{d=1}^{D} \left(w_d z_{do}^u - \theta_d\left(z_{do}^u - z_{do}^{m_2}\right)\right) \le 0,$$

$$\sum_{r=1}^{s} \left(u_r y_{rj}^u - \eta_r\left(y_{rj}^u - y_{rj}^{m_2}\right)\right) - \sum_{d=1}^{D} \left(w_d z_{dj}^l + \theta_d\left(z_{dj}^{m_1} - z_{dj}^l\right)\right) \le 0,$$

$$j = 1, \ldots, n, \ j \ne o,$$

$$\sum_{d=1}^{D} \left(w_d z_{do}^u - \theta_d\left(z_{do}^u - z_{do}^{m_2}\right)\right) = 1,$$

$$u_r \ge \eta_r \ge 0, \quad r = 1, \ldots, n,$$

$$w_d \ge \theta_d \ge 0, \quad d = 1, \ldots, D,$$

$$v_i \ge \lambda_i \ge 0, \quad i = 1, \ldots, m.$$

(9.129)

Then, using the obtained total efficiency scores, we classify the supply chains into three groups: E^-, E^-, and E^{++}. So that E^{++} includes the index of the units that the upper and lower bounds of the efficiency interval are equal to 1. It is necessary to explain that $e_j^l \le e_j^u$, therefore, if $e_j^l = 1$, then necessarily $e_j^u = 1$. In other words $E^{++} = \left\{j \in J \middle| e_j^l = 1\right\}$. E^+ includes units that are not efficient from above but are efficient from below. Therefore, $E^+ = \left\{j \in J \middle| e_j^l < 1, e_j^u < 1\right\}$ and finally E^- include the units that $E^- = \left\{j \in J \middle| e_j^l < 1, e_j^u < 1\right\}$. In this way, E^{++}, E^+, and E^- can be defined for steps 1 and 2.

9.8   Determining the Type of Return to Scale in the Supply Chain

Based on market conditions, economic policies, etc., managers must constantly change the network under their management. Therefore, they tend to have information related to relative changes in output compared to relative changes in input. In such a case, some questions arise for a supply chain manager. For example:

- Are the proportional changes in the inputs of a network more/less than the proportional changes in the network outputs?
- Do proportional changes in the internal productions' size affect the network's overall performance?

We need to discuss return to scale in the network to answer the above questions. Return to scale is one of the critical issues in management decisions, and it can provide helpful information regarding the optimal size of the under-evaluation unit. Therefore, Saleh et al. [44] investigated the return to scale in the network.

Suppose we have n two-stage network, as shown in Fig. 9.1. So that each network in stage 1 consumes the input x and produces the output z. Then, in the next stage, using the input z, it has the final output y. Therefore, by using the definition of return to scale provided by Tone [53] and Soleimani-damaneh [50], we define the return to scale in the network as follows:

Definition 9.2 Network o is located on the efficient frontier of T_{SC}^{VRS} in this case:

1. RTS of Network o is increasing if and only if exists $\delta^* > 0$ such that:

$$\forall \delta, 0 < \delta \leq \delta^* \Rightarrow ((1+\delta)x_o, (1+\delta)y_o) \in \mathrm{int}T_{SC}^{VRS} \qquad (9.130)$$

2. RTS of Network o is decreasing if and only if exists $\delta^* > 0$ such that:

$$\forall \delta, 0 < \delta \leq \delta^* \Rightarrow ((1-\delta)x_o, (1-\delta)y_o) \in \mathrm{int}T_{SC}^{VRS} \qquad (9.131)$$

3. RTS of Network o is constant if and only if one of the following four conditions is true.

   (a)   There exists $\delta^* > 0$ such that:

   $$\forall \delta, \ 0 < \delta \leq \delta^*$$
   $$\Rightarrow ((1+\delta)x_o, (1+\delta)y_o) \in \partial T_{SC}^{VRS} \qquad (9.132)$$
   $$\text{and } ((1-\delta)x_o, (1-\delta)y_o) \in \partial T_{SC}^{VRS}$$

   (b)   For each $\delta > 0$:

   $$((1+\delta)x_o, (1+\delta)y_o) \notin T_{SC}^{VRS}$$
   $$\text{and } ((1-\delta)x_o, (1-\delta)y_o) \notin T_{SC}^{VRS} \qquad (9.133)$$

Fig. 9.22 Geometric
interpretation of the RTS of
two-stage units

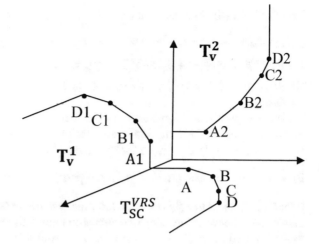

(c)   There exists $\delta^* > 0$ such that:

$$\forall \delta, \ 0 < \delta \leq \delta^*$$
$$\Rightarrow ((1 + \delta)x_o, (1 + \delta)y_o) \in \partial T_{SC}^{VRS} \tag{9.134}$$
$$\text{and } ((1 - \delta)x_o, (1 - \delta)y_o) \notin T_{SC}^{VRS}$$

(d)   There exists $\delta^* > 0$ such that:

$$\forall \delta, \ 0 < \delta \leq \delta^*$$
$$\Rightarrow ((1 - \delta)x_o, (1 - \delta)y_o) \in \partial T_{SC}^{VRS} \tag{9.135}$$
$$\text{and } ((1 + \delta)x_o, (1 + \delta)y_o) \notin T_{SC}^{VRS}$$

The geometric interpretation of the Definition 9.2 is shown in Fig. 9.22.

In Fig. 9.22, the diagrams on the left and right show the PPS of stage 1 $\left(T_v^1\right)$ and the PPS of stage 2 $\left(T_v^2\right)$. Also, the diagram at the bottom of the figure illustrates the PPS of the network without considering the principle of constant return to scale. Using the Definition 9.2, network A has IRS, networks D and C have DRS, and network B has CRS.

Theorem 9.5 Suppose the network o is located on the frontier T_{SC}^{VRS} and $\alpha > 0$ and $\beta > 0$ are proportional changes in the inputs and outputs of the network, respectively. In this case:

(a)   Network o has increasing return to scale if only if $\beta/\alpha > 1$.
(b)   Network o has decreasing return to scale if only if $\beta/\alpha < 1$.
(c)   Network o has constant return to scale if only if $\beta/\alpha = 1$.

Proof of Part (a) Assumes that $(\alpha x_o, \beta y_o) \in T_{SC}^{VRS}$ so $\sum_{j=1}^{n} \left(\theta_{Sj}^* x_{ij} \right) \lambda_j \leq \alpha x_{io}$ and $\sum_{j=1}^{n} \left(\varphi_{Mj}^* y_{rj} \right) \lambda_j \geq \beta y_{ro}$. We know that $\beta > \alpha$ thus $\beta = \alpha + \delta_1$. We define $\alpha = 1 + \delta$ and $\beta = (1 + \delta) + \delta_1$. So $\sum_{j=1}^{n} \left(\varphi_{Mj}^* y_{rj} \right) \lambda_j \geq ((1 + \delta) + \delta_1) y_{ro} > (1 + \delta) y_{ro}$. Consequently: $((1 + \delta) x_o, (1 + \delta) y_o) \in \text{int} T_{SC}^{VRS}$. Therefore, returns to scale are increasing.

Now suppose that returns to scale are increasing. So: $((1 + \delta) x_o, (1 + \delta) y_o) \in \text{int} T_{SC}^{VRS}$. Based on the definition of an interior point, There exists $\delta_1 > 0$ such that: $((1 + \delta) x_o, ((1 + \delta) + \delta_1) y_o) \in T_{SC}^{VRS}$. Thus $\beta / \alpha = ((1 + \delta) + \delta_1) / (1 + \delta) > 1$.

Proof of Part (b) The proof is similar to part (a). So it is ignored.

Proof (c) Assume that $\beta / \alpha = 1$ and returns to scale are not constant. As a result, the network o either has increasing or decreasing returns to scale. In this case, according to the last two parts, $\beta / \alpha < 1$ or $\beta / \alpha > 1$. And this is a contradiction. Thus the return to scale is constant.

Now assume that network o has a constant return to scale but $\beta / \alpha \neq 1$, therefore $\beta / \alpha < 1$ or $\beta / \alpha > 1$. As a result, according to parts (a) and (b), the return to scale of network o is not constant, which is a contradiction. So $\beta / \alpha = 1$.

Also, to examine the relationship between the RTS of the network and the RTS of each stage, the Theorem 9.6 is provided.

Theorem 9.6 Consider the two-stage network o.

(a) If stages 1 and 2 have increasing return to scale, then the network o has increasing return to scale.
(b) If stages 1 and 2 have decreasing return to scale, then the network o has decreasing return to scale.
(c) If stages 1 and 2 have constant return to scale, then the network o has constant return to scale.

Proof of Part (a) Let (x, z, y) be the network's initial input, intermediate, and final output, respectively. And α, β, and γ are proportional changes in x, y, and z, respectively. We know that stage 1 has increasing returns to scale. Therefore, $\gamma > \alpha$. Similarly, it can be shown that $\beta > \gamma$. Consequently: $\beta > \alpha$. Hence $\beta / \alpha > 1$. Thus network o has increasing return to scale.

Proof of Part (b) The proof is similar to the Proof of Part (a).

Proof of Part (c) The proof is similar to the Proof of Part (a).

Consider Fig. 9.22 again. As seen, the return to scale of $D1$ and $D2$ is DRS. Also, network D has a decreasing return to scale. In the same way, $B1$ and $B2$ have constant returns to scale, so according to Theorem 9.6, network B also has constant returns to scale.

Remark 9.1 It is essential to say that the converse of the above theorems is not necessarily true. For example, in Fig. 9.22, network A has an increasing return to scale, while stage 1 has an increasing return to scale and stage 2 has a constant return to scale.

The Definition 9.2 is not suitable for practical applications. To use it in real-world issues, we suggest the Theorem 9.7.

Theorem 9.7 Suppose network o is located on the efficient frontier of T_{SC}^{VRS} and (v^*, u^*, ω_o^*) is the optimal solution of model (9.17), then:

(a) If in every optimal solution $\omega_o^* > 0$, the network has increasing return to scale.
(b) If in every optimal solution $\omega_o^* < 0$, the network has decreasing return to scale.
(c) If at least in one of the optimal solutions $\omega_o^* = 0$, the network has constant return to scale.

Proof of Part (a) Let network o is under evaluation and (v^*, u^*, ω_o^*) is optimal solution of the model (9.17). So: $\sum_{r=1}^{s} u_r^* \left(\varphi_{mj}^* y_{ro} \right) - \sum_{i=1}^{m} v_i^* \left(\theta_{sj}^* x_{io} \right) + \omega_o^* = 0.$ We define:

$$z = \sum_{r=1}^{s} u_r^* \left(\varphi_j^* (1+\delta) y_{ro} \right) - \sum_{i=1}^{m} v_i^* \left(\theta_j^* (1+\delta) x_{io} \right) + \omega_o^* \qquad (9.136)$$

As a result:

$$z = (1+\delta) \left(\sum_{r=1}^{s} u_r^* \left(\varphi_j^* y_{ro} \right) - \sum_{i=1}^{m} v_i^* \left(\theta_j^* x_{io} \right) + \omega_o^* \right) - \delta \omega_o^*$$
$$= 0 - \delta \omega_o^* < 0 \qquad (9.137)$$

Therefore: $((1+\delta) x_o, (1+\delta) y_o) \in int T_{SC}^{VRS}$ and using Definition 9.2, network o has an increasing return to scale.

Proof of Part (b) The proof is similar to the Proof of Part (a).

Proof of part (c) Considering that (v^*, u^*, ω_o^*) is the optimal solution of the model (9.17) and $\omega_o^* = 0$, therefore: $\sum_{r=1}^{s} u_r^* \left(\varphi_j^* y_{ro} \right) - \sum_{i=1}^{m} v_i^* \left(\theta_j^* x_{io} \right) = 0.$ As a result, network o is located on the joint frontier of T_{SC}^{CRS} and T_{SC}^{VRS}. So the network o has constant returns to scale.

We present the following algorithm using the Theorem 9.7 to specify the type of return to scale.

Algorithm 1

1. Obtain the optimal solution of the model (9.17).
2. If $\omega_o^* = 0$ then returns to scale is constant.

3. If $\omega_o^* > 0$, solve the model (9.138). If the optimal solution of the model (9.138) is equal to zero, the return to scale is constant. Otherwise, the return to scale is increasing.

$$\min \ \omega_o$$
$$s.t.$$
$$\sum_{i=1}^{m} v_i x_{io} = 1,$$
$$\sum_{r=1}^{s} u_r y_{ro} - \sum_{i=1}^{m} v_i x_{io} + \omega_o = 0, \tag{9.138}$$
$$\sum_{r=1}^{s} u_r \left(\varphi_{mj}^* y_{rj} \right) - \sum_{i=1}^{m} v_i \left(\theta_{sj}^* x_{ij} \right) + \omega_o \leq 0, \ j = 1,...,n, \ j \neq o,$$
$$u_r, v_i, \omega_o \geq 0, \ r = 1, ..., s, \ i = 1, ..., m.$$

4. If $\omega_o^* < 0$, solve the model (9.139). If the optimal solution of the model (9.139) is equal to zero, the return to scale is constant. Otherwise, the return to scale is decreasing.

$$\max \ \omega_o$$
$$s.t.$$
$$\sum_{i=1}^{m} v_i x_{io} = 1,$$
$$\sum_{r=1}^{s} u_r y_{ro} - \sum_{i=1}^{m} v_i x_{io} + \omega_o = 0, \tag{9.139}$$
$$\sum_{r=1}^{s} u_r \left(\varphi_{mj}^* y_{rj} \right) - \sum_{i=1}^{m} v_i \left(\theta_{sj}^* x_{ij} \right) + \omega_o \leq 0, \ j = 1,...,n, \ j \neq o,$$
$$u_r, v_i \geq 0, \ \omega_o \leq 0, \ r = 1, ..., s, \ i = 1, ..., m.$$

In the above discussion, it is assumed that the network is efficient. In the following, this assumption is removed. Using the Definition 9.1 and Theorem 9.3, Algorithm 2 is presented to determine the return to scale in the inefficient network.

Algorithm 2

1. Obtain $\left(\hat{x}_o, \hat{y}_o \right)$ using the model (9.18).
2. Calculate the optimal solution of model (9.17) when $\left(\hat{x}_o, \hat{y}_o \right)$ is under evaluation, $\left(v^*, u^*, \omega_o^* \right)$.
3. If $\omega_o^* = 0$ then returns to scale is constant.
4. If $\omega_o^* > 0$, solve the model (9.140). If the optimal solution of the model (9.140) is equal to zero, the return to scale is constant. Otherwise, the return to scale is

increasing.

$$\min \ \omega_o$$

$$s.t.$$

$$\sum_{i=1}^{m} v_i \hat{x}_{io} = 1,$$

$$\sum_{r=1}^{s} u_r (\hat{y}_{ro}) - \sum_{i=1}^{m} v_i (\hat{x}_{io}) + \omega_o = 0, \tag{9.140}$$

$$\sum_{r=1}^{s} u_r (\varphi_j^* y_{rj}) - \sum_{i=1}^{m} v_i (\theta_j^* x_{ij}) + \omega_o \leq 0, \ j = 1,...,n, \ j \neq o$$

$$u_r, v_i \geq 0, \ \omega_o \geq 0, \ r = 1, ..., s, \ i = 1, ..., m.$$

5. If $\omega_o^* < 0$, solve the model (9.141). If the optimal solution of the model (9.141) is equal to zero, the return to scale is constant. Otherwise, the return to scale is decreasing.

$$\max \ \omega_o$$

$$s.t.$$

$$\sum_{i=1}^{m} v_i \hat{x}_{io} = 1,$$

$$\sum_{r=1}^{s} u_r (\hat{y}_{ro}) - \sum_{i=1}^{m} v_i (\hat{x}_{io}) + \omega_o = 0, \tag{9.141}$$

$$\sum_{r=1}^{s} u_r (\varphi_j^* y_{rj}) - \sum_{i=1}^{m} v_i (\theta_j^* x_{ij}) + \omega_o \leq 0, \ j = 1,...,n, \ j \neq o,$$

$$u_r, v_i \geq 0, \ r = 1, ..., s, \ i = 1, ..., m, \ \omega_o \leq 0.$$

9.9 Conclusion

As stated in this chapter, the supply chain is a type of decision-making unit that not only has input and output indicators but also uses intermediary indicators that flow from the previous stage to the next stage. Each stage may have its inputs and outputs. Therefore, due to the networked or multi-stage nature of the supply chain, the traditional data envelopment analysis models cannot wholly and correctly evaluate the supply chain, provide reliable results and accurately identify the sources of inefficiency in a supply chain. Therefore, identifying sources of inefficiency in units under evaluation using DEA may not be done correctly. Thus, to solve this problem, researchers used NDEA models. When using NDEA models, paying attention to the following points is necessary.

- Supply chain structure.

  Before starting any work to use NDEA models, the number of supply chain components and how they relate to each other should be determined. In other words, it should be specified that internal connections are in series, parallel, or mixed. Accuracy in this step is critical. Otherwise, the results obtained for supply chain evaluation will be discredited.
- Specify the appropriate NDEA model.

  As seen in this chapter, various envelopment and multiplier models exist for each parallel, series, and mixed supply chain group. Appropriate envelopment or multiplier models can be chosen based on the decision maker's objectives. For example, some economic interpretations are based only on having optimal weights in multiplier forms. So in such a situation, using envelopment models does not provide the required results to managers. It should also be noted that, unlike DEA models, some NDEA models are presented only by considering the principle of constant return to scale and cannot be presented without this principle. Also, in some NDEA models, relatively efficient units are not necessarily introduced.

  And another point that should be noted is that some NDEA models are only used to measure two-stage units. And they cannot be expanded to measure the performance of multi-member supply chains. Therefore, all aspects must be weighed carefully before choosing the suitable model for evaluation.
- Type of data

  Contrary to the initial assumption in NDEA modeling, in some applied studies, the inputs and outputs are not greater than or equal to 0. We will face negative, fuzzy, integer, and interval data in the supply chains under evaluation. So precision to the data type is emphasized before starting the assessment and selecting the model. Otherwise, the evaluation results will not be accurate enough. And even in some cases, it will be discredited.

References

1. Abbasbandy, S., Asady, B.: Ranking of fuzzy numbers by sign distance. Inf. Sci. **176**(16), 2405–2416 (2006)
2. Akyar, E., Akyar, H., Düzce, S.A.: A new method for ranking triangular fuzzy numbers. Int. J. Uncertainty Fuzziness Knowl. Based Syst. **20**(5), 729–740 (2012)
3. Allahviranloo, T., Lotfi, F.H., Firozja, M.A.: Fuzzy efficiency measure with fuzzy production possibility set. Appl. Appl. Math. Int. J. (AAM) **2**(2), 8 (2007)
4. Allahviranloo, T., Lotfi, F.H., Firozja, M.A.: Efficiency in fuzzy production possibility set. Iran. J. Fuzzy Syst. **9**(4), 17–30 (2012)

5. Bagheri, M., Ebrahimnejad, A., Razavyan, S., Hosseinzadeh, L.F., Malekmohammadi, N.: Fuzzy arithmetic DEA approach for fuzzy multi-objective transportation problem. Oper. Res. **22**(4), 1–31 (2020)
6. Chen, C., Yan, H.: Network DEA model for supply chain performance evaluation. Eur. J. Oper. Res. **213**(1), 147–155 (2011)
7. Chen, Y., Zhu, J.: Measuring information technology's indirect impact on firm performance. Inf. Technol. Manage. **5**(1), 9–22 (2004)
8. Cook, W.D., Zhu, J., Bi, G., Yang, F.: Network DEA: additive efficiency decomposition. Eur. J. Oper. Res. **207**(2), 1122–1129 (2010)
9. Despotis, D.K., Koronakos, G., Sotiros, D.: Composition versus decomposition in two-stage network DEA: a reverse approach. J. Prod. Anal. **45**, 71–87 (2016)
10. Despotis, D.K., Smirlis, Y.G.: Data envelopment analysis with imprecise data. Eur. J. Oper. Res. **140**(1), 24–36 (2002)
11. Ebrahimnejad, A., Tavana, M., Lotfi, F.H., Shahverdi, R., Yousefpour, M.: A three-stage data envelopment analysis model with application to banking industry. Measurement **49**, 308–319 (2014)
12. Ezzati, R., Allahviranloo, T., Khezerloo, S., Khezerloo, M.: An approach for ranking of fuzzy numbers. Expert Syst. Appl. **39**(1), 690–695 (2012)
13. Färe, R., Grosskopf, S.: Network DEA. Socio-Econ. Plann. Sci. **34**(1), 35–49 (2000)
14. Ghobadi, S., Jahanshahloo, G.R., Lotfi, F.H., Rostamy-Malkhalifeh, M.: Dynamic inverse DEA in the presence of fuzzy data. Adv. Environ. Biol. **8**(24), 139–151 (2014)
15. Huang, C.-W.: Assessing the performance of tourism supply chains by using the hybrid network data envelopment analysis model. Tour. Manage. **65**, 303–316 (2018)
16. Izadikhah, M., Saen, R.F.: Evaluating sustainability of supply chains by two-stage range directional measure in the presence of negative data. Transp. Res. Part D: Transp. Environ. **49**, 110–126 (2016)
17. Jahanshahloo, G.R., Lotfi, F.H., Malkhalifeh, M.R., Namin, M.A.: A generalized model for data envelopment analysis with interval data. Appl. Math. Model. **33**(7), 3237–3244 (2009)
18. Jahanshahloo, G.R., Lotfi, F.H., Rezaie, V., Khanmohammadi, M.: Ranking DMUs by ideal points with interval data in DEA. Appl. Math. Model. **35**(1), 218–229 (2011)
19. Jahanshahloo, G.R., Lotfi, F.H., Shahverdi, R., Adabitabar, M., Rostamy-Malkhalifeh, M., Sohraiee, S.: Ranking DMUs by l_1-norm with fuzzy data in DEA. Chaos, Solitons Fractals **39**(5), 2294–2302 (2009)
20. Kao, C.: Efficiency decomposition in network data envelopment analysis: a relational model. Eur. J. Oper. Res. **192**(3), 949–962 (2009)
21. Kao, C., Hwang, S.-N.: Efficiency decomposition in two-stage data envelopment analysis: an application to non-life insurance companies in Taiwan. Eur. J. Oper. Res. **185**(1), 418–429 (2008)
22. Kao, C., Hwang, S.-N.: Efficiency measurement for network systems: IT impact on firm performance. Decis. Support Syst. **48**(3), 437–446 (2010)
23. Khalili-Damghani, K., Taghavi-Fard, M., Abtahi, A.-R.: A fuzzy two-stage DEA approach for performance measurement: real case of agility performance in dairy supply chains. Int. J. Appl. Decis. Sci. **5**(4), 293–317 (2012)
24. Khalili-Damghani, K., Sadi-Nezhad, S., Hosseinzadeh-Lotfi, F: Imprecise DEA models to assess the agility of supply chains. In: Supply Chain Management under Fuzziness, pp. 167–198. Springer, Berlin, Heidelberg (2014)
25. Kleindorfer, P.R., Kunreuther, H.C., Schoemaker, P.J.H.: Decision sciences: an integrative perspective. Cambridge University Press (1993)
26. Lewis, H.F., Sexton, T.R.: Network DEA: efficiency analysis of organizations with complex internal structure. Comput. Oper. Res. **31**(9), 1365–1410 (2004)
27. Li, S.X., Huang, Z., Ashley, A.: Seller—buyer system cooperation in a monopolistic market. J. Oper. Res. Soc. **46**(12), 1456–1470 (1995)
28. Liang, L., Cook, W.D., Zhu, J.: DEA models for two-stage processes: game approach and efficiency decomposition. Naval Res. Logistics **55**(7), 643–653 (2008)

29. Liang, L., Yang, F., Cook, W.D., Zhu, J.: DEA models for supply chain efficiency evaluation. Ann. Oper. Res. **145**(1), 35–49 (2006)
30. Liu, S.T.: A fuzzy DEA/AR approach to the selection of flexible manufacturing systems. Comput. Ind. Eng. **54**(1), 66–76 (2008)
31. Lotfi, F.H., Eshlaghy, A.T., Shafiee, M., Nikoomaram, H., Seyedhoseini, S.M.: A new two-stage data envelopment analysis (DEA) model for evaluating the branch performance of banks. Afr. J. Bus. Manage. **6**(24), 7230–7241 (2012)
32. Lozano, S.: Information sharing in DEA: a cooperative game theory approach. Eur. J. Oper. Res. **222**(3), 558–565 (2012)
33. Lozano, S., Hinojosa, M.Á., Mármol, A.M., Borrero, D.V.: DEA and cooperative game theory. In: Handbook of Operations Analytics Using Data Envelopment Analysis, pp. 215–239. Springer, Boston, MA (2016)
34. Momeni, E., Tavana, M., Mirzagoltabar, H., Mirhedayatian, S.M.: A new fuzzy network slacks-based DEA model for evaluating performance of supply chains with reverse logistics. J. Intell. Fuzzy Syst. **27**(2), 793–804 (2014)
35. Neumann, J.V.: On the theory of game of strategy. In: Contributions to thetheory of games, vol. IV. Annals of Mathematic Studies, (40), 13-42.
36. Nikfarjam, H., Rostamy-Malkhalifeh, M., Mamizadeh-Chatghayeh, S.: Measuring supply chain efficiency based on a hybrid approach. Transp. Res. Part D: Transp. Environ. **39**, 141–150 (2015)
37. Omrani, H., Keshavarz, M.: A performance evaluation model for supply chain of shipping company in Iran: an application of the relational network DEA. Marit. Policy Manag. **43**(1), 121–135 (2016)
38. Ramzi, S.: Modeling the education supply chain with network DEA Model: the case of tunisia. J. Quant. Econ. **17**(3), 525–540 (2019)
39. Rezaee, M.J., Izadbakhsh, H., Yousefi, S.: An improvement approach based on DEA-game theory for comparison of operational and spatial efficiencies in urban transportation systems. KSCE J. Civ. Eng. **20**(4), 1526–1531 (2016)
40. Rostamy-Malkhalifeh, M., Sanei, M., Saleh, H.: A new method for solving fuzzy DEA models by trapezoidal approximation. J. Math. Extension **4**(1), 115–126 (2009)
41. Rostamy-Malkhlifeh, M., Ebrahimkhani, G.S., Sale, H., Ebrahimkhani, G.N.: Congestion in DEA model with fuzzy data. Int. J. Appl. Oper. Res. **1**(2), 49–56 (2011)
42. Saaty, T.L.: Rank from comparisons and from ratings in the analytic hierarchy/network processes. Eur. J. Oper. Res. **168**(2), 557–570 (2006)
43. Sahoo, B.K., Saleh, H., Shafiee, M., Tone, K., Zhu, J.: An alternative approach to dealing with the composition approach for series network production processes. Asia-Pacific J. Oper. Res. **38**(06), 2150004 (2021)
44. Saleh, H., Hosseinzadeh, F., Rostamy, M., Shafiee, M.: Performance evaluation and specifying of return to scale in network DEA. J. Adv. Math. Model. **10**(2), 309–340 (2020)
45. Sanei, M., Rostami-Malkhalifeh, M., Saleh, H.: A new method for solving fuzzy DEA models. Int. J. Ind. Math. **1**(4), 307–313 (2009)
46. Shafiee, M., Saleh, H.: Evaluation of strategic performance with fuzzy data envelopment analysis. Int. J. Data Envelopment Anal. **7**(4), 1–20 (2019)
47. Shafiee, M., Saleh, H., Ghaderi, M.: Benchmarking in the supply chain using data envelopment analysis and system dynamics simulations. Iran. J. Supply Chain Manag. **23**(70), 55–70 (2021)
48. Shafiee, M., Lotfi, F. H., Saleh, H.: Supply chain performance evaluation with data envelopment analysis and balanced scorecard approach. Appl. Math. Model. **38**(21–22), 5092–5112 (2014)
49. Shoja, M., Hosseinzadeh, L.F., Gholam, A.A., Rashidi, K.A.: Efficiency of green supply chain in the presence of non-discretionary and undesirable factors, using data envelopment analysis. Bus. Inf. **15**(3), 78–96 (2021)
50. Soleimani-damaneh, M.: On a basic definition of returns to scale. Oper. Res. Lett. **40**(2), 144–147 (2012)
51. Tamaddon, L., Jahanshahloo, G.R., Lotfi, F.H., Mozaffari, M.R., Gholami, K.: Data envelopment analysis of missing data in crisp and interval cases. Int. J. Math. Anal. **3**(20), 955–969 (2009)

52. Tavana, M., Mirzagoltabar, H., Mirhedayatian, S.M., Saen, R.F., Azadi, M.: A new network epsilon-based DEA model for supply chain performance evaluation. Comput. Ind. Eng. **66**(2), 501–513 (2013)
53. Tone, K.: On returns to scale under weight restrictions in data envelopment analysis. J. Prod. Anal. **16**, 31–47 (2001)
54. Tone, K., Tsutsui, M.: An epsilon-based measure of efficiency in DEA – A third pole of technical efficiency. Eur. J. Oper. Res. **207**(3), 1554–1563 (2010)
55. Torabi, N., Tavakkoli-Moghaddam, R., Najafi, E., Hosseinzadeh-Lotfi, F.: A two-stage green supply chain network with a carbon emission price by a multi-objective interior search algorithm. Int. J. Eng. **32**(6), 828–834 (2019)
56. Wagner, B.A., Macbeth, D.K., Boddy, D: Improving supply chain relations: an empirical case study. Supply Chain Manag. Int. J. **7**, 253–264 (2002)
57. Wang, Y.-M., Greatbanks, R., Yang, J.-B.: Interval efficiency assessment using data envelopment analysis. Fuzzy Sets Syst. **153**(3), 347–370 (2005)
58. Yang, F., Wu, D., Liang, L., Bi, G., Wu, D.: Supply chain DEA: production possibility set and performance evaluation model. Ann. Oper. Res. **185**(1), 195–211 (2011)
59. Yaya, S., Xi, C., Xiaoyang, Z., Meixia, Z.: Evaluating the efficiency of China's healthcare service: a weighted DEA-game theory in a competitive environment. J. Clean. Prod. **270**, 122431 (2020)
60. Zhai, D., Shang, J., Yang, F., Ang, S.: Measuring energy supply chains' efficiency with emission trading: a two-stage frontier-shift data envelopment analysis. J. Clean. Prod. **210**, 1462–1474 (2019)

Chapter 10
Performance Evaluation of Supply Chains by Bi-Level DEA

10.1 Introduction

In management decisions in real-world scenarios, companies are mainly composed of different interconnected departments and levels with many interactions. In other words, in decentralized decision-making, companies are a set of decisions at different levels, along with independent decision variables, and sometimes related to other departments, which are often taken by the managers of each department, who usually only control their level. Therefore, these decisions, in some cases, even contradict each other. Therefore, organizations are considered a set of shared resources and subgroups, and each organization competes with others to achieve the overall goals of that organization and the independent goals of their subgroups [15]. As a result, these subgroups usually act as a hierarchical structure at different levels. So that the leading managers in each organization as leaders at the top level of operations and other subgroups as followers at a lower level by making decisions seeking to optimize their individual goals and the overall goal of the respective organization. In other words, the hierarchical process means that the leader first determines specific decisions and plans by considering the followers' reactions to these decisions. Then the follower provides the appropriate response and strategy considering the leader's decisions and goals.

In many supply chain management studies, a supply chain is considered a black box or, in some cases, a multi-stage structure for a more detailed analysis. In other words, the supply chain members are often regarded as independent, so this issue makes the obtained results doubtful in some applied studies. Because in many supply chains, each of the components has its own goals and decision variables on the one hand, and on the other hand, they have effects, limitations, and mutual reactions from other members. For example, consider a manufacturing supply chain that includes a factory and distribution center (DC). In this system, the factory is the leader and assigns the products to the follower, i.e., the DC, although there may be a follower or several followers at this level. At the next level, the DCS distributes the products to the customers. Thus DC acts as a follower for the factory and a leader for the customers.

© The Author(s), under exclusive license to Springer Nature Switzerland AG 2023
F. Hosseinzadeh Lotfi et al., *Supply Chain Performance Evaluation*,
Studies in Big Data 122, https://doi.org/10.1007/978-3-031-28247-8_10

Since the leader's information at different levels is not necessarily complete, leaders should take their decisions, such as product pricing, into consideration of the possible reactions of followers (customers) [25]. Therefore, the performance of a supply chain should be evaluated in a way that fully considers the structures or natural characteristics of the system. Thus, in macro-level supply chain management, supply chain planning and management actually is decision-making about a multi-level decision-making network. Therefore, using DEA and NDEA models in such conditions does not provide acceptable results.

Although the evaluation of the performance of bi-level supply chains through the combination of NDEA and game theory was explained in Chap. 8, in this chapter, we examine this issue with another approach. Internal interactions and a hierarchical structure in supply chains are easily considered using multi-level mathematical programming and DEA. Therefore, in the following, we will explain the types of models in the multi-level data envelopment analysis technique and, in a particular case, bi-level DEA (BLPDEA) as a valuable tool for providing information to the manager when evaluating the performance of supply chains.

10.2   Bi-Level Programming

Multi-level programming problems refer to problems that optimize various goals in a hierarchical structure. In such issues, there are several levels of decision-making, each level has its objective function, and each objective function in each hierarchical level is associated with its constraints. In a particular case, if the problem has two levels, it is called a bi-level programming problem. In simpler terms, a bi-level programming problem is mathematical programming in which an optimization problem is within the constraints of another optimization problem. The higher level decision maker (HLDM) is called the leader, and the lower level decision maker (LLDM) is called the follower.

Suppose that $x \in X \subseteq \mathbb{R}^n$, $y \in Y \subseteq \mathbb{R}^m$ are the variables related to levels 1 and 2 in a BPL, so the general form of a BPL problem is shown as the model (10.1).

$$\min_x c_1 x + d_1 y$$

$$s.t$$

$$A_1 x + B_1 y \le b_1 y,$$

$$\min_y c_2 x + d_2 y,$$

$$s.t$$

$$A_2 x + B_2 y \le b_2,$$

$$x \ge 0, y \ge 0. \tag{10.1}$$

So that: $c_1, c_2 \subseteq \mathbb{R}^n, d_1, d_2 \subseteq \mathbb{R}^m, b_1 \subseteq \mathbb{R}^e, b_2 \subseteq \mathbb{R}^f, A_1 \subseteq \mathbb{R}^{e \times n}, A_2 \subseteq \mathbb{R}^{f \times n},$ $B_1 \subseteq \mathbb{R}^{e \times m},$ and $B_2 \subseteq \mathbb{R}^{f \times m}.$. The model (10.1) includes two optimization problems. In the first problem (higher level), the leader seeks to minimize the objective function under the constraints specified in the model (10.1). In the second level, the follower seeks to find the optimal solution of $\min c_2 x + d_2 y$ considering the constraints $A_2 x + B_2 y \leq b_2$ and $x \geq 0, y \geq 0.$ Different methods have been presented to solve bi-level programming problems. In general, these methods can be categorized as follows:

(1) Search methods such as the k-best algorithm
(2) Transfer methods such as branch and bound
(3) Heuristic methods such as genetic algorithm
(4) Evolutionary methods such as the optimization algorithm
(5) Methods based on the fuzzy concept.

10.3 Cost Efficiency in the Two-Member Supply Chain (Leader–Follower) Using Bi-Level DEA

As one of the critical management issues, company managers, factory owners, and so on, considering the limited resources, always seek to maintain or improve the quality of their services by minimizing the overall cost. In other words, managers using resources and allocating costs for resources are always looking for ways to reduce resource-cost consumption in competition with other competitors while maintaining or improving performance.

These challenges, "how to spend" and "using available resources," has become a part of their management practices. Therefore, the management of the companies is constantly trying to identify and eliminate the reasons for inefficiency in the under-evaluation system in competition with others. Although Fare et al. [4], for the first time, expressed the concept of cost efficiency in DEA in the form of the model (6.91), this model is not suitable for dealing with multi-stage and multi-level problems. For example, suppose the goal is to evaluate the cost efficiency in the branches of a bank. The performance evaluation of banks is considered one of the complex issues in the field of performance evaluation due to the existence of extensive internal activities. Therefore, due to the importance of this issue, many studies have been conducted in this field. Sun et al. [20] suggested that to improve the performance of banks, each bank branch should be considered as a supply chain and supply chain management (SCM) methods are used to improve the performance of banks. Because this can be beneficial in providing timely and reliable quality products and services while keeping costs low. Thus, in this approach, the deposit system is considered as the manufacturer and the loan payment system is the retailer. Because the deposit collection department produces products (capital) through costs (labor costs, operating costs, etc.) and employees. In the next stage, the loan disbursement department sells the products (collected capital) to borrowers through loan disbursement.

Wu et al. [22] believe the type and amount of the loan disbursement and the bank's investment in the second stage are based on deposits collected in the first stage, so it is evident that in the supply chain shown in Fig. 10.1, the first stage is the leader, and the second stage is the follower. As a result, using conventional efficiency models in DEA and NDEA is not recommended in such cases, so [21] proposed a bi-level cost efficiency calculation model. For this purpose, consider the general bi-level supply chain shown in Fig. 10.2.

So that x_j^{D1} and x_j^{D2} are direct inputs for the leader and follower, x_j is the shared input between the leader and the follower where x_j^1 and x_j^2 are the input values used for the leader and the follower, respectively, $x_j^1 + x_j^2 = E$, $E \in \mathbb{R}$. Also, c^1 and D^1 correspond to the cost of one unit of shared input in the leader (follower) stage and one unit of intermediate products (z). Similarly, c^2 and D^2 correspond to the cost of a unit of direct input in the leader and follower stages. Therefore, the model (10.1) was introduced by ref. [21] to calculate cost efficiency in bi-level supply chains.

$$\min\left(c^1 \overline{x}^1 + c^2 \overline{x}^{D1}\right) + \left(c^1 \overline{x}^2 + D^1 \overline{z} + D^2 \overline{x}^{D2}\right)$$

$s.t$

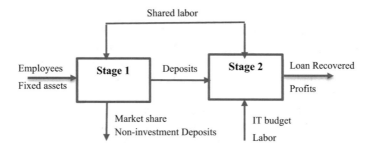

Fig. 10.1 Bi-level system for bank efficiency evaluation

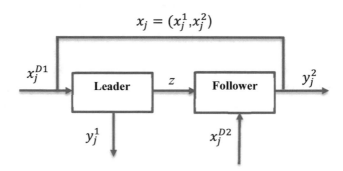

Fig. 10.2 Bi-Level system with constrained resource

$$\sum_{j=1}^{n} \lambda_j x_j^1 \leq \overline{x}^1, \qquad \sum_{j=1}^{n} \lambda_j x_j^{D1} \leq \overline{x}^{D1},$$

$$\sum_{j=1}^{n} \lambda_j z_j \geq z_o, \qquad \sum_{j=1}^{n} \lambda_j y_j^1 \geq y_o^1,$$

$$\overline{x}^1 + \overline{x}^2 = E,$$

$$\min c^1 \overline{x}^2 + D^1 \overline{z} + D^2 \overline{x}^{D2}$$

s.t

$$\sum_{j=1}^{n} \pi_j x_j^2 \leq \overline{x}^2, \qquad \sum_{j=1}^{n} \pi_j x_j^{D2} \leq \overline{x}^{D2},$$

$$\sum_{j=1}^{n} \pi_j z_j \leq \overline{z}, \qquad \sum_{j=1}^{n} \pi_j y_j^2 \geq y_o^2,$$

$$\left(\overline{x}^1, \overline{x}^2, \overline{x}^{D1}, \overline{x}^{D2}, \overline{z} \right) \geq 0,$$

$$\pi_j, \lambda_j \geq 0, \qquad j = 1, \ldots, n. \tag{10.2}$$

Thus, in the first stage, the leader obtains the optimal value of the cost in the objective function by determining the optimal input consumption (direct input and shared input). Then in the next stage, the follower, by considering the optimal values obtained from stage 1 ($\overline{x}^2 = E - \overline{x}^1$), management of other resources, and consumption costs, gets the optimal value of the objective function.

Suppose that $\left(\overline{x}^{1*}, \overline{x}^{2*}, \overline{x}^{D1*}, \overline{x}^{D2*}, z^* \right)$ is the optimal solution of model (10.2) when DMU_o, is under evaluation. As a result, DMU_o is cost-efficient, if only if $CE_o = 1$ so that: $CE_o = \frac{c^1 \overline{x}^{1*} + c^2 \overline{x}^{D1*} + c^1 \overline{x}^{2*} + D^1 \overline{z}^* + D^2 \overline{x}^{D2*}}{c^1 x_o^1 + c^2 x_o^{D1} + c^1 x_o^2 + D^1 z_o + D^2 x_o^{D2}}$. Also, in DMU_o, leader and follower are cost-efficient, if only if $CE_o^L = 1$ and $CE_o^F = 1$, respectively. So that: $CE_o^L = \frac{c^1 \overline{x}^{1*} + c^2 \overline{x}^{D1*}}{c^1 x_o^1 + c^2 x_o^{D1}}$ and $CE_o^F = \frac{c^1 \overline{x}^{2*} + D^1 \overline{z}^* + D^2 \overline{x}^{D2*}}{c^1 x_o^2 + D^1 z_o + D^2 x_o^{D2}}$.

It can be easily shown in the model (10.2): $c^1 \overline{x}^1 + c^1 \overline{x}^2 = c^1 E$ and $\sum_{i=1}^{n} \pi_j x_j^2 \leq \overline{x}^2 = E - \overline{x}^1$. Therefore, we rewrite the model (10.2) as the BPL (10.3).

$$\min(c^2 \overline{x}^{D1} + D^2 \overline{x}^{D2} + D^1 \overline{z} + c^1 E)$$

s.t

$$\sum_{i=1}^{n} \lambda_j x_j^1 \leq \overline{x}^1, \qquad \sum_{j=1}^{n} \lambda_j x_j^{D1} \leq \overline{x}^{D1},$$

$$\sum_{j=1}^{n} \lambda_j y_j^1 \geq y_o^1, \qquad \sum_{j=1}^{n} \lambda_j z_j \geq z_o,$$

$$\min c^1 E - c^1 \overline{x}^1 + D^2 \overline{x}^{D2} + D^1 \overline{z}$$

s.t

$$\sum_{j=1}^{n} \pi_j x_j^2 \leq E - \bar{x}^1, \qquad \sum_{j=1}^{n} \pi_j x_j^{D2} \leq \bar{x}^{D2},$$

$$\sum_{j=1}^{n} \pi_j z_j \leq \bar{z}, \qquad \sum_{j=1}^{n} \pi_j y_j^2 \geq y_o^2,$$

$$\left(\bar{x}^1, \bar{x}^{D1}, \bar{x}^{D2}, \bar{z}\right) \geq 0,$$

$$\pi_j, \lambda_j \geq 0, \qquad j = 1, \dots, n. \tag{10.3}$$

It is necessary to explain that in some cases, according to the conditions of the problem and the opinion of the supply chain manager, the constraint $\bar{x}^1 + \bar{x}^2 = E$ in the model (10.2) is replaced by constraint $\bar{x}^1 + \bar{x}^2 \leq E$. Therefore, model (10.2) is rewritten as the model (10.4).

$$\min\left(c^1 \bar{x}^1 + c^2 \bar{x}^{D1}\right) + \left(c^1 \bar{x}^2 + D^1 \bar{z} + D^2 \bar{x}^{D2}\right)$$

$$s.t$$

$$\bar{x}^1 + \bar{x}^2 \leq E,$$

same constraints in (10.2)

$$\min c^1 \bar{x}^2 + D^1 z + D^2 \bar{x}^{D2}$$

$$s.t$$

same constraints in (10.2). $\tag{10.4}$

10.3.1   BPL Problem-Solving Method

Reference [19] stated the theorem (10.1) to solve the BPL problem introduced in (10.1).

Theorem 10.1 *Suppose that $u \in \mathbb{R}^p$, $w \in \mathbb{R}^m$, $v \in \mathbb{R}^q$ are dual variables corresponding to the model's constraints (10.1). In this case, the necessary and sufficient condition for (x^*, y^*) to be the optimal solution to the medel (10.1) is that (x^*, y^*) is the optimal solution to the problem (10.5).*

$$\min c_1 x + d_1 y$$

$$s.t$$

$$A_1 x + B_1 y \leq b_1,$$

$$A_2 x + B_2 y \leq b_2, \tag{10.5}$$

$$u B_1 + v B_2 - w = -d_2,$$

$$u(b_1 - A_1 x + B_1 y) + v(b_2 - A_2 x + B_2 y) + wy = 0,$$

$$x \geq 0, y \geq 0, u \geq 0, v \geq 0, w \geq 0.$$

Proof See ref. [19].

Because of the constraint $u(b_1 - A_1 x + B_1 y) + v(b_2 - A_2 x + B_2 y) + wy = 0$, model (10.5) is a nonlinear programming problem. Therefore, Shi et al. used the extended branch and bound method to solve the model (10.5).

To solve model (10.3) based on the theorem 10.1, model (10.3) needs to be converted to a standard BPL similar to model (10.1). So model (10.6) is presented.

Therefore, based on theorem 10.1, instead of solving BPL (10.6), model (10.5) can be solved so that: $c_1 = (0, c^2, 0)$, $d_1 = (D^2, D^1, 0)$, $c_2 = (-c^1, 0, 0)$, $d_2 = (D^2, D^1, 0)$,

$$
x = \begin{pmatrix} \bar{x}^1 \\ \bar{x}^{D1} \\ \lambda \end{pmatrix}, \quad
y = \begin{pmatrix} \bar{x}^{D2} \\ \bar{z} \\ \pi \end{pmatrix}, \quad
b_1 = \begin{pmatrix} 0 \\ 0 \\ -y_o^1 \\ -z_o \end{pmatrix},
$$

$$
b_2 = \begin{pmatrix} E \\ 0 \\ 0 \\ -y_o^2 \end{pmatrix}, \quad
A_1 = \begin{bmatrix}
-1 & 0 \, x_1^1 & \cdots & x_n^1 \\
0 & -1 \, x_1^{D1} & \cdots & x_n^{D1} \\
0 & 0 -y_1^1 & \cdots & -y_n^1 \\
0 & 0 -z_1^1 & \cdots & -z_n^1
\end{bmatrix}, \quad B_1 = 0
$$

Similarly, model (10.4) can also be solved.

$$
\min (0, c^2, 0) \begin{pmatrix} \bar{x}^1 \\ \bar{x}^{D1} \\ \lambda \end{pmatrix} + (D^2, D^1, 0) \begin{pmatrix} \bar{x}^{D2} \\ \bar{z} \\ \pi \end{pmatrix}
$$

s.t

$$
(-1, 0, x_1^1, \dots, x_n^1) \begin{pmatrix} \bar{x}^1 \\ \bar{x}^{D1} \\ \lambda_1 \\ \vdots \\ \lambda_n \end{pmatrix} \le 0, \quad
(0, -1, x_1^{D1}, \dots, x_n^{D1}) \begin{pmatrix} \bar{x}^1 \\ \bar{x}^{D1} \\ \lambda_1 \\ \vdots \\ \lambda_n \end{pmatrix} \le 0,
$$

$$
(0, 0, -y_1^1, \dots, -y_n^1) \begin{pmatrix} \bar{x}^1 \\ \bar{x}^{D1} \\ \lambda_1 \\ \vdots \\ \lambda_n \end{pmatrix} \le -y_o^1, \quad
(0, 0, -z_1, \dots, -z_n) \begin{pmatrix} \bar{x}^1 \\ \bar{x}^{D1} \\ \lambda_1 \\ \vdots \\ \lambda_n \end{pmatrix} \le -z_o,
$$

$$
\min(-c^1, 0, 0) \begin{pmatrix} \overline{x}^1 \\ \overline{x}^{D1} \\ \lambda \end{pmatrix} + (D^2, D^1, 0) \begin{pmatrix} \overline{x}^{D2} \\ \overline{z} \\ \pi \end{pmatrix}
$$

$s.t$

$$
(1, 0, 0) \begin{pmatrix} \overline{x}^1 \\ \overline{x}^{D1} \\ \lambda \end{pmatrix} + (0, 0, x_1^2, \ldots, x_n^2) \begin{pmatrix} \overline{x}^{D2} \\ \overline{z} \\ \pi_1 \\ \vdots \\ \pi_n \end{pmatrix} \leq E,
$$

$$
(-1, 0, x_1^{D2}, \ldots, x_n^{D2}) \begin{pmatrix} \overline{x}^{D2} \\ \overline{z} \\ \pi_1 \\ \vdots \\ \pi_n \end{pmatrix} \leq 0,
$$

$$
(0, -1, z_1, \ldots, z_n) \begin{pmatrix} \overline{x}^{D2} \\ \overline{z} \\ \pi_1 \\ \vdots \\ \pi_n \end{pmatrix} \leq 0, \quad (0, 0, -y_1^2, \ldots, -y_n^2) \begin{pmatrix} \overline{x}^{D2} \\ \overline{z} \\ \pi_1 \\ \vdots \\ \pi_n \end{pmatrix} \leq -y_o^2,
$$

$$
\overline{x}^1, \overline{x}^2, \overline{x}^{D1}, \overline{x}^{D2}, \overline{z} \geq 0,
$$

$$
\pi_j, \lambda_j \geq 0 \quad j = 1, \ldots, n. \tag{10.6}
$$

10.4 Cost Efficiency Assessment in a Supply Chain with Multiple Followers Using Bi-Level DEA

Although the cost efficiency model presented in the previous subsection partially solves the problem of calculating the cost efficiency in supply chains with the leader and follower approach, this model is only applicable for two-member supply chains. In reality, due to the complexities we face in various industries, supply chains are usually multi-member. For example, consider Fig. 10.3, which is the development of Fig. 10.2. As seen in Fig. 10.3 unlike the first stage (leader), the second stage (follower) includes several members.

Suppose that x_{Lj}^D and $x_{F_k j}^D$, $k = 1, \ldots, K$ are direct inputs for the k-th follower. Also, the vector $x = (x_L, x_{F_1 j}, \ldots, x_{F_K j})$ is shared input between the leader and

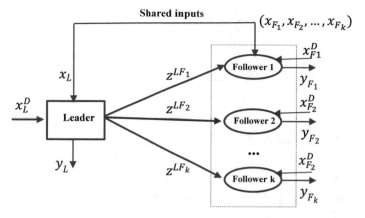

Fig. 10.3 Bi-Level system with constrained resources and multiple followers

followers, x_L and $x_{F_k j}$ are input for leader and follower k respectively, so that: $\bar{x}_L + \sum_{k=1}^{K} \bar{x}_{F_k} \le E, E \in \mathbb{R}$. Also c_L and c_{F_k} correspond to the cost of a unit of shared input for the leader and follower stage, and c_{LF_k} is the price of one unit of intermediate products from the leader to the follower $(z_{LF_k j})$. In the same way, c_L^D and $c_{F_k}^D$ are the costs corresponding to a unit of direct input in the leader and follower stages. And finally, $y_{F_k j}$ is the independent output of the first stage (leader) and follower k. Therefore, to calculate cost efficiency in bi-level supply chains with several followers, model (10.7) was introduced by Zhou et al. [24].

$$\min\left(c_L \bar{x}_L + c_L^D \bar{x}_L^D\right) + \left(\sum_{k=1}^{k} c_{F_k} \bar{x}_{F_k} + \sum_{k=1}^{k} c_{F_k}^D \bar{x}_{F_k}^D + \sum_{k=1}^{k} c_{LF_k} \bar{z}_{LF_k}\right)$$

$s.t$

$$\sum_{j=1}^{n} \lambda_j x_{Lj} \le \bar{x}_L, \quad \sum_{j=1}^{n} \lambda_j x_{Lj}^D \le \bar{x}_L^D,$$

$$\sum_{j=1}^{n} \lambda_j y_{Lj} \ge y_{Lo},$$

$$\sum_{j=1}^{n} \lambda_j z_{LF_k j} \ge z_{LF_k o}, \quad k = 1, \ldots, K,$$

$$\bar{x}_L + \sum_{k=1}^{K} \bar{x}_{F_k} \le E,$$

$$\min f_1 = c_{F_1} \bar{x}_{F_1} + c_{F_1}^D \bar{x}_{F_1}^D + c_{LF_1} \bar{z}_{LF_1}$$

$s.t$

$$\sum_{j=1}^{n} \pi_j^1 x_{F_1 j} \leq \overline{x}_{F_1}, \quad \sum_{j=1}^{n} \pi_j^1 x_{F_1 j}^D \leq \overline{x}_{F_1}^D,$$

$$\sum_{j=1}^{n} \pi_j^1 z_{LF_1 j} \leq \overline{z}_{LF_1}, \quad \sum_{j=1}^{n} \pi_j^1 y_{F_1 j} \geq y_{F_1 o},$$

$$\vdots$$

$$\min f_K = c_{F_1} \overline{x}_{F_1} + c_{F_1}^D \overline{x}_{F_1}^D + c_{LF_1} \overline{z}_{LF_1}$$

$s.t$

$$\sum_{j=1}^{n} \pi_j^1 x_{F_K j} \leq \overline{x}_{F_K}, \quad \sum_{j=1}^{n} \pi_j^1 x_{F_K j}^D \leq \overline{x}_{F_K}^D,$$

$$\sum_{j=1}^{n} \pi_j^1 z_{LF_K j} \leq \overline{z}_{LF_K}, \quad \sum_{j=1}^{n} \pi_j^1 y_{F_K j} \geq y_{F_K o},$$

$$\left(\overline{x}_L^D, \overline{x}_L, \overline{x}_{F_k}^D, \overline{x}_{F_k}, \overline{z}_{LF_k} \right) \geq 0,$$

$$\pi_j^K, \lambda_j \geq 0, \quad j = 1, \ldots, n, \quad k = 1, \ldots, K. \tag{10.7}$$

Thus, in the first stage, the leader obtains the optimal amount of cost in the objective function by determining the optimal input consumption (direct input and shared input). Then, in the next stage, the follower manages the input price by considering the optimal values obtained from stage 1 and other sources and gets the optimal value of the objective function. $z_{LF_k j}$ plays a dual role among the two levels of the supply chain. At the first level, it is output, and the constraint $\sum_{j=1}^{n} \lambda_j z_{LF_k j} \leq z_{LF_k o}$ indicates this issue. In contrast, at the second level, according to the constraint $\sum_{j=1}^{n} \pi_j^1 z_{LF_k j} \leq \overline{z}_{LF_k}$ in the k-th optimization problem, this indicator is in the input role for each component of the second level.

Suppose that $\left(\overline{x}_L^{s*}, \overline{x}_L^{D*}, \overline{x}_{F_1}^{*}, \ldots, \overline{x}_{F_k}^{*}, \overline{x}_{F_1}^{D*}, \ldots, \overline{x}_{F_K}^{D*}, \overline{z}_{LF_1}^{*}, \ldots, \overline{z}_{LF_K}^{*} \right)$ is the optimal solution of model (10.7) when DMU_o, is under evaluation as a result, DMU_o is cost-efficient, if only if $CE_o = 1$ so that: $CE_o = \dfrac{[c_L \overline{x}_L^* + c_L^D \overline{x}_L^{D*}] + \left[\sum_{k=1}^{k} c_{F_k} \overline{x}_{F_k}^* + \sum_{k=1}^{k} c_{F_k}^D \overline{x}_{F_k}^{D*} + \sum_{k=1}^{k} c_{LF_k} \overline{z}_{LF_k}^* \right]}{[c_L x_{Lo} + c_L^D x_{Lo}^D] + \left[\sum_{k=1}^{k} c_{F_k} x_{F_k o} + \sum_{k=1}^{k} c_{F_k}^D x_{F_k o}^D + \sum_{k=1}^{k} c_{LF_k} z_{LF_k o} \right]}$. Also, in DMU_o, the leader and follower k are cost-efficient, if only if $CE_o^L = 1$ and $CE_o^{F_k} = 1$, respectively. So that: $CE_o^L = \dfrac{c_L \overline{x}_L^* + c_L^D \overline{x}_L^{D*}}{c_L x_{Lo} + c_L^D x_{Lo}^D}$ and $CE_o^{F_k} = \dfrac{c_{F_k} \overline{x}_{F_k}^* + c_{F_k}^D \overline{x}_{F_k}^{D*} + c_{LF_k} \overline{z}_{LF_k}^*}{c_{F_k} x_{F_k o} + c_{F_k}^D x_{F_k o}^D + c_{LF_k} z_{LF_k o}}$. So the lower level of DMU_o is the cost-efficient if only if $CE_o^F = \dfrac{\sum_{k=1}^{k} c_{F_k} \overline{x}_{F_k}^* + \sum_{k=1}^{k} c_{F_k}^D \overline{x}_{F_k}^{D*} + \sum_{k=1}^{k} c_{LF_k} \overline{z}_{LF_k}^*}{\sum_{k=1}^{k} c_{F_k} x_{F_k o} + \sum_{k=1}^{k} c_{F_k}^D x_{F_k o}^D + \sum_{k=1}^{k} c_{LF_k} z_{LF_k o}} = 1$.

10.4.1 BPL Problem-Solving Method

By generalizing the solving method presented by [19], Zhou et al. [24] introduced the theorem 10.2 to solve bi-level problems with several followers.

Theorem 10.2 *Let* $(x^*, y_1^*, \ldots, y_K^*)$ *be the optimal solution to the problem* (10.8):

$$F^* = \min c_1 x + \sum_{k=1}^{K} d_k y_k$$

s.t

$$A_1 x + \sum_{k=1}^{K} B_{1k} y_k \leq b$$

$$f_k^* = \min c_2 x + d_k y_k$$

s.t

$$A_{2k} x + B_{2k} y_k \leq b_k$$

$$x \geq 0, y_k \geq 0, k = 1, \ldots, K. \tag{10.8}$$

In this case, the necessary and sufficient condition for (x^*, y_k^*) to be the optimal solution of the model (10.7) is that $(x^*, u_k^*, v_k^*, y_k^*, w_k^*)$ is the optimal solution to the problem (10.9). So that: $u_k^* \in \mathbb{R}^p$, $v_k^* \in \mathbb{R}^{q_k}$, and $w_k^* \in \mathbb{R}^{m_k}$ are the dual variables corresponding to the constraints of $A_1 x + \sum_{k=1}^{K} B_{1k} y_k \leq b_1, A_{2k} x + B_{2k} y_k \leq b_{2k}$, and $y_k \geq 0$ in the model (10.9).

$$\min c_1 x + \sum_{k=1}^{K} d_k y_k$$

s.t

$$A_1 x + \sum_{k=1}^{K} B_{1k} y_k \leq b_1,$$

$$A_{2k} x + B_{2k} y_k \leq b_{2k},$$

$$u_k B_{1k} + v_k B_{2k} - w_k = -d_k,$$

$$u_k \left(b_1 - A_1 x - \sum_{k=1}^{K} B_{1k} y_k \right) + v_k (b_{2k} - A_{2k} x - B_{2k} y_k) + w_k y_k = 0,$$

$$x \geq 0, y_k \geq 0, u_k \geq 0, v_k \geq 0, w_k \geq 0, k = 1, \ldots, K. \tag{10.9}$$

Proof See ref. [24].

We define $x = \begin{pmatrix} \overline{x}_L \\ \overline{x}_L^D \\ \lambda \end{pmatrix}$, $y_k = \begin{pmatrix} \overline{x}_{F_k} \\ \overline{x}_{F_k}^{D*} \\ \overline{z}_{LF_k} \\ \pi^k \end{pmatrix}$, Also, if the coefficients of the objective

functions are $c_1 = (c_L, c_L^D, 0)$, and $d_k = (c_{F_k}, c_{F_k}^D, c_{LF_k}, 0)$ and the right-hand side

is considered equal to $b_1 = \begin{pmatrix} 0 \\ 0 \\ -y_{Lo} \\ -z_{LF_1} \\ \vdots \\ -z_{LF_K} \\ E \end{pmatrix}$ and $b_k = \begin{pmatrix} 0 \\ 0 \\ 0 \\ -y_{F_k} \end{pmatrix}$. And finally, describing

the technology matrixes as follows:

$$A_1 = \begin{bmatrix} -1 & 0 & x_{L1} & \cdots & x_{Ln} \\ 0 & -1 & x_{L1}^D & \cdots & x_{Ln}^D \\ 0 & 0 & -y_{L1}^D & \cdots & -y_{Ln}^D \\ 0 & 0 & -z_{LF_11} & \cdots & -z_{LF_1n} \\ \vdots & \vdots & \vdots & \ddots & \vdots \\ 0 & 0 & -z_{LF_k1} & \cdots & -z_{LF_kn} \\ 1 & 0 & 0 & \cdots & 0 \end{bmatrix}, \quad B_{1k} = \begin{bmatrix} 0 & 0 & 0 & 0 & \cdots & 0 \\ 0 & 0 & 0 & 0 & \cdots & 0 \\ \vdots & \vdots & \vdots & \vdots & \ddots & \vdots \\ 0 & 0 & 0 & 0 & \cdots & 0 \\ 1 & 0 & 0 & 0 & \cdots & 0 \end{bmatrix}, \quad B_{2K} = $$

$$\begin{bmatrix} -1 & 0 & 0 & x_{F_K1} & \cdots & x_{F_Kn} \\ 0 & -1 & 0 & x_{F_K1}^D & \cdots & x_{F_Kn}^D \\ 0 & 0 & -1 & z_{LF_k1} & \cdots & z_{LF_kn} \\ 0 & 0 & 0 & -y_{F_k1} & \cdots & -y_{F_kn} \end{bmatrix}$$, and $A_{2k} = [0]$. So by using variable changes

explained, model (10.7) is transformed into the standard BPL with several followers like the model (10.9), which can be solved using the extended branch and bound method.

10.5 Profit Efficiency Using Bi-Level DEA

Another essential concept in supply chain management is the calculation of profit efficiency. Although several studies have been presented to calculate the profit efficiency in the supply chain using DEA, and NDEA, including Saleh et al. [14], all chain members are considered at the same level in the proposed models. Therefore, these models are not suitable for evaluating the profit efficiency in a bi-level supply chain. Reference [23] presented a new bi-level DEA model to solve this problem.

Consider Fig. 10.4, unlike Fig. 10.2, the intermediate products sent from level one are not necessarily used entirely in level 2. Hence, z_j^1 and z_j^2 represent the

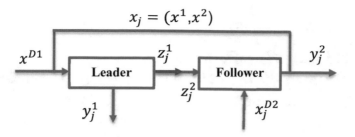

$$x_j = (x^1, x^2)$$

Fig. 10.4 Bi-Level system with constrained resource and independent output

outputs produced at the higher level and the consumption of these products at level 2, respectively. Also, if the income from the sale of one unit of intermediate products, the independent output of level 1 and the independent output of level 2 are indicated by the symbols P^1, P^2, and P^3, respectively, also, c^1, c^2, and c^3 is correspond to the cost of one unit of shared input, the direct input of the higher level, and the direct input of the lower level. So, the current profit of the leader and follower levels is calculated using (10.10) and (10.10) relations.

$$\pi_o^L = \left(P^1 z_o^1 + P^2 y_o^1 \right) - \left(c^1 x_o^1 + c^2 x_o^{D1} \right) \tag{10.10}$$

$$\pi_o^F = P^3 y_o^2 - \left(c^1 x_o^2 + c^3 x_o^{D2} + P^1 z_o^2 \right) \tag{10.11}$$

Therefore, supply chain managers at both levels seek to achieve a combination of inputs and outputs that will improve the relative efficiency of the chain and gain more profit. In this case, the values of the variables should be calculated in such a way that:

$$\overline{\pi}_o^L = \left(P^1 \overline{z}^1 + P^2 \overline{y}^1 \right) - \left(c^1 \overline{x}^1 + c^2 \overline{x}^{D1} \right) \geq \pi_o^L = \left(P^1 z_o^1 + P^2 y_o^1 \right) - \left(c^1 x_o^1 + c^2 x_o^{D1} \right) \tag{10.12}$$

$$\overline{\pi}_o^F = P^3 \overline{y}^2 - \left(c^1 \overline{x}^2 + c^3 \overline{x}^{D2} + P^1 \overline{z}^2 \right) \geq \pi_o^F = P^3 y_o^2 - \left(c^1 x_o^2 + c^3 x_o^{D2} + P^1 z_o^2 \right) \tag{10.13}$$

In this case, the model (10.14) is used to calculate the profit efficiency of a bi-level supply chain.

$$\max \left(P^1 \overline{z}^1 + P^2 \overline{y}^1 \right) - \left(c^1 \overline{x}^1 + c^2 \overline{x}^{D1} \right)$$

$$s.t$$

$$\sum_{j=1}^n \lambda_j x_j^1 + \sum_{j=1}^n \pi_j x_j^2 \leq \overline{x}^1 + \overline{x}^2, \quad \sum_{j=1}^n \lambda_j x_j^{D1} \leq \overline{x}^{D1},$$

$$\sum_{i=1}^{n} \lambda_j y_j^1 \geq \overline{y}^1, \quad \sum_{i=1}^{n} \lambda_j z_j^1 \geq \overline{z}^1,$$

$$\overline{x}^1 + \overline{x}^2 \leq E, \quad \overline{x}^{D1} \leq x_o^{D1},$$

$$\left(P^1 \overline{z}^1 + P^2 \overline{y}^1\right) - \left(c^1 \overline{x}^1 + c^2 \overline{x}^{D1}\right) \geq \left(P^1 z_o^1 + P^2 y_o^1\right) - \left(c^1 x_o^1 + c^2 x_o^{D1}\right),$$

$$\max P^3 \overline{y}^2 - \left(c^1 \overline{x}^2 + c^3 \overline{x}^{D2} + P^1 \overline{z}^2\right)$$

s.t

$$\sum_{i=1}^{n} \pi_j x_j^{D2} \leq \overline{x}^{D2},$$

$$\sum_{i=1}^{n} \pi_j z_j^2 \leq \overline{z}^2,$$

$$\sum_{i=1}^{n} \pi_j y_j^2 \geq \overline{y}^2,$$

$$\overline{z}^2 \leq \overline{z}_o^2, \quad \overline{x}^{D2} \leq x_o^{D2},$$

$$P^3 \overline{y}^2 - \left(c^1 \overline{x}^2 + c^2 \overline{x}^{D2} + P^1 \overline{z}^2\right) \geq P^3 y_o^2 - \left(c^1 x_o^2 + c^3 x_o^{D2} + P^1 z_o^2\right),$$

$$\overline{x}^{D1}, \overline{x}^{D2}, \overline{x}^1, \overline{x}^2, \overline{y}^1, \overline{y}^2, \overline{z}^1, \overline{z}^2, \pi_j, \lambda_j \geq 0, \qquad j = 1, \dots, n. \qquad (10.14)$$

The constraint $\overline{z}^2 \leq \overline{z}^1$ emphasizes that the intermediate products sent from level one are not necessarily completely used at level 2. Also, due to the existing limitations in the use of shared input, the use of this input is limited by the adverb $\overline{x}^1 + \overline{x}^2 \leq E$. Therefore, after determining $\overline{x}^1$ in the first level, the optimal value for $\overline{x}^2$ is selected in the interval $\left[0, E - \overline{x}^1\right]$.

Suppose that $\left(\overline{x}^{D1*}, \overline{x}^{D2*}, \overline{x}^{1*}, \overline{x}^{2*}, \overline{y}^{1*}, \overline{y}^{2*}, \overline{z}^{1*}, \overline{z}^{2*}, \pi^*, \lambda^*\right)$ is the optimal solution of model (10.14), so DMU_o is profit efficient if only if $PE_o = \frac{\left(P^3 \overline{y}^{2*} + P^1 \overline{z}^{1*} + P^2 \overline{y}^{1*}\right) - \left(c^1 \overline{x}^{1*} + c^2 \overline{x}^{D1*} + c^1 \overline{x}^{2*} + c^3 \overline{x}^{D2*} + P^1 \overline{z}^{2*}\right)}{\left(P^3 y_o^2 + P^1 z_o^1 + P^2 y_o^1\right) - \left(c^1 x_o^1 + c^2 x_o^{D1} + c^1 x_o^2 + c^3 x_o^{D2} + P^1 z_o^2\right)} = 1$. Also, in DMU_o, leader and follower are profit efficient if only if $PE_o^L = \frac{\left(P^1 \overline{z}^{1*} + P^2 \overline{y}^{1*}\right) - \left(c^1 \overline{x}^{1*} + c^2 \overline{x}^{D1*}\right)}{\left(P^1 z_o^1 + P^2 y_o^1\right) - \left(c^1 x_o^1 + c^2 x_o^{D1}\right)} = 1$ and

$$PE_o^F = \frac{P^3 \overline{y}^{2*} - \left(c^1 \overline{x}^{2*} + c^3 \overline{x}^{D2*} + P^1 \overline{z}^{2*}\right)}{P^3 y_o^2 - \left(c^1 x_o^2 + c^3 x_o^{D2} + P^1 z_o^2\right)} = 1.$$

10.5.1 BPL Problem-Solving Method

To solve BPL (10.14) using theorem (10.1), it is enough to convert the model (10.14) into the standard BPL. Therefore, the model (10.15) is obtained.

$$\min\left(-p^1, -p^2, c^1, c^2, 0\right)x$$

s.t

$$\left(0, 0, -1, 0, x_1^1, \dots, x_n^1\right)x + \left(0, -1, 0, 0, x_1^2, \dots, x_n^2\right)y \leq 0,$$

$$\left(0, 0, 0, -1, x_1^{D1}, \ldots, x_n^{D1}\right)x \leq 0,$$

$$\left(0, 1, 0, 0, -y_1^1, \ldots, -y_n^1\right)x \leq 0,$$

$$\left(-1, 0, 0, 0, z_1^1, \ldots, z_n^1\right)x \leq 0,$$

$$(0, 0, 1, 0, 0)x + (0, 1, 0, 0, 0)y \leq E,$$

$$(0, 0, 0, 1, 0)x \leq x_o^{D1},$$

$$\left(-p^1, -p^2, c^1, c^2, 0\right)x \leq \left(c^1 x_o^1 + c^2 x_o^{D1}\right) - \left(P^1 z_o^1 + P^2 y_o^1\right)$$

$$\min\left(-p^3, c^1, c^3, p^1, 0\right)y$$

s.t

$$\left(0, 0, -1, 0, x_1^{D2}, \ldots, x_n^{D2}\right)y \leq 0,$$

$$\left(0, 0, 0, -1, z_1^2, \ldots, z_n^2\right)y \leq 0,$$

$$\left(1, 0, 0, 0, -y_1^2, \ldots, -y_n^2\right)y \leq 0,$$

$$(0, 0, 0, 1, 0)y \leq z_o^2,$$

$$(0, 0, 1, 0, 0)y \leq x_o^{D2},$$

$$(0, -1, 0, 0, 0)x + (0, 0, 0, 1, 0)y \leq 0,$$

$$\left(-p^3, c^1, c^3, p^1, 0\right)y \leq \left(c^1 x_o^2 + c^2 x_o^{D2} + P^1 z_o^2\right) - P^3 y_o^2,$$

$$x \geq 0, y \geq 0. \tag{10.15}$$

As a result, if:

$$x = \begin{pmatrix} \bar{z}^1 \\ \bar{y}^1 \\ \bar{x}^1 \\ \bar{x}^{D1} \\ \lambda \end{pmatrix}, y = \begin{pmatrix} \bar{y}^2 \\ \bar{x}^2 \\ \bar{x}^{D2} \\ \bar{z}^2 \\ \pi \end{pmatrix}, A_1 = \begin{pmatrix} 0 & 0 & -1 & 0 & x_1^1 & \cdots & x_n^1 \\ 0 & 0 & 0 & -1 & x_1^A & \cdots & x_n^{n_1} \\ 0 & 1 & 0 & 0 & -y_1^1 & \cdots & -y_n^1 \\ -1 & 0 & 0 & 0 & z_1^1 & \cdots & z_n^1 \\ 0 & 0 & 1 & 0 & 0 & \cdots & 0 \\ 0 & 0 & 0 & 1 & 0 & \cdots & 0 \\ -p^1 & -p^2 & c^1 & c^2 & 0 & \cdots & 0 \end{pmatrix},$$

$$B_1 = \begin{pmatrix} 0 & -1 & 0 & 0 & x_1^2 & \cdots & x_n^2 \\ 0 & 0 & 0 & 0 & 0 & \cdots & 0 \\ 0 & 0 & 0 & 0 & 0 & \cdots & 0 \\ 0 & 0 & 0 & 0 & 0 & \cdots & 0 \\ 0 & 1 & 0 & 0 & 0 & \cdots & 0 \\ 0 & 0 & 0 & 0 & 0 & \cdots & 0 \\ 0 & 0 & 0 & 0 & 0 & \cdots & 0 \end{pmatrix},$$

$$A_2 = \begin{pmatrix} 0 & 0 & 0 & 0 & 0 & \cdots & 0 \\ 0 & 0 & 0 & 0 & 0 & \cdots & 0 \\ 0 & 0 & 0 & 0 & 0 & \cdots & 0 \\ 0 & 0 & 0 & 0 & 0 & \cdots & 0 \\ 0 & 0 & 0 & 0 & 0 & \cdots & 0 \\ 0 & 0 & 0 & 0 & 0 & \cdots & 0 \\ 0 & -1 & 0 & 0 & 0 & \cdots & 0 \end{pmatrix}, B_2 = \begin{pmatrix} 0 & 0 & -1 & 0 & x_1^{D2} & \cdots & x_n^{D2} \\ 0 & 0 & 0 & -1 & z_1^2 & \cdots & z_n^2 \\ 1 & 0 & 0 & 0 & -y_1^2 & \cdots & -y_n^2 \\ 0 & 0 & 0 & 1 & 0 & \cdots & 0 \\ 0 & 0 & 1 & 0 & 0 & \cdots & 0 \\ 0 & 0 & 0 & 1 & 0 & \cdots & 0 \\ -p^3 & c^1 & c^3 & p^1 & 0 & \cdots & 0 \end{pmatrix}$$

$$b_1 = \begin{pmatrix} 0 \\ 0 \\ 0 \\ 0 \\ E \\ x_o^{D1} \\ (c^1 x_o^1 + c^2 x_o^{D1}) - (P^1 z_o^1 + P^2 y_o^1) \end{pmatrix},$$

$$b_2 = \begin{pmatrix} 0 \\ 0 \\ 0 \\ z_o^2 \\ x_o^{D2} \\ 0 \\ (c^1 x_o^2 + c^2 x_o^{D2} + P^1 z_o^2) - P^3 y_o^2 \end{pmatrix},$$

$$c_1 = (-p^1, -p^2, c^1, c^2, 0), d_2 = (-p^3, c^1, c^3, p^1, 0), d_1 = [0], c_2 = [0]$$

10.6 Resources Reallocation in the Supply Chain Using Bi-Level DEA

Fair resource allocation, including budget, labor, demand, primary resources, etc., is one of the essential and influential issues in managers' decisions in any organization. In general, the resource allocation problem refers to a process that allows managers to fairly distribute limited resources among the subgroups under their management to achieve a final goal. Therefore, fair resource allocation is one of the critical issues in management science, and many researchers have studied this field and introduced new techniques. Among these techniques is DEA. As an example, the studies conducted by refs. [9, 8, 9, 10, 11, 12, 18], are examples of providing a fair allocation by DEA models.

Researchers have used the allocation of resources using DEA in various industries until now. One of the most important supply chains in the industry is the supply chain in the petrochemical industry because the products produced by petrochemicals are an essential part of production factories. Also, in addition to financial criteria, petrochemical industries face many non-financial measures. Therefore, the evaluation of petrochemical industries always needs a comprehensive performance evaluation system that, in addition to maintaining financial criteria, also considers their numerous non-financial criteria. Therefore, the evaluation of petrochemical industries always needs a comprehensive performance evaluation system that, in addition to maintaining financial criteria, also considers their numerous non-financial measures. Due to the weaknesses of the current performance measurement systems, many leading organizations have designed and established more comprehensive approaches that link organizational tasks with strategic goals. And by providing a balanced set of financial and non-financial indicators, the organization's activities are directed toward the needs and wants of customers.

Also, the importance of petrochemical companies in the economy is emphasized in the following two points:

1. This industry's significant role is providing other industries with raw materials.
2. The main inputs in the petrochemical industry are oil and gas. Due to the depletion of fossil fuels and the non-renewability of these resources, it can be said that the optimal allocation of the resources above causes double sensitivity in the petrochemical industry.

For example, consider an industrial group's activity in 5 different regions. Each area contains three production units. The production operation in these five production areas is such that it uses seven primary sources to produce a final product. In other words, the production lines transfer circulating cooling water, deionized water, compressed air, nitrogen, electricity, feed gas, and fuel gas. Finally, each production line uses its independent production method to produce ammonia (final output). Figure 10.5 shows how these production lines work. x^o is the standard input distributed among the three production units in each area. Then each production unit produces the desired outputs using its input and the technology used in that production unit.

Fig. 10.5 Three-stage parallel network

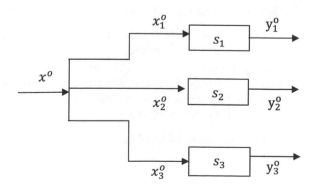

Fig. 10.6 Parallel network

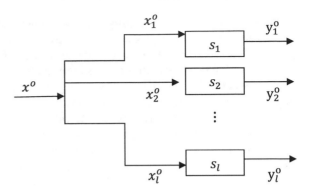

Although using DEA models for resource and cost allocation has brought good results, ignoring an organization's internal processes in fair distribution is a fundamental problem that classical DEA models seriously face. In other words, in these studies, each of the DMUs under evaluation has been considered as a unit with one stage that includes primary inputs and final outputs, while many organizations operate multi-stage in practice. Fare and Groskoff [1] proposed the network DEA, for the first time. After that, the use of NDEA in providing an optimal and fair allocation by refs. [25, 1] as a powerful tool was used.

Shafiee et al. [16] proposed a new resource allocation approach for multi-members supply chain with parallel structure. In general, consider a production unit with m parallel production line as shown in Fig. 10.6.

Suppose that the vector of the initial resources $x^j = \left(x_1^j, \ldots, x_m^j \right)$ is available and a share of as much as $x_k^j = \left(x_{k1}^j, \ldots, x_{km}^j \right)$ is allocated to each of the production lines so that $\sum_k x_k^j = x^j$ Thus, each production line produces the output vector $y_k^j = \left(y_{k1}^j, \ldots, y_{ks}^j \right)$ by the consumption of the vector x_k^j.

This section provides a method for the reallocating of the initial resources in production lines to improve the network performance. Of course, before starting the discussion, it is necessary to pay attention to a few points:

1. Resource reallocation has been considered an internal flow, and foreign exchanges have been omitted.
2. Reallocation may have costs. Initially, the costs are considered equal to zero.
3. Reallocation may have some limitations. Initially, we ignore these constraints.
4. In some organizations, some inputs may not have reallocation capabilities. Initially, this issue is ignored.

We introduce the model (10.16) to obtain the optimal allocation flow.

$$\min \ t$$

$$s.t \quad \sum_k x_{ki} = x_i^o, \qquad i = 1, \ldots, m,$$

$$x_{ki} \geq 0, \qquad i = 1, \ldots, m, k = 1, \ldots, d,$$

$$t = \max \sum_k \varphi_k$$

$$\qquad\qquad\qquad\qquad\qquad\qquad\qquad\qquad\qquad\qquad (10.16)$$

$$s.t \quad \sum_j \lambda_k^j x_{ki}^j \leq x_{ki}, \qquad i = 1, \ldots, m, k = 1, \ldots, d,$$

$$\sum_j \lambda_k^j y_{kr}^j \geq \varphi_k y_{kr}^o, \ r = 1, \ldots, s, k = 1, \ldots, d,$$

$$\lambda_k^j \geq 0, \qquad j = 1, \ldots, n, k = 1, \ldots, d,$$

The model (10.16) is a bi-level model. At level 1, using the constraint $\sum_k x_{ki} = x_i^o$, a new feasible resource allocation is introduced for level 2.

Henceforth, the allocation obtained from level 1 is considered a constant value for level 2. At level 2 for this new allocation, φ_k^* is calculated for each stage, and $\sum_k \varphi_k^*$ is calculated subsequently for the network. Finally, among all $\sum_k \varphi_k^*$ calculated for each feasible allocation, the best solution is selected at level 1.

Suppose there are restrictions on the allocation of input x_{ki} in some production lines. For example, suppose that based on the organization's internal policies, at least a_{ki}^o and at most b_{ki}^o specialized workers are required per production line. Therefore, we always have $a_{ki}^o \leq x_{ki} \leq b_{ki}^o$. Note that a_{ki}^o and b_{ki}^o are constant values specified by the network manager.

In the next stage, suppose each transfer imposes a cost c_{ki}^o to the organization (For example, the cost of training newly arrived workers in the production line). On the other hand, the company has been considered a budget equal to $\bar{c}$ for the reallocation, and no more costs are not possible for the company. To consider this condition, the constraint $\sum_k \sum_i c_{ki}^o \leq \bar{c}$ is added to the model (10.16).

Now assume that the cost of using one unit of the i-th initial input in the DMU_o is equal to c_i^o, then DMU_o must pay a fee of $c_i^o x_i^o$ to use the i-th initial resources. Therefore, the initial cost in the DMU_o is equal to $c^o x^o$. Assume also that the revenue from the production of one unit of the r-th output is equal to p_r^o. In this case, DMU_o will have the revenue of $p^o y^o$ by producing the output y^o, and the profit of the company is equal to $p^o y^o - c^o x^o$. Here, a new question arises for the managers as follows:

"How to select a reallocate initial resources so that the profit efficiency of DMU_o is increased?" To answer this question, model (10.17) is suggested as follows.

$$\max p^o y - c^o x$$
$$s.t \qquad \sum_k x_{ik} = x_i^o, \qquad i = 1, \ldots, m,$$
$$x_{ik} \geq 0, \qquad i = 1, \ldots, m, k = 1, \ldots, d,$$
$$\max p^o y$$
$$s.t \qquad \sum_j \lambda_k^j x_{ki}^j \leq x_{ik}, \quad i = 1, \ldots, m, k = 1, \ldots, d, \qquad (10.17)$$
$$\sum_j \lambda_k^j y_{kr}^j \geq y_{rk}, \quad r = 1, \ldots, s, k = 1, \ldots, d,$$
$$\lambda_k^j \geq 0, \qquad j = 1, \ldots, n, k = 1, \ldots, d,$$
$$y_{rk} \geq 0, \qquad r = 1, \ldots, s, k = 1, \ldots, d.$$

Thus, at level one, a feasible allocation is initially proposed. Then at level two, the new outputs and the revenue from the new production are calculated for this new allocation.

10.6.1 BPL Problem-Solving Method

Bi-level DEA is a particular type of bi-level linear programming. Due to the specific complexities in these issues, hierarchical structures with the DEA technique have been less evaluated. Among the studies carried out, we can refer to the research of [7] and Hajiagha et al. [6] In this research, we evaluate bi-level structures using the method provided by Abo-Sinna [2]. First, suppose:

$$G = \left\{ (x, y) \mid \sum_k x_k = x^o, \sum_j \lambda_k^j x_k^j \leq x_k, \sum_j \lambda_k^j y_{kr}^j \geq y_{rk}, x_{ik} \geq 0, \right.$$
$$\left. y_{rk} \geq 0, \lambda_k^j \geq 0, j = 1, \ldots, n, k = 1, \ldots, d \right\} \qquad (10.18)$$

Also, suppose that f_o^* and $\overline{f}_o$ are the best and worst values of the objective function of the level H on G, respectively. In other words, f_o^* is the optimal solution of the model (10.19):

$$\max p^o y - c^o x$$
$$s.t \sum_k x_{ik} = x_i^o, \ i = 1, \ldots, m,$$
$$x_{ik} \geq 0, \qquad i = 1, \ldots, m, k = 1, \ldots, d$$
$$\sum_j \lambda_k^j x_{ki}^j \leq x_{ik}, \ i = 1, \ldots, m, k = 1, \ldots, d \qquad (10.19)$$
$$\sum_j \lambda_k^j y_{kr}^j \geq y_{rk}, \ r = 1, \ldots, s, k = 1, \ldots, d,$$
$$\lambda_k^j \geq 0, \qquad j = 1, \ldots, n, k = 1, \ldots, d,$$
$$y_{rk} \geq 0, \qquad r = 1, \ldots, s, k = 1, \ldots, d.$$

Also, to calculate $\overline{f}_o$, we obtain the optimal solution of the model (10.20):

$$\min \; p^o y - c^o x$$

s.t

$$\sum_k x_{ik} = x_i^o, \quad i = 1, \dots, m,$$

$$x_{ik} \geq 0, \quad i = 1, \dots, m, k = 1, \dots, d,$$

$$\sum_j \lambda_k^j x_{ki}^j \leq x_{ik}, \quad i = 1, \dots, m, k = 1, \dots, d, \qquad (10.20)$$

$$\sum_j \lambda_k^j y_{kr}^j \geq y_{rk}, \quad r = 1, \dots, s, k = 1, \dots, d,$$

$$\lambda_k^j \geq 0, \quad j = 1, \dots, n, k = 1, \dots, d,$$

$$y_{rk} \geq 0, \quad r = 1, \dots, s, k = 1, \dots, d.$$

Now, to calculate the level of satisfaction from the optimal solution, we define the following satisfaction function using the concept of membership function in fuzzy theory as follows:

$$\mu_{f_o}[f_o(x, y)] = \begin{cases} 1, & f_o(x, y) \geq f_o^* \\ \frac{f_o(x,y) - \overline{f}_o}{f_o^* - \overline{f}_o}, & \overline{f}_o < f_o(x, y) < f_o^* \qquad (10.21) \\ 0, & f_o(x, y) \leq \overline{f}_o \end{cases}$$

Now, to find the solution that has the smallest distance from the satisfactory solution at level H, we solve the model (10.22).

$$\max \; \lambda$$

s.t

$$(x, y) \in G,$$

$$\mu_{f_o}[f_o(x, y)] \geq \lambda,$$

$$0 \leq \lambda \leq 1. \qquad (10.22)$$

We show the optimal solution of model (10.22) as (λ^H, x^H, y^H). In the next step, similarly, we obtain $\mu_{f_1}[f_1(\overline{x})]$ using the relation (10.23):

$$\mu_{f_1}[f_1(x, y)] = \begin{cases} 1, & f_1(x, y) > f_1^* \\ \frac{f_1(x,y) - \overline{f}_1}{f_1^* - \overline{f}_1}, & \overline{f}_1 < f_1(x, y) < f_1^* \qquad (10.23) \\ 0, & f_1(x, y) \leq \overline{f}_1 \end{cases}$$

So that f_1^* and $\overline{f}_1$ are respectively the best and worst values of the objective function of level L on G. To calculate the solution that has the smallest distance from the satisfactory solution at the level L, we solve (10.24). The optimal solution of model (10.24) is illustrated as (λ^L, x^L, y^L).

$$\max \ \beta$$
$$s.t$$
$$(x, y) \in G,$$
$$\mu_{f_1}[f_1(x, y)] \geq \beta,$$
$$0 \leq \lambda \leq 1. \tag{10.24}$$

Since the nature of both levels is different, the answers obtained by solving models (10.22) and (10.24) usually conflict. Therefore, to achieve a solution accepted by both levels, at first, HLDM defines tolerance for the solution of level H. As a result, $\mu_x(x)$ is defined as (10.25).

$$\mu_x(x) = \begin{cases} \frac{x - (x^H - t_1)}{t_1}, & (x^H - t_1) \leq x \leq x^H \\ \frac{(x^H + t_1) - x}{t_1}, & x^H \leq x \leq (x^H + t_1) \\ 0, & o.w \end{cases} \tag{10.25}$$

From the view of HLDM, if the value is greater than f_o^H, entirely acceptable. But if $f_o < f_o' = f_o(x^l, y^l)$, this solution is not satisfactory. So the level of satisfaction with the optimal solution for HLDM can be determined using the following function define:

$$\mu_{f_o}[f_o(x, y)] = \begin{cases} 1, & f_o(x, y) \geq f_o^H \\ \frac{f_o(x,y) - f_o'}{f_o^H - f_o'}, & f_o' < f_o(x, y) < f_o^H \\ 0, & f_o(x, y) \leq f_o' \end{cases} \tag{10.26}$$

Similarly, with HLDM, the satisfaction level of the optimal solution for LLDM is defined as follows:

$$\mu_{f_o}[f_o(x, y)] = \begin{cases} 1, & f_1(x, y) > f_1^L \\ \frac{f_1(x,y) - f_1'}{f_1^L - f_1'}, & f_1' < f_1(x, y) < f_1^L \\ 0, & f_1(x, y) \leq f_1' \end{cases} \tag{10.27}$$

Such that: $f_1' = f_1(x^H, y^H)$. Finally, the model (10.28) is solved to obtain an acceptable answer for both decision-makers.

$$\max \ \delta$$
$$s.t$$
$$(x, y) \in G$$
$$\frac{x - (x^H - t_1)}{t_1} \geq \delta,$$
$$\frac{(x^H + t_1) - x}{t_1} \geq \delta,$$

$$\frac{f_0(x, y) - f_0'}{f_0^H - f_0'} \geq \delta,$$

$$\frac{f_1(x, y) - f_1'}{f_1^L - f_1'} \geq \delta,$$

$$x, y \geq 0,$$

$$0 \leq \delta \leq 1 \tag{10.28}$$

If both levels accept the obtained answer, the work is finished. Otherwise, the process must be repeated by introducing a new tolerance.

10.7 Conclusion

Unlike the supply chains shown in the previous chapters, in some supply chains, the chain's components are not at the same level; in other words, one member is under the leadership of another. While in NDEA models, it is assumed that either all components of the chain are under unified supervision or they act entirely independently of each other. Therefore, this chapter shows that bi-level DEA models are used in such situations. But solving bi-level DEA models is somewhat complicated and requires using special methods that have been explained. To evaluate multi-level supply chains, it should be noted that, like multi-stage chains, at first, the number of chain stages, the number of chain members in each stage, and the type of serial or parallel linking among the chain members should be determined. In addition to these cases, another vital point to pay attention to is identifying the member of the leader and the follower. This step has such importance that if there is not enough accuracy in determining the leader and the follower, the results will not be valid.

References

1. Fare, R., & Grosskopf, S. Network DEA–Socio-Economic Planning Sciences. Issue, **49**, 34-35 (2000)
2. Abo-Sinna, M.A.: A bi-level non-linear multi-objective decision-making under fuzziness. Opsearch **38**(5), 484–495 (2001)
3. Ding, T., Zhu, Q., Zhang, B., Liang, L.: Centralized fixed cost allocation for generalized two-stage network DEA. INFOR: Inf. Syst. Oper. Res. **57**(2), 123–140 (2019)
4. Fare, R., Grosskopf, S., Lovell, C.A.K.: The measurement of efficiency of production, Kluwer Academic Publisher (1985)
5. Ghazi, A., Lotfi, F.H.: Assessment and budget allocation of Iranian natural gas distribution company-A CSW DEA-based model. Socioecon. Plann. Sci. **66**, 112–118 (2019)
6. Hajiagha, S.H.R., Mahdiraji, H.A., Tavana, M.: A new bi-level data envelopment analysis model for efficiency measurement and target setting. Measurement **147**, 106877 (2019)
7. Hakim, S., Seifi, A., Ghaemi, A.: A bi-level formulation for DEA-based centralized resource allocation under efficiency constraints. Comput. Ind. Eng. **93**, 28–35 (2016)

8. Jahanshahloo, G.R., Lotfi, F.H., Moradi, M.: A DEA approach for fair allocation of common revenue. Appl. Math. Comput. **160**(3), 719–724 (2005)
9. Jahanshahloo, G.R., Lotfi, F.H., Shoja, N., Sanei, M.: An alternative approach for equitable allocation of shared costs by using DEA. Appl. Math. Comput. **153**(1), 267–274 (2004)
10. Lotfi, F.H., Noora, A.A., Jahanshahloo, G.R., Gerami, J., Mozaffari, M.R.: Centralized resource allocation for enhanced Russell models. J. Comput. Appl. Math. **235**(1), 1–10 (2010)
11. Malekmohammadi, N., Lotfi, F.H., Jaafar, A.B.: Centralized resource allocation in DEA with interval data: an application to commercial banks in Malaysia. Int. J. Math. Anal. **3**(13–16), 757–764 (2009)
12. Mirsalehy, A., Abu Bakar, M.R., Jahanshahloo, G.R., Hosseinzadeh Lotfi, F., Lee, L.S.: Centralized resource allocation for connecting radial and nonradial models. J. Appl. Math. **2014** (2014)
13. Momeni, E., Lotfi, F.H., Saen, R.F., Najafi, E.: Centralized DEA-based reallocation of emission permits under cap and trade regulation. J. Clean. Prod. **234**, 306–314 (2019)
14. Saleh, H., Hosseinzadeh Lotfi, F., Rostmay-Malkhalifeh, M., Shafiee, M.: Provide a mathematical model for selecting suppliers in the supply chain based on profit efficiency calculations. J. New Res. Math. **7**(32), 177–186 (2021)
15. Shafiee, M., Lotfi, F.H., Saleh, H., Ghaderi, M.: A mixed integer bi-level DEA model for bank branch performance evaluation by Stackelberg approach. J. Ind. Eng. Int. **12**(1), 81–91 (2016)
16. Shafiee, M., Rostamy- Malkhalifeh, M., Saleh, H.: Resource re-allocation at network units by using bi-level data envelopment analysis. J. Oper. Res. Appl. In Press (2023)
17. Shamsi, R., Jahanshahloo, G.R., Mozaffari, M.R., Lotfi, F.H.: Centralized resource allocation with MOLP structure. Indian J. Sci. Technol. **7**(9), 1297–1306 (2014)
18. Sharafi, H., Lotfi, F.H., Jahanshahloo, G.R., Razipour-GhalehJough, S.: Fair allocation fixed cost using cross-efficiency based on Pareto concept. Asia-Pacific J. Oper. Res. **37**(01), 1950036 (2020)
19. Shi, C., Lu, J., Zhang, G., Zhou, H.: An extended branch and bound algorithm for linear bilevel programming. Appl. Math. Comput. **180**(2), 529–537 (2006)
20. Sun, J., Wang, C., Ji, X., Wu, J.: Performance evaluation of heterogeneous bank supply chain systems from the perspective of measurement and decomposition. Comput. Ind. Eng. **113**, 891–903 (2017)
21. Wu, D.D.: Bilevel programming data envelopment analysis with constrained resource. Eur. J. Oper. Res. **207**(2), 856–864 (2010)
22. Wu, D.D., Birge, J.R.: Serial chain merger evaluation model and application to mortgage banking. Decis. Sci. **43**(1), 5–36 (2012)
23. Wu, D.D., Luo, C., Wang, H., Birge, J.R.: Bi-level programming merger evaluation and application to banking operations. Prod. Oper. Manag. **25**(3), 498–515 (2016)
24. Zhou, X., Luo, R., Tu, Y., Lev, B., Pedrycz, W.: Data envelopment analysis for bi-level systems with multiple followers. Omega **77**, 180–188 (2018)
25. Zhu, W., Zhang, Q., Wang, H.: Fixed costs and shared resources allocation in two-stage network DEA. Ann. Oper. Res. **278**(1), 177–194 (2019)

Printed in the United States
by Baker & Taylor Publisher Services